B.S. Beckett

Biology

A modern introduction

Second edition

Oxford University Press

Oxford University Press, Walton Street, Oxford OX2 6DP
London Glasgow New York Toronto
Delhi Bombay Calcutta Madras Karachi
Kuala Lumpur Singapore Hong Kong Tokyo
Nairobi Dar es Salaam Cape Town
Melbourne Auckland
and associates in
Beirut Berlin Ibadan Mexico City Nicosia

ISBN 0 19 914088 X

First published 1976
Reprinted 1976, 1977 (twice), and 1978
Reprinted with corrections 1981
Second edition 1982

Notice to readers:
The author welcomes constructive criticism of his books, suggested improvements, and general correspondence related to topics covered in the text. Correspondence should be addressed to B. S. Beckett, care of the Oxford University Press.

Typeset by Advanced Filmsetters (Glasgow) Ltd

Printed in Great Britain at the University Press, Oxford
by Eric Buckley Printer to the University

To the teacher

Biology is an established and successful O-level textbook. For the second edition, the text has been altered to cater for syllabus changes and recent developments in scientific thinking. New drawings have been included, and many of these are designed for copying by the student.

The organization and methodology of the new edition remain unaltered. Plant and animal biology are still treated together, with special emphasis on the concepts and principles common to both. The interrelationships between groups of organisms are underlined by means of comparative studies. Groups such as mammals, insects, and flowering plants are discussed together, and chapters are arranged so that characteristics like transport, reproduction, and sensitivity can be compared. Chapters are arranged sequentially so that some topics like cells and photosynthesis can be developed gradually until their biological significance is fully appreciated.

There are two different types of illustration. Some are an attempt to represent subjects as they appear in nature, and from these the student can see how specimens appear when viewed under ideal conditions. Other illustrations are schematic representations of the realistic drawings, or diagrammatic summaries of biological principles described in the text. Important structures such as the heart and lungs are drawn both ways so that by comparison the student can see how to reduce complex information to a simple diagram. Almost all illustrations have been drawn by the author and all are carefully integrated into the text and referred to wherever they will aid understanding.

The book tries to achieve a balance between the traditional and more progressive approaches to science teaching by selecting the best from both methodologies. Each chapter includes a body of knowledge which the student is expected to learn and understand. This can be tested by the factual recall and comprehension tests included in most chapters. Most chapters also include experiments showing how this knowledge can be verified. This is considered important even though the outcome may be beyond doubt and known beforehand. In addition there are inquiry exercises, which invite the student to expand and extend his knowledge by exploring ideas in more detail and experiencing for himself the scientific method and the excitement of discovery.

Of course, no book at this level, however modern, can be any more than an introduction. This one will serve its purpose if it helps the teacher to interest his students in that vast, rich, and exciting world which is the subject of *Biology*.

To the student

A scientist approaches his subject in the following way. First, he *studies* the existing knowledge that other scientists have already discovered. Second, he *learns* this body of knowledge including all its technical terms. Third, he makes sure he *understands* what he learns. Fourth, he may repeat some experiments that others have already done to *verify* the results. Fifth, and most important, he devises new experiments to *investigate* new subjects and perhaps create new knowledge.

This book is written in such a way that you can follow roughly the same five steps yourself. This is possible because most chapters consist of the following parts.

The text

The text of each chapter introduces the existing knowledge on a biological topic. These accounts are up to date, fully illustrated, simply worded, and contain explanations of the more important technical terms. Technical terms are a vital part of science. They are a kind of shorthand which is used to summarize ideas which might otherwise require several pages of explanation.

Summary and factual recall tests

Each chapter ends with a brief summary of the basic facts contained in tne text. You can take part in constructing these summaries by supplying missing words and phrases, and sometimes by summarizing ideas in your own words. You should try to supply these missing parts from memory before referring to the chapter.

Comprehension tests

These tests provide you with an opportunity to test your understanding of certain information contained in the chapter.

Verification exercises

Most chapters contain exercises which show you how to verify the accuracy of factual statements contained in the text. You can do this by repeating some of the experiments from which these facts were discovered.

Inquiry exercises

Most chapters also contain an opportunity for you to make inquiries into knowledge which is not in the chapter, but which follows naturally from it. Sometimes these exercises state problems for you to solve and give detailed instructions on how to solve them; but in other cases there is no more than a problem, a list of necessary apparatus, and a few clues on how to proceed. Work on an inquiry exercise is obviously not the same as the work of a scientist who is solving a problem and creating new knowledge. But it is like the work of a scientist in some ways: it gives you a chance to use knowledge gained from a chapter to discover knowledge which is new *to you*. At the same time it allows you to experience what it is like to solve a problem making use of controlled experiments which, in some cases, you have devised yourself.

This book introduces biology and the way biologists work. This involves more than memorizing facts. It involves *understanding* the facts by reading them, summarizing them, checking them, and inquiring beyond them.

Contents

Part 1 The nature and variety of life

1
The characteristics and variety of living things **1**

1.1 The characteristics of life
1.2 Metabolism
1.3 The variety and classification of living things

2
Cells **11**

2.1 Cell structure
2.2 Plant and animal cells compared
2.3 Cells, tissues, and organs
2.4 Cell size
2.5 Cell division and growth
2.6 Cell division by mitosis

Part 2 Nutrition

3
Food and diet **24**

3.1 Energy value of food
3.2 Types of food and their composition
3.3 A balanced diet

4
Feeding, digestion, and absorption **35**

4.1 Digestion and absorption
4.2 Amoeba and Paramecium
4.3 Hydra
4.4 Insects
4.5 Mammals

5
Photosynthesis **52**

5.1 The discovery of photosynthesis
5.2 The mechanism of photosynthesis
5.3 Leaf structure in relation to photosynthesis
5.4 The importance of photosynthesis
5.5 Plant mineral requirements

Part 3 The maintenance and organization of life

6
Transport in mammals 66

6.1 Composition of blood
6.2 Circulation of blood in mammals
6.3 Tissue fluid and the lymphatic system
6.4 The main functions of blood

7
Support and transport in plants 80

7.1 Vascular tissue in flowering plants
7.2 Osmosis and diffusion
7.3 Osmosis in plants
7.4 Flow of water through plants
7.5 Absorption of minerals by roots
7.6 Transpiration rate
7.7 Stomata
7.8 Translocation

8
Respiration 94

8.1 The release of energy
8.2 The utilization of respiratory energy

9
Breathing and gaseous exchange 101

9.1 Respiratory organs of mammals
9.2 Respiratory organs of fish
9.3 Respiratory organs of insects

10
Homeostasis: the maintenance of a constant internal environment 112

10.1 The meaning of homeostasis
10.2 The liver
10.3 Temperature control and the skin
10.4 Excretion and the kidneys
10.5 Osmoregulation

Part 4 Sensitivity, movement, and co-ordination

11
Sensitivity and movement in plants 125

11.1 Tropisms
11.2 Phototropism
11.3 Geotropism

12
Support and movement in animals 133

12.1 Mammals
12.2 Birds
12.3 Fish
12.4 Insects
12.5 Earthworms

13
Co-ordination in animals **153**

13.1 The nervous system
13.2 The endocrine system

14
Sensitivity in animals **165**

14.1 Internal receptors
14.2 Skin receptors
14.3 Taste and smell
14.4 Insect antennae (and other surface receptors)
14.5 Vision
14.6 Hearing
14.7 Sense of balance

Part 5 Reproduction, evolution, and heredity

15
The reproductive process and reproduction in protists **182**

15.1 The reproductive process
15.2 Amoeba
15.3 Yeast and Mucor
15.4 Spirogyra

16
Reproduction in plants **189**

16.1 Mosses
16.2 Flowering plants
16.3 Life cycles of flowering plants
16.4 Vegetative reproduction
16.5 Artificial propagation

17
Reproduction in animals **207**

17.1 Hydra
17.2 Insects
17.3 Fish
17.4 Amphibia
17.5 Birds
17.6 Mammals
17.7 Birth control

18
Evolution and natural selection **228**

18.1 The theory of evolution
18.2 Evidence in support of evolution
18.3 The search for a mechanism of evolution
18.4 The theory of natural selection
18.5 Evidence in support of natural selection
18.6 Scientific critics of Darwinism

Contents

19
Variation, heredity, and genetics **239**

19.1 Mendel's experiments
19.2 Chromosomes and heredity
19.3 Modern genetics

Part 6 Interrelationships between organisms and their environment

20
Parasites and the body's defences against them **254**

20.1 Bacteria and viruses
20.2 The spread and prevention of infection
20.3 The tapeworm
20.4 The body's defences against parasites

21
Soil **268**

21.1 Composition and origin of soil
21.2 Types of soil
21.3 Cultivation of soil

22
The balance of nature **276**

22.1 The nitrogen cycle
22.2 The carbon cycle
22.3 Food chains and food webs
22.4 Pollution

Glossary **286**

Index **295**

Answers to factual recall tests **303**

Acknowledgements

The publishers would like to thank the following for permission to reproduce photographs:

A. C. Allison, p. 67; Anatomical Institute, Bern, p. 104; American Museum of Natural History, p. 229; A. Bajer, p. 21; British Museum (Natural History), p. 236; Camera Press, p. 248; J. Allan Cash, pp. 282, 283; Bruce Coleman, pp. 142, 212, 213, 215, 218; J. Free, p. 213; Philip Harris Biological Ltd., pp. 117, 175, 187; D. M. Kendal, Centre for Overseas Pest Research, and British Museum (Natural History), p. 168; Keystone Press, p. 235; J. H. Kugler, p. 11; K. R. Lewis, p. 19; D. B. Moffat, p. 123; Oxfam, p. 24; J. J. Pritchard, p. 136; Radio Times Hulton Picture Library, pp. 31, 264; Royal College of Surgeons, p. 224; Sport and General Press Agency, p. 95; Toner and Carr (1969) *Journal of Pathological Bacteriology* 97, p. 49; M. P. Whitehouse, p. 190; Sir Vincent Wigglesworth, p. 109; R. H. J. Williams, p. 282.

The author would like to thank Mr. I. D. Watson, Head of Biology at Winchester College, whose comments and criticisms on the manuscript of this book were invaluable.

1 The characteristics and variety of living things

1.1 The characteristics of life

Biology is the study of living things, but it has not yet produced a definition of life. It has, however, produced a list of the characteristics of life. It has done this by studying the differences between living and non-living things. It is not difficult, for instance, to see the differences between a man and a waxwork dummy. The man walks, eats, sees, etc., whereas the dummy does not. However, characteristics such as these cannot be applied to all living things. Most plants do not move about from place to place, eat, or perform breathing movements, but this does not mean they are not alive. Furthermore, some objects such as plant seeds show no signs of life whatsoever until they begin to grow. Lotus seeds are an interesting example: some were once grown after being stored for 160 years.

When listing the characteristics of life, therefore, only those which are common to all living things must be included. It must also be remembered that under certain circumstances all these characteristics can be suspended for a period, as in seeds, and reappear later.

It is now generally accepted that a thing is alive if it exhibits, or is capable of exhibiting, the seven characteristics listed below. From a biological standpoint, these are the features which distinguish living from non-living things:

1. *Movement*

Living things move in a directed and controlled way. In other words they move of their own accord, whereas non-living things move only if pushed or pulled by something else. Animals usually move their whole bodies and often have special organs which do this, such as fins, wings, and legs. These are called **locomotory organs** because they move the animal from place to place. Plants also move in a directed and controlled manner but generally only parts of their bodies move. Consequently, their movements are not locomotory. Leaves which turn towards the light, and roots which grow down into the soil are examples of plant movements. Plant movements are generally very slow and not always obvious to the casual observer.

2. *Sensitivity*

Living things are sensitive to their environment. This means that they detect and respond to events in the world around them. Simple organisms such as *Amoeba* (page 5) have limited sensitivity, while higher animals such as humans are more sensitive and can react to small changes in light, sound, touch, smell, taste, temperature, etc. Humans have highly developed sense organs and a complex nervous system through which responses are co-ordinated. Plants, on the other hand, have no sense organs, but the way they move shows that certain regions of their bodies are sensitive to light, gravity, water, and various chemicals.

3. *Feeding*

Living things feed. Food is the material from which organisms obtain energy for movement, and the raw materials necessary for growth and repair of the body. The scientific study of food and the different ways in which organisms feed is called **nutrition**. There are two types of nutrition:

a) *Autotrophic nutrition* Autotrophic organisms make their own food. Green plants, for example, manufacture sugar and starch from carbon dioxide and water, using the energy of sunlight to drive the necessary chemical reactions. This process is called **photosynthesis.**

b) *Heterotrophic nutrition* Heterotrophic organisms obtain food from the bodies of other organisms. This is done in various ways. **Carnivores,** such as cats and foxes, eat the flesh of animals. **Herbivores,** such as cattle, horses, and rabbits, eat plants. **Omnivores,** such as humans, eat both plants and animals. **Parasites,** such as fleas, mosquitoes, and tapeworms, live on or in another living organism called the **host** from which they obtain food. **Saprotrophs** (sometimes called saprophytes), which include many types of fungi and bacteria, obtain their food in liquid form from the decaying remains of dead organisms.

4. *Respiration*
Living things respire. Respiration is a complex sequence of chemical reactions which results in the release of energy from food. These reactions are vital to life because they provide the power for the numerous chemical and physical processes within living organisms. Most organisms respire using oxygen which they absorb from the surrounding air or water. The oxygen reacts with food substances in the body of the organism, releasing energy and producing carbon dioxide gas and water as waste substances. Most animals obtain oxygen by means of respiratory organs, such as gills and lungs. These animals carry out breathing movements to ensure that oxygen reaches these organs. These movements also ensure that waste carbon dioxide is removed from the body. Plant respiration is less obvious, but it can be detected using methods described in chapter 8.

5. *Excretion*
Living things excrete. Excretion is the removal from the body of waste products which result from normal life processes. Waste products such as carbon dioxide must be excreted. If they accumulate in the body they cause poisoning which slows down vital chemical reactions. Excretion must not be confused with **egestion**, which is the removal from the body of substances with no food value that have passed unused through the digestive system. Most animals have special excretory organs, such as kidneys.

6. *Reproduction*
Living things are able to reproduce. Unless reproduction occurs populations of organisms will diminish and eventually disappear as their members die from old age, disease, accidents, attacks from other organisms, etc. It is a fundamental law of nature that living things can only be produced by other living things, i.e. every living organism owes its existence to the reproductive activities of other living organisms. This fact has not always been accepted. At one time it was believed that life could develop spontaneously. People believed, for example, that mould formed out of decaying bread, that cockroaches were formed out of the crumbs and dust on a bakery floor, and that rotting sacks of grain turned into rats and mice. These beliefs contributed to the theory that living things can arise spontaneously out of non-living material. This theory of spontaneous generation has now been totally rejected in its original form.

7. *Growth*
Living things grow. Most animals grow until they reach maturity. Plants usually continue to increase in size throughout their life span.

1.2 Metabolism

Metabolism is a word used to describe all the chemical and physical changes within an organism which are necessary for life. It is one of the most useful technical words in biology because it summarizes the most vital processes of life. Apart from its use in this general sense, metabolism also refers to specific processes within part of an organism, such as **muscle metabolism**. In addition, it is used to refer to the part played by a particular substance within an organism's life processes, such as **protein metabolism**.

The word **metabolite** refers to substances which undergo various changes during metabolism. For example, carbon dioxide and water are metabolites used in the process of photosynthesis.

There are two different types of metabolism:

Catabolism
Catabolism is a process in which complex substances are *broken down* into simpler ones, resulting in the release of energy. Respiration is an important example of catabolism in living things.

Anabolism
Anabolism is a process in which simple raw materials are used to *build up* complex compounds. This type of reaction requires a supply of energy. Examples of anabolism are photosynthesis, and all cases of growth and repair in the bodies of organisms.

Both catabolic (breaking down) and anabolic (building up) processes occur in living things. In growing organisms anabolism proceeds at a faster rate than catabolism, whereas in most healthy adult animals the two are more or less balanced. Neither process stops until the organism dies. Furthermore, new experimental techniques have shown that the bodies of living organisms are constantly changing. Hardly any body substance remains for long in a stable condition. The idea that the adult human body is a relatively permanent structure is an illusion. Almost every part, even the solid parts such as bones, are constantly and simultaneously being broken down and remade. The raw materials for these changes come from food, and the energy needed comes from respiration.

1.3 The variety and classification of living things

Approximately 1 500 000 different organisms are known to science and more are being discovered all the time. The task of studying this enormous variety is made a little easier by the fact that organisms can be sorted, or **classified**, into groups according to similarities and differences.

Man has always used 'group words' in ordinary speech, such as dog, horse, bird, fish, and tree. These words make it easier to communicate facts and ideas. When talking about horses, for example, it is not necessary to describe these animals in detail; use of the group name 'horse' tells the listener all he needs to know. The scientific classification of living things is no more than an elaboration of this simple every-day naming and sorting process.

Something which often dismays students of biology is the use of Latin and 'latinized' words in the scientific naming of organisms. Even though some of these names are difficult to pronounce and remember they are nevertheless of great value. The names are agreed on at regular meetings of scientists from several different countries. The main object of their work is to give every organism an 'official' name for use in scientific literature. Consider the alternative. Certain organisms live in many different countries and often have a name in the language of each country. If some of the countries are large the organisms may have more than one name in the same language. If all these names were used in books and scientific papers the result would be hopeless confusion. It is far simpler to stick to official scientific names. In this way people from different countries can communicate with each other and be sure they are talking about the same organism.

Classifying organisms

Scientific classification begins by sorting organisms into very large groups called **kingdoms**. Kingdoms are divided into smaller groups called **phyla** in the case of animals (singular **phylum**), and **divisions** in the case of plants. Each phylum or division is divided into **classes**, the classes are divided into **orders**, and the orders are divided into **families**. Each family is divided into **genera** (singular **genus**), and genera are divided into **species**.

Tables 1 and 2 give examples of how various organisms are classified according to this system.

A kingdom contains a large number of organisms which have comparatively few common features. The animal kingdom, for example, contains an enormous number of different organisms but they share only

Table 1 Classification of amoeba, earthworm, and man

Group	*Amoeba*	*Earthworm*	*Man*
Kingdom	Protista	Animal	Animal
Phylum	Protozoa	Annelida	Chordata
Class	Sarcodina	Chaetopoda	Mammalia
Order	Amoebina	Oligochaeta	Primate
Family	Amoebidae	Lumbricidae	Hominidae
Genus	Amoeba	Lumbricus	Homo
Species	A. proteus	L. terrestris	H. sapiens

Table 2 Classification of bracken, pine, and buttercup

Group	*Bracken fern*	*Scots pine*	*Creeping buttercup*
Kingdom	Plant	Plant	Plant
Division	Pteridophyta	Spermatophyta	Spermatophyta
Class	Filicineae	Gymnospermae	Angiospermae
Order	Filicales	Coniferales	Ranales
Family	Polypodiaceae	Pinaceae	Ranunculaceae
Genus	Pteridium	Pinus	Ranunculus
Species	P. aquilinum	P. sylvestris	R. repens

two common features: they are all multicellular and their nutrition is heterotrophic.

Progressive sub-division of kingdoms all the way to species level produces sub-groups containing fewer and fewer organisms, but the members of the sub-groups have progressively more and more common features. For example, the only living representative of the family Hominidae is Man, but all members of this family, living or extinct, have many features in common: a large brain, a prominent chin, small canine teeth, and the ability to use tools.

Naming organisms

Every species of organism known to science, living and extinct, has been given a double scientific name: one name for its genus (the generic name) and one for its species (the specific name). For example, the scientific name for man is *Homo sapiens*. This is usually written *H. sapiens* for short. The generic name is always written with a capital letter and the specific name with a small letter.

Definition of species

The word species describes any group of organisms which exhibit two major characteristics. First, they all share the same general physical shape. Second, and of far greater importance, members of a species have the ability to mate together and produce young similar to themselves, which in turn have the same reproductive capacity. In short, the fundamental characteristic of a species is the ability of its members to mate and produce fertile offspring.

The fact that this is only a rough convenient definition is shown by the reproductive behaviour of different yet very similar species. Horses and asses (donkeys) are different species, yet they can interbreed. The offspring, depending upon which species provides the father, are called mules or hinnies. It is important to note, however, that such offspring are infertile.

The problem of identifying organisms

It is often necessary to identify various specimens during the course of a biological investigation. This task is quite easy when it involves telling the difference between major groups. For instance, there is little difficulty in distinguishing between beetles and butterflies. In detailed studies, however, there remains the problem of identifying organisms all the way to genus or species level. Having found a beetle, for example, how is it to be identified from among the 300 000 different species in the world? First, it is almost certain that the beetle's country of origin will be known, which reduces the range of possibilities a little—there are about 3700 British beetles. Second, there is the possibility of making an identification by looking through books with accurate illustrations of living things from various countries, but in the case of organisms which look very much alike it is difficult to know where to begin.

Table 3 Key to vertebrate animals

1	Hair present	Mammals
	Hair absent	2
2	Feathers present	Birds
	Feathers absent	3
3	Breathe with lungs	4
	Breathe with gills	5
4	Dry scaly skin	Reptiles
	Moist scaleless skin	Amphibians
5	Fins and cartilage skeleton	Chondrichthyes (cartilage fish)
	Fins and bony skeleton	Osteichthyes (bony fish)

Devices called **keys** are used to make this task easier. Table 3 shows one of the simplest types of key. It is made up of brief descriptions arranged in numbered pairs. To use this type of key you begin at the first pair of descriptions and decide which one fits the organism to be identified. The key either names the organism or gives the number of the next pair of descriptions which must be consulted. This procedure is continued until an identification is made.

Table 4 shows an alternative form of the same key. Here, the descriptions are arranged as a diagram. Starting at the top, follow the branches by deciding which descriptions fit the organism to be identified.

Table 4 An alternative form of Table 3

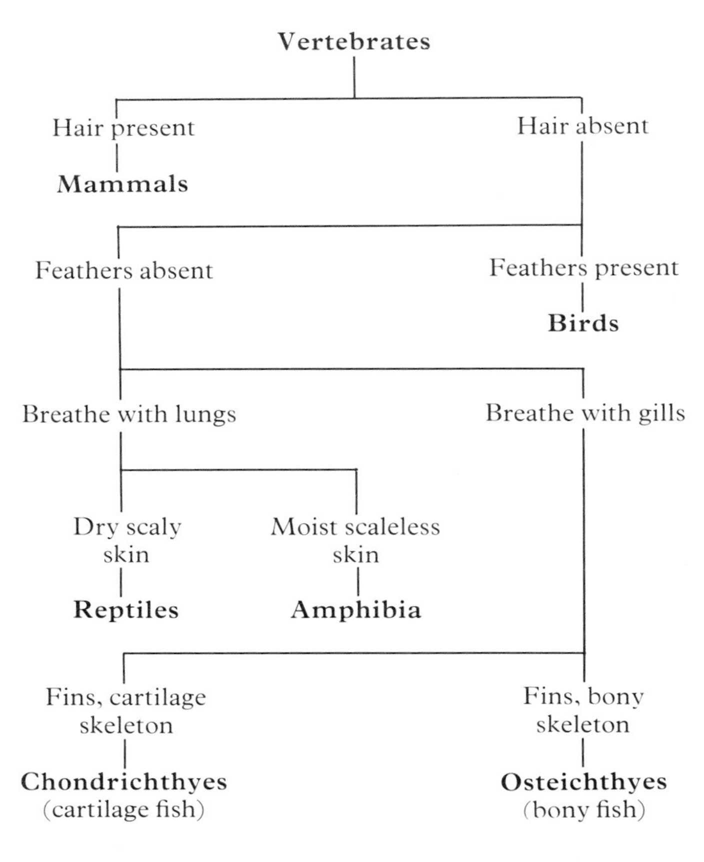

The following pages show an illustrated outline classification of some of the main phyla. It must be remembered that this is only one of several ways in which living things can be classified.

Protista kingdom

Protists are organisms whose bodies are relatively simple in structure. They consist of cells with a well-defined nucleus.

Phylum Protozoa
Protozoa are microscopic unicellular organisms. The main classes are:

Class Rhizopoda Protozoa which move by means of pseudopodia. Some are enclosed in shells made of silica (e.g. *Radiolarians*) or lime (e.g. *Foraminiferans*). Others consist of naked protoplasm (e.g. *Amoeba*).

Class Ciliophora (ciliates) Protozoa which possess cilia used in movement and feeding. They usually have a large and a small nucleus (e.g. *Vorticella*).

Class Mastigophora (flagellates) Protozoa which move by means of whip-like flagella. Many are green (e.g. *Euglena*) and capable of photosynthesis.

Phylum Mycophyta (fungi)
Fungi are mostly multicellular, but yeast is an exception. They are usually made up of fine threads called hyphae, which are known collectively as the mycelium of the fungus. Many are saprotrophic (e.g. *Mucor*, mushrooms). Others are parasitic, causing plant and animal diseases such as potato blight and ringworm.

Phylum Chlorophyta (green algae)
These algae live in the sea, in fresh water, and in damp places on land. They occur as single cells, hollow balls of cells, fine threads (e.g. *Spirogyra*), and as hollow tubes.

Phylum Phaeophyta (brown algae)
These algae live in the sea. They often reach several metres in length (e.g. *Fucus*, kelp).

Phylum Bacilliarophyta (diatoms)
These are unicellular algae with cell walls made of silica. They consist of two halves which fit together like a box and lid.

Plant kingdom

Plants are multicellular autotrophic organisms.

Division Bryophyta
Bryophytes are plants without true roots, stems, or leaves, but they often possess structures which resemble these parts of higher plants. Their life cycle consists of two alternating generations: a sexual or gametophyte generation, and an asexual or sporophyte generation. The gametophyte is dominant, but often bears the sporophyte as a parasite. The main classes are:

Class Hepaticae (liverworts) Bryophytes in which the plant body is either flattened and liver-like in shape, or consists of a 'stem' with two parallel rows of leaf-like structures. The sporophyte is short-lived (e.g. *Pellia*, *Marchantia*).

Class Musci (mosses) Bryophytes with structures resembling roots, stems, and leaves. The sporophyte is long-lived (e.g. *Polytrichum*, *Sphagnum*).

Division Pteridophyta (ferns)
These plants have true roots, stems, and leaves, but no flowers. The sporophyte generation is dominant and bears sporangia on the lower surface of the leaves (e.g. male fern, hart's-tongue fern, horsetail).

Division Spermatophyta (seed plants)
These are plants with a complex and highly organized plant body. They produce seeds. The main classes are:

Class Gymnospermae Seeds not enclosed within a fruit but within a cone (e.g. pine, fir, spruce).

Class Angiospermae Flowering plants. Seeds develop within an ovary, the walls of which become a fruit. *Monocotyledons* are types which germinate with one seed leaf, or cotyledon (e.g. grasses, iris). *Dicotyledons* germinate with two seed leaves (e.g. oak, buttercup).

Animal kingdom

Animals are multicellular heterotrophic organisms.

Phylum Coelenterata
Coelenterates are aquatic animals found mainly in the sea. They have a hollow sac-like body with a single opening, the mouth, at one end which is typically surrounded by tentacles. The tentacles and body are armed with sting cells, or nematocysts. The body wall consists of two layers of cells, illustrated in Figure 2.4 (e.g. *Hydra*, jellyfish, sea anemones).

Phylum Annelida (true worms)
Annelid worms have bodies consisting of many similar segments, the boundaries of which are marked by external grooves around the body. Some live in the soil (e.g. earthworms), some in the sea under sand or mud (e.g. sandworms) and others are parasites (e.g. leeches).

Phylum Arthropoda (arthropods)
The largest animal phylum. Arthropods possess a segmented body enclosed in a jointed exoskeleton, with paired jointed legs. Many have compound eyes and antennae. The main classes are:

Class Crustacea Arthropods with two pairs of antennae. Many have a hard chalky exoskeleton (e.g. crab, lobster); others have a more delicate exoskeleton (e.g. woodlouse, water flea).

Class Arachnida (spiders) Arthropods with four pairs of legs and no antennae (e.g. spiders, scorpions, ticks).

(continued on page 10)

Protista kingdom

Protozoa

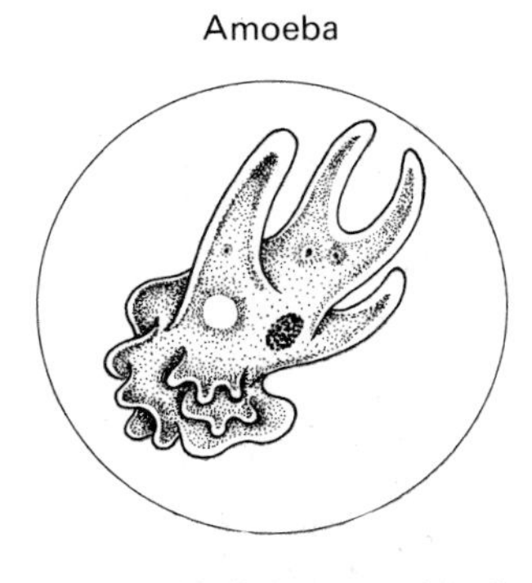
Amoeba

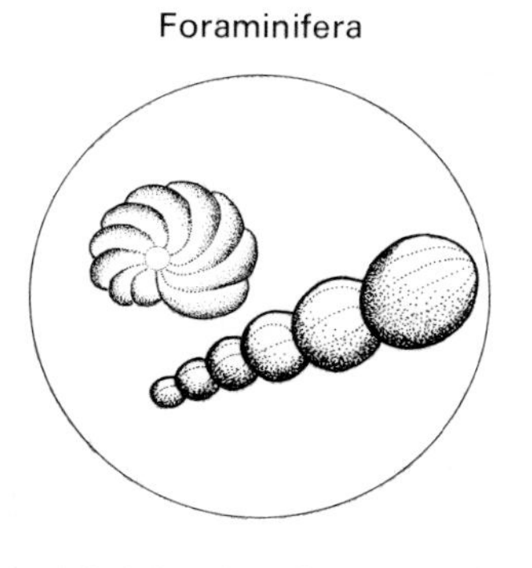
Foraminifera

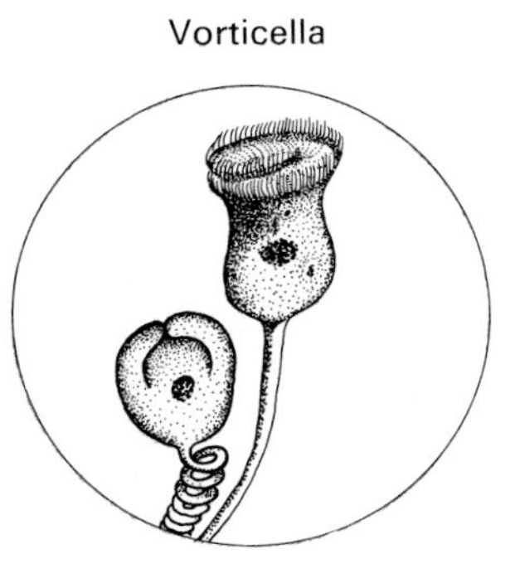
Vorticella

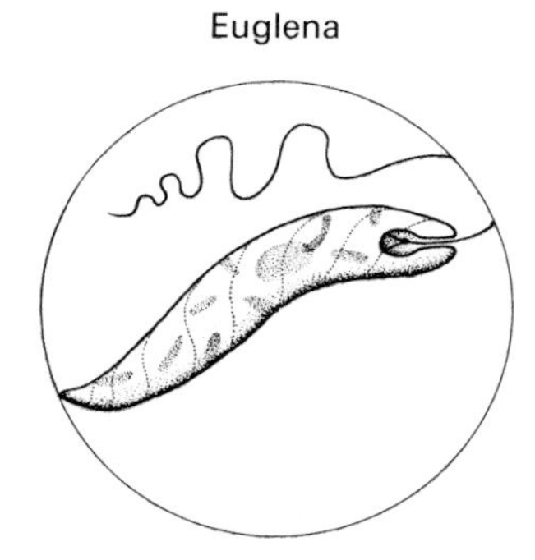
Euglena

Mycophyta (fungi)

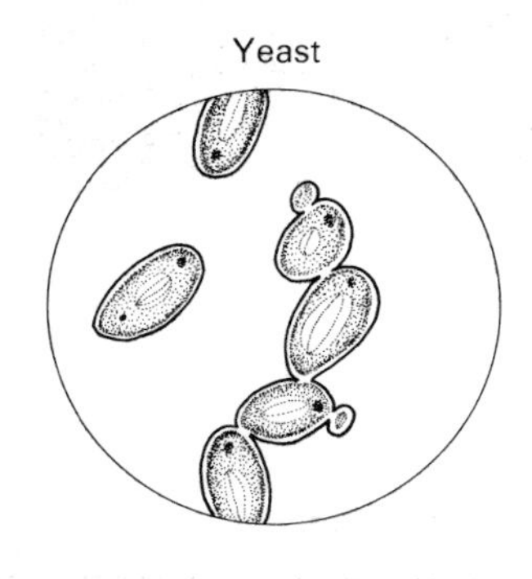
Yeast

Mushroom

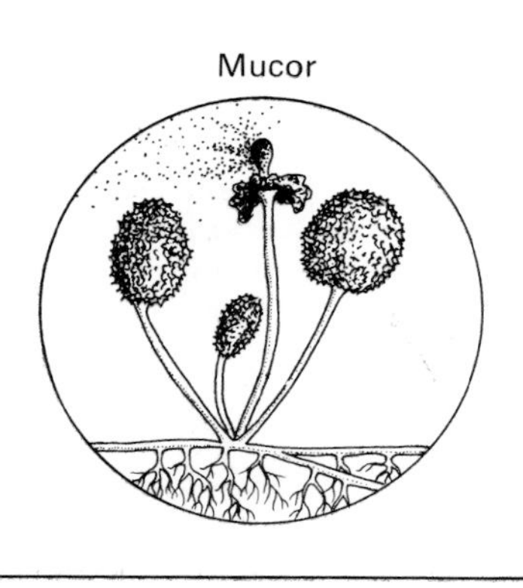
Mucor

Chlorophyta (green algae)

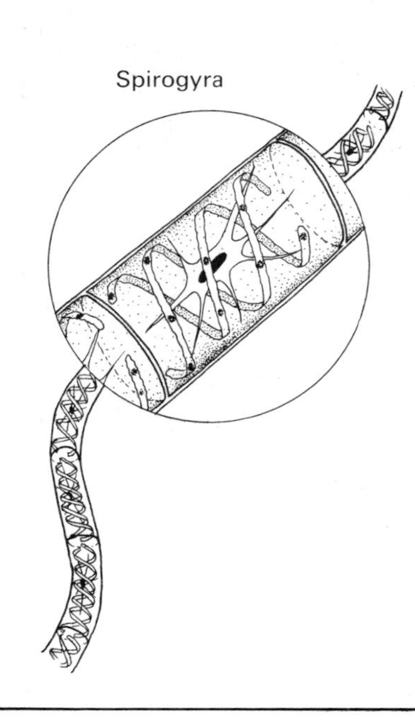
Spirogyra

Phaeophyta (brown algae)

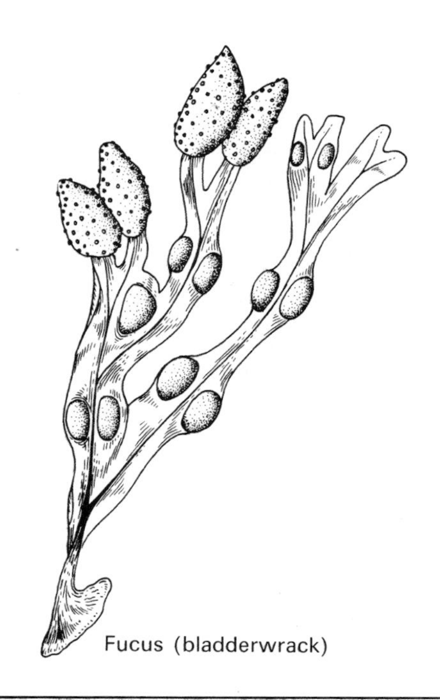
Fucus (bladderwrack)

Bacilliarophyta

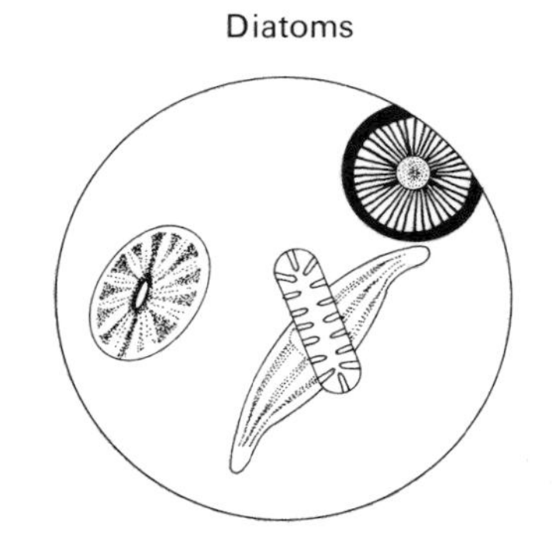
Diatoms

Bryophyta (mosses and liverworts)

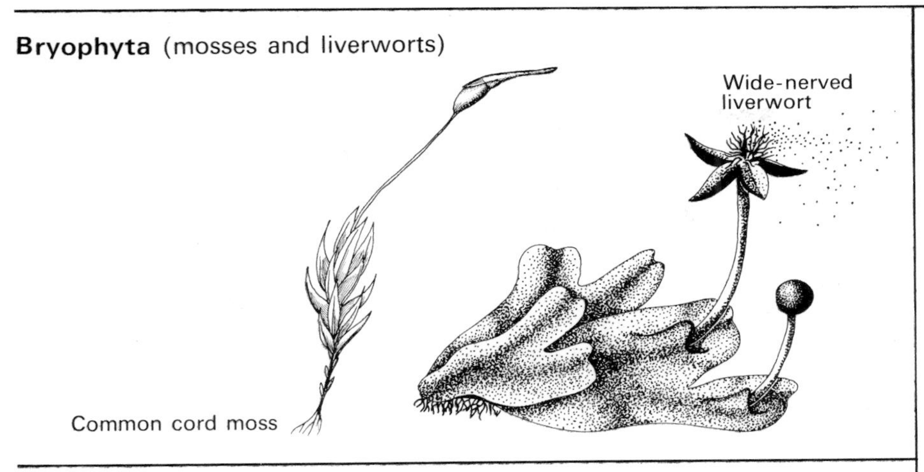

Gymnospermae (pines, firs, and spruces)

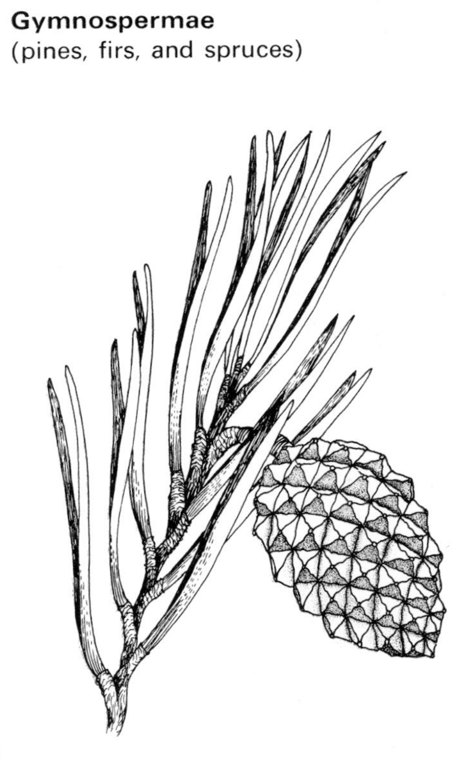

Scots pine (branch with female cone)

Pteridophyta (ferns)

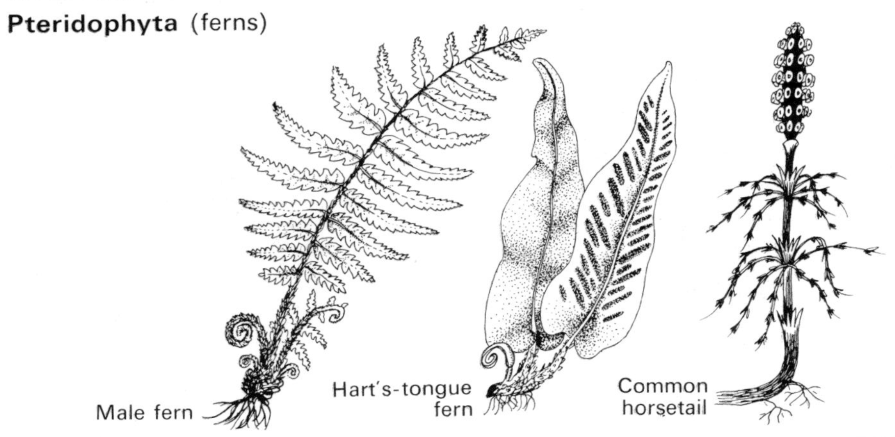

Angiospermae (flowering plants)

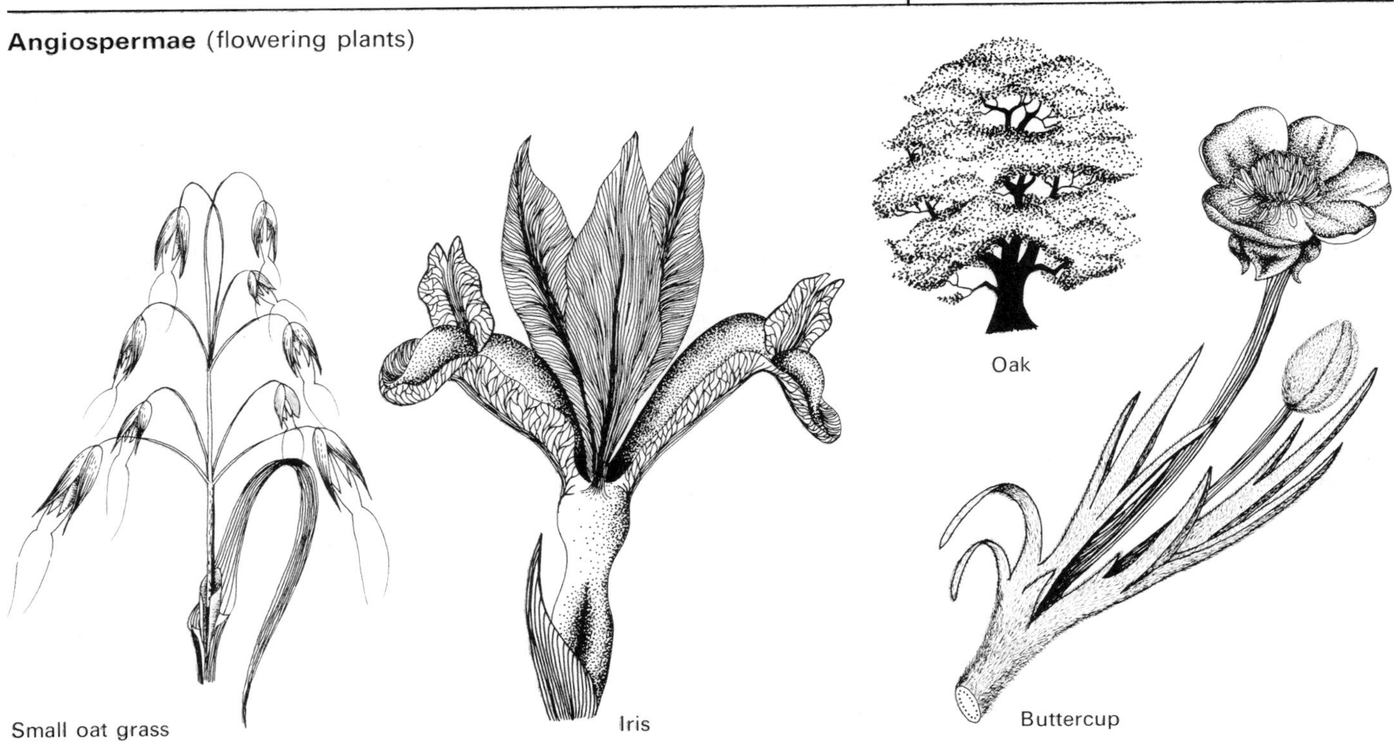

Coelenterata

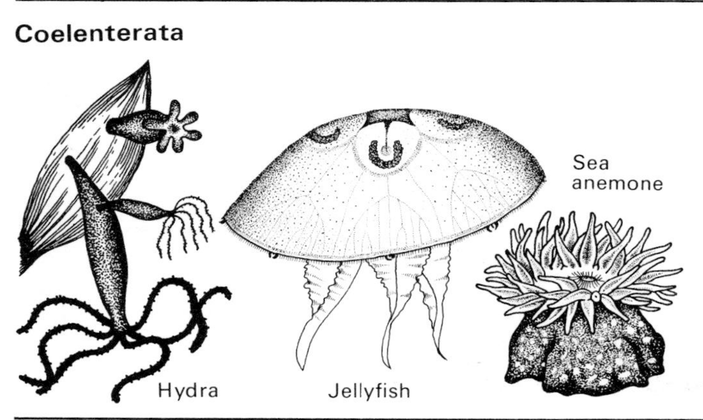

Annelida

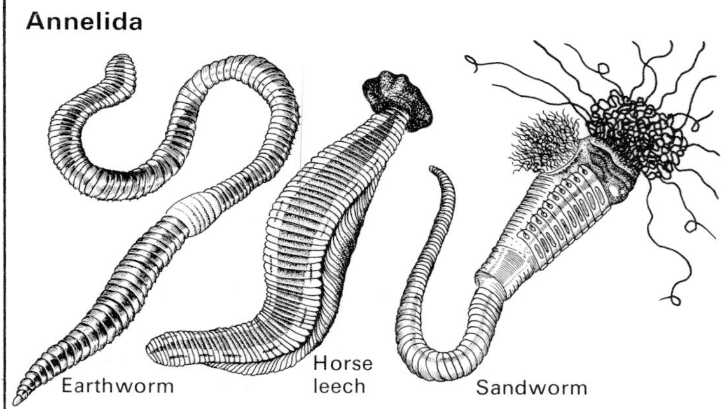

Arthropoda

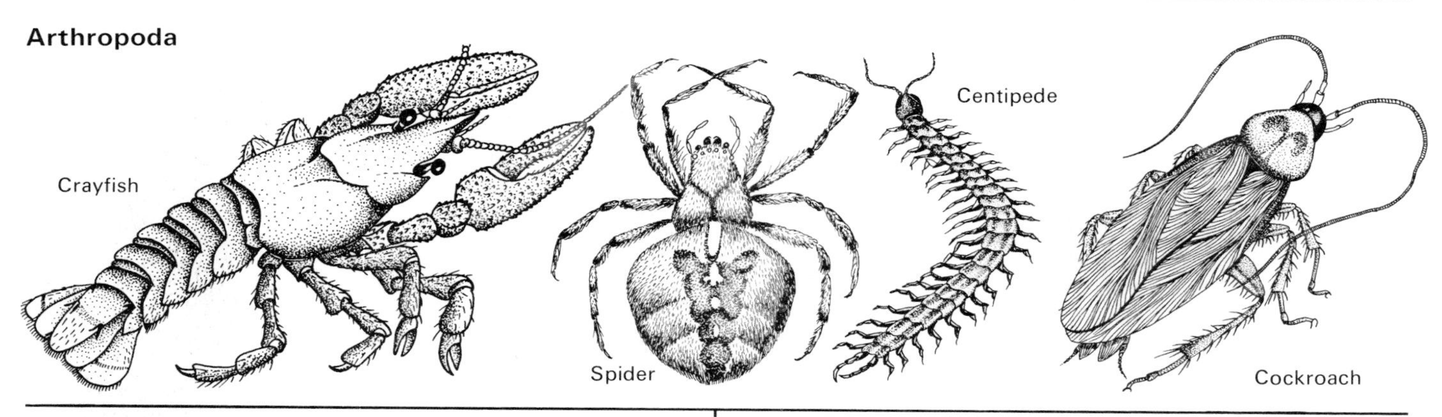

Mollusca

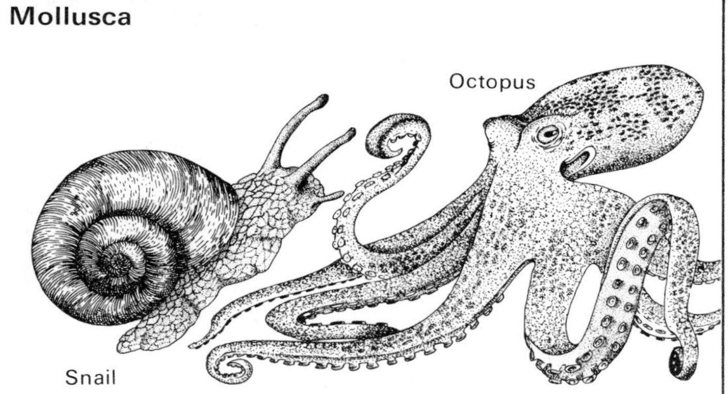

Echinodermata

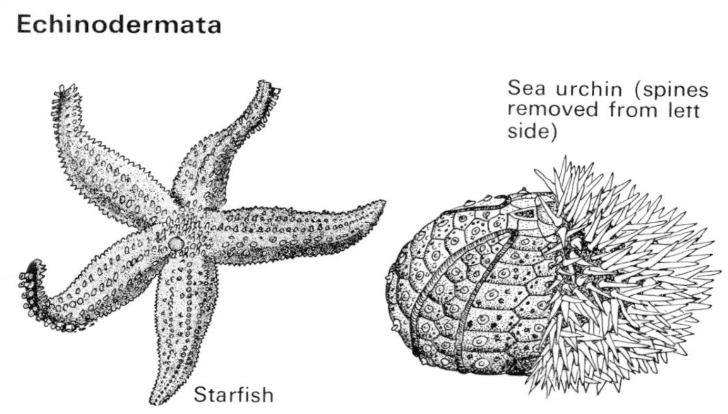

Chordata

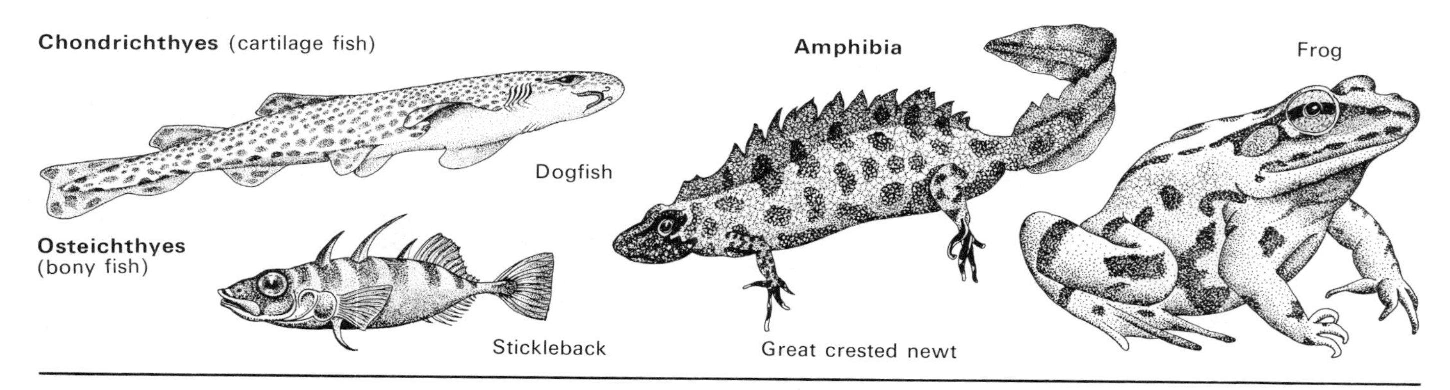

Chordata (continued)

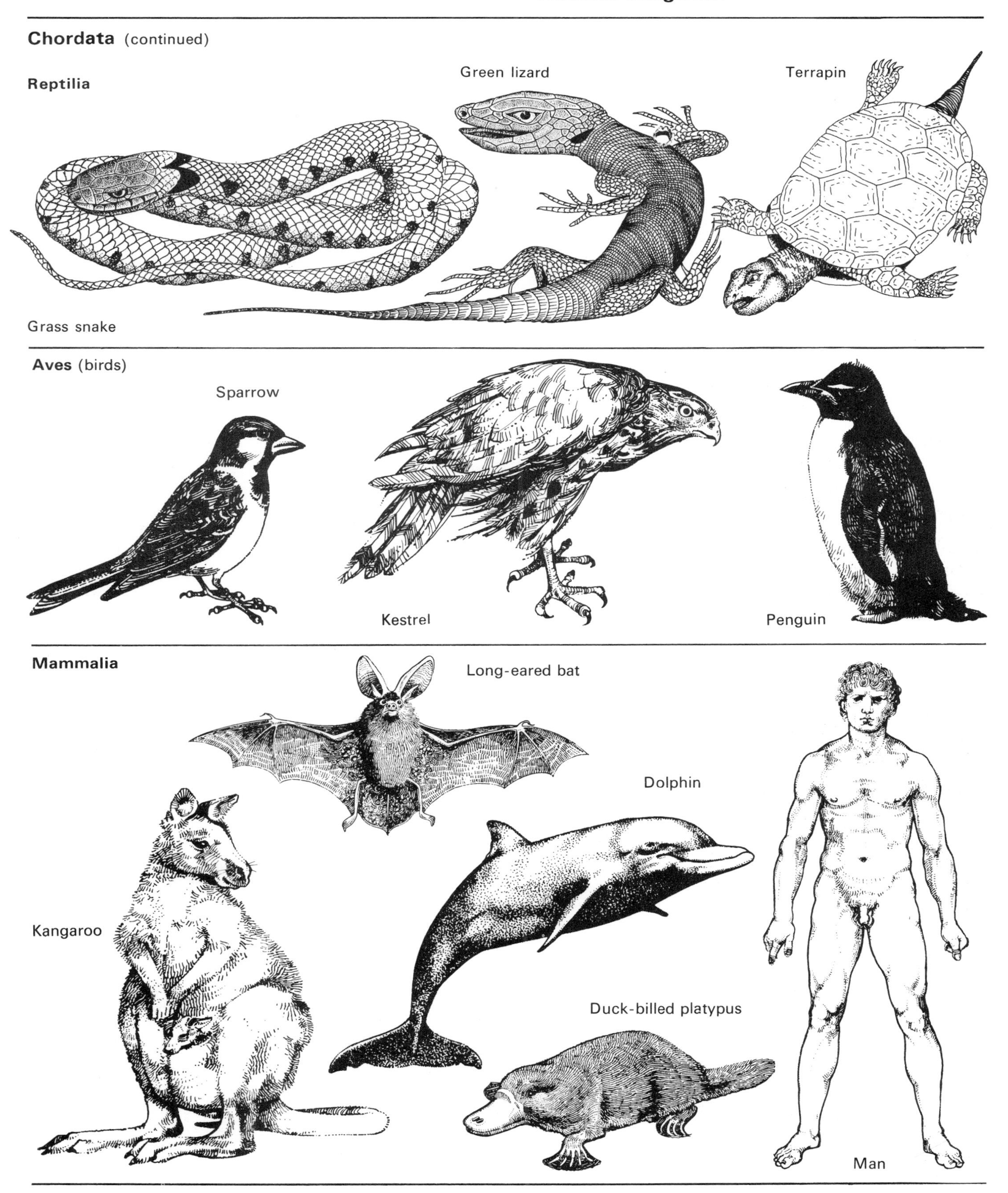

Animal Kingdom *(continued)*

Class Insecta (insects) Arthropods with three pairs of legs, and often wings. The body is divided into a head (with eyes and antennae), a thorax (with legs and wings), and an abdomen (e.g. cockroach, butterfly, bee, earwig, flea).

Class Myriapoda Arthropods with a long body consisting of many segments each bearing one or two pairs of legs (e.g. centipede, millipede).

Phylum Mollusca (molluscs)
Molluscs have a soft body enclosed in one or two shells. Some have a coiled shell and move about on a soft slimy 'foot' (e.g. snails); some have two shells (e.g. mussel, oyster); and in some the shell is inside the body (e.g. slug, octopus).

Phylum Echinodermata
Echinoderms are marine animals with a spiny skin and a body usually divided into five radially arranged portions, such as the 'arms' of a starfish, which bear tube feet (e.g. starfish, sea urchins, sea cucumbers).

Phylum Chordata
Chordates are animals with a notochord which becomes the backbone in higher types, a hollow dorsal nerve cord, a ventral heart, and gill slits at some stage in their development.

Sub-Phylum Vertebrata (vertebrates)
These are chordates with a backbone, or vertebral column. The main classes are:

Class Chondrichthyes Fish with a skeleton of cartilage (e.g. dogfish, sharks, rays).

Class Osteichthyes Fish with a skeleton of bone and a swim-bladder (e.g. stickleback, cod).

Class Amphibia Vertebrates with a moist, scaleless skin. They live on land, and have lungs. They lay eggs, mostly in water, which develop into larvae with gills (e.g. frogs, newts, salamanders).

Class Reptilia (reptiles) Vertebrates adapted to life on land (e.g. snakes, lizards) although some live in water (e.g. terrapin). They have a dry scaly skin, breathe with lungs, and lay eggs with a tough leathery shell.

Class Aves (birds) Vertebrates with a constant warm body temperature, feathers, and forelimbs adapted as organs of flight (e.g. kestrel, sparrow). Some are flightless (e.g. penguin, kiwi).

Class Mammalia (mammals) Vertebrates with a constant warm body temperature and a hairy skin. The young are born alive and suckled with milk from the mother's mammary glands. *Monotremes* are primitive egg-laying mammals (e.g. duck-billed platypus). *Marsupials* are mammals which carry their young in pouches (e.g. kangaroos, koalas). *Primates* are the most advanced animals, with large brains and eyes positioned at the front of the head (e.g. monkeys, apes, man).

Summary and factual recall test

Examples of locomotory organs are (1–name three). Plant movements are not locomotory because (2). Examples of plant movements are (3–name two).

Higher animals are sensitive to (4–name five things to which animals are sensitive), and they have complex nervous systems which (5) their responses. Plants are sensitive to (6–name three things to which plants are sensitive).

Food provides (7) for movement, and raw materials necessary for (8) and (9) of the body. (10) nutrition is typical of green plants, in which (11) and (12) are absorbed and made into food using the energy of (13). This process is called (14). The main types of heterotrophic organism are (15–name five).

During respiration (16) is absorbed and reacts with food in the body releasing (17), and producing (18) gas as waste. Examples of respiratory organs are (19–name two).

(20) is the removal of waste substances produced by the body, and must not be confused with (21) which is the removal of undigested substances from the intestine.

Living organisms arise only from (22). This is the opposite of the now disproved theory of (23) generation which proposed that (24).

One difference between animal and plant growth is that (25).

Metabolism is the word which stands for (26). An example of catabolism is (27), and (28) is an example of anabolism. In a healthy adult animal catabolism and anabolism are more or less (29).

The scientific name for a human being is (30). The first of these names denotes the (31) and the second the (32). Scientific names are useful because (33).

A species is a group of organisms which can mate and produce (34) young.

2
Cells

In 1665 an English naturalist called Robert Hooke made a chance observation while using a microscope which he had designed himself. When examining a thin slice of cork, a substance which comes from the bark of a tree, he saw that it looked 'much like a honey-comb' consisting of 'a great many little boxes'. Hooke called these boxes **cells**, from the Latin for 'a little room'.

This seemingly trivial incident is important because it was the first time anyone had noticed that living things are not necessarily made up of continuous material, but sometimes appear to consist of separate units. Furthermore, Hooke's use of the word 'cell' to describe these units has survived to this day and has become a fundamental part of the language of biology.

It must be remembered, however, that Hooke was a long way from understanding the true nature of cells. To him a cell was no more than an empty space surrounded by dead walls of cork. He probably never suspected that these spaces once contained living material. Today the word 'cell' refers to this living material rather than an empty space. Cells are alive, and are not concerned only with the formation of cork; they are a vital part of almost every type of living thing.

Two of the most popular ways of describing cells are to call them 'the units of life', or 'the building-blocks of which living things are made'. These descriptions are useful because they emphasize that cells are the structural units of life. In simple terms cells are like the bricks which make up a wall. But bricks are dead, identical in shape, and quite large; cells are living, of many different shapes, and microscopic in size.

The human body is made up of several million million cells. These are invisible except under high magnification because they measure on average between 0.005 mm and 0.02 mm in diameter. If it were possible to increase the size of a man to two hundred times his normal volume his cells would still be only the size of a pin-head. But if his size could be increased a million times his cells would then be the size of a cricket ball and it might be possible to see a little of their structure with the naked eye. The electron-micrograph below shows an animal cell magnified 7500 times. Figure 2.1A is an attempt to illustrate the appearance of a cell, separated from an animal and magnified about a million times. Figure 2.1B is a diagrammatic cross-section of a cell showing the internal structure.

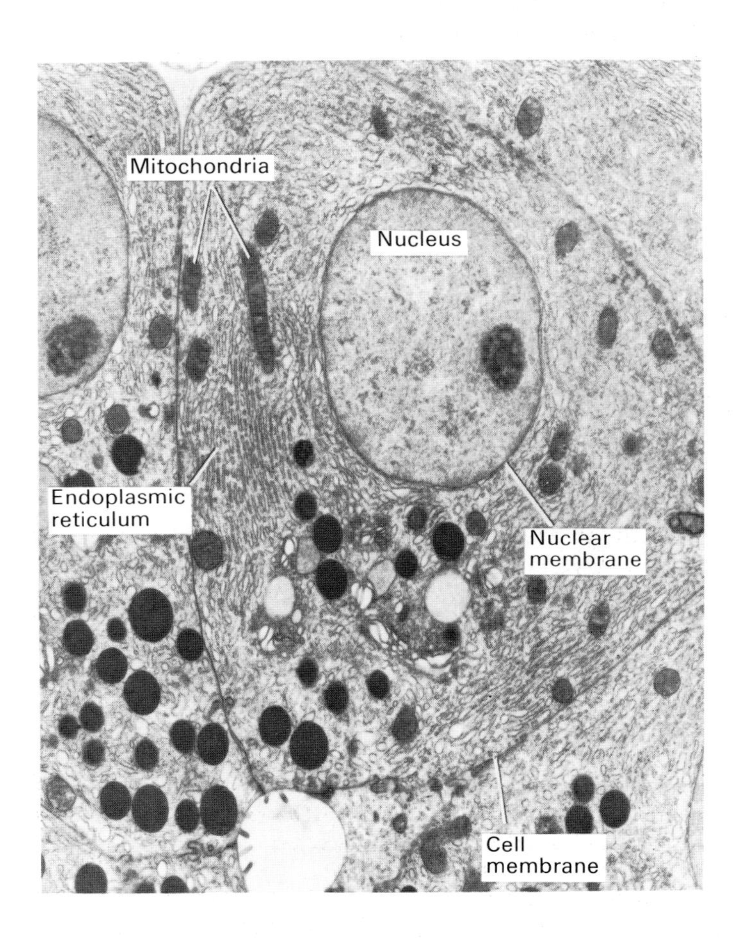

Electronmicrograph of an animal cell × 7500

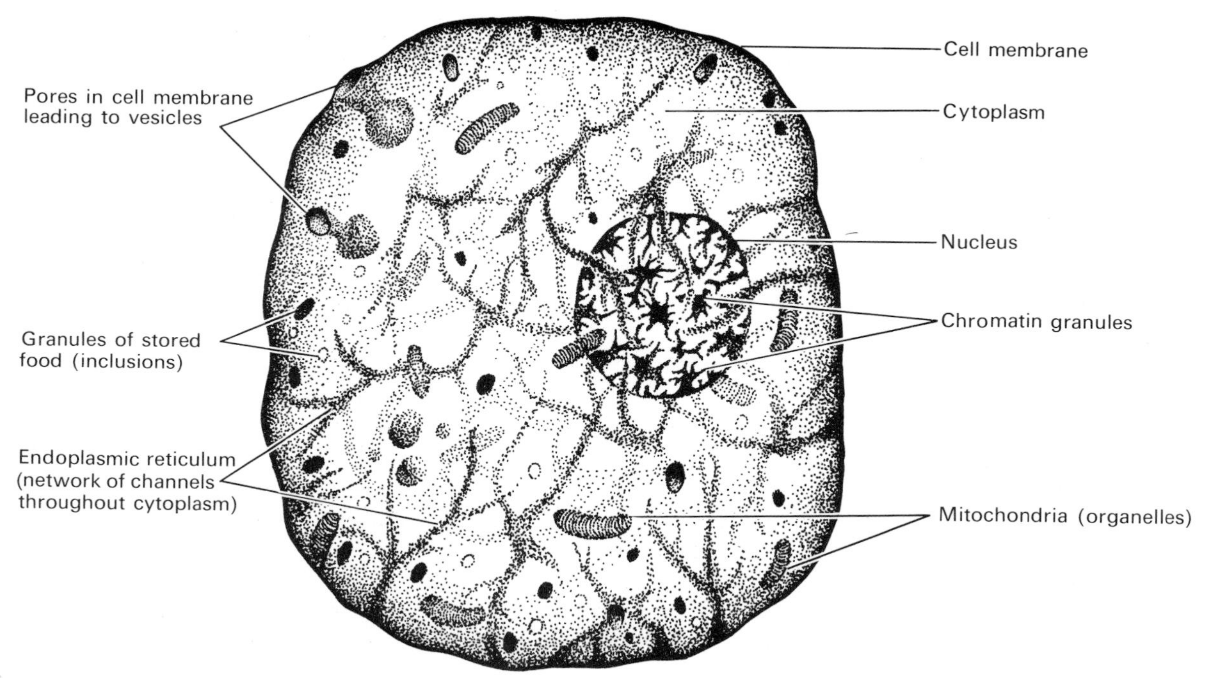

Fig. 2.1A An unspecialized animal cell magnified approximately one million times

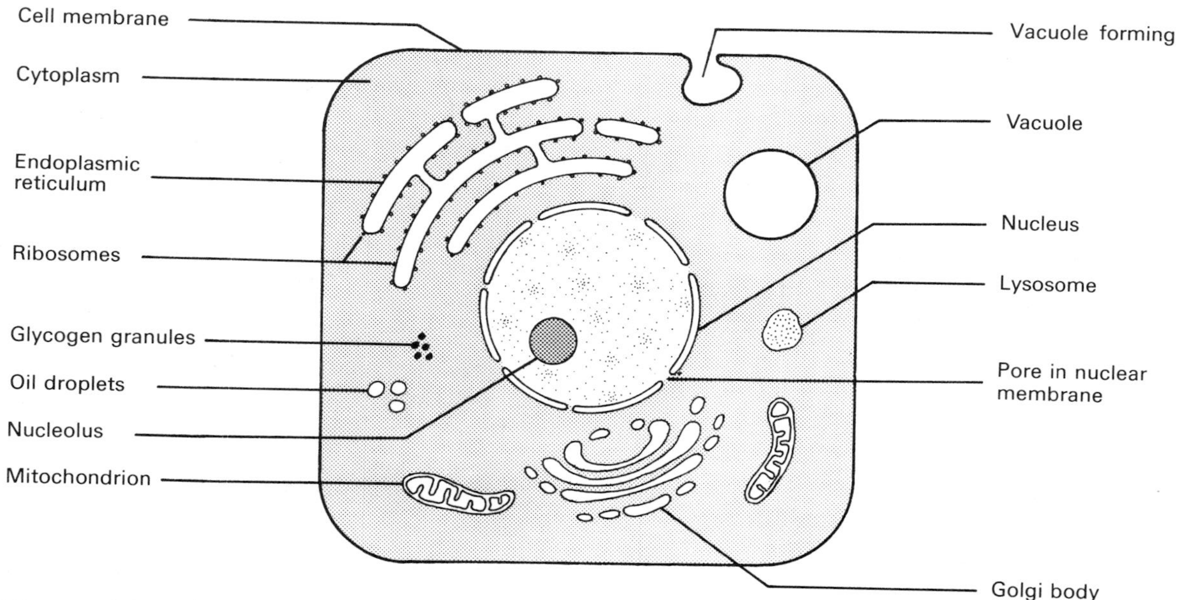

Fig. 2.1B Diagram of the main parts of an animal cell

2.1 Cell structure

One of the most astonishing things about cells is that they nearly always have the same basic structure, no matter what their function is or what organism they are found in. The single cell which forms the body of an amoeba, a brain cell of a frog, and a leaf cell of a buttercup all have certain features in common. All cells contain a round or oval object called a **nucleus**, surrounded by a jelly-like substance called **cytoplasm**, both of which are enclosed within a very thin skin known as the **cell membrane**.

Cell membrane

The cell membrane is 0.000 01 mm thick and forms the outer boundary of the cell. It is here that all exchanges take place between a cell and its surrounding environment. In a way which is not yet fully understood this membrane allows certain chemicals to pass in and out of the cell, but prevents the passage of others. Hence, cell membranes are said to be **semi-permeable** or, to be more accurate, **selectively permeable**.

Cytoplasm

The term cytoplasm refers to all the living substance of a cell except the nucleus. Cytoplasm contains a number of sub-units of which there are two types: **inclusions** and **organelles**.

Inclusions consist of a variety of materials, such as starch grains, fat globules, and crystals of excretory substances, which are stored temporarily in the cell. Organelles are more permanent structures and are active in the life of the cell. Here are some examples:

Mitochondria These are organelles found in all animal and plant cells. Mitochondria are generally rod-shaped and contain substances called **enzymes** which, in this case, are concerned with respiration. By means of these enzymes, mitochondria control the release of energy within a cell.

Endoplasmic reticulum This is a system of membranes which runs throughout the cytoplasm. Between the membranes are channels which are thought to allow soluble materials to become distributed around the cell. **Ribosomes** are organelles attached to one surface of the reticulum membranes. Ribosomes are concerned with the manufacture of proteins.

The nucleus

At least one nucleus is found in the cells of all animals, plants, and protists.

Using very tiny knives and needles, manipulated under a microscope, it is possible to remove the nucleus from large cells, particularly of organisms such as *Amoeba*. If this is done the cytoplasm of the organism dies after about one week unless the nucleus is replaced by one from another amoeba. Since the replacement nucleus restores the organism's normal life processes it is assumed that nuclei are vital to the continuation of cell metabolism.

It has also been observed that after the removal of its nucleus a cell can no longer reproduce: a process which involves division of the cell into two halves. This, along with other evidence, shows that nuclei play an essential part in cell division.

Nuclei of living cells are almost transparent. They are usually examined by first killing the cell and then treating it with coloured stains. After this, a nucleus appears to consist of a darkly-stained irregular network composed of tiny granules which are made of **chromatin.** During cell division the chromatin granules are reorganized into rod-shaped objects called **chromosomes**. The structure and significance of chromatin and chromosomes are described in section 2.5 of this chapter, and in chapter 19.

2.2 Plant and animal cells compared

Having described features common to both plant and animal cells it is now necessary to discuss those which are found only in one group or the other.

Cellulose

A tough and fairly rigid layer of a substance called cellulose completely surrounds the cell membrane in plant cells. The cellulose forms the **cell wall** (Fig. 2.2B). The presence of this cell wall around every cell in a plant gives a great deal of support to the plant.

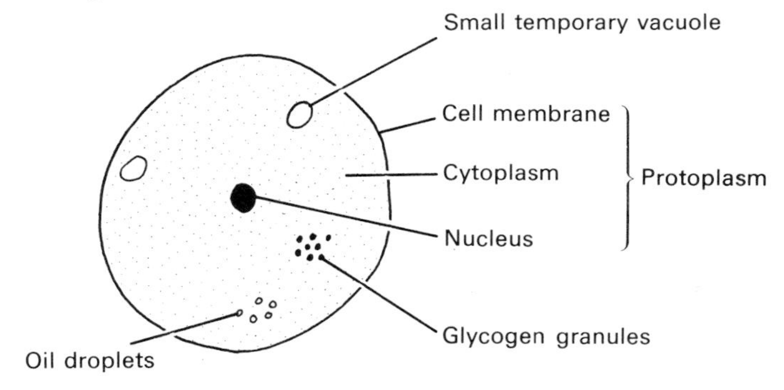

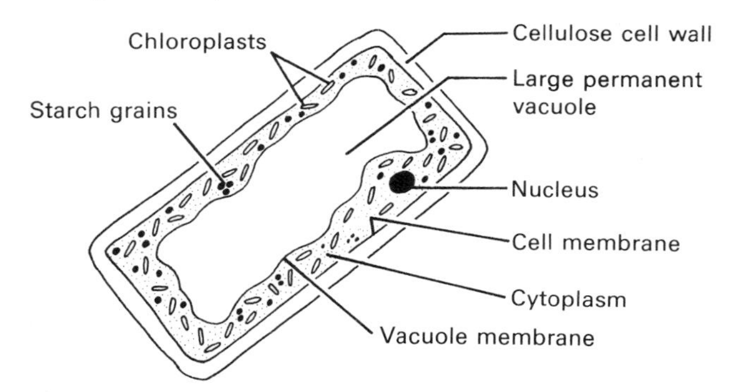

Fig. 2.2 Animal and plant cells compared

Plant cells are not isolated from each other by their cellulose cell walls because cellulose is permeable to all fluids and is perforated at intervals by tiny holes through which cells are interconnected by fine cytoplasmic threads. The presence of a cell wall makes plant cells clearly visible as distinct units when viewed under a microscope.

Almost all types of animal cell are naked: that is, there is no wall of any kind outside the cell membrane. Animal cells have a less distinct outline when seen under a microscope.

Vacuoles

A vacuole is a space filled with fluid in the cytoplasm of a cell. Vacuoles are lined with a semi-permeable membrane similar to the cell membrane. Mature plant cells usually have a large permanent vacuole at their centre (Fig. 2.2B). Small temporary vacuoles, or **vesicles**, are often found in animal cells (Fig. 2.2A), and are particularly common in fresh-water protists such as *Paramecium* (Fig. 2.3).

Chlorophyll

Chlorophyll is a green substance which absorbs light for use as a source of energy in the chemical reactions of photosynthesis. Chlorophyll is present in most plants, and is usually though not exclusively found in the cells of leaves. It is always contained within disc-shaped organelles called **chloroplasts**. Animal cells never contain chlorophyll.

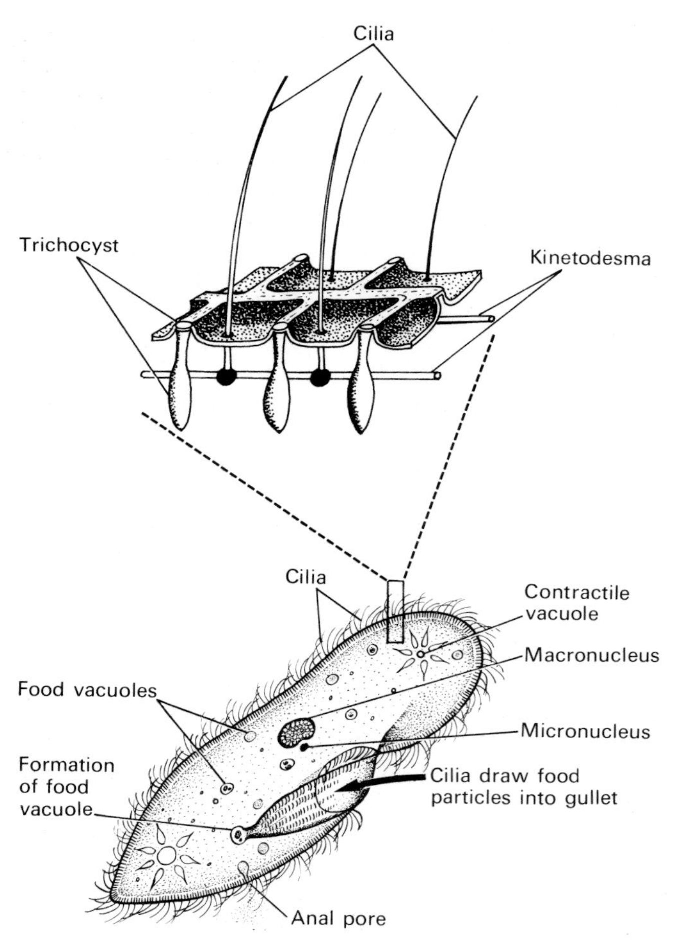

Fig. 2.3 Structure of *Paramecium* to show specialization within a unicellular organism

2.3 Cells, tissues, and organs

It has already been pointed out in chapter 1 that many organisms in the protista kingdom consist of only one cell: i.e. they are **unicellular**. Certain other protists and all plants and animals consist of many cells, and so they are called **multicellular** organisms.

One of the most fundamental characteristics of both unicellular and multicellular organisms is that their bodies consist of parts specialized to perform particular functions, such as locomotion, feeding, or digestion. Great advantages arise from having specialized parts. The following analogy may help to explain why.

Compare two imaginary villages. In village A every person tries to be independent of every other person by growing his own food, building his own house, making his own clothes, and so forth. In village B some people are trained to spend their working lives growing food, others to build houses, and others to make clothes. Under these circumstances the various groups of workers in village B serve the community as a whole in addition to looking after their own needs. The outcome is that village B has more and better quality food, houses, clothes, etc., because it is much more efficient to have groups of workers specially trained to do one particular job.

In a sense village B is like the body of an organism. The groups of specialized workers are roughly equivalent to specialized parts of the body. The workers serve the village as a whole. In a similar way, the parts of the body serve the body as a whole. The efficiency which results from the division of labour in village B is comparable to the efficiency which results from 'dividing' the 'labour' involved in the processes of life among different parts specialized for one particular function in an organism.

Division of labour in unicellular organisms

Figure 2.3 illustrates the structure of a unicellular protist called *Paramecium*. Its body consists of organelles which are specialized regions of cytoplasm with one particular function. It is possible to compare some of these organelles with parts of the human body.

Cilia, for example, move the *Paramecium* by beating rhythmically; they act rather like human limbs. The beats are co-ordinated by another organelle called the **kinetodesma**, which corresponds to the nervous system in man. *Paramecium's* **gullet** and its associated cilia, together with the **food vacuoles** and **anal pore**, can be compared with human feeding and the digestive system, and in certain minor respects the **contractile vacuoles** of *Paramecium* compare with the human excretory system.

Thus, even in microscopic creatures there can be a degree of specialization and division of labour; the single cell which makes up the body of a unicellular organism is often biologically equivalent to the whole body of a multicellular organism.

Division of labour in multicellular organisms

The cells of a multicellular organism lead a double life, similar to the people living in the imaginary village B. Consider the workers for a moment. For part of the time they act as individuals: they wash themselves, feed themselves, or read a book. At other times they work together in groups which serve the community. The cells of an organism also act as individuals: they absorb food, respire to obtain energy, or manufacture proteins. In addition, the cells work together in groups which 'serve' the whole organism, and these groups show special structural adaptations enabling them to perform their particular function. Groups of specialized cells are called **tissues**. Muscle tissue, for example, consists of cells adapted in a way which enables them to contract and cause movement, and nerve cells in nervous tissue are so adapted that they can transmit nerve impulses around the body.

When different tissues work together they form an **organ** of the body. The heart is an example of an organ which is composed of muscle and connective tissue. Several organs working in conjunction form an **organ system**. The circulatory system, which is composed of the heart, blood vessels, and blood, is an example.

Animal tissues Figure 2.4 illustrates the structure of *Hydra*, an animal belonging to the Coelenterates (page 5). Animals of this group are of interest because their cells show some specialization into tissues, but not into organs. Coelenterates may be thought of as intermediate between unicellular organisms and advanced multicellular animals such as man. Some human tissues are illustrated in Figure 2.5.

The body wall of a hydra consists of an inner and an outer layer of cells with a non-cellular, jelly-like layer called **mesogloea** sandwiched between them. The two layers of cells are classed as tissues because

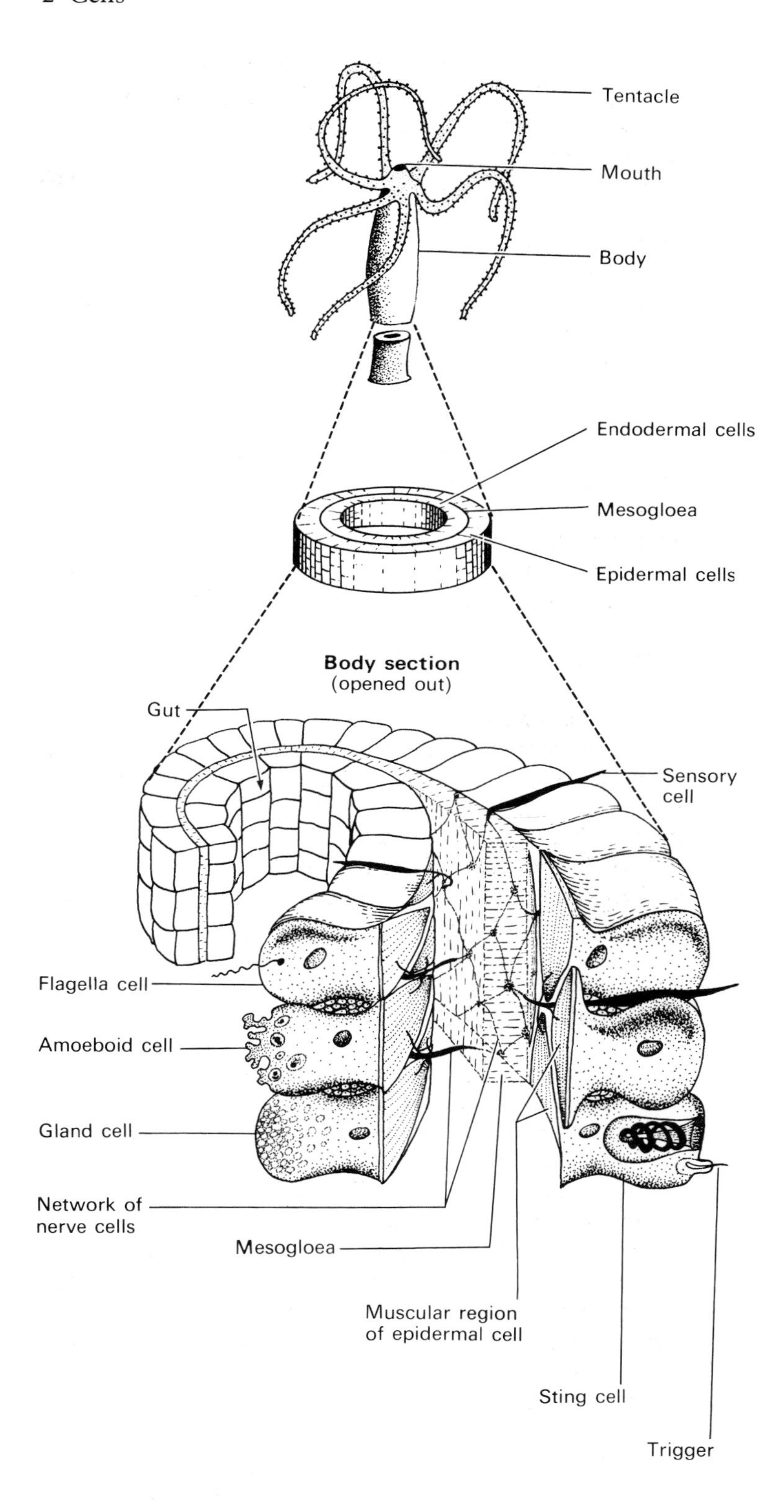

Fig. 2.4 Structure of *Hydra* to show specialization within a primitive multicellular organism

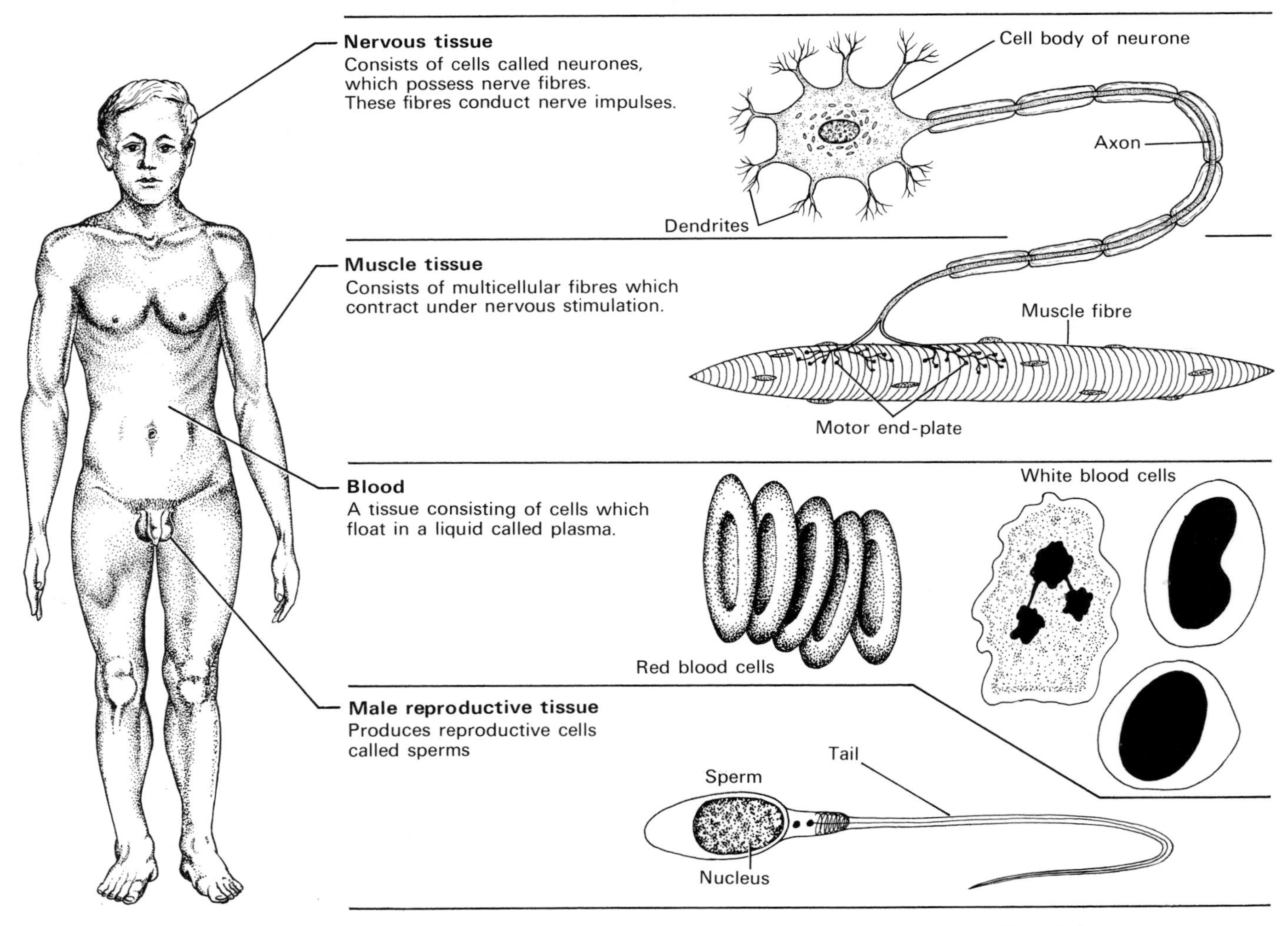

Fig. 2.5 A selection of human cells and tissues

their cells are basically similar in shape and have a common function: muscular contraction. Each cell has a flat muscular region at its base which interlocks with muscular regions of adjacent cells, thereby forming a muscle-like sheet. The muscle fibres in this sheet run vertically on the outer layer of cells, and around the animal in the inner layer. In other words, they form longitudinal and circular muscles.

Besides contraction, the cells of a hydra's inner and outer layers show a considerable variety of other functions. For instance, there are amoeboid, flagellate, glandular, and sting cells, all of which are concerned with feeding and digestion. A hydra's tissues are not the same as human tissues, which are mostly confined to the performance of only one function.

Plant tissues The different types of plant tissue are quite easy to observe under a microscope and should be investigated by methods given in exercise A. Some examples of plant tissue are illustrated in Figure 2.6.

Epidermal tissue forms the outer skin of a plant, and consists of cells which are regularly shaped, and sometimes flat like paving stones. **Vascular tissue** is a system of tubes within a plant which carry water and food substances in solution from place to place. Vascular tissues can be seen as 'veins' in a leaf. Figure 2.6A-D shows how these tubes are formed from cylindrically-shaped cells joined end to end. **Photosynthetic tissue** is characterized by the presence of chlorophyll in its cells.

2.4 Cell size

The majority of cells are between 0.005 mm and 0.02 mm in diameter. A human hair is about 0.1 mm thick.

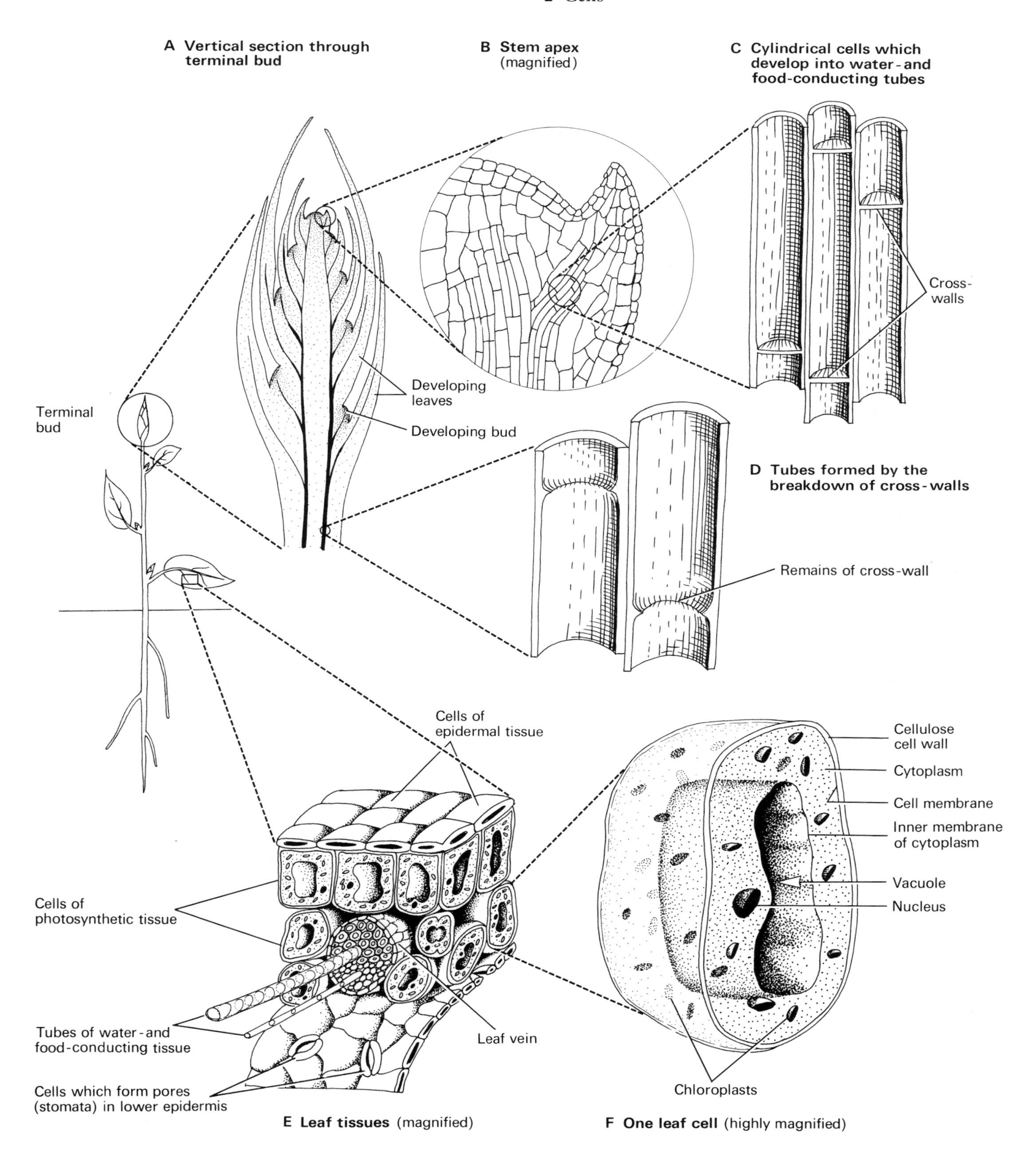

Fig. 2.6 A selection of plant cells and tissues

All cells absorb food and oxygen and remove waste products through their cell membranes, and so they cannot grow so large that the surface area of the membrane is insufficient to support the volume of the cell. For this reason, when cells reach a size at which their surface area is just sufficient to support their volume – a point known as **optimum size** – they either stop growing or divide. Therefore, the upper limit of cell growth is determined by the ratio of volume to surface area. Exercise D explores this relationship and investigates its implications.

2.5 Cell division and growth

In certain circumstances a cell, usually referred to as the **parent cell**, divides into two cells which are called **daughter cells**. The nature, occurrence, and location in the body of cell division differ according to the species and age of an organism.

Cell division in multicellular organisms

When a cell divides in a multicellular organism the daughter cells remain attached to each other and grow, usually without much delay, to the size of the original parent cell. As a result, the organism grows in size. Growth can be defined as an increase in the over-all size of an organism, and an increase in the number of its cells.

By far the greatest amount of cell division and growth takes place as a multicellular organism develops from a single fertilized egg cell called the **zygote**. The zygote grows into an organism composed of billions of cells specialized into tissues, organs, and organ systems. During this growth and development period the organism is called an **embryo**.

The cells making up an embryo in its early stages of growth are unspecialized, and when isolated from each other they look something like Figure 2.1A. As growth proceeds, groups of cells adopt the shape and form of the different types of permanent tissue. This is called **cell differentiation** because the cells develop their *different* shapes and functions. When a cell is fully differentiated it usually loses the power of division. As growth proceeds cell division consequently becomes localized to restricted areas of the embryo.

When an organism reaches its adult state cell division does not stop altogether. Cells in certain areas remain unspecialized and have the power to divide. These areas supply cells which can, theoretically, differentiate into any kind of tissue and bring about further growth, or repair worn and damaged tissues.

Cell division in plants Regions of active cell division in plants are called **meristems**. Examples of these occur at the root and shoot tips. Meristems are retained in these areas throughout the life span of plants and produce new tissues by growth and differentiation whenever conditions are favourable. Figure 2.6A-D illustrates cell differentiation behind the shoot apex meristem.

Cell division in animals Unlike plants, most animals do not grow throughout their whole life span. In adults regions of cell division occur only where new cells are required to replace damaged or worn-out tissues. (The word meristem is not used to describe regions of cell division in animals.) In humans, for example, the skin surface is rubbed away whenever solid objects are touched. It is replaced from below by cell division within what is called the **Malpighian layer** (Fig. 10.3). This layer also produces cells which repair cuts and abrasions. An example of cell replacement is found in the life cycle of red blood cells. These have a useful life of only a few months, after which they are destroyed by the spleen, and replaced by new cells produced in the bone marrow.

The hereditary significance of cell division

There is a close relationship between the cell nucleus and an organism's hereditary features; that is, those bodily features which an organism inherits from its parents. In order to understand the significance of this fact consider again the zygote (fertilized egg cell) which is the first stage of growth in all multicellular organisms.

Zygotes of most animals, apart from the amount of yolk which they possess, are very similar in that they all consist of cytoplasm and a nucleus. Yet an elephant zygote will develop only into an elephant, and never into a cat or a man. Furthermore, it will not develop into just any kind of elephant, but into one with all the characteristic features of its particular species down to the last detail. Obviously these features, which are the hereditary characteristics of the species, have in some way been transmitted to the zygote by the parent elephants which produced it.

A simple way of describing how this happens is to say that a zygote contains all the 'instructions' which are necessary for it to 'build' the one species of organism which produced it, and no other. The zygote receives these instructions from the parent organisms, and the same instructions will eventually be passed on to future generations when the zygote's own development is complete.

The nature of these instructions and where they are stored in a cell was once one of the greatest puzzles of biological science. Parts of the puzzle are now solved.

It has been discovered that the instructions are contained within the chromosomes of the cell nucleus. In other words, chromosomes contain the hereditary information from which an organism is made. This information is now called the **genetic code**, and **genetics** is the study of how the information is passed to future generations.

The genetic code has a chemical nature. It takes the form of a specific pattern of molecules that make up a complex substance called **deoxyribonucleic acid**, DNA for short, and this chemical is the chief constituent of the substance of which chromosomes are made.

To summarize so far: the DNA within the chromosomes of a zygote contains all the instructions to build an adult organism with the characteristic features of the species which produced the zygote.

Chapter 19 describes DNA and the genetic code in greater detail.

2.6 Cell division by mitosis

Mitosis is the mechanism by which a cell passes a copy of its genetic code to each of the daughter cells produced when it divides. Mitosis is therefore the type of division responsible for processes such as growth and repair of tissues within adult organisms, and it is particularly important for its role in the development of a zygote into an adult organism.

The nature of mitosis follows from what has already been said about the genetic code: the code is contained within chromosomes, and chromosomes are in the nucleus of a cell. Consequently, mitosis is the process by which the nucleus of a cell divides in such a way that each daughter cell receives exactly the same number and type of chromosomes as were present in the parent cell.

Division of plant cells differs in certain ways from animal cell division, but the following description applies equally well to both. Plant cells have been illustrated in Figure 2.7 as an aid to the study of mitosis using methods given in exercise B, and to help in the interpretation of commercially prepared slides of a root tip.

Although mitosis is a continuous process, it is divided for ease of description into four phases: prophase, metaphase, anaphase, and telophase. Each phase represents a distinct episode as division takes place. A cell which is not undergoing mitosis is said to be in **interphase**. The chromosomes of an interphase cell are thought to exist as long, fine threads, invisible except for the chromatin granules, which contain DNA, situated along their length like beads on a string. In an interphase cell which is about to undergo mitosis, a perfect replica is created of each DNA molecule within each chromosome; the replica is exactly the same as the original down to atomic level. In other words a perfect copy is made of the genetic code. When the copy is complete mitosis begins.

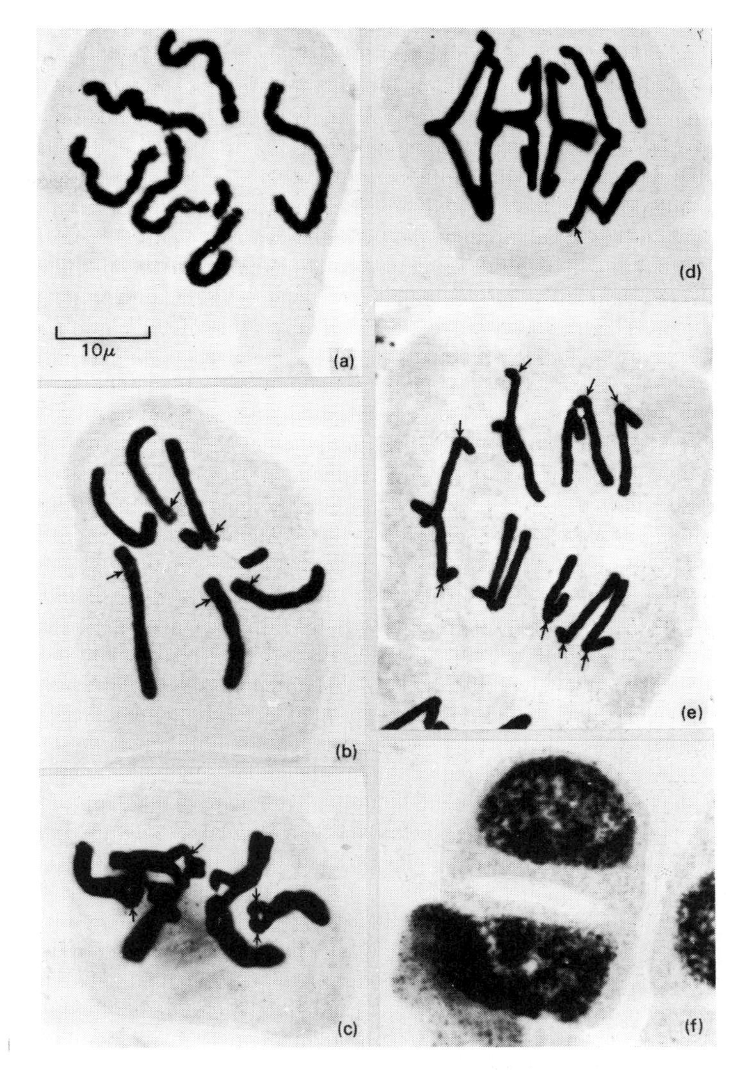

Stages of mitosis from the root apex of a crocus. Arrows point to the centromeres ($1\mu = 0.001$ mm). (a) Prophase. Each chromosome consists of two chromatids. (b) Early metaphase. (c) Metaphase. (d) Middle anaphase. The chromatids are being drawn apart. (e) Late anaphase. (f) Telophase. A nucleus has formed in each daughter cell

Prophase

During prophase, the first stage of mitosis, the chromosomes become shorter and thicker until they are clearly visible in the cell. Owing to the DNA duplication process already mentioned the chromo-

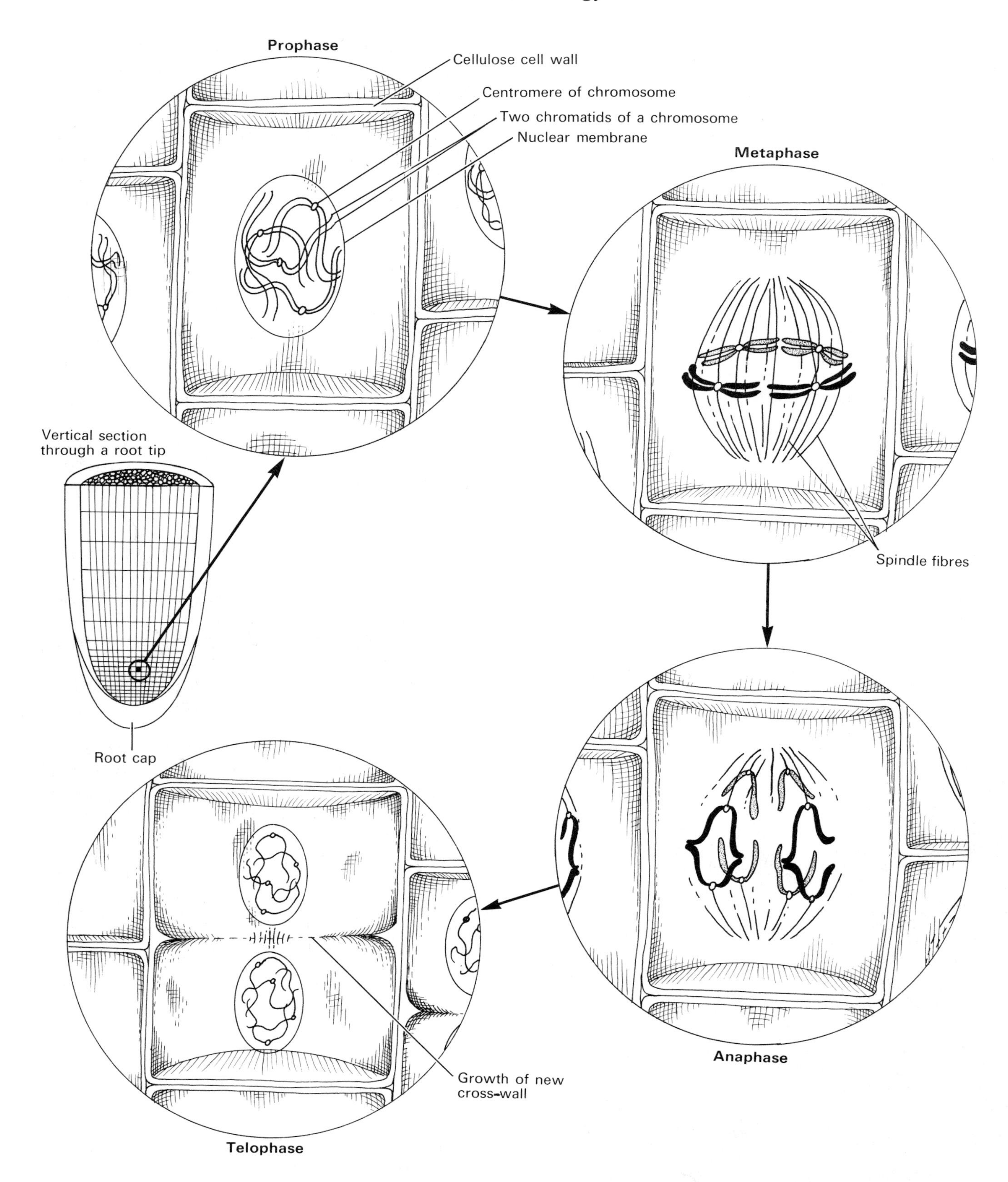

Fig. 2.7 Mitosis in a cell from the root tip of a plant

somes, which were originally single threads, are seen to be *double* structures. As they become thick enough to be visible during prophase they are seen to consist of two **chromatids** joined at a point called the **centromere**. Chromatids are considered to be chromosomes because each contains a complete replica of the DNA molecules contained in the original single chromosomes which produced them. In other words, a prophase nucleus contains twice as many chromosomes as usual.

Metaphase

During metaphase, the second phase of mitosis, the nuclear membrane disappears and a structure called the **spindle apparatus** is established. The spindle is an arrangement of very fine fibres to which each chromatid pair becomes attached at the centromere. Metaphase is completed when all the chromatid pairs are positioned at the 'equator', or middle, of the cell.

Anaphase

During anaphase, the third phase of mitosis, the two chromatids of each double chromosome separate.

Metaphase of mitosis. Arrows point to fibres of the spindle apparatus. The two chromatids of each chromosome are clearly visible

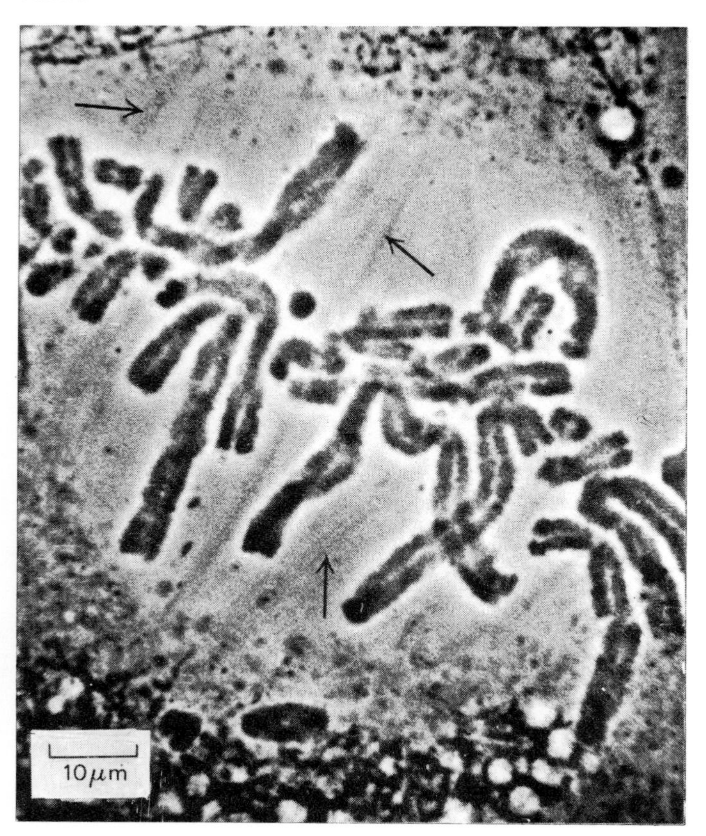

Initially the centromere splits and separation of the chromatids begins, progressing outwards until the members of each pair of chromosomes are completely separated. The chromatids of each pair then move apart in opposite directions. Remember that throughout this process each chromatid pair is attached to a separate fibre of the spindle apparatus. It is thought that the spindle fibres contract, and draw the chromatids apart.

Telophase

During telophase, the final phase of mitosis, a group of chromatids – now true chromosomes – assembles at opposite ends of the cell. Each of these two groups, through a process of 'reverse prophase', becomes an interphase nucleus with long, fine, invisible chromosomes and a nuclear membrane. In plants a cross-wall of cellulose grows between them. Animal cells begin to constrict about their equator into an hour-glass shape. This process continues until the two cells are nipped apart.

Summary of mitosis

1. A cell which is about to divide produces a replica of all the DNA contained in its nucleus. This results in its chromosomes becoming double structures consisting of two chromatids.

2. The chromatids of each chromosome separate and move to opposite ends of the cell.

3. This results in two cells being formed, each with the *same* number of chromosomes as the original parent cell.

4. Mitosis is important because it is the mechanism which ensures that the information contained within an organism's genetic code reaches all the cells of its body, where it is decoded and used to build tissues, organs, and organ systems.

There is a more or less fixed number of chromosomes for each species of organism. This varies only under very special circumstances, which are explained in chapter 19. For example, normal human cells contain 46 chromosomes, and mice cells 40. The mechanism which ensures that this number remains constant when plants and animals reproduce sexually is also described in chapter 19.

Verification and inquiry exercises

A *Looking at cells*

The following techniques can be used to prepare slides for examination under a microscope.

Onion epidermis Cut up an onion bulb to obtain a

piece of one of the white, fleshy 'leaves' from inside the bulb. A piece of epidermis may be stripped from this by breaking and tearing it into smaller pieces. Mount the epidermis, without folding it, on a slide and add a few drops of iodine. Place a cover slip over it and observe the cell shapes and nuclei.

Section cutting Cut thin slices with a razor-blade from bottle cork, carrot, potato, soft plant stems and roots. Small pieces of tissue may be cut using the method illustrated in Figure 5.10. Select the thinnest slices and mount them on a slide in 25% glycerine solution (which reduces air bubbles in the tissue). Observe the variety of cell shapes. Compare stem and root sections with the illustrations in chapter 7.

Animal cells Scrape the inside of the cheek with a finger-nail and mix the collected tissue with a drop of methylene blue stain on a slide using mounted needles. Observe individual epidermal cells noting their shapes (where they are not distorted by folding), and the presence of nuclei.

B *Observation of chromosomes during mitosis, and subsequent cell growth*

1. Cut off about 15-20 mm from a root of an onion bulb which has been suspended with its base just touching water for several days. (This will have encouraged root growth.)

2. Place the piece of root on millimetre graph paper and slice off the terminal 5 mm with a razor-blade, and then a further 5 mm length.

3. Place these two pieces in separate watch-glasses containing 10 parts of acetic orcein stain to one part normal hydrochloric acid.

4. *Warm* (do not boil) the preparation for five minutes and then place the root pieces on separate labelled slides with a few drops of the same solution.

5. Break up the tissues of each preparation with the tip of a mounted needle without altering the relative arrangement of cells.

6. Place a cover slip on the preparation and cover with a paper towel. Press down with the thumb on the cover slip through the towel taking care not to move the cover slip sideways.

7. Observe cells from the root-tip portion for the various stages of mitosis, using Figure 2.7 as a guide to the appearance of chromosomes at each stage.

8. Are there any signs of mitosis in cells from the *upper* 5 mm root segment? Compare the size and shape of cells in both preparations to confirm that cells do in fact grow after the stage of rapid cell division at the tip of the root.

C *An investigation into the effects of pressure on cell shape*

Within tissues, the pressure of all the cells against each other has a great effect upon cell shape. The shapes can be investigated in the following way.

1. Make at least thirteen plasticine spheres of equal

Fig. 2.8 Comparison between animal and plant growth patterns

A Animal growth pattern (human)

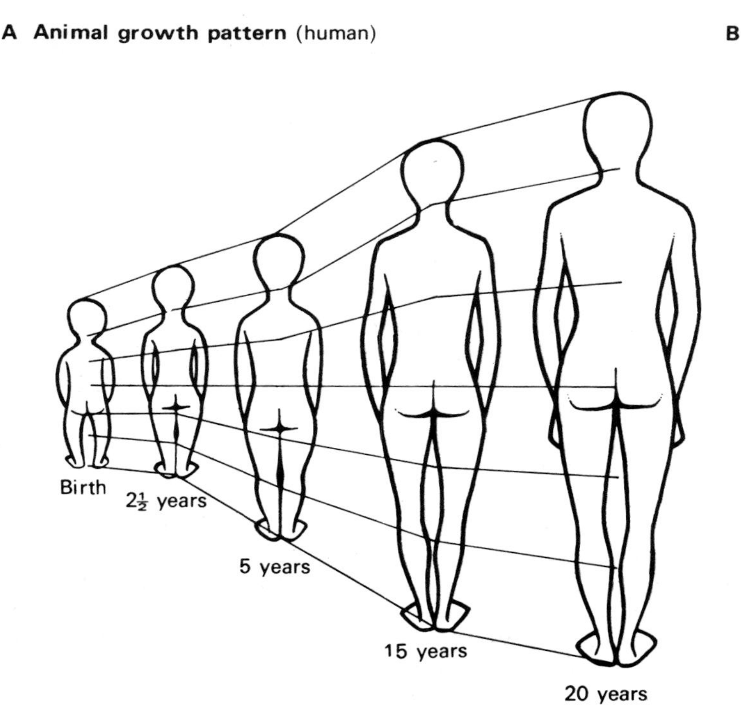

B Plant growth pattern (broad bean)

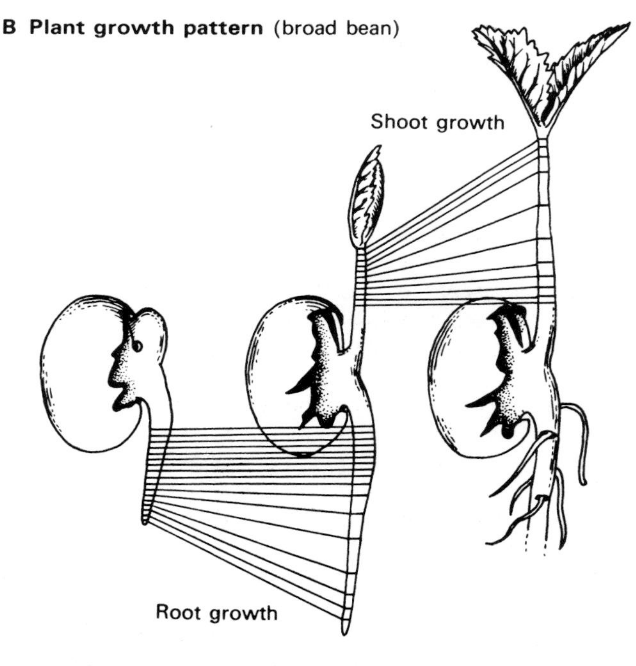

diameter (about 1.5 cm). Surround one sphere with the other twelve in a roughly symmetrical, spherical arrangement and roll the whole collection between the palms until the spheres are squashed firmly together into a large composite ball. Pull this ball carefully apart and observe the different shapes which have resulted from the spheres pressing against each other. Observe especially the shape of the sphere at the centre of the ball.

2. Arrange six plasticine spheres around a central one on a desk top so that they are all touching. Squash them flat with the hand. From Figure 2.6, find the cells in the leaf which the resulting shapes resemble.

3. Arrange six cylindrical plasticine rods around a central rod and roll them on a desk top to squash them together. From Figure 7.2 find the plant tissue which these shapes resemble.

D *An investigation into the relationship between the volume and surface area of a cell*

1. Weigh out 10 g of plasticine (or clay). Either roll it into a sphere and estimate its volume ($1/6\pi d^3$) and surface area (πd^2); or more simply, mould it roughly into a cube and calculate these values. Repeat this procedure at least three times, doubling the volume of plasticine (by adding an equal weight) on each occasion.

2. Does the surface area increase slower, faster, or at the same rate as volume? How does the answer to this question relate to what is said in section 2.4 of this chapter?

3. Flatten the plasticine from 1 above to make a thin rectangular solid and calculate its surface area. How does this affect the ratio of surface area to volume? How does this help to explain why large cells are often flattened in shape?

E *An investigation into growth patterns of leaves*

1. Grow a bean seedling to the last stage shown in Figure 2.8B. Make an ink marker by stretching cotton thread across the gap between the ends of a piece of wire bent into a U-shape. Use this, dipped in finger-print ink (or indian ink), to print a 2 mm grid on a young leaf attached to the plant. Copy the leaf outline and its grid on tracing paper. On the next and subsequent days for at least a week, compare the leaf with its tracing, making new tracings if a difference is observed.

2. Does the leaf grow evenly all over? If not, discover where the major growth regions are by studying the sequence of tracings.

3. Repeat with different plants.

Comprehension test

Growth is an increase in both the number and the size of cells. This process is distributed unevenly throughout the bodies of animals and plants, giving each a different growth pattern. The diverging lines on each sequence of drawings in Figures 2.8A and B show how growth occurs in animals and plants. The youngest organism in each sequence is marked off into regions of equal length, but these regions soon show different rates of development.

1. Find the human body region in Figure 2.8A which grows very little compared with the rest, and suggest reasons why this is so.

2. The diverging lines in Figure 2.8B indicate the major growth regions in plants. Where are they located?

3. The vertical section of a root tip illustrated in Figure 2.7 shows what happens to the size of cells produced by mitosis at the tip of the root. Using this information try to explain the pattern of lines over the root in Figure 2.8B.

4. From the answers to the above questions try to explain the basic difference between plant and animal growth patterns.

Summary and factual recall test

The outer boundary of a cell is called the (1) and it is (2) permeable. All the living substance in a cell except the nucleus is called (3). This substance contains temporarily stored materials such as (4–name three) known collectively as (5). In comparison (6) are more permanent and play an active part in metabolism. Examples are (7–name four). Nuclei contain (8) granules, which are reorganized into rod-shaped (9) during cell division. These objects contain a complex chemical called (10), known to carry an organism's (11) information or (12) code.

Plant cells differ from animal cells in the following ways (13–name three).

Multicellular organisms consist of groups of cells called (14) which are specialized in performing a particular function and thereby achieve greater efficiency through the division of (15). Examples of such cell groups are (16–name two). Where several of these groups work together they form an (17), an example of which is (18).

Cell size is limited by the ratio of (19) to (20), because (21).

Growth results from an increase in both the (22) and (23) of cells.

Before a cell undergoes mitosis it first produces a perfect (24) of all the DNA in its (25), which results in them becoming double structures consisting of two (26), joined together at a (27). These separate at (28), and move to (29) ends of the cell where they undergo reverse (30) to form a nucleus. Eventually two cells are formed with the (31) number of chromosomes as the parent cell. Mitosis is important because (32).

3
Food and diet

Every cell of every tissue and organ in the body has been made from, and is maintained in a healthy state by the food which is eaten.

If a person eats too little food his body becomes weak and thin because the tissues are used up as a source of energy; growth and repair processes slow down so that wounds fail to heal properly; and the body loses the ability to fight off infections. On the other hand, if a person eats too much his body becomes inflated with stored food. This condition, known as **obesity**, may result in damage to the heart and circulatory system. Finally, if a person eats only a moderate amount of food but of the wrong type to suit his body's requirements, he is likely to suffer from diseases such as scurvy, beri-beri, and pellagra, described later in the chapter.

To remain healthy fairly precise amounts of the right kinds of food must be consumed. This is known as living on a **balanced diet**. Failure to do so results in **malnutrition**, a term which describes the effects on the body of eating too little, too much, or the wrong kinds of food.

This chapter describes various types of food and their different functions in the body. It then shows how these foods can be put together to make meals which form a balanced diet. But first it is necessary to describe a function common to all types of food, which is to supply energy.

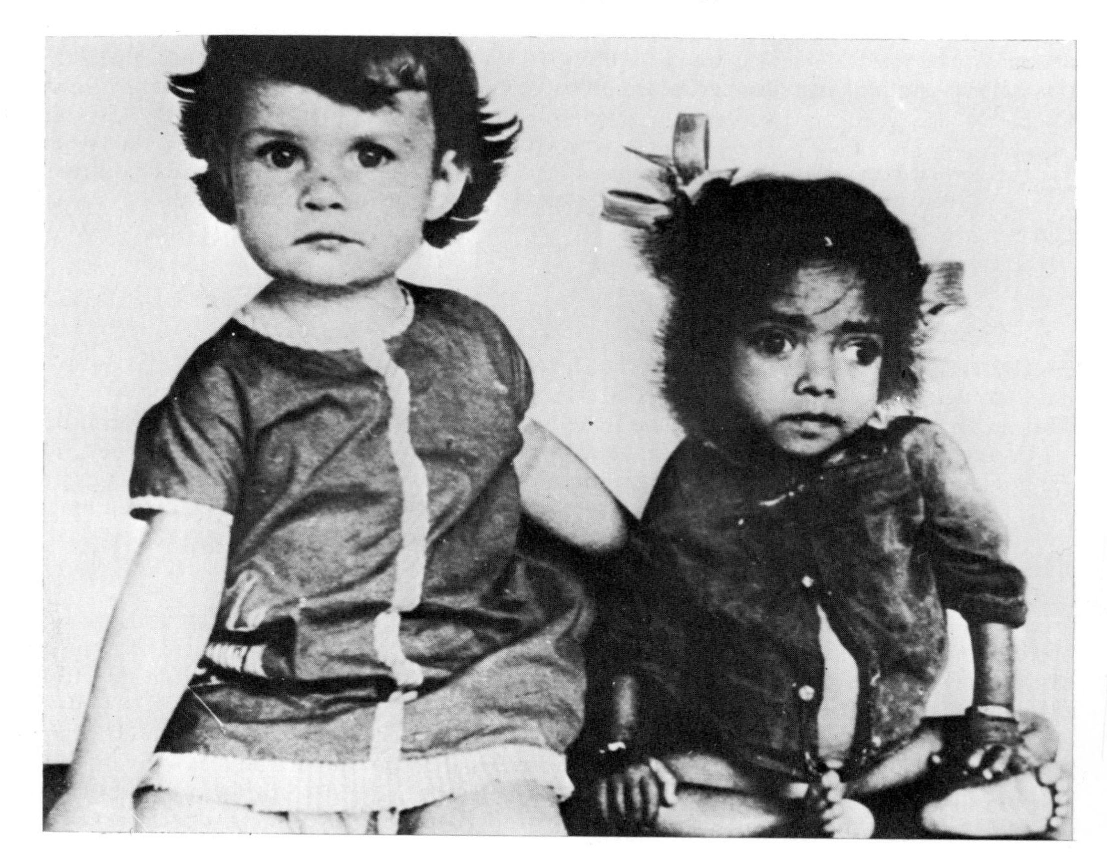

Effects of malnutrition. Compare the well-fed child on the left who is 18 months old and weighs 11 kg, with the under-fed child on the right who is 3 years old and weighs 6.5 kg

3.1 Energy value of food

Most types of food contain stored energy. But this energy cannot be used by the body until it is released by the chemical reactions of respiration, described in chapter 8.

There is a simple method of measuring the amount of energy in each type of food so that its energy value may be stated in so many units. When a measured quantity of food is burned in air it gives off exactly the same amount of energy, in the form of heat, as it does when the same quantity is respired inside an organism. The food is burned inside a device called a bomb calorimeter. This ensures that all the heat released during burning is transmitted to a known quantity of water, thereby raising its temperature. All forms of energy are now measured in units called **joules.** It takes 4.2 joules of heat energy to raise the temperature of 1 g of water by 1 °C. After a food has finished burning, its energy value is calculated by the formula: temperature rise × 4.2 × mass of the water in the calorimeter.

The energy value of food is usually measured in **kilojoules** (abbreviated to kJ). 1 kilojoule = 1000 joules. Table 1 in this chapter gives the kJ energy values for many foods. It must be emphasized that these figures indicate only the *potential* energy value of these foods to any organism which eats them. The *actual* energy yield per gram of a food depends upon the extent to which it is digested and assimilated into the body. In addition, the body cannot make full use of all the energy released from a food by respiration since a lot of the energy is lost as heat. In humans, for instance, up to 85 per cent of respiratory energy is lost in this way.

3.2 Types of food and their composition

Chemical analysis shows that there are five main types of food: carbohydrates, fats and oils, proteins, minerals, and vitamins. They each have different chemical compositions, different properties, and different functions in living organisms.

Carbohydrates

The most familiar carbohydrates are sugars and starches. Sweet fruits, honey, jam, and treacle are examples of sugary foods. Examples of starchy foods are bread, potatoes, rice, and spaghetti. Cellulose is a less familiar carbohydrate. It is a major constituent of plants, where it forms a wall around each cell. But cellulose can only be digested by herbivores, and consequently the enormous supplies of cellulose in the world are only available to carnivores and omnivores indirectly, i.e. when they eat herbivores.

Functions of carbohydrates Carbohydrates are the chief source of energy for living things, which is why they are sometimes called the 'fuel' of life. On average, 1 g of carbohydrate can yield 17 kJ of energy when respired. The principal carbohydrate used in respiration is glucose sugar. In fact, most carbohydrate foods are converted by the body into glucose sugar before they are consumed in respiration.

Carbohydrates are also important as food reserves which are stored within organisms. Many plants store large quantities of starch: in wheat and maize seeds, and in potato tubers. In animals, the main carbohydrate food reserve is glycogen, which is very similar to starch in its chemical composition. If more carbohydrates are eaten than are necessary to satisfy the body's immediate energy requirements the excess food is converted into glycogen and stored in the liver and muscles. The body can store only limited amounts of glycogen, however, and when this limit has been reached, any excess carbohydrate in the diet is converted into fat or oil and stored in special tissues described below.

Structure of carbohydrates Carbohydrates are made of carbon, hydrogen, and oxygen atoms, which are joined together so that the hydrogen and oxygen atoms are always present in the ratio of 2 : 1. Glucose sugar, for example, which is one of the simplest carbohydrates, has the chemical formula $C_6H_{12}O_6$. Glucose and other simple sugars such as fructose, or fruit sugar, are known collectively as **monosaccharides**, or simple sugars.

Disaccharides, or double sugars, are made up of monosaccharide sugar molecules joined together in pairs. Sucrose, commonly known as cane sugar, is an example of a disaccharide. It consists of pairs of glucose and fructose molecules. Lactose, or milk sugar, is another disaccharide.

Polysaccharides, or multiple sugars, are made up of monosaccharide molecules joined together in long, often branching chains. Starch and cellulose are examples of polysaccharides. The chain-like structure of cellulose molecules is illustrated in Figure 5.3.

Fats and oils

Examples of fatty and oily foods are butter, lard, suet, dripping, olive oil, and cod-liver oil. The main difference between fats and oils is that oils are liquid at 20 °C, while fats are solid at that temperature.

Functions of fats and oils On average, 1 g of fat or oil can yield up to 38 kJ of energy when respired,

Table 1 Composition of food per 100 g. (Percentage not accounted for represents inedible matter, e.g. bone, shell, gristle, skin, water.)

Figures from *Manual of Nutrition*, Ministry of Agriculture, Fisheries and Food, H.M.S.O., 1978

Type of food	*Kilojoules*	*Carbo-hydrate* g	*Fat* g	*Protein* g	*Vitamins* (*a rich source) A	B	C	D	E	K
Food from animals										
Meat										
Bacon, cooked	1851	—	48	11		*				
Beef, stewing, cooked	932	—	11	30		*				
Beef, corned	905	—	15	22						
Chicken, roast	599	—	7	30						
Kidney, grilled	375	—	4	17						
Lamb, roast	1209	—	22	23						
Liver, fried	1016	2	10	30	*	*		*	*	*
Luncheon meat	1298	5	29	11		*				
Mutton, roast	1554	—	33	15						
Pork chop, grilled	1308	—	20	35		*				
Sausage, pork	1520	13	31	10		*				
Steak and kidney pie	1195	16	21	13		*				
Tripe	252	—	1	12						
Dairy produce										
Butter	3041	—	82	0.5	*			*	*	
Cheese	1700	—	35	25	*			*	*	
Eggs raw, one	612	—	12	12	*	*		*		
Ice cream	698	21	13	4						
Milk, liquid whole	273	5	4	3	*	*				
Yogurt low fat, fruit	405	18	1	5	*					
Fish										
Kipper fillets	770	—	12	19				*		
White fish, baked	399	0.5	1	20		*				
White fish, fried	840	8	10	20						
Salmon, tinned	649	—	6	20				*		
Sardine, tinned	906	—	23	21	*			*		
Food from plants										
Cereals										
Bread, white	991	55	2	8						
Bread, wholemeal	918	47	3	10		*			*	*
Biscuit, chocolate	2197	65	25	7						
Cornflakes	1567	88	0.5	8		*				
Rice	1536	87	1	6						
Spaghetti	1612	84	1	10						
Vegetables										
Beans, tinned	270	17	0.5	6						
Beetroot, boiled	189	10	—	2						
Brussels sprouts, boiled	75	2	—	3			*		*	*
Cabbage, raw	92	6	—	2	*		*		*	*
Cabbage, boiled	66	1	—	1					*	*
Carrot, raw	84	5	—	1	*					
Cauliflower	56	3	—	1			*			
Lettuce	36	2	—	1	*		*		*	
Onion	99	5	—	1						
Peas, boiled	161	8	—	5						
Potatoes, raw	369	18	—	2		*	*			
Potatoes, boiled	339	20	—	1						

Table 1 *(continued)*

Type of food	*Kilojoules*	*Carbo-hydrate* g	*Fat* g	*Protein* g	*Vitamins* (*a rich source) A	B	C	D	E	K
Vegetables *(continued)*										
Potatoes, fried	1065	37	9	4						
Tomatoes, fresh	60	3	—	1	*		*			*
Watercress	61	1	—	3	*		*		*	
Fruit										
Apple	196	12	—	0.5						
Banana	326	19	—	1						
Blackcurrants	121	7	—	1			*			
Cherries	201	12	—	1						
Gooseberries	73	6	—	1			*			
Grapefruit	95	5	—	0.5			*			
Melon	97	5	—	1	*		*			
Orange, peeled	150	9	—	1			*			
Peaches, fresh	155	9	—	0.5						
Peaches, canned	373	23	—	0.5						
Pear, fresh	175	11	—	0.5						
Raspberries	105	6	—	1			*			
Strawberries	109	6	—	0.5			*			*
Nuts										
Almonds, shelled	2336	4	54	21						
Coconut, desiccated	2492	6	62	7						
Peanuts, shelled	2364	7	49	28						
Miscellaneous										
Apple pie	1179	40	15	3						
Beer, keg bitter	129	2	—	0.5						
Cakes and pastry	1680	55	20	6						
Chocolate, milk	2214	54	38	9						
Coffee, instant powder	424	0.5	0.5	—						
Custard	496	18	4	3						
Jams, assorted	1116	69	—	1						
Sugar, white	1680	105	—	—						
Tea with milk (1 cup)	84	0.5	0.5	—						

which makes them a very important source of energy. But they are less easily digested than most other foods, which offsets their energy value a little.

Fats and oils are also important as food reserves in the body. In mammals, for instance, they are stored in special cells known collectively as **adipose tissue**. This tissue occurs under the skin, and around muscles, the heart, the kidneys, and several other body organs. These fat deposits have two main functions. First, they form a food reserve with an extremely high energy yield per gram and are therefore less bulky than carbohydrates with the same energy value. Second, the layer of fat under the skin insulates the body against loss of heat and is especially well developed in arctic animals such as seals and polar bears.

Structure of fats and oils Like carbohydrates, fats and oils consist of the elements carbon, hydrogen, and oxygen, but their molecules differ from carbohydrates in that they contain relatively little oxygen. Beef and mutton fat, for example, both contain a fatty substance called tristearin which has the chemical formula (greatly simplified) $C_{57}H_{110}O_6$.

The wide variety of fats and oils found in living things are made up of substances called **fatty acids**, which are linked in various combinations with **glycerol** (glycerine).

Proteins

Meat, liver, kidney, eggs, and fish contain a high proportion of protein.

Functions of protein Protein is often described as a body-building food because it supplies the building materials from which the tissues and organs of the body are made. Muscles and the liver, for example, consist of about 30% protein, kidneys nearly 20%, and the brain about 10%.

The importance of proteins in living matter is demonstrated by the fact that their energy value, which is up to 17 kJ per gram, is used only in extreme emergencies. When an animal is deprived of food, stored glycogen and fats are the first food reserves to be respired. Only when all these reserves are exhausted is protein broken down as fuel for energy. When this happens cell structure soon degenerates and metabolism slows down and finally stops.

In addition to proteins which form part of body structure, there are other protein substances called **enzymes** which play a vital role in body chemistry. Enzymes are described more fully in chapter 4. Here it is sufficient to say that they are chemicals which, in minute amounts, control the speed and direction of all the chemical reactions of metabolism. In technical terms enzymes are **catalysts**. This means that they speed up reactions which would otherwise proceed very slowly. Without enzymes, life processes which normally take seconds would take months or even years. For this and many other reasons enzymes are vital to life.

A simple comparison between a living cell and a motor-car engine may illustrate the points made so far about the importance of various foodstuffs. The metal structural parts of an engine such as the cylinder block, pistons, and valves are comparable to the protein structures of a cell. The fuel, petrol, which is burned in the cylinders to liberate energy to drive the engine is like the carbohydrates and fats which are respired in a cell for energy to drive metabolism. The throttle or accelerator of an engine, which controls its speed, is in certain respects like the controlling influence which enzymes have over the speed of metabolism.

Structure of proteins All proteins consist of carbon, hydrogen, oxygen, and nitrogen, and some types contain sulphur. Like polysaccharides, proteins consist of chemical units linked together in chains. In proteins these units are called **amino acids**. Unlike the simple sugars which make up polysaccharides, the amino acids in proteins are of many different types. This means that the structure of proteins is vastly more complex. Polysaccharides have at most only one or two functions in organisms, whereas proteins as a group serve thousands, and possibly even millions of different functions in living things. It is their complexity which makes proteins so biologically versatile.

A clue to the nature of this complexity comes from a study of amino acids and how they are put together by cell metabolism in the manufacture of proteins. There are about twenty different types of amino acid and they are joined together in different sequences and in different numbers to make protein molecules. A simple analogy shows how an immense variety of protein structures is possible using these raw materials.

Imagine a large box full of coloured glass beads of which there are just twenty different colours. Then think of a jeweller who is set the task of making necklaces with twenty beads in each. It is a statistical fact that if he has enough patience, and enough beads, he could make more than two million million million necklaces without repeating any sequence of the twenty beads. In other words, twenty different beads can be arranged in this number of different combinations.

Reverting to biology again, imagine a cell which is making proteins and using twenty different amino acids. Like the jeweller, it can make at least this number of different protein molecules, but there is an added complication. Cells are not restricted to chains of only twenty amino acid units. In fact, the average protein molecule consists of about 500 units, and some contain thousands of units. Under these circumstances it is clear that the number of different proteins is almost limitless.

The main point is that the properties of proteins, and their functions in organisms, vary according to the sequence and number of amino acids in their molecules. Furthermore, it is almost certain that living things owe their immense and ever changing variety to the almost infinite number of different proteins which are possible.

Proteins and diet Autotrophic organisms, such as green plants, can synthesize proteins from carbon, hydrogen, oxygen, and nitrogen atoms. But heterotrophic organisms, such as animals, cannot do this. At most they can change one type of amino acid into another, and this ability is fairly limited. Consequently, heterotrophs must obtain amino acids by eating plants and other heterotrophs.

Humans, for example, must obtain sufficient quantities of eight **essential amino acids**. From these all their protein requirements are synthesized. Any non-essential amino acids in the diet are either converted into glycogen and fat or excreted. Protein foods which contain these essential amino acids are said to be **adequate**. Examples of adequate proteins are eggs, meat, and milk. When these and other adequate proteins are eaten, enzymes in the digestive system split them into individual amino acid units. When these units reach the body cells they are put together again, but this time in the sequence which is specific to human proteins. This break-up and recombination process is illustrated in Figure 3.1.

Foods from plants contain relatively little protein; this can be seen from Table 1. Moreover, few of these plant proteins are adequate in the way described above. But plant proteins lack different essential amino acids, and if they are eaten in sufficient variety, deficiencies can be avoided. Unfortunately this variety of food is unobtainable in certain countries and the people who live there have to rely on one type of plant food. For example, rice is the staple food in the Far East, and maize in South America. In such areas protein deficiency disorders such as kwashiorkor, which involves degeneration of the liver and pancreas, are common.

Since proteins are body-building foods it is important that growing children, pregnant women, and those recovering from severe illness have a diet rich in protein foods.

Minerals

Humans require about fifteen different mineral elements in their diet. Most of these are supplied by

Fig. 3.1 Diagram to show how amino acids are split up and recombined during digestion

Amino acid sequence in food

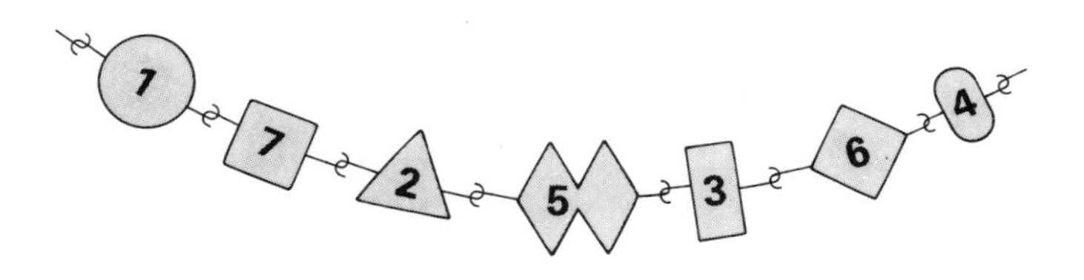

Amino acids split apart by digestive enzymes

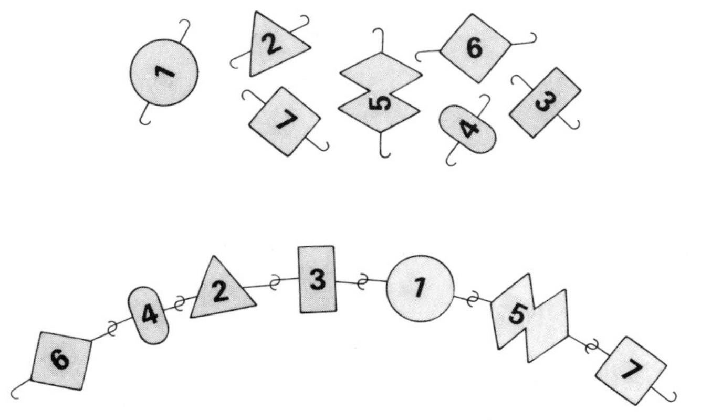

Amino acids rearranged to form human protein

Table 2 Mineral requirements of Man (1 mg = 0.001 g)

Mineral	*Daily requirement* mg	*Function in the body*
Sodium chloride (common salt)	5–10	Blood plasma is almost 1% salt. Needed for digestion and passage of impulses along the nerve fibres.
Potassium	2	Necessary for muscle contraction.
Magnesium	0.3	Necessary for muscle contraction.
Phosphorus	1.5	A major constituent of bones and teeth. Part of DNA molecule. Needed for the chemical reactions of respiration.
Calcium	0.8	A major constituent of bones and teeth.
Iron	0.01	A major constituent of haemoglobin (red substance of blood needed to transport oxygen to body tissues).
Copper	0.001	Needed for the utilization of iron by the body, and for growth.
Manganese	0.003	Needed for growth.
Iodine	0.000 03	Needed by the thyroid gland. Deficiency causes goitre.

meat, eggs, milk, green vegetables, and fruits. Minerals have no energy value but they do have important functions in the body, which are listed in Table 2.

Sodium chloride, or common salt, deserves special mention not just because of its functions but because it is constantly and rapidly lost from the body in perspiration, urine, and faecal matter. In temperate climates an intake of at least 4 g daily is required to make up for this loss. Some of this salt is already in food when it is eaten; the rest must be added to the diet if health is to be maintained. More salt is needed on a hot day because perspiration rate is higher. Strenuous exercise under these conditions can result in salt loss at such a high rate that muscle cramp and heat exhaustion may occur unless extra salt is taken into the body.

Vitamins

Vitamins are complex chemicals which are required in minute amounts in the diet of all heterotrophic organisms. Autotrophs manufacture their own vitamins from simple raw materials.

Vitamins have no energy value. They play a vital role in many chemical reactions of metabolism, for the most part in conjunction with enzymes. Lack of any vitamin in the diet causes the reaction in which it takes part to slow down, and since most metabolic reactions are part of a long sequence of events an alteration in the rate of any one of them can have widespread effects on the body. These effects are called **vitamin deficiency diseases**, to distinguish them from the ill effects caused by disease organisms. Stunted growth results from lack of almost any vitamin, while other disorders are specific to particular vitamins.

Long before the discovery of vitamins there was some evidence to suggest that certain diseases could be cured by improving or adjusting the victim's diet in various ways. An example of such a disease is scurvy, the symptoms of which are easily bruised skin, wounds which fail to heal, and gums which become so soft and pulpy that the teeth fall out. This disease was particularly common among seamen on long voyages whose diet was restricted to biscuits and salt pork. In the middle of the eighteenth century a naval doctor called Lind discovered that scurvy could be cured by eating oranges and lemons. But despite the spectacular results of this treatment the naval authorities did not take action until 1800, when they ordered regular doses of lemon juice to be given to British naval crews. This completely eliminated the disease. The British have been nicknamed 'limeys' ever since, because American sailors mistakenly believed that their counterparts were forced to drink lime juice every day.

Further evidence that certain foods are essential for health was discovered in the late nineteenth century by Admiral Takaki of the Imperial Japanese Navy. Admiral Takaki was alarmed by the fact that most of his men were practically useless as a fighting force owing to the ravages of beri-beri, a disease which causes paralysis of the limbs. However, the disease was completely eradicated when he added meat and vegetables to the monotonous diet of rice usually fed to his men. But this, and other evidence was not enough to convince a Dutch doctor called Eijkman. Determined to find the 'germ' responsible for beri-beri, he set out for the Dutch East Indies where the disease was prevalent, and discovered the truth in a rather unexpected way. Eijkman used chickens as experimental animals and was completely baffled by the incidence of beri-beri among them until, quite by chance, he discovered that the man responsible for looking after the birds was pocketing the money intended for their feed and giving them kitchen scraps instead. The disease disappeared as soon as a proper diet was provided.

Ultimately laboratory investigations, notably those of the English scientist Gowland Hopkins, led to the discovery of what were called 'accessory food factors', the absence of which in the diet caused certain diseases. These factors were eventually renamed vitamins. At first it was thought there were two types of vitamin: the A vitamins which dissolve in fat (fat-soluble), and the B vitamins which dissolve in water (water-soluble). Later, many different types of vitamin were discovered, and today vitamins A, D, E, and K are placed in the fat-soluble group, and B and C are placed in the water-soluble group. The modern tendency is to use the chemical name of each vitamin rather than its alphabetical label.

Vitamin deficiency diseases are not likely to occur when a balanced diet is obtainable, and it should be noted that excessive intake of vitamins, particularly of A and D, can result in poisoning. Vitamin pills, therefore, should never be taken in large quantities.

The following notes describe vitamin requirements in humans. Table 1 must be consulted to find out the foods which contain particular vitamins.

Vitamin A (retinol, or retinene) Adults require only 0.001 g of vitamin A daily. Amounts greatly in excess of this are harmful to the liver. Vitamin A is essential for proper growth and for efficient functioning of the eyes, especially in conditions of low illumination. Deficiency symptoms include poor night vision; drying up of the skin, the cornea of the eyes, and the moist membranes of the nose and bronchial tubes; a tendency to get skin and throat infections; and in severe cases damage to the retina of the eye. Carotene, the yellow substance in many plant foods such as carrots, can be converted by the body into vitamin A. This vitamin is stored in the liver.

Vitamin B complex Vitamin B was originally thought to be a single substance, but it is now known that there are twelve different types, of which seven are essential to human nutrition. These are of particular importance: thiamine (B1), riboflavine (B2), nicotinic acid (niacin), and cyanocobalamine (B12).

Thiamine is unstable at high temperatures. For this reason it is destroyed when foods are cooked or canned. It plays a part in the chemical reactions that release energy from carbohydrate during respiration. Deficiency symptoms include stunted growth and the disease beri-beri, which is characterized by paralysis of the limbs. Beri-beri is common in the Far East

The child on the right has rickets, a vitamin-deficiency disease caused by an inadequate supply of vitamin D in her diet

owing to an unfortunate set of circumstances associated with reliance on rice as a staple food. The outer husk of rice grains contains oil which quickly goes rancid and spoils the grains when the food is stored. Peasant farmers overcome this problem by removing the husks prior to storage, but the offending husks contain virtually their only source of thiamine.

Riboflavine is also necessary for the release of energy from food in respiration. Lack of this vitamin causes stunted growth, cracks at the mouth corners, an inflamed tongue, and damage to the cornea of the eye.

Nicotinic acid is sometimes called **niacin** to avoid confusion with nicotine in tobacco and thereby stop people from getting the impression that cigarettes are beneficial to health. This vitamin is also necessary for the release of energy from food, and lack of it causes stunted growth, rough red patches on the skin of face, neck and hands, and the disease pellagra which involves digestive upsets, and eventual mental disorder.

Cyanocobalamine, usually abbreviated to **cobalamine,** is involved in the formation of protein, fat, and glycogen in the body. A deficiency symptom is the disease pernicious anaemia, which results when the body fails to produce sufficient of the red pigment haemoglobin in red blood cells. The work of the English biologist William Castle and others has shown that this disease occurs in people whose stomach digestive juice lacks a substance now called **Castle's intrinsic factor**. This substance is necessary for the absorption of cobalamine by the intestine wall, and pernicious anaemia can be cured by including a little intrinsic factor in the diet. Only one-thousandth as much cobalamine is required by the body as of other B vitamins.

Vitamin C (ascorbic acid) Almost all animals except man can manufacture their own vitamin C. Adult humans need up to 0.06 g daily in their diet. This vitamin is very soluble in water and is destroyed by prolonged cooking. It is also lost when food is grated or minced because these treatments release enzymes from the food which destroy the vitamin. When food is stored for long periods this vitamin can deteriorate and disappear altogether. Very little is known about the functions of vitamin C. A lack of it, however, results in stunted growth, and the disease scurvy, the symptoms of which are easily bruised skin, soft pulpy gums with loose teeth, and failure of wounds to heal.

Vitamin D (calciferol) Adults require only 0.000 01 g of this vitamin daily, and amounts greatly in excess of this can cause a painful stiffness of the joints. A lack of vitamin D may result in rickets, a deficiency disease characterized by soft weak bones which easily become deformed. Vitamin D is manufactured in the skin during exposure to sunlight. There is a theory that this is why people in northern latitudes have evolved 'white' skin: this feature may permit the weaker sunlight of these regions to penetrate the skin far enough for vitamin D formation. Calciferol is necessary for the proper absorption of calcium and phosphorus by the upper part of the small intestine. Both of these minerals are constituents of bone. The modern opinion is that calciferol is more properly described as a hormone than a vitamin. Hormones are substances manufactured in one part of the body which have effects on other, distant regions. This description certainly fits calciferol, which is made in the skin and has effects on the intestine.

Vitamin E There is no conclusive evidence that humans need a supply of vitamin E, although experimental evidence shows it to be necessary for fertility in rats.

Vitamin K (phylloquinone) Only very small amounts of this vitamin are required by humans. It is necessary in the complicated sequence of reactions leading to the clotting of blood in wounds. Lack of vitamin K can cause a haemorrhage, or excessive bleeding, when wounding occurs. This vitamin does not have to be present in the diet because it is manufactured by bacteria which live in the human intestine. The blood of newly-born babies will not clot in wounds until their intestines become infected with the necessary bacteria.

Water

At least two-thirds of the human body consists of water. Water has no food value, but it is still one of the most essential components of living matter. One of the most important functions of water is to act as the medium in which all the chemical reactions of metabolism take place, and for this reason life cannot continue in the absence of water. Water is also the medium in which soluble foods and excretory wastes are transported throughout the body. In higher animals the function of transporting such substances is carried out by the blood, which is mostly water. In temperate climates at least two litres of water are lost daily by the human body as urine and perspiration. This must be replaced by drinking, and by eating foods which contain water.

Roughage

Roughage is indigestible material in food, and consists mostly of cellulose and plant fibres. Its presence in food is important because it stimulates the muscular movements called **peristalsis** which propel food through the digestive system. Lack of roughage is a major cause of constipation.

3.3 A balanced diet

A diet is balanced when it contains sufficient carbohydrates, fats, proteins, minerals, and vitamins to provide for the body's energy, growth, and repair requirements.

Unfortunately, despite extensive research, it is not yet possible to say exactly how much of each type of food a person must eat in order to satisfy all of these needs. The problem of doing this is complicated by the fact that the body's food requirements vary according to age, body size, sex, occupation, state of health, and perhaps race. It has even been suggested that every person may have unique individual requirements which vary from day to day. At the moment all that can be done is to list the average requirements which experiments have shown to be necessary in various age groups.

Table 3 *Approximate total energy requirements per day (in kilojoules)*

Age	*Male*	*Both sexes*	*Female*
Birth to 1 year		3 360	
1–5 years		6 300	
5–10 years		8 400	
10–15 years		11 760	
15–20 years		13 450	
20 + years			
At rest	7 560		6 300
Light work	11 550		9 450
Heavy work	14 700		12 600
Very heavy work	21 000		15 750

Table 3 gives the daily intake of food that is necessary to satisfy the requirements of various age groups and occupations. The figures are adapted from 'Daily Dietary Allowances as recommended by the Committee on Nutrition of the British Medical Association' (1950).

The figures which refer to adults were obtained by studying the rate at which the human body expends energy in various situations. During the deepest sleep, for example, an adult male expends energy at a rate of 4.2 kJ per minute. When awake and resting in a chair he expends about 10 kJ per minute; walking, he expends 21 kJ per minute; and running very fast he expends 42 kJ per minute.

It is possible to meet total daily energy requirements by eating only pure white sugar. This would mean eating 760 g a day in order to meet the average adult's daily requirement of 12 600 kJ. But anyone who did this would not remain healthy, or alive, for long since his mineral, vitamin, and protein reserves would quickly be exhausted. To remain healthy, it is necessary to make up the total energy requirement of the body by eating balanced proportions of carbohydrate, fat, and protein.

A list of recommended weights and proportions of the various foods is found in 'The Nutritional Standard of the School Dinner' by the Department of Education and Science (1966). This book recommends that the average school dinner should have an energy value of 8700 kJ, and be made up of: 29 g protein, including 18.5 g animal protein; 32 g of fat, and the remainder carbohydrate.

Provided specific amounts and types of protein and fat are eaten the remainder of the total energy requirement may consist of almost any carbohydrate food. The details are as follows:

Fats and oils

A certain amount of fat and oil is necessary in order to supply essential fatty acids, including linoleic and linolenic acids. In addition, fats are important because they contain the fat-soluble vitamins A, D, E, and K. Beyond these requirements, fat is not essential but it does make food more palatable, which is one reason why fats are used to bake and fry many foodstuffs. Fat also reduces the bulk of food which must be eaten since, weight for weight, fat contains twice the energy value of carbohydrate.

Proteins and carbohydrates

A Department of Health and Social Security pamphlet (1969) recommends that at least 2 g of protein are needed per kg of body weight from 0–6 months of age, and 1.6 g per kg from 6–9 months, gradually reducing to 1 g per kg. The pamphlet recommends that adults should eat 80–100 g of protein per day, and that 60 per cent of this should be first-class (animal) protein since this contains a higher proportion of the essential amino acids than plant protein. Pregnant women should eat at least 85 g of protein per day, increasing to 100 g during breast feeding.

Adults should eat about 300 g of carbohydrate per day, unless they are doing very heavy work when they should eat far more. Sugar should not be eaten in large amounts as it increases tooth decay and contributes to obesity.

Verification and inquiry exercises

A *Food tests*

1. *Range of food substances required*

Solutions of glucose, sucrose, and starch; ground-up suspensions in water of pea, bean, carrot, and grape; milk; castor-oil beans, peanuts, and cooking oil; pieces of potato, bread, and boiled white of egg.

2. *Tests for carbohydrates*

a) Test for glucose Place equal quantities of Benedict's solution and glucose solution (about 2 cm³ of each) in a test-tube, lower the tube into a beaker of boiling water and note the brick-red precipitate of cuprous oxide which appears. Repeat the test with carrot and grape juice.

b) Test for starch Add a few drops of iodine solution to a starch solution and note the blue-black colouration which appears. Repeat with bread, and add iodine to the cut surface of bean, maize, and potato. Scrape stained tissue from the last three substances on to separate microscope slides and observe the shape of starch grains under high magnification.

3. *Test to verify that sucrose, starch, and cellulose are made up of glucose molecules*

a) Sucrose Verify that sucrose gives a negative result with glucose and starch tests, then add a few drops of dilute hydrochloric acid to 2 cm³ of sucrose, and boil the mixture for one minute. After neutralization with sodium hydroxide the resulting liquid will show a positive result with the glucose test.

b) Starch Add an equal volume of dilute hydrochloric acid to starch solution in a 50 cm³ beaker and boil gently for ten minutes, stirring all the time. Neutralize as in (a) above and test for glucose.

c) Cellulose Verify that filter paper, which is made of cellulose, does not contain starch or glucose. Place a drop of concentrated sulphuric acid on to iodine-stained filter paper and note the blue-black colouration which appears. Next, dissolve as much filter paper as possible in 5 cm³ of concentrated sulphuric acid and pour into 100 cm³ of distilled water. Boil the liquid in a round-bottomed flask for one hour, cool, and neutralize with solid calcium carbonate. Filter, and test for glucose.

These tests depend upon the ability of acids to break up complex molecules.

4. *Tests for fat and oil*

a) Emulsion test Add a few cm³ of ethanol to a small amount of fatty food, such as ground-up castor-oil bean, peanut, suet, or cooking oil, in a test-tube and shake the mixture. After allowing it to settle pour off the ethanol into an equal quantity of water noting that a white emulsion is formed.

b) Translucent paper mark Press pieces of peanut and castor-oil bean on to paper and note that the translucent mark which appears will wash out with acetone but not with water.

5. *Tests for proteins*

a) Millon's test Add a few drops of Millon's reagent to protein, such as pea and bean suspensions, egg white, cut maize grains, and finger-nail clippings, in a test-tube and boil for one minute. Note that the protein becomes deep red in colour.

b) Biuret test Add 2 cm³ of 2% sodium hydroxide solution to milk and mix. Add a few drops of 1% copper sulphate solution and note the violet colouration which appears. It is not necessary to boil the mixture. This is a test for soluble protein. Repeat it with pea and bean suspensions.

6. *Test for vitamin C*

Place 2 cm³ of the dye Phenol-indo-2:6-dichlorophenol in a test tube. Add 1% vitamin C solution a drop at a time from a graduated pipette. Note the quantity of vitamin needed to decolorize the dye. This result can be used as a standard for estimating the vitamin C content of fruit juices (e.g. lemon), and of extracts from fresh and old vegetables.

B *An investigation of diet*

Note that *all* the information required to complete these exercises is provided in the text and tables of this chapter.

1. *A favourite meal* Using only the items listed in Table 1, design the menu for a favourite meal, and include approximate weights of each food. Calculate the energy value of the meal in kilojoules, and see how this figure compares with the daily energy requirements listed in Table 3.

2. *A balanced diet* A balanced meal should consist of approximately one part protein, one part fat, and five parts carbohydrate. With this in mind, use the information in Table 1 to design balanced meals for one day to provide approximately 12 600 kJ of energy, essential amino acids, minerals, and vitamins.

3. *A slimming diet* Obesity, or overweight, can be defined as 10% above normal weight for one's age and physical build. People who are obese should not simply starve themselves because this will make them weak and unhealthy. Consider the following information:

Approximate daily food intake, in kilojoules, for slow healthy weight loss in moderately active people

Man	*Woman*	*Boy 13–15 years*	*Girl 13–15 years*
10 500	8 400	11 350	9 250

Using the foods listed in Table 1 design a daily slimming diet, keeping protein intake at approximately 80–85 g, with adequate supplies of minerals and vitamins. The remainder of the diet can consist of any fat or carbohydrate food up to the energy limit of the chosen group.

Comprehension test

1. 'We are what we eat.' Discuss the meaning and significance of this statement.

2. Europeans tend to eat 80 g of fatty food, while Eskimos eat up to 300 g daily. Account for the difference.

3. In the school dinner described on page 32 it is recommended that at least 18.5 g of the protein content should be made up of animal proteins. Why is this advisable?

4. Between 1906 and 1912 a biochemist called Frederick Gowland Hopkins performed many experiments on rats. The following is the most famous. He used two groups of young rats. *Group A* rats were fed on a diet of purified casein (cheese protein), starch, glucose, lard, minerals, and water. *Group B* rats were fed the same diet with the addition of 3 cm^3 of milk daily. *Group A* rats stopped growing and lost weight, while *Group B* rats gained steadily in weight and size. After eighteen days the milk was given to *Group A* rats and removed from the diet of *Group B*. *Group A* rats now recommenced growth and gained in weight, while *Group B* rats stopped growing and lost weight.

Gowland Hopkins knew that 3 cm^3 of milk has an insignificant food value in terms of carbohydrate, fat, protein, and minerals. If you were in his position with *only* the information given so far, what conclusions could you reach about the presence in milk of substances other than these four foods, the possible functions of the hypothetical substances, and the quantities needed by the rat?

Why was it necessary to transfer the milk from Group B rats to Group A half-way through the experiment?

The rats received only one type of protein (casein). Why can protein starvation be ruled out as a possible cause of growth stoppage and weight loss?

Summary and factual recall test

Examples of carbohydrate foods are (1–name six). These foods are the body's chief source of (2). They yield (3) kJ of energy per gram when respired and if eaten in excess they are either converted into (4), which is similar to starch, and stored in the (5) and (6), or are converted into (7).

Examples of fatty foods are (8–name five). These foods yield (9) kJ of energy per gram. Fat is stored in special cells which make up (10) tissue. This tissue occurs in the following body regions (11–name three). The fat under the (12) insulates the body against (13) loss. Fat is generally a better food reserve material than carbohydrate because (14–give two reasons).

Examples of protein foods are (15–name five). Proteins are described as body-(16) foods because (17). In addition, proteins form substances called (18), which are examples of catalysts because in (19) amounts they (20). All proteins are made of the elements (21–name all four). These are combined into units called (22) acids of which there are (23) different types. Examples of adequate protein foods are (24–name three). They are called adequate proteins because (25). It is especially important that (26), (27), and (28) receive a diet rich in protein because (29).

Humans require about (30) different minerals in their diet, which are found in foods such as (31–name four).

Vitamins are required in (32) quantities by the body, and lack of them causes diseases such as (33–name four).

Two-thirds of the human body consists of (34). This is the medium in which the chemical reactions of (35) take place, and in which (36) and (37) are transported to their proper locations in the body.

A balanced diet is one which contains (38). It is difficult to say exactly how much a person should eat because food requirements vary according to (39–name five factors).

4

Feeding, digestion, and absorption

Animals are the killers and thieves of this world. But they do not kill and steal out of malice; they do it to obtain food because, unlike plants, they cannot sit in the sun and make food by photosynthesis. Herbivores, carnivores, and omnivores all obtain food at the expense of other organisms.

When hungry animals find food they take it into their bodies by means of a feeding mechanism of some kind. This is called **ingestion**. But most of the food which animals consume is of no use whatsoever to their bodies in the form in which it is eaten. In the first place, most foods are insoluble and in this state cannot pass through cell membranes into body cells. In the second place, most foods are chemically different from substances which make up the body tissues and so cannot take part in metabolism directly.

Imagine a man who has just eaten a ham sandwich. Bread from the sandwich contains the carbohydrate food starch, which is a source of energy. But starch is both insoluble and chemically different from glucose and glycogen, which are the carbohydrates used by human body cells as fuel for the energy-releasing process of respiration. Similarly, fats and oils contained in the butter and ham of the sandwich cannot immediately be used by the human body because they are insoluble and differ in chemical structure from human fats and oils. Again, proteins in the ham are useless to the human body in the form in which they are eaten mainly because the number and sequence of amino acids in their molecules differ from any type of human protein. Only the salt used to flavour the sandwich and the few vitamins in butter, ham, and bread can be used directly by the man's body. Nevertheless, his body is able to make full use of all the foods in a sandwich by means of the following events.

First, the man's digestive system produces chemicals which make the food soluble. This process is called **digestion**. Second, the soluble food passes through the intestine walls into the blood stream. This is called **absorption**. Finally, soluble food is transported by the blood to all parts of the body where it enters cells and is transformed into substances which can take part in human metabolism. This is called **assimilation**.

This chapter begins with a fuller explanation of digestion and absorption. It then describes some of the ways in which animals feed, giving details, in certain cases, of their digestive and absorptive processes.

4.1 Digestion and absorption

Digestion is the process in which food is broken down into chemically simpler materials which are soluble in water. Carbohydrate foods are broken down into monosaccharide sugars such as glucose; fats and oils are broken down into fatty acids and glycerol; and proteins are broken down into amino acids.

In animals, and all other heterotrophic organisms, food is broken down by digestive juices which contain chemicals called **digestive enzymes.**

Digestive enzymes

All enzymes are **catalysts**; that is, they speed up chemical reactions which would otherwise proceed very slowly. Digestive enzymes are only one example of many types of enzyme which exist in living things. The reactions which these enzymes speed up involve splitting complicated molecules into simpler ones. Figure 4.1 illustrates one theory of how this may happen.

It is thought that the enzyme combines briefly with molecules of food and while in this state the food undergoes a rapid chemical change in which its molecules are split apart into chemically simpler substances. These substances separate from the enzyme leaving it immediately available for another identical reaction. In other words, enzymes are not used up in the reactions which they control but are used countless times in rapid succession.

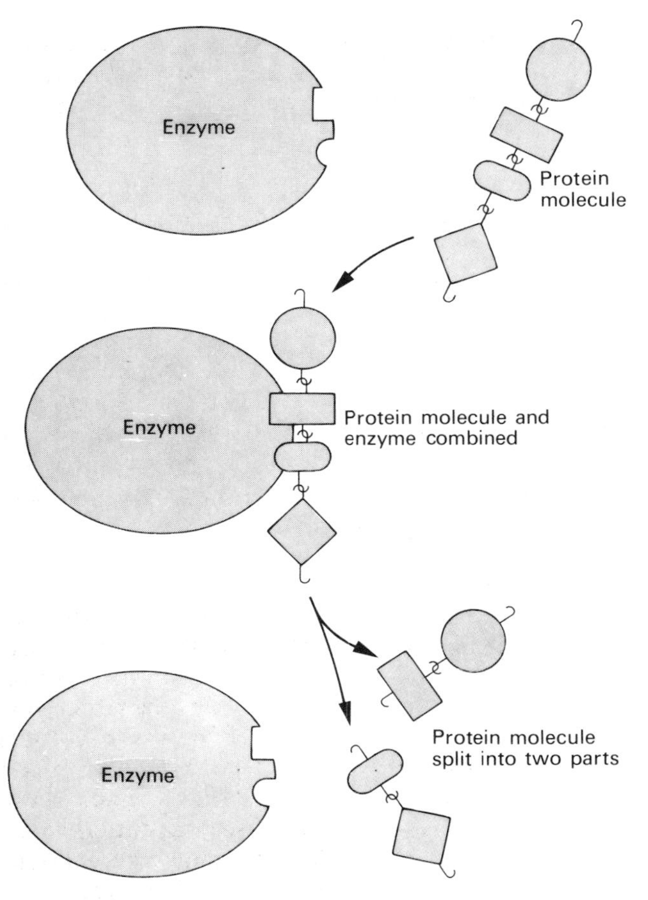

Fig. 4.1 Diagram of how digestive enzymes work

In this example a protein molecule, made up of linked amino acids, combines with an enzyme molecule and undergoes a chemical change which breaks the link between two of the amino acids. This splits the protein into two smaller molecules. Protein and enzyme molecules are not actually shaped like those in the diagram. The shapes indicate that each enzyme combines with only one specific part of the food molecule.

The following list describes the more important features of digestive enzymes, most of which they share with enzymes in general.

1. Small amounts of enzyme can bring about a chemical change in relatively large amounts of another substance.

2. Enzymes are specific. This means that each enzyme is limited to reactions with only one, or very few, types of food substance. Because of this digestive enzymes can be divided into groups according to the foods which they digest. **Amylases**, sometimes called **diastases**, are a group of enzymes which break down starchy foods into sugars such as glucose. **Lipases** are a group of enzymes which break down fats and oils into their component fatty acids and glycerol. **Proteases** are enzymes which break down proteins into their component amino acids.

3. The speed of reactions involving enzymes is influenced by temperature. A rise in temperature causes an increase in reaction speed, up to what is called the enzyme's **optimum efficiency**. At temperatures above this point the reaction goes faster for a time but eventually stops because excessive heat destroys enzymes.

4. Some enzymes work best in acid conditions, others in neutral conditions, and others in conditions which are alkaline. In technical terms, each enzyme requires a specific pH level for optimum efficiency.

5. Certain enzymes require the presence of vitamins of the B complex before they can function. These vitamins are known as **co-enzymes**.

Types of digestion

In general, there are two types of digestion: intracellular and extracellular.

Intracellular digestion This type of digestion takes place inside a cell. It is typical of unicellular organisms such as *Amoeba* and, to a limited extent, of simple multicellular animals such as *Hydra*, both of which are described below. The sequence of events typical of intracellular digestion is illustrated in Figure 4.2. Food particles enter a cell, are broken down by digestive juices into soluble form, and are then absorbed directly into the cytoplasm. Indigestible substances, i.e. those which cannot be made soluble, are expelled from the cell.

Extracellular digestion This type of digestion occurs outside the cells of the body and is typical of most animals and many other heterotrophic organisms. The process is illustrated diagrammatically in Figure 4.3. Typically, extracellular digestion occurs within a tube called the **intestine**, or **alimentary canal**, which runs through the organism from mouth to anus. In certain specialized regions the walls of the intestine produce digestive juices which break down the food into soluble form. In larger animals, such as humans, the soluble food is absorbed in two stages. First, it passes through the intestine wall and is absorbed into the blood-stream which transports it to all parts of the body. Second, it passes out of the blood-stream and is absorbed into body cells.

4.2 Amoeba and Paramecium

Both of these unicellular organisms carry out intracellular digestion. They take food into their cytoplasm, where it is enclosed in a space called a **food vacuole** which contains water and digestive juice. After diges-

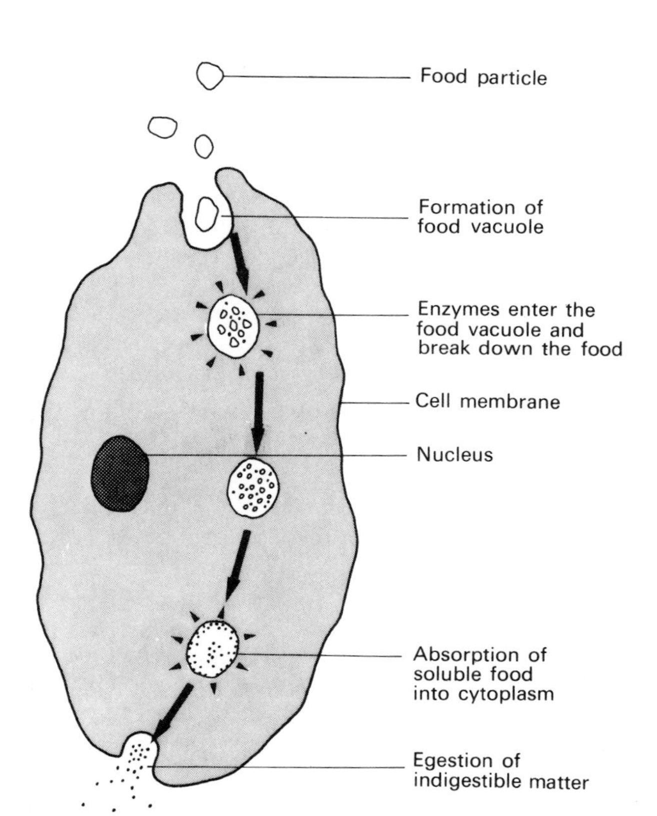

Fig. 4.2 Diagram of intracellular digestion

Digestion takes place within food vacuoles, from which soluble food is absorbed directly into the cytoplasm of the cell.

Fig. 4.3 Diagram of extracellular digestion

Digestion takes place in an intestine, shown here as a straight tube through the animal from mouth to anus. Enzymes are released into this tube, and food is absorbed into the body through the walls of this tube.

tion, soluble food passes out of the vacuole and is absorbed into the cytoplasm. Indigestible matter is egested through a temporary hole in the cell membrane.

Amoeba

Amoeba feeds mainly on protozoa smaller than itself. It does this by producing outgrowths from its body called **pseudopodia,** which surround and then completely enclose the food particle (Fig. 4.4). The food is engulfed along with a droplet of water so that a food vacuole is automatically formed as the pseudopodia close and fuse together. *Amoeba* has no permanent mouth, digestive system, or anus. Its digestive mechanism is no more than a sequence of food vacuoles scattered throughout its cytoplasm, containing food in various stages of digestion.

Paramecium

Unlike *Amoeba*, *Paramecium* has a permanent feeding mechanism consisting of a funnel-shaped **gullet** into which food is drawn by the combined action of cilia which cover the body and other cilia which line the gullet's inner surface (Fig. 2.3). Food vacuoles form at the base of the gullet from where they circulate through the cytoplasm on a more or less fixed route during which digestion takes place. Eventually the vacuoles reach an area of cell membrane where a temporary pore opens to let out indigestible material.

4.3 Hydra

Hydra, and all other Coelenterates, carry out both intracellular and extracellular digestion. Many types

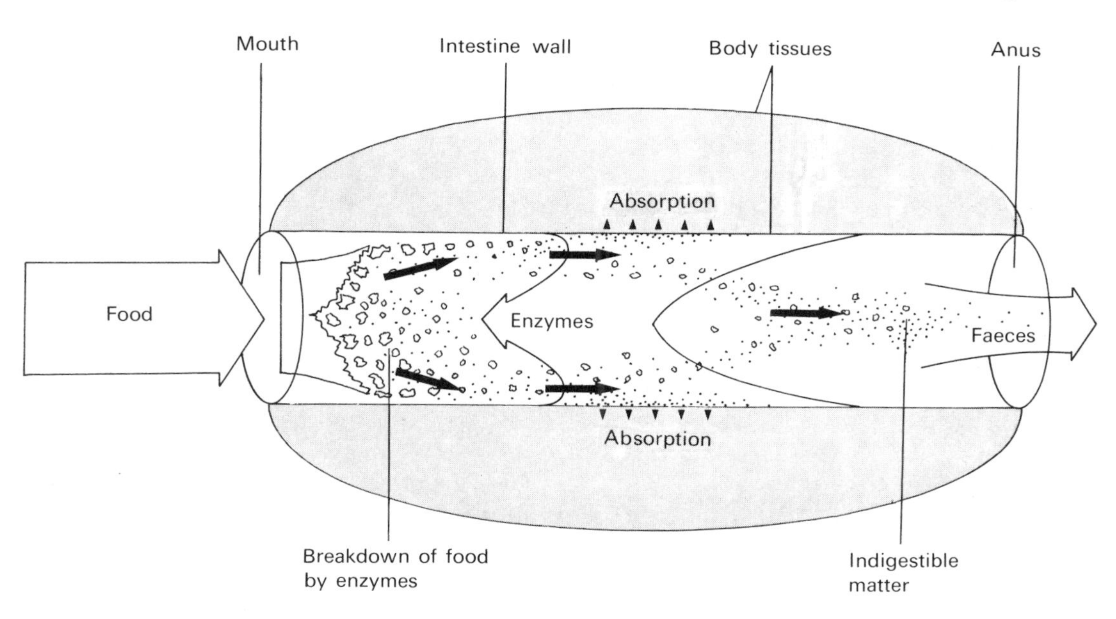

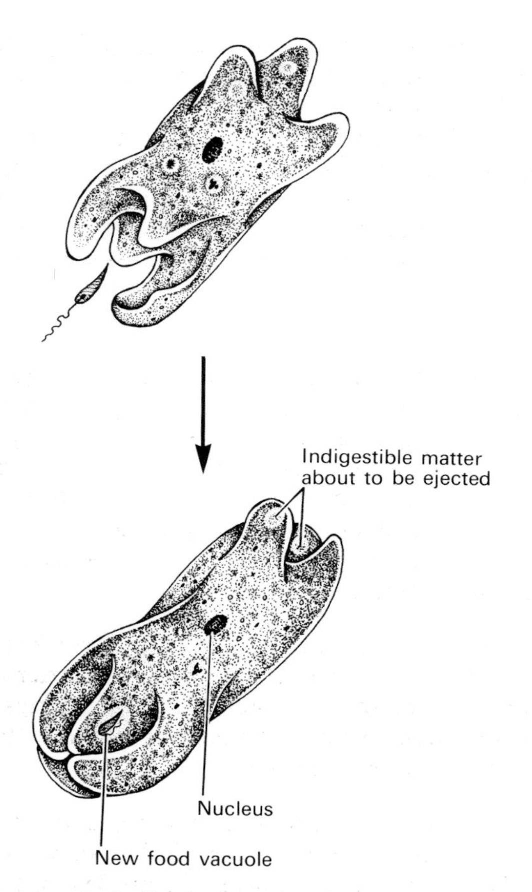

Fig. 4.4 *Amoeba* feeding

of *Hydra* are found living in ponds and slow-moving streams where their slender tube-like bodies with six or more tentacles are found attached to pond weed, or hanging upside-down from the water surface. These animals are carnivorous and live on small crustaceans such as *Daphnia* and *Cyclops*. Figure 2.4 illustrates a hydra's body structure, and Figure 4.5 shows how they catch prey. Hydras do not actively hunt. They wait, often for long periods, in one position with tentacles fully extended. When prey accidentally brushes against a tentacle several things happen. First, the prey is riddled by poison-injecting threads from specialized tentacle cells called **nematocysts** (Fig. 4.6). The poison injected by these cells paralyses the prey, and other types of nematocyst send out threads which coil around the prey holding it temporarily. Second, several tentacles coil around the prey and thrust it whole into the animal's mouth. Third, a hydra's gut cavity is lined with different types of cell which now take part in digesting the prey. Extracellular digestion begins when gland cells in the gut wall produce digestive enzymes. The whip-like action of flagella cells mixes food and enzymes together and eventually the food breaks up into small partly-digested fragments. Intracellular digestion begins when amoeboid cells in the gut wall engulf these food fragments into food vacuoles where digestion is completed. Finally, sudden contraction of circular muscle fibres in the hydra's body wall ejects indigestible matter through the animal's mouth.

Fig. 4.5 *Hydra* feeding

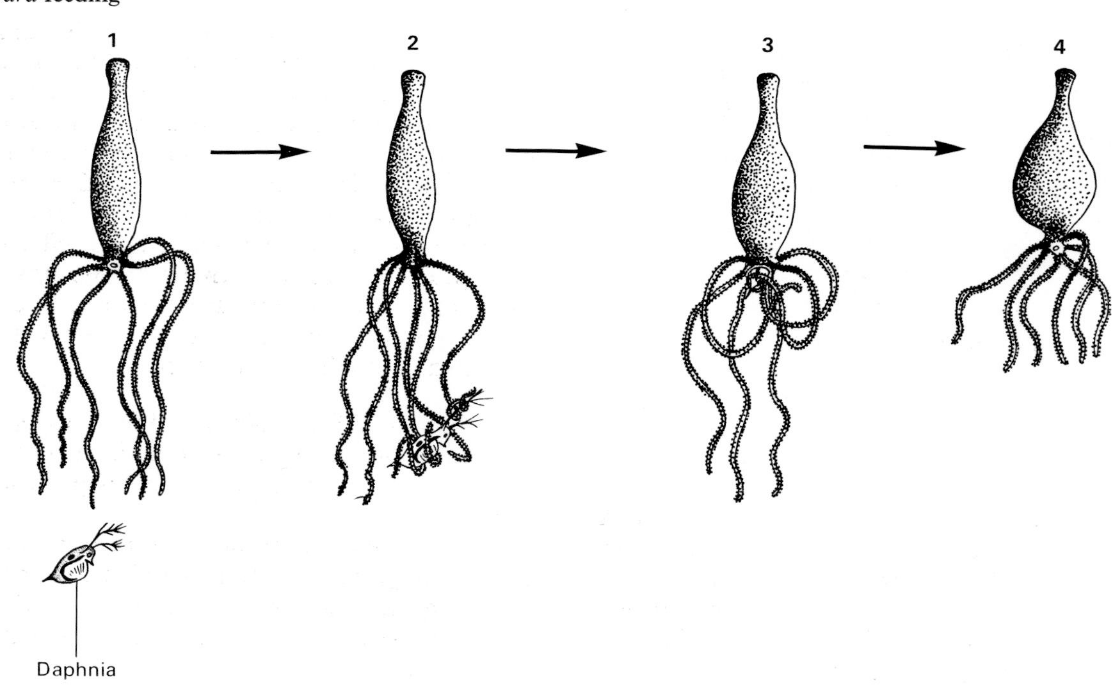

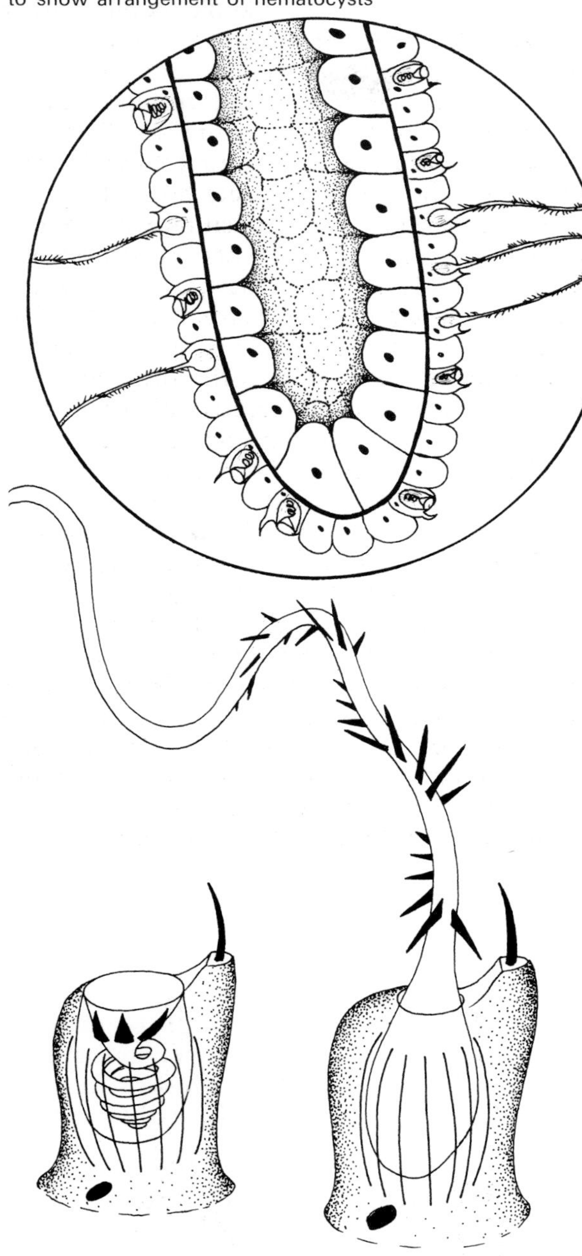

Fig. 4.6 Nematocysts

4.4 Insects

The mouth of an insect is surrounded by a number of different structures called **mouth parts** which are used in feeding. The shape of these mouth parts varies from species to species and enables each type of insect to exploit a particular source of food. Houseflies, for instance, feed on rotting animal and vegetable matter and their mouth parts are very different from those of butterflies which are so adapted that they can suck nectar from flowers.

Comparison between fossil insects and living specimens suggests that species such as cockroaches, grasshoppers, and locusts, all of which have biting and cutting mouth parts, most closely resemble the extinct species from which all insects are thought to have evolved.

Locust

Locust mouth parts, illustrated in Figure 4.7, enable them to eat a wide variety of plant materials, but especially grasses and all other cereals. In parts of Africa swarms of locusts, sometimes measuring hundreds of square kilometers in extent, will often eat every scrap of vegetation over huge areas.

Starting from the front of the animal the mouth parts are arranged in the following sequence.

Labrum The labrum is a large flap which conceals the other mouth parts from view. Its inner surface bears minute bristles containing nerve endings sensitive to various chemicals. These function like the taste buds on the human tongue.

Mandibles Behind the labrum, and on either side of the mouth are the mandibles, or jaws. Each mandible is moved by powerful muscles and has an extremely hard biting edge. During feeding these edges are brought together in a precise shearing action, which is used to cut food into small, easily manageable pieces.

Maxillae Behind the mandibles are a pair of maxillae which have three main functions. First, the tooth-like edges of each maxilla are used to assist the mandibles in cutting and breaking up food. Second, maxillae are capable of a wider range of movements than the mandibles and this enables them to be used in manipulating food and guiding it into the mouth. Third, each maxilla bears a long finger-like organ called a **palp**. Palps possess many nerve endings sensitive to touch and chemicals and are used, in conjunction with the eyes and other sense organs on the feet and antennae, to detect edible substances.

Labium The labium, or lower lip, appears to be formed by the fusion of two maxilla-like structures. It bears no teeth but has two palps with sensory functions.

Locusts detect food at a distance mainly by smell using sense organs located on their antennae, described in chapter 14. Their sense of sight is limited. Exercise C explains ways of investigating feeding and sensitivity in locusts.

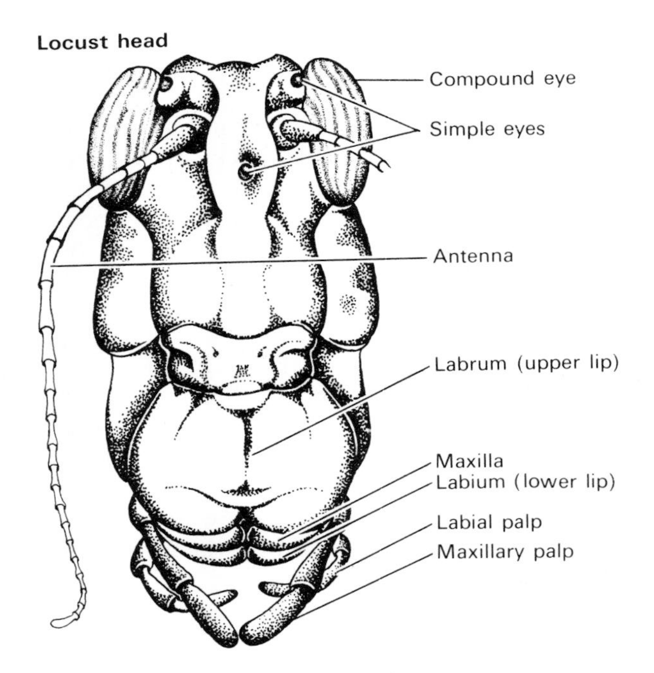

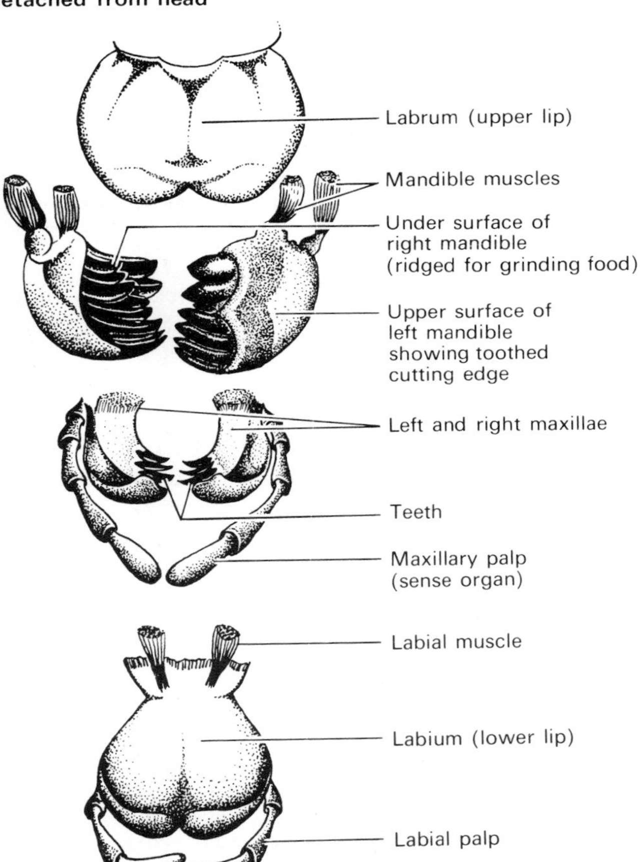

Fig. 4.7 Locust head and mouth parts

Mosquito

Mosquitoes are parasites, and have mouth parts which enable them to cut into and deeply pierce the skin and soft tissues of their hosts. The female *Anopheles* mosquito, for example, feeds on the blood of mammals and has mouth parts which resemble the needle of a hypodermic syringe. She uses these mouth parts to suck blood from her hosts, as illustrated in Figure 4.8.

In mosquitoes the basic arrangement of insect mouth parts as illustrated by the locust has been adapted to such an extent in the course of evolution that the ancestral pattern is no longer recognizable. The mosquito has six long needle-like **stylets** enclosed in a sheath formed by the labium. The largest stylet, the labrum, is deeply grooved on its underside to form the roof of a tube which is completed when another stylet, called the **hypopharynx**, is fitted closely against it. This arrangement is used to suck food into the insect's digestive system, and so it is called the **food-tube**. There are a further two pairs of stylets: the maxillae which have saw-like barbs at their tips, and the needle-like mandibles. The maxillae and mandibles work together to penetrate host tissues and form a hole through which the food-tube is passed. During this last operation the sheath-like labium holds the stylets together.

In blood-sucking mosquitoes there is a channel through the centre of the hypopharynx which is used to pump saliva into the wound. Saliva contains chemicals which prevent the formation of blood clots which would block the food-tube.

Housefly

The mouth parts of a housefly, illustrated in Figure 4.9, consist of a long **proboscis**, or tongue, through which passes a food channel and a salivary channel. Both these channels open into many tiny grooves in the surface of a pair of flaps at the base of the proboscis.

It is easy to watch the 'lapping' action of a housefly's proboscis by putting moist sugar cubes as bait near an open window in warm weather. During feeding saliva containing digestive enzymes is squirted on to the food through the salivary channel. The saliva digests the food and turns it into a liquid which is sucked up the food channel into the fly's intestine where the last stages of digestion are completed.

Houseflies live and breed in places such as refuse tips, where there is decaying animal and vegetable matter. Germs which cause typhoid, dysentery, and many other diseases also live in decaying matter, which suggests that houseflies are responsible for the spread of certain diseases.

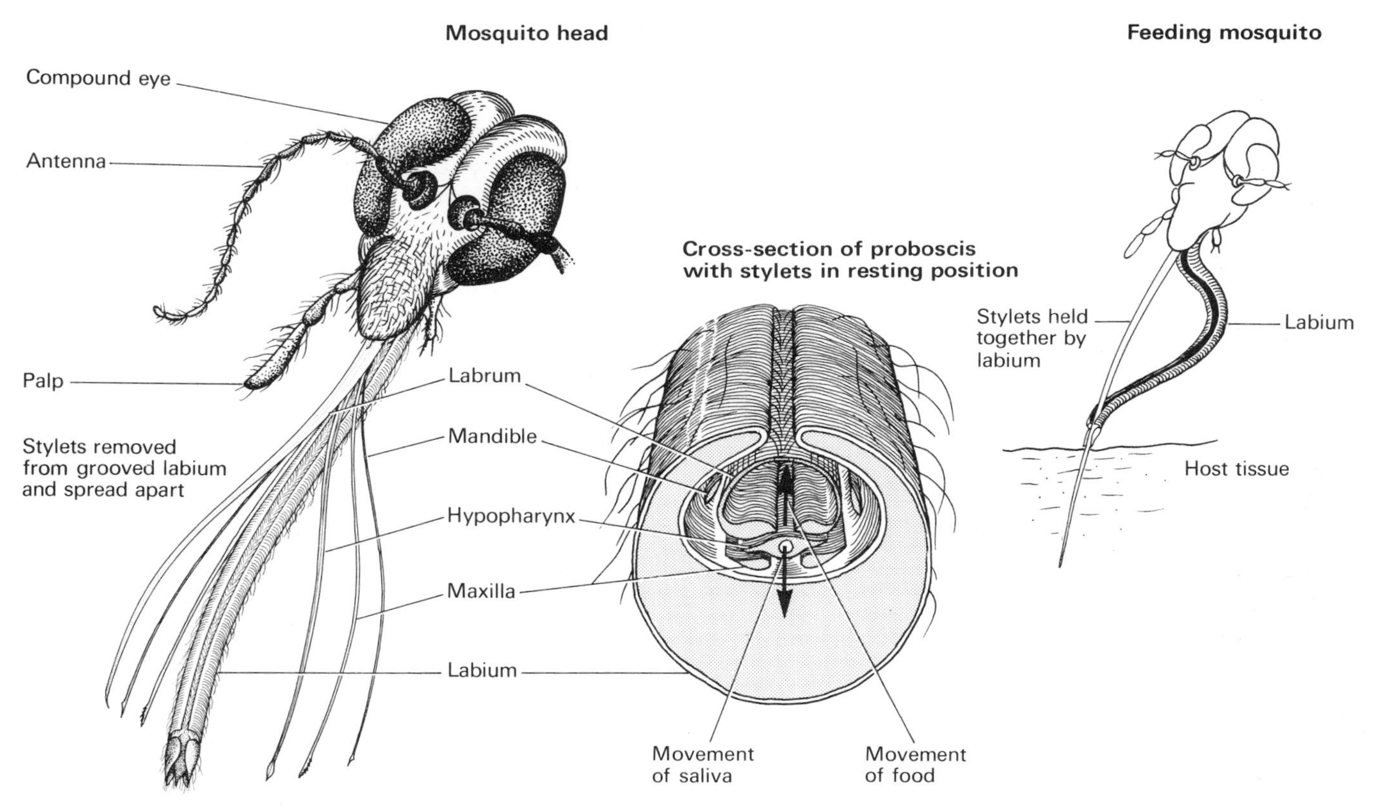

Fig. 4.8 Mosquito head and mouth parts

Fig. 4.9 Housefly head and mouth parts

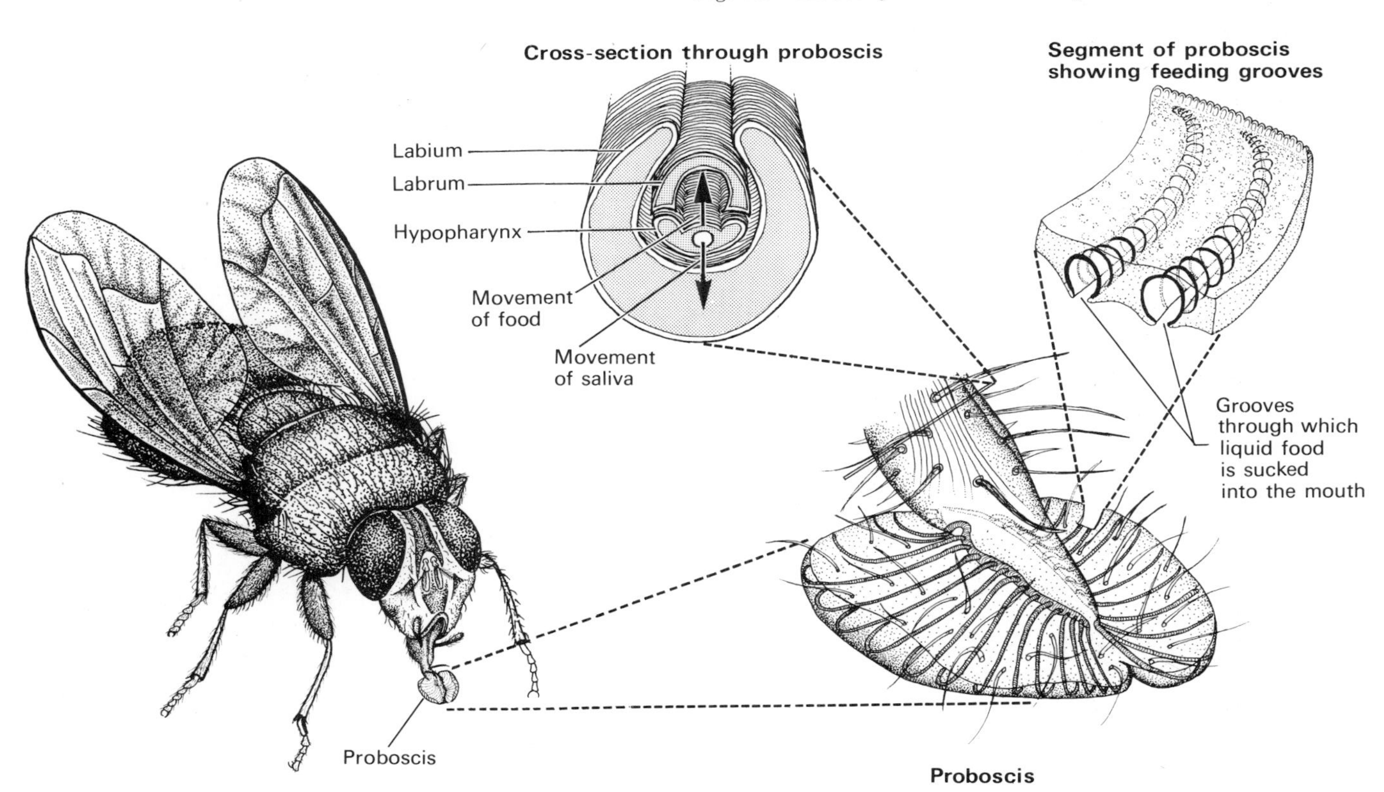

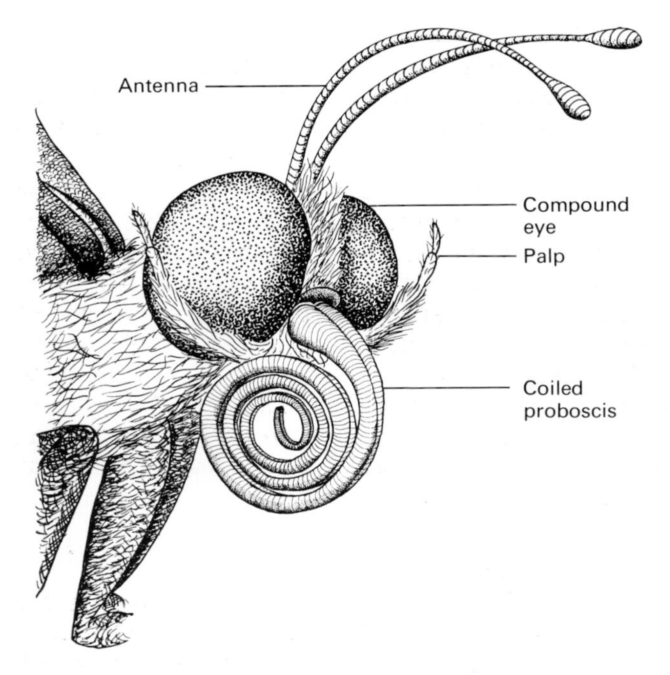

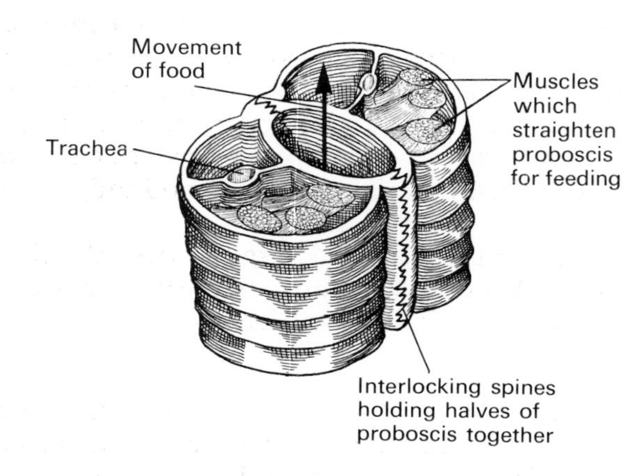

Fig. 4.10 Butterfly head and mouth parts

Butterfly

A butterfly has a proboscis through which it sucks nectar from flowers. The proboscis is formed by two modified maxillae grooved along their inner surfaces and fitted together by interlocking spines to form a tube (Fig. 4.10).

This complex structure is tightly coiled like a watch-spring when not in use. It is uncoiled during feeding by the contraction of tiny muscles along its length which work against the curvature of the coil. When uncoiled the proboscis is dipped into the nectary of a flower and used like a drinking-straw to suck food into the insect's mouth and digestive system.

4.5 Mammals

In mammals food is first broken down by the mechanical action of teeth. Chewing is a physical process of reducing food to small particles. Thorough chewing of food is important for at least three reasons. First, small pieces of food are more easily swallowed than large lumps. Second, the surface area of food is vastly increased when it is broken into fragments, and since digestive juices work from the surface of food inwards, thorough chewing speeds up the rate of digestion. Third, chewing mixes food with saliva, the functions of which are described later in this chapter.

Structure of a tooth

Teeth of mammals are set within sockets in the jaw-bone (Fig. 4.11). Each tooth consists of: a **crown** the part above gum level; a **neck**, the part surrounded by gum; and a **root**, the part embedded in bone. The crown is the biting surface. Its outer layer consists of **enamel**, which is the hardest substance found in animals. The root and inner portion of the crown consist of a bone-like substance called **dentine**, except for a central **pulp cavity**. This contains nerves, and blood vessels that supply the growing tooth with food and oxygen.

Fig. 4.11 Structure of a tooth

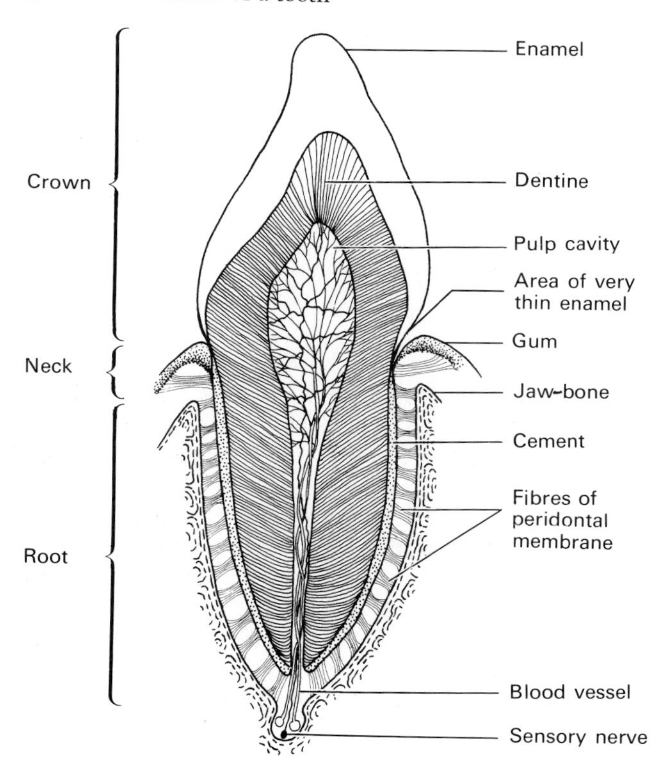

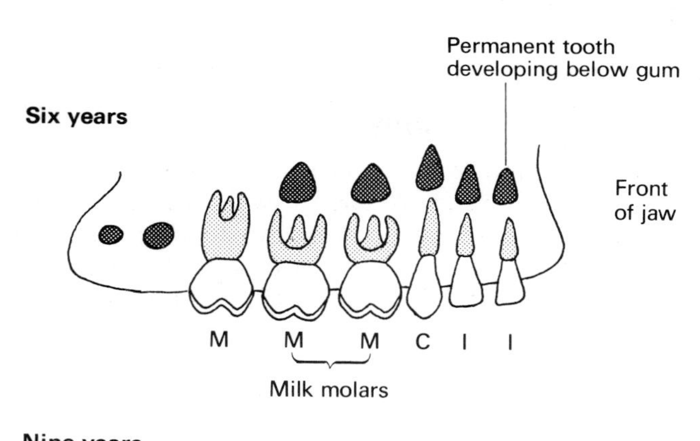

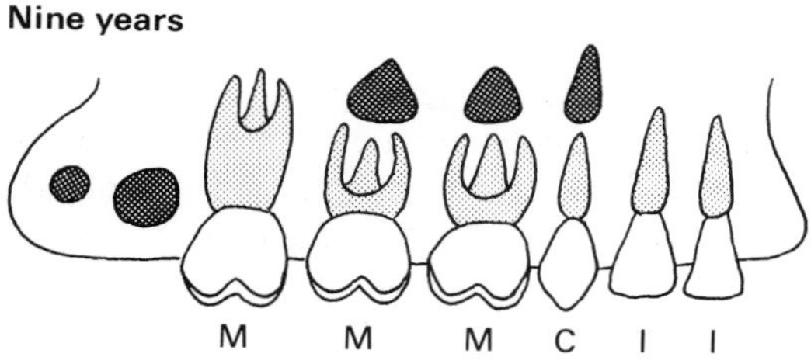

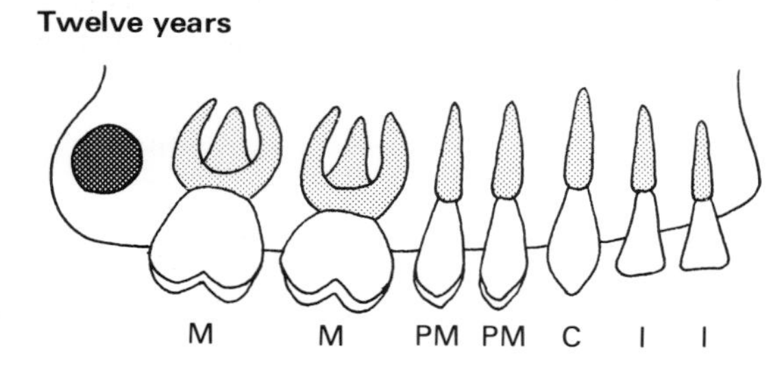

Fig. 4.12 Development of permanent teeth (The diagram shows only half of one jaw.)

In carnivores and omnivores the teeth stop growing when they reach a certain size. In herbivores, however, the teeth continue to grow throughout the life of the animal.

The root of each tooth is covered with **cement**, another bone-like material, which extends upwards as far as the enamel of the crown. Tough fibres of the **peridontal membrane** attach the cement to the jaw bone so that each tooth is fixed firmly, but can move a little in its socket during chewing. This slight flexibility reduces the risk of teeth breaking when hard food is chewed.

Milk and permanent teeth

Most mammals have two sets of teeth, the first of which are known as the **milk set** (Fig. 4.12). In humans, teeth form in the jaws before birth and begin to appear, or erupt, at about five months of age. By approximately six years most children have twenty-four teeth. Only twenty of these, however, are milk teeth. They are shed between the ages of seven and eleven years to be replaced by larger permanent teeth. By eighteen years of age most people have a total of thirty-two teeth. Twenty of these are replacement teeth (in place of milk teeth), and twelve are non-replacement teeth. The non-replacement teeth consist of molars. These appear at approximately six, twelve, and eighteen years of age. The last molars to appear are the wisdom teeth.

Types of teeth and their functions

The variety of teeth found in mammals is the result of evolution from four basic types (Fig. 4.13).

At the front of the jaw are a set of **incisors** which are typically chisel-shaped and used to cut food into pieces of manageable size. Next are two **canines**, one at the side of each jaw. Canines are characteristically long and pointed and are prominent in carnivores. Last are the cheek-teeth, the **premolars** and **molars**, which have projections on their surface called **cusps.** These teeth are used to pulverize food and, sometimes, grind it into small particles.

The word **dentition** is used to describe the shape and number of teeth. The pattern of an animal's dentition varies according to diet.

Carnivore dentition (Fig. 4.14A) In the course of evolution carnivores have become highly specialized and efficient at catching and killing other animals, particularly herbivores. A typical carnivore has small closely-fitting incisors. These are used to clean fur, and to cut away pieces of flesh close to the bones. The long dagger-like canines are used to pierce the prey when it is captured, thus preventing its escape, and often killing it. The huge **carnassials**, and other similar but smaller premolars, have a shape which is especially well adapted to the task of shearing flesh from bones and cracking them open so that bone marrow is exposed. Carnivores have comparatively small molars and spend little time chewing food into small pieces. Cats, for instance, have only one tiny molar at the side of each jaw. Chewing is therefore restricted to cutting food into pieces which are small enough to swallow.

In biology the word **articulation** refers to a moveable joint in the skeleton. The jaw articulation in carnivores is a closely fitting hinge-joint which permits

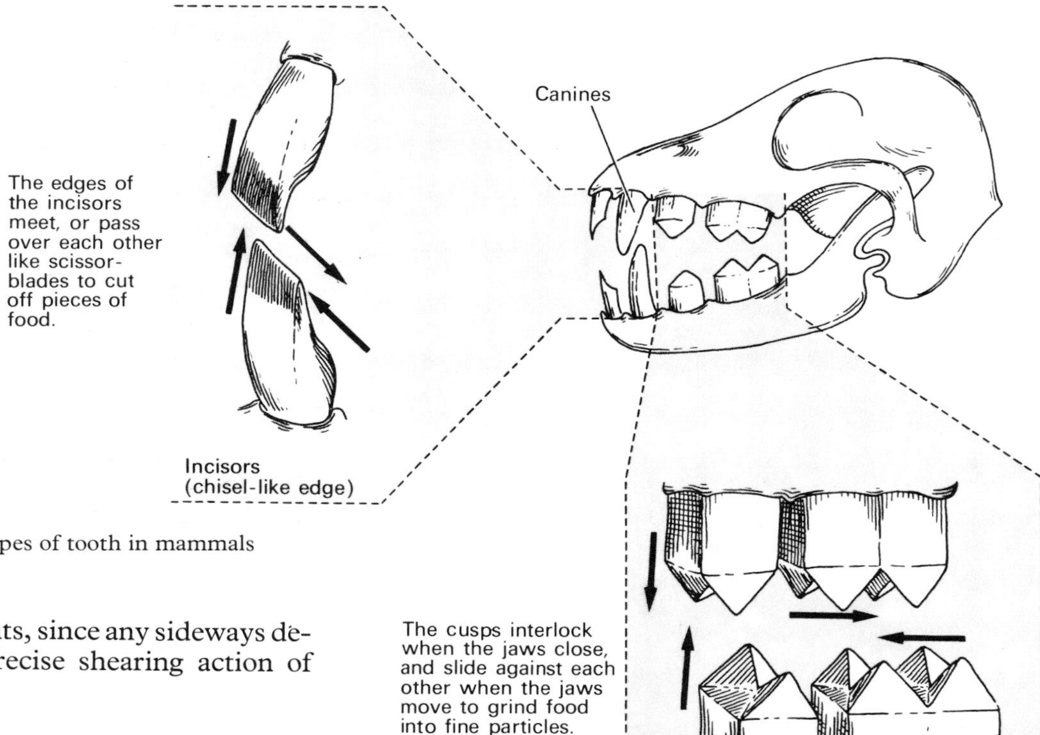

Fig. 4.13 Diagram of the four types of tooth in mammals

only up-and-down movements, since any sideways deviation would hinder the precise shearing action of carnassial teeth.

Herbivore dentition (Fig. 4.14B) Herbivore teeth are adapted in such a way that the animals can eat large quantities of vegetable matter, especially grass. The following description applies mainly to a group of herbivores called **ruminants** which includes sheep, cattle, oxen, goats, deer, antelope, and giraffes.

Ruminants have no incisors in the upper jaw. This region consists of a thick horny pad against which the lower incisors bite. Canine teeth are either absent, very small, or present as tusks. The most noticeable and important teeth are premolars and molars. These teeth wear down into alternating ridges of hard enamel and furrows of softer dentine, but the teeth do not wear away altogether since they continue to grow throughout the animal's life. The ridges of the teeth in the upper jaw have a shape which fits exactly against the ridges of the teeth in the lower jaw. Jaw articulation in ruminants is very loose and permits extensive movements forwards, backwards, and sideways. The function of each of these features is best understood by watching a ruminant eating.

A sheep or cow, for instance, uses its long flexible tongue to place a tuft of grass between the lower incisors and horny pad of the upper jaw. A sideways jaw movement, and a simultaneous upward jerk of the head, shears off the grass close to the soil. The food is not chewed immediately but swallowed whole and stored in the **rumen**. The rumen is the first part of the four-chambered stomach which gives the group of animals its name. At the next stage grass is moved from the rumen to the **reticulum**, the second chamber of the stomach, where it is moulded into round balls of 'cud' which are returned to the mouth for chewing. Continuous and prolonged side-to-side movements of the jaws, made possible by loose jaw articulation, draw the animal's ridged teeth closely across each other. This action grinds up the cellulose walls of plant cells and reduces the material to a fine consistency.

In wild ruminants these feeding habits make it possible for the animals to eat very quickly and then retire to a safe place from which they can keep watch for predatory carnivores while chewing their cud.

Omnivore dentition All that need be said about omnivorous animals is that their teeth show very little specialization of any kind. Humans, for example, can eat a wide variety of foods – everything from rare steak to soft pulpy fruit.

Digestion in the mouth

Chewing mixes food with **saliva**. This is a neutral or slightly alkaline fluid produced by three pairs of salivary glands (Fig. 4.15). In humans these glands produce about 1.5 litres of saliva daily. Saliva consists of water, the digestive enzyme **salivary amylase** (sometimes called ptyalin), **mucin**, and several other substances including sodium, potassium, bicarbonate, and chloride.

Functions of saliva

1. The water and mucin in saliva moisten, soften, and lubricate dry food so that it is more easily chewed and swallowed.

2. The enzyme salivary amylase breaks down starch into the soluble sugar maltose. This reaction is the first stage of carbohydrate digestion. Food remains in the mouth for only a few seconds and little digestion occurs here, but after swallowing the action of saliva continues for some time in the stomach. It is finally stopped when the acid in stomach digestive juice penetrates the food and destroys the amylase.

3. Amylase, and another enzyme in saliva called **lysozyme**, help to remove carbohydrate food and bacteria from between the teeth and thus help to prevent tooth decay.

4. Saliva moistens the mouth, tongue, and lips, which facilitates talking. Saliva production stops during nervousness or severe illness, making speech difficult.

5. By dissolving food, saliva makes it possible for the chemicals within it to reach the taste buds in the tongue. This is important because taste buds are not stimulated by dry food.

6. Bicarbonate in saliva acts as a **buffer**, which means that it keeps saliva at a more or less constant level of weak alkalinity. This helps prevent tooth decay by reducing the strength of mouth acids that dissolve tooth enamel. During sleep saliva production slows down considerably.

Fig. 4.14 Comparison between carnivore and herbivore dentition

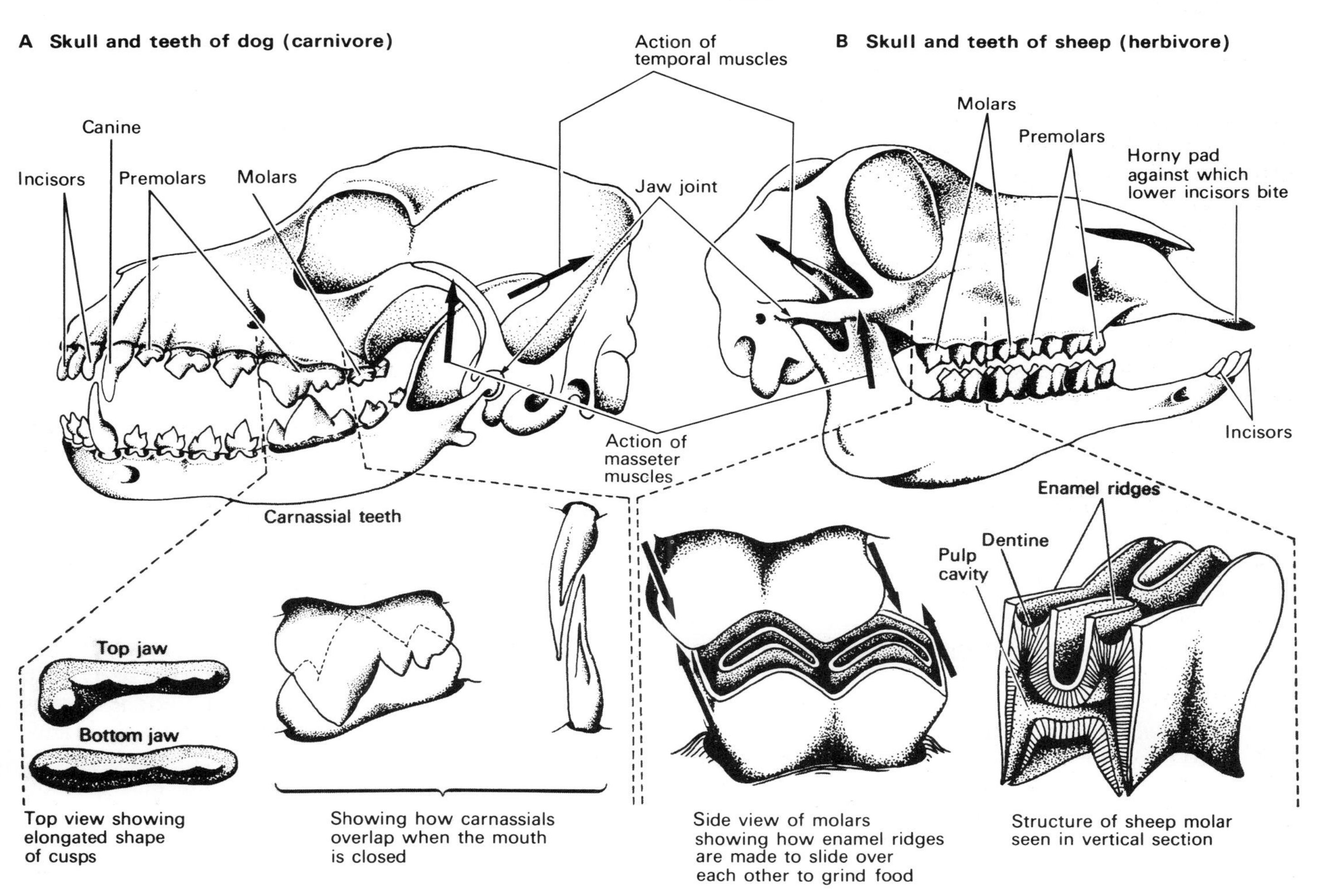

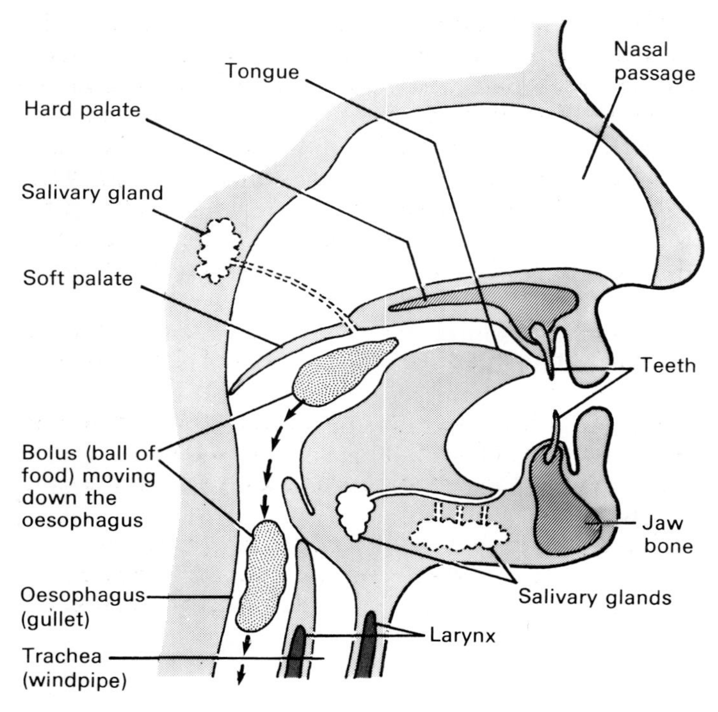

Fig. 4.15 Swallowing

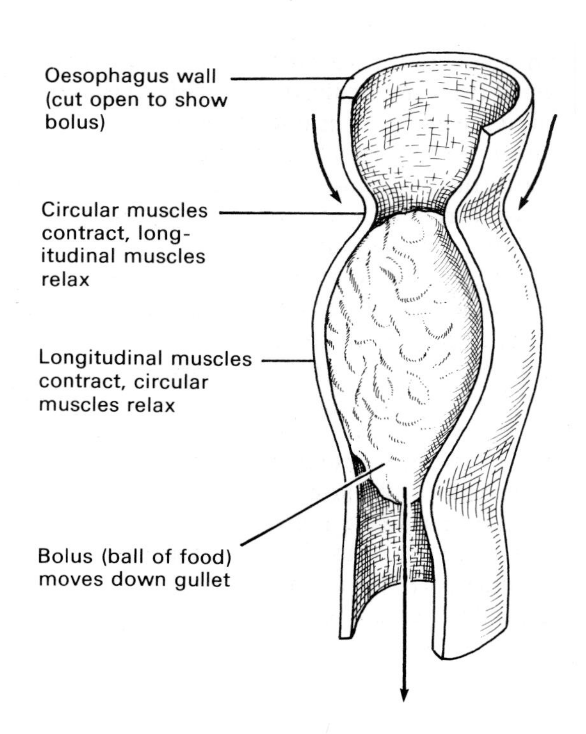

Fig. 4.16 Peristalsis (muscular contractions which move food along the gut)

Swallowing

Immediately before swallowing, food is rolled into a ball, or **bolus**, by the tongue and is then thrust to the back of the mouth. When this is done clumsily, or while breathing in through the mouth, food can be drawn into the windpipe, or **trachea**, causing a fit of coughing which usually clears any blockage. Normally, the following sequence of events prevents the risk of a blocked trachea.

First, the soft palate is pushed upwards to shut off entry to the nasal cavity. Second, muscles pull the top of the trachea upwards and under the back of the tongue. This action also pulls the larynx upwards causing the familiar 'bobbing' action of the 'Adam's apple', another name for the larynx. Third, the entrance to the trachea is reduced in size by the contraction of a ring of muscle around its circumference. Fourth, the **epiglottis**, a flat piece of cartilage and skin, drops over the tracheal entrance, forming a bridge over which food passes as it goes into the **oesophagus**, a tube leading to the stomach (Fig. 4.15).

Movement of food along the intestine

Food does not simply fall through the intestine under the influence of gravity. In fact a person can drink and eat while standing on his head, because food is moved along the intestine by wave-like muscular contractions known as **peristalsis** (Fig. 4.16).

These contractions start as soon as food enters the oesophagus. Circular muscles in the oesophagus wall contract immediately behind the bolus pushing it towards the stomach. These waves of contraction continue throughout the whole intestine at speeds of up to 20 cm per second. Movement of food is assisted first by the lubricating action of salivary mucin and then by **mucus**, a similar substance produced by **goblet cells** in the intestine wall (Fig. 4.19).

Digestion in the stomach

Perhaps the first person to investigate the stomach was an Italian called Spallanzani who in 1770 was curious enough, and brave enough, to swallow pieces of sponge tied to lengths of string. The sponges were pulled back up again, squeezed out, and the liquid from them examined. (This procedure could be dangerous and is not recommended to readers.)

In 1822 an army doctor called Beaumont treated a gunshot wound in one of his patients. The man's injury healed, but left a small hole right through the body wall into his stomach. Over an eleven year period the doctor took samples of stomach contents through this hole and from them discovered a great deal about what happens to food in the stomach. Eventually, and perhaps not surprisingly, the 'patient' grew tired of the experiments and refused to let them continue.

The experiments of Spallanzani and Beaumont were

among the first to provide evidence that the stomach produces chemicals, now called digestive juices, which dissolve food. Beaumont went on to investigate the conditions in which these juices work and the types of food that they digest. His results, and those of more advanced studies carried out by other workers, can be summarized as follows.

The stomach is an enlarged, bag-like region of the intestine. In humans it is J-shaped when standing and U-shaped when lying down. There are rings of muscle called **sphincters** around the entrance and exit to the stomach. These muscles act like valves: when they contract movement of food is prevented and when they relax food movements are resumed. As swallowed food approaches the stomach its uppermost sphincter relaxes and its lower sphincter contracts. The food enters the stomach, and the stomach walls expand until it reaches its normal capacity. In humans this is approximately 1 litre. More than this amount of food can be eaten at one meal but it produces an over-full sensation.

Once food is in the stomach the circular muscles in its walls begin wave-like contractions similar to peristalsis which pass down the stomach from the oesophagus end about three times a minute. These contractions have the important effect of churning up food and mixing it thoroughly with a substance called **gastric juice** which flows out of thousands of tiny hollow pits, the **gastric glands**, in the stomach lining. The semi-liquid result of this churning process is called **chyme**.

Gastric juice Adult humans produce about three litres of gastric juice daily. It consists of the following substances:

Hydrochloric acid is produced by the stomach in concentrations of up to 0.4%. Its function is to produce acid conditions which are necessary for the action of gastric enzymes.

Pepsin is an enzyme which breaks down protein molecules. It requires the conditions created by hydrochloric acid to do this. To understand the action of

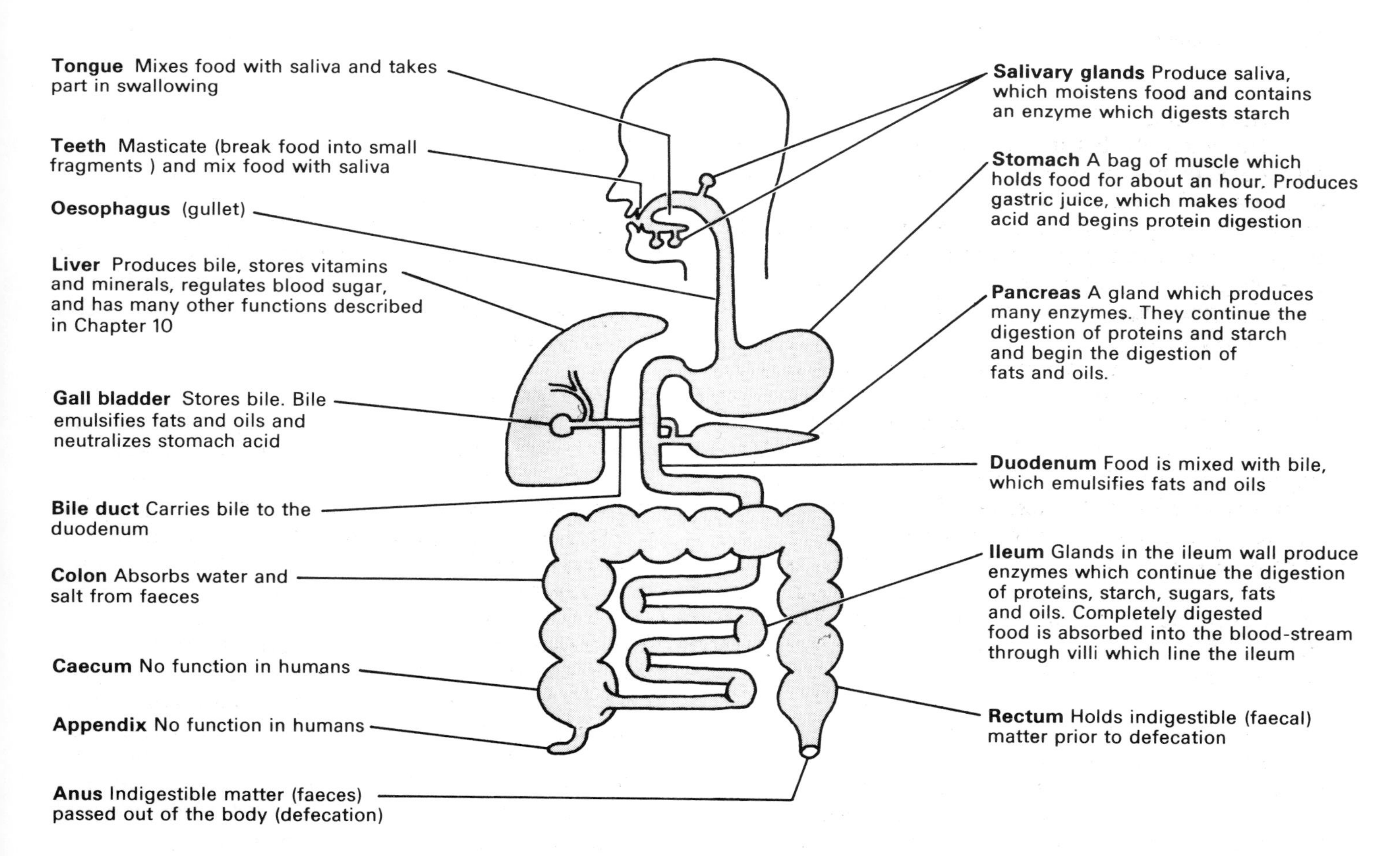

Fig. 4.17 Diagram of the gut and its functions

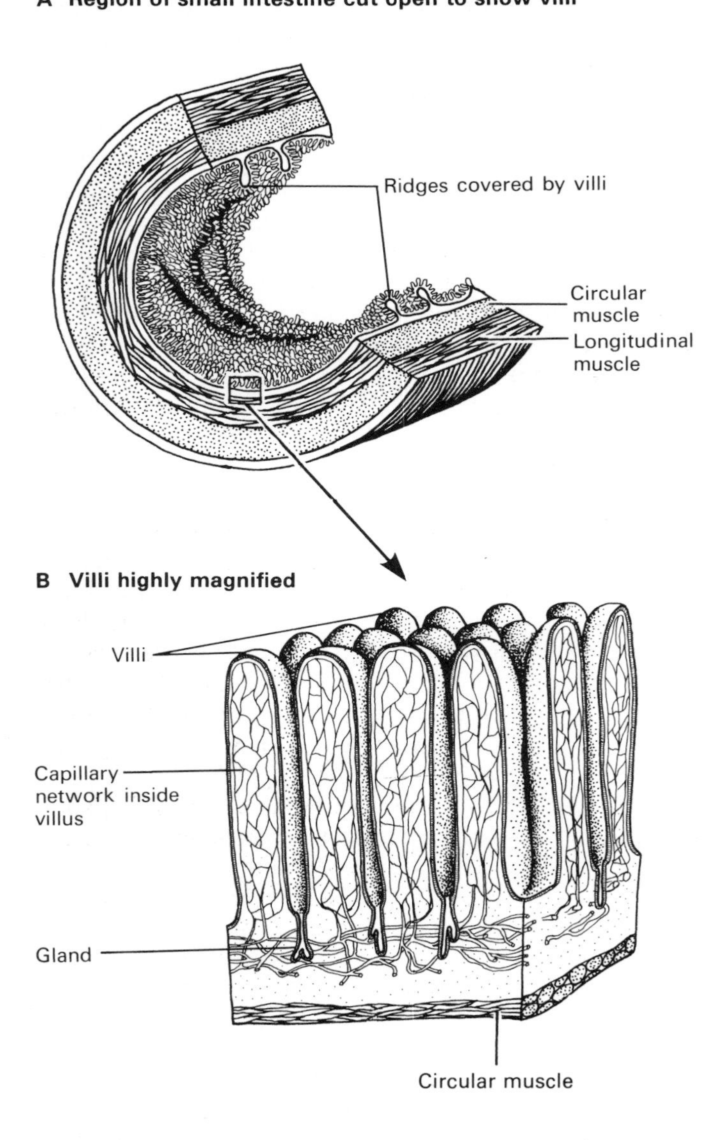

Fig. 4.18 Structure of the small intestine wall

pepsin it is necessary to remember that proteins consist of long chains of linked amino acids. Pepsin is an enzyme of a type which breaks these chains only at the specific point where certain amino acids appear together. The result is that proteins are broken into large fragments called **peptides**.

Rennin is an enzyme produced only in the stomachs of young mammals. It solidifies the proteins in milk and in this state they are retained in the stomach long enough for pepsin to act upon them.

Castle's intrinsic factor is a substance produced by the stomach which is necessary for the proper absorption of cobalamine (vitamin B12) by the intestine (section 3.2).

When food reaches the liquid consistency of chyme the lower sphincter of the stomach relaxes periodically releasing small quantities of food at a time into the next region of the intestine. Different types of food pass through the stomach at different speeds. Liquids and starchy foods pass through quickly: they begin to leave the stomach about ten minutes after entering the mouth. Meat and vegetables are slower, and fatty foods are slowest of all, taking up to 30 hours to drain through the stomach. In general, however, the stomach is empty about six hours after a fair sized meal, and then its walls contract, which stimulates the nervous system and results in the sensation of hunger.

In humans, the digestive system below the stomach consists of two regions: the **small intestine** and the **large intestine**. The small intestine has a smaller diameter than the large intestine. The small intestine consists of two regions: the **duodenum** which is about 30 cm long, and the **ileum** which is about 8 metres long.

Digestion in the duodenum

The duodenum receives fluid called **bile** through the **bile duct** from the **gall-bladder** in the liver. It also receives **pancreatic juice** through the **pancreatic duct** from the **pancreas**.

Bile Bile is a greenish-yellow liquid which is manufactured in the liver partly from substances resulting from the breakdown of old red blood cells. Bile is stored in the gall bladder, and is released whenever food enters the duodenum (Fig. 4.17).

Bile contains dissolved substances including: yellow-green bile pigments (2.5%) which have no digestive function; organic bile salts (6%); and inorganic substances of which sodium bicarbonate (0.8%) is the most important.

Organic bile salts have several functions. They react with fat-soluble vitamins (i.e. A, D, E, and K) and cholesterol, an important constituent of cell membranes. This makes them water-soluble and therefore easier for the intestine to absorb. Bile salts assist in the digestion of fats and oils in two ways. First, they reduce the surface tension of fats and oils so they disintegrate into a mass of tiny oil droplets, called an **emulsion**. This increases the surface area of fat exposed to digestive enzymes. Second, bile salts activate the fat-digesting enzyme **lipase** produced by the pancreas and the small intestine wall. Up to 95% of bile salts are absorbed by the small intestine and returned to the liver for re-use.

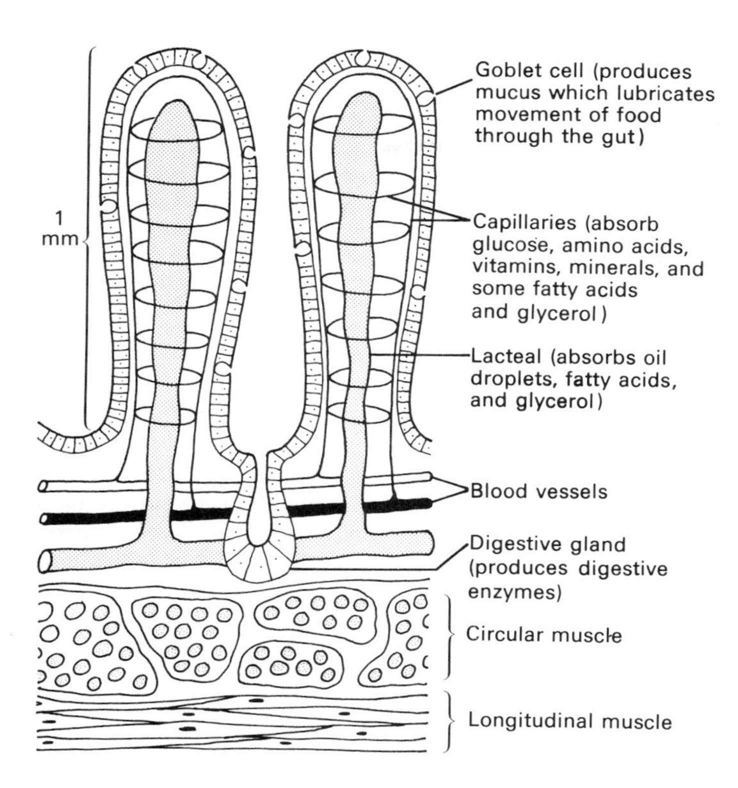

Fig. 4.19 Diagram of the ileum wall and villi

A variety of shapes of villi from the ileum. **F** is finger-like; **L** is leaf-like; and **C** is long and curved, a form known as convoluted

The **sodium bicarbonate** in bile is extremely important because it neutralizes stomach acid and creates an alkaline medium which is necessary for efficient functioning of all enzymes produced in the small intestine.

Pancreatic juice The pancreas is a long, narrow organ situated between the duodenum and the stomach. The pancreatic juice which it produces is colourless and contains sodium bicarbonate, which makes it alkaline. It also contains many powerful enzymes, in fact they are so powerful that in combination with bile and other enzymes from the ileum wall they are capable of completing the whole digestive process without the aid of mouth and stomach enzymes. The main pancreatic enzymes are listed below:

Trypsin is an enzyme which digests protein. It is released from the pancreas as an inactive substance called **trypsinogen** which becomes digestive trypsin when acted upon by another substance called **enteropeptidase** which is produced from glands in the duodenum wall. Trypsin, like pepsin, breaks protein molecules into large fragments called peptides, but it also breaks large peptides into smaller peptides. This process often results in molecules consisting of only two amino acids, which are called **dipeptides**.

Pancreatic amylase is an enzyme which is probably identical to salivary amylase, and shares its function of breaking down starch to maltose.

Pancreatic lipase is an enzyme which breaks down fats and oils to fatty acids and glycerol.

All of these enzymes continue their digestive activities within food as it passes into the ileum.

Digestion in the ileum

Within the ileum all digestive processes are completed and the soluble products are absorbed into the blood-stream. The ileum is therefore both a digestive and an absorptive organ.

Glands situated in the duodenum and ileum walls secrete many important enzymes which together form a fluid called **succus entericus**, or simply intestinal juice (Figs. 4.18 and 4.19). The following enzymes are present in this juice.

Erepsin is a mixture of several enzymes all of which break peptides into single amino acid molecules, thereby completing the digestion of protein foods.

Maltase, **sucrase**, and **lactase** break down maltose, sucrose, and lactose sugars respectively into simpler monosaccharide sugars such as glucose, fructose, and galactose. This process completes carbohydrate digestion.

Various lipases continue the digestion of fats and oils to fatty acids and glycerol.

Absorption in the duodenum and ileum

The end product of enzyme activity upon food is a liquid called **chyle**. This liquid is a mixture of many soluble substances including amino acids, fatty acids, glycerol, monosaccharide sugars, vitamins, and minerals together with insoluble but finely emulsified fats and oils. All of these are absorbed into the blood-stream through the duodenum and ileum walls.

The whole of the ileum's internal surface is covered by finger-like projections called **villi** (singular villus) which are about 1 mm long (Figs. 4.18 and 4.19). Each square millimetre of ileum has up to forty villi and there are about five million in the ileum as a whole. The presence of villi gives the ileum a far greater internal surface area available for absorption than if it had a smooth lining. Under very high magnification ($\times$ 40 000) the surface of each individual cell lining the ileum is seen to be folded into **micro-villi**, giving the ileum an estimated total internal surface area of thirty square metres.

Inside each villus there is a dense network of blood capillaries, and a single **lacteal**, or lymph vessel, which is closed at its upper end (Fig. 4.19). It is into these two kinds of vessels that digested food passes after it has been absorbed into the cells covering the outer surface of each villus. In general the soluble foods, amino acids, sugars, vitamins, minerals, and some fatty acids and glycerol, pass into the blood capillaries of the villi. These capillaries join together to form a larger blood vessel called the **hepatic portal vein** which carries the food to the liver. The bulk of fatty acids and glycerol passes through the villi walls into the lacteals. The remaining emulsion of fat and oil droplets is absorbed into the lacteals of the villi in the following manner. Fat and oil droplets pass between the micro-villi of the cells covering each villus, and are absorbed whole into these cells by a process like amoeboid feeding (section 4.2). The droplets then pass through the cytoplasm of the cells and out the other side into the lacteals. They then pass into the main lymphatic system which eventually discharges them into the blood (section 6.3).

Movement of absorbed food out of the villi into the blood-stream and lymphatic system is assisted by periodic contractions of villi during which they suddenly become shorter and fatter, then relax slowly. This occurs about six times per minute in each villus.

Functions of the large intestine

The large intestine consists of the **colon, rectum,** and **anus**. The colon receives indigestible material from the ileum. This material consists of plant fibres, cellulose, bacteria, dead cells dislodged by friction from the intestine walls, mucus, and considerable quantities of water. It remains in the colon for about thirty-six hours, during which time most of the water and salt are absorbed from it. The semi-solid residue, called **faecal matter** or **faeces**, is lubricated by mucus and periodically ejected from the body via the rectum and through the anus.

Caecum and appendix

At the point where the ileum joins the colon there is a tube called the **caecum**, from which arises a sac called the **appendix**. In ruminants and other herbivores such as rabbits and horses these organs are large and contain the bacteria which help digest cellulose. In humans these organs have no function and are referred to as **vestigial structures**. This term is used to describe organs which in the course of evolution have become reduced in size and have lost their original functions. The presence of a vestigial caecum and appendix may indicate that long-extinct ancestors of the human species were capable of digesting cellulose material from a vegetable diet.

Verification and inquiry exercises

A *Feeding and digestion in Paramecium*

1. On a microscope slide prepare a mixture of: 1 drop *Paramecium* culture, 1 drop methyl cellulose solution ('Polycell' paste), and 1 drop of yeast suspension previously stained in Congo Red dye. (The methyl cellulose slows down the movement of *Paramecium*.)

2. Observe how *Paramecium* feeds on the yeast cells. Note that the yeast cells change colour as they pass through the organism in food vacuoles. Congo Red is blue-violet in acid and red in alkaline conditions. What conclusions can be made about changing conditions inside a food vacuole? Note the disintegration of yeast cells as digestion and eventual egestion occurs.

B *Feeding in Hydra*

1. Observe *Hydra* feeding by placing one in a watch-glass of water with a few *Daphnia*.

2. Use fine forceps to hold the following objects within a hydra's reach: a small glass bead; a bead which has previously been coated with saliva and allowed to dry; and a small piece of uncooked meat. What conclusions can be reached about sensitivity in *Hydra*?

3. Place a hydra either on a cavity slide under a cover slip, or under a supported cover slip. Add a drop of 1% acetic acid to the preparation while observing the animal under high magnification. Watch for the discharge of nematocysts, of which there are four types.

C *Feeding in insects*

1. Using Figure 4.7 as a guide and fine forceps as a tool remove and study the mouthparts from a locust or cockroach head, starting with the labrum. This can be made a little easier by first boiling the head in 5% sodium

hydroxide solution for one or two minutes to dissolve away muscles and soften the cuticle.

2. After starving them for twelve hours wrap a number of live locusts separately in cotton wool so that only the head is free. Using a different locust each time hold one of the following objects near, but not touching, its labial and maxillary palps and observe the action of mouth parts under a lens: dry filter paper; wet filter paper; filter paper soaked in strong sugar solution. Repeat with other food substances. What conclusions can be made about chemical sensitivity in locusts?

D *An investigation of enzymes*

1. Materials: saliva, obtained by rinsing the mouth with water and collecting it in a test-tube; 10% pepsin solution; 10% pancreatin solution; 5% starch solution; coagulated egg white, obtained by mixing the white of one egg with 500 cm^3 of water and heating it, while stirring, until a white suspension is obtained; Benedict's reagent; dilute hydrochloric acid; dilute sodium hydroxide solution. Refer to the food tests on page 33 whenever necessary.

2. Verify that saliva digests starch into maltose as follows. Mix 2 cm^3 each of starch solution and saliva in a test-tube and place the tube in a beaker of water warmed to about 20°C. After five minutes, during which the tube must be shaken several times, test the contents for starch, and use the glucose test to detect maltose.

3. Verify that pepsin digests protein as follows. Mix 2 cm^3 each of pepsin solution and coagulated egg white in a test-tube with one drop of dilute hydrochloric acid. Observe that the white suspension of egg becomes a clear liquid when the tube is left in a beaker of warm water for a few minutes and shaken from time to time.

4. Devise and carry out controlled experiments to prove that it was the pepsin and not the dilute hydrochloric acid which digested the egg white within a few minutes in 3 above.

5. Devise and carry out controlled experiments to provide evidence in support of the following statements:

a) An enzyme will digest only one type of food.

b) A small amount of enzyme will digest a large amount of food.

c) Finely ground food digests faster than large solid lumps with the same volume.

d) Enzymes work with greatest efficiency in warm conditions (e.g. around 20°C) but most are destroyed by temperatures above 60°C.

e) Pepsin digests only in acid conditions, and pancreatin only in alkaline conditions.

Comprehension test

Consider the following information and use it to answer the questions below:

It has been estimated that each year British children lose 4 000 000 teeth due to neglect of proper dental hygiene. One of the major causes of tooth decay is failure when cleaning the teeth to remove dental plaque, an invisible layer of bacteria and decaying food on and in between the teeth. Another important cause of decay is eating sugary foods. Within one and a half minutes of eating sugar it combines with dental plaque to form acids that dissolve tooth enamel.

1. Why is it better to clean the teeth after the last meal of the day rather than first thing in the morning?

2. After a meal clean the teeth, and immediately chew a *Ceplac Disclosing Tablet* (obtainable at chemists). This will stain any dental plaque remaining on the teeth and gums bright red if they have not been cleaned properly. What is the most effective brushing technique for removing all dental plaque from teeth?

3. Make a list of foods likely to cause tooth decay.

4. From Figure 4.11 find the area where tooth decay is most likely to begin.

Summary and factual recall test

Digestion is the process by which food is broken down into chemically (1) molecules which are (2) in water. Digestion is necessary because most foods are (3) when eaten and so cannot be (4) into cells.

Amylases are a group of (5) which break down (6) foods into (7) such as (8); lipases break down (9) and (10) into (11) and (12); and proteases break down (13) into (14).

Intracellular digestion takes place (15) cells and is typical of (16) organisms such as (17). Extracellular digestion takes place (18) cells and typically occurs within a tube called the (19) which runs from the (20) to the (21).

Thorough chewing of food is important because (22– three reasons). The main action of saliva is to break down (23) into (24). Food is moved along the intestine by (25)-like (26) contractions called (27). Movement of food in and out of the stomach is controlled by muscles called (28). The stomach wall produces (29) juice, which contains (30) acid, and the enzyme (31) which breaks down proteins into (32). The duodenum receives (33) from the liver. Its main function is to break down fats and oils into tiny (34) which are known as (35).

The main enzymes in pancreatic juice are (36) which breaks down protein; (37) which breaks down starch; and (38) which breaks down fats and oils. The main enzymes of the ileum are (39) which breaks down (40) into amino acids; (41 – name three) which complete the digestion of carbohydrates; and (42) which digests fats and oils into (43) and (44).

The ileum wall is covered with millions of (45). These (46) the surface area available for (47) of soluble food. The main function of the large intestine is to remove (48) and (49) from indigestible substances, called (50) matter.

5
Photosynthesis

Photosynthesis is the process by which green plants make carbohydrates from carbon dioxide and water using sunlight energy absorbed by chlorophyll. Oxygen gas is a by-product of the process and is released into the atmosphere.

This chapter describes the mechanism of photosynthesis and the internal structure of leaves: the 'factories' in which photosynthesis takes place. It then goes on to show that the products of photosynthesis are the raw materials not only for the manufacture of all the substances which make up a plant body, but for the foodstuffs used by the rest of the living world.

As an introduction, however, it is interesting to study the events leading to the discovery of how photosynthesis works. These events read like a detective story in which the mystery cannot be solved until the detectives (or in this case the biologists) learn a new way of solving problems: i.e. until they develop scientific methods.

5.1 The discovery of photosynthesis

The mystery is perhaps first referred to in the works of Aristotle, a Greek philosopher who lived over two thousand years ago. He wanted to know how plants are able to produce food for their own use, as well as for man and other animals, and yet never apparently eat anything themselves. Since scientific knowledge and methods were very limited in Aristotle's time he could do little more than give an hypothesis, which is a reasoned opinion based on observations. He ob-

Fig. 5.1 Diagram of Priestley's experiment in August 1771

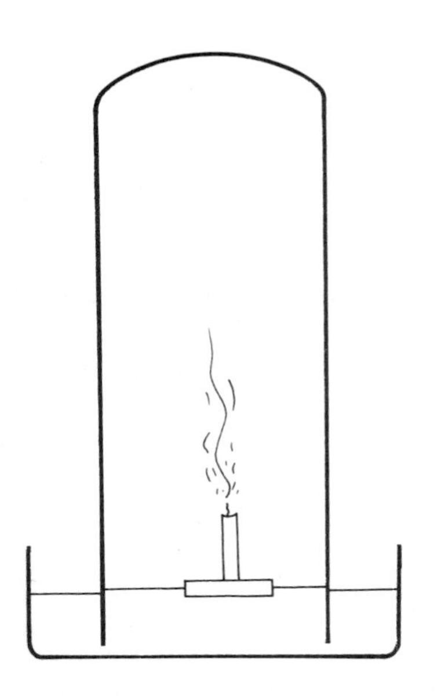

1 Candle allowed to burn out.

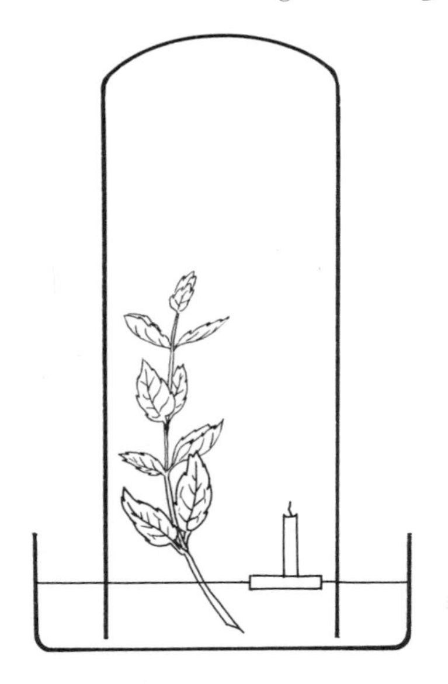

2 Sprig of mint inserted without disturbing gas in the bell jar. Apparatus left for ten days.

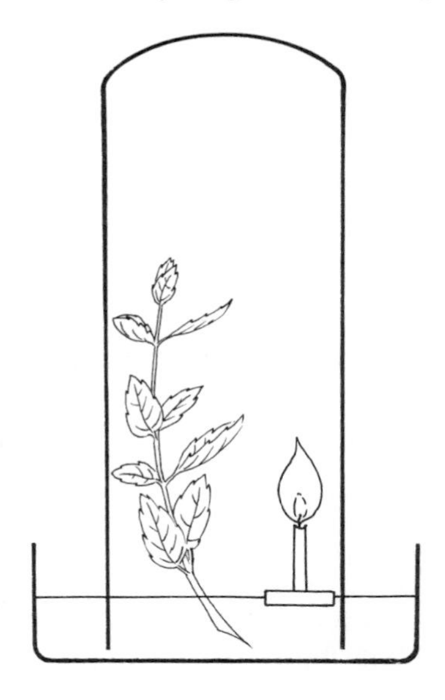

3 Candle can be relighted.

served that when animals and plants die their bodies decay and pass into the soil, and that the soil is penetrated by plant roots. He put these two observations together and formed the hypothesis that plant roots extract decayed materials from soil, and that these materials are used by the rest of the plant for the production of food.

Observation and hypothesis formation are the first stages of scientific method. They are normally followed by experiments designed to test the hypothesis under controlled conditions. Aristotle did not test his hypothesis in this way; perhaps he was satisfied with intelligent guesswork or perhaps it never occurred to him that experimental testing was the next logical step. One thing is certain, however: he could never have solved the mystery with the equipment available in his age.

For whatever reason, the idea that plants extract substances from the soil remained untested until the late seventeenth century. In 1692, a Dutchman called van Helmont planted a willow tree seedling in 90 kg of soil and left it to grow for five years, giving it nothing but water. After this time he discovered that the tree had gained about 74 kg in weight while the soil had lost only 56 g. Van Helmont concluded that plants take nothing at all from the soil. He discounted the lost 56 g as experimental error, and assumed that wood, bark, roots, and leaves are somehow 'transmuted out of water which falls on the soil as rain'. At first sight this work seems to disprove Aristotle's hypothesis. However, van Helmont's technique was not precise enough to prove, as he believed, that plants take nothing from soil except water. What his work does suggest is that the bulk of raw materials used by plants come from some other source than the soil. Whether or not water is the main source was still open to question.

It was not until the mid-nineteenth century that the true roles of water and soil in plant growth became clear. At this time Julius Sachs demonstrated that plants develop very poorly when grown in distilled water compared with others whose roots are in solutions of various soil minerals. This work is described more fully in section 5.5.

Meanwhile the development of the microscope in the seventeenth century led to an interest in the structure of plants. In particular it was discovered that the stem and leaves of plants contain pores, later called **stomata**. The fact that stomata are the openings of extensive air spaces within the leaves gave rise to speculation about the possible role of atmospheric gases in plant growth. The experiments of Joseph Priestley (1733–1804) eventually provided evidence that there is a link between plants and the atmosphere.

Priestley discovered that flames will quickly go out, and that animals will suffocate when enclosed in a limited amount of air. He concluded that animals and flames 'damage' the air in some way so that it can no longer support combustion or life. What puzzled Priestley was the fact that the earth's atmosphere does not become permanently damaged in this way.

This puzzle was partly solved in August 1771 when Priestley made a remarkable discovery. He put a sprig of mint into a quantity of air in which a candle flame had burned out. Ten days later he found that a candle would again burn in this air (Fig. 5.1). He concluded that the mint had 'purified' the air, and proposed that if all plants had this capacity then they must be 'nature's restorative' of the earth's atmosphere.

It was soon apparent, however, that Priestley had revealed only part of the truth. Other experimenters often failed to get the same results after repeating his work. The reason for these failures was discovered by Jan Ingenhousz (1730–99) when he decided to investigate two factors completely overlooked by Priestley and his followers: light, and the green pigment in plants. His experiments clearly demonstrated that only the green parts of plants can, as he put it, 'purify air', and then only in the light. Furthermore, he discovered that in the dark all parts of plants make air 'impure'.

Later in the eighteenth century oxygen and carbon dioxide were isolated, named, and their properties investigated. It was then realized that plants 'restore' the atmosphere in daylight hours by absorbing carbon dioxide and releasing oxygen, while in the dark the reverse happens.

Despite the immense progress which has been made since Aristotle's time it is still not possible to say that the mystery of how plants live on light, water, and carbon dioxide is completely solved. Some of what is known, however, is described in the next section.

5.2 The mechanism of photosynthesis

There are four main factors involved in photosynthesis. These are water, carbon dioxide gas, the green pigment chlorophyll, and the energy of sunlight.

During photosynthesis plants make sugars such as glucose. To do this, the energy of light is absorbed by chlorophyll and used to combine carbon dioxide and water. Oxygen is released as a by-product and diffuses out of the plant into the atmosphere. Traditionally this process has been summarized in the following equations but, as will be explained below, modern discoveries have led to certain modifications.

$$6CO_2 + 6H_2O + \text{Light energy} \xrightarrow{\text{Chlorophyll}} \underset{\text{Glucose}}{C_6H_{12}O_6} + 6O_2$$

This can be simplified further:

$$CO_2 + H_2O \longrightarrow C(H_2O) + O_2$$

These equations are not only out of date. They are also misleading because they do not indicate the immense complexity of the mechanism. Photosynthesis actually consists of at least fifty separate reactions which follow in sequence, each catalysed by a different enzyme system, and connected with many side reactions. A detailed study is beyond the scope of this book, but the following account gives some impression of its depth and importance.

Photosynthesis as an energy conversion system

When plants make sugar they convert light energy into chemical energy. This can be verified to a certain extent by demonstrating that sugar contains energy. This can be done by burning sugar, which releases heat energy at the rate of 17 kJ per gram (section 3.2). The heat so released is equivalent to the light energy which a plant requires to make sugar out of carbon dioxide and water.

In simple terms, plants use light energy to transform carbon dioxide and water, which contain little energy, into sugar which is very rich in energy. This process is important because no organism can make direct use of light as a source of energy for metabolism, whereas all organisms can use the chemical energy in foods such as sugar for this purpose.

The events briefly described so far have been investigated in detail with a variety of techniques. Using isotopes, for example, it is now possible to study the fate of carbon, hydrogen, and oxygen atoms as they take part in the food-making process.

Use of isotopes in the study of photosynthesis

Isotopes are atoms of an element with different atomic weights but identical chemical properties. For example, the ^{18}O isotope of oxygen has an atomic weight of 18 whereas normal oxygen atoms have an atomic weight of 16. The abnormal atomic weight of an isotope provides one means by which it may be detected after it has taken part in a chemical reaction.

Isotope techniques have led to a revision of early theories about the source of oxygen from illuminated plants. Early investigators assumed, incorrectly, that the source of oxygen was the carbon dioxide that plants absorb. Their theory was that oxygen splits off the carbon dioxide molecules, leaving carbon free to react with water. This *incorrect* idea is summarized as follows:

$$CO_2 \begin{cases} \rightarrow O_2 \text{ (Released oxygen)} \\ \rightarrow C + H_2O \xrightarrow{\text{Chlorophyll}} C(H_2O) \end{cases}$$

In 1941, a team of American scientists used the ^{18}O isotope to make a special kind of carbon dioxide and water; i.e. $C^{18}O_2$ and $H_2{}^{18}O$ respectively. They supplied plants with $C^{18}O_2$ under controlled conditions and found that the oxygen which they produced did not contain any isotope. However, when $H_2{}^{18}O$ was supplied to the plants they did release the ^{18}O isotope. From these results it was possible to conclude, firstly, that the oxygen liberated in photosynthesis originates from water and not carbon dioxide, and secondly that water molecules must be ‘split’ up into hydrogen and oxygen during photosynthesis.

This work has led to a revision of the equation for photosynthesis given earlier. Consider this equation again, rewritten to incorporate the isotopic oxygen:

Traditional equation

$$CO_2 + H_2{}^{18}O \longrightarrow C(H_2O) + {}^{18}O_2$$

This is no longer a balanced equation: it shows the production of *two* ^{18}O atoms whereas only *one* is available from the single water molecule. Water as the source of oxygen in photosynthesis is correctly indicated as follows:

Corrected equation

$$CO_2 + 2H_2O \longrightarrow C(H_2O) + H_2O + O_2$$

The function of light in photosynthesis

The experiments just described and others have shown that light energy is responsible for splitting water molecules into their component hydrogen and oxygen atoms. This is called the **light reaction**. The oxygen is released as gas, but the hydrogen atoms are retained for combination with carbon dioxide in the manufacture of sugar. Additional experiments have provided more precise details of these reactions.

When a plant is illuminated in a carbon dioxide-free atmosphere no oxygen is produced. But when the light is switched off and carbon dioxide is immediately supplied oxygen is produced for a short time *in the dark*. If this experiment is repeated, withholding carbon dioxide for increasing lengths of time after switching off the light, less and less oxygen, and finally none

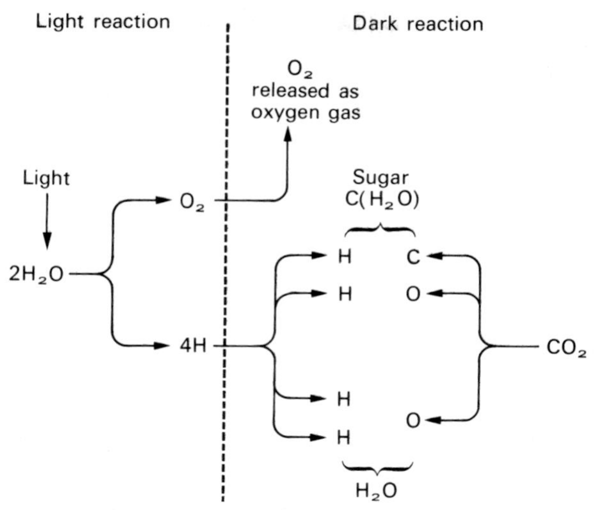

Fig. 5.2 Diagram of the 'light' and 'dark' reactions of photosynthesis (greatly simplified)

at all, is evolved. These results are interpreted in the following way.

First, it is concluded that an illuminated plant in carbon dioxide-free air forms a 'substance' which can react with carbon dioxide in the dark to form sugar and release oxygen. Second, this 'substance' can be thought of as a temporary store of converted light energy. The evidence also shows that it is unstable since it cannot be stored for more than a few minutes in the dark. Third, other experiments show that this 'substance' consists of water held in a state from which its hydrogen atoms can be rapidly released for the transformation of carbon dioxide into sugar, and its oxygen atoms released to the atmosphere.

These experiments provide evidence for what is known as the **dark reaction**. They suggest that light energy is not required for the ultimate transformation of carbon dioxide and hydrogen, with oxygen release, into sugar molecules. The reaction is called 'dark' merely because it requires no light, and not because it must take place in the dark. Light and dark reactions are summarized diagrammatically in Figure 5.2.

The elaboration of photosynthetic products

The reactions of photosynthesis are only the first step in the manufacture of all the substances that make up a plant body. Simple (monosaccharide) sugars such as glucose, which are produced by photosynthesis, are the raw materials out of which plants manufacture many other substances, such as complex (polysaccharide) carbohydrates like starch and cellulose.

The manufacture of cellulose, for example, involves the linking together of glucose molecules into long chains. Millions of these chains make up a single cellulose fibre, and millions of fibres go to make up the cellulose wall of one cell. Rayon (artificial silk), cotton, and flax (which is made into linen) are all examples of cellulose fibres used by man.

5.3 Leaf structure in relation to photosynthesis

All the green parts of a plant carry out photosynthesis in daylight hours, but leaves are the principal photosynthetic organs. The following description refers mainly to the structure of leaves found on the group of flowering plants known as dicotyledons (section 1.3). Fig. 5.3 illustrates the external features and the arrangement of leaves on a plant typical of this group.

External features of a leaf

A leaf usually consists of the following parts:

Petiole The petiole is the narrow stalk of a leaf by which it is attached to the stem. In some leaves it is almost non-existent and the lamina extends to the stem.

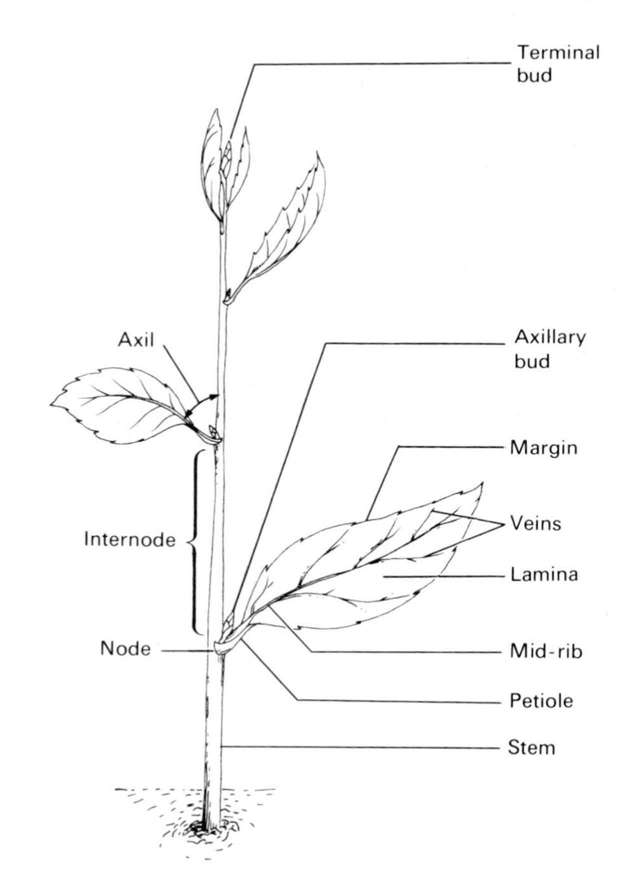

Fig. 5.3 Arrangement and external features of leaves on a typical dicotyledonous plant

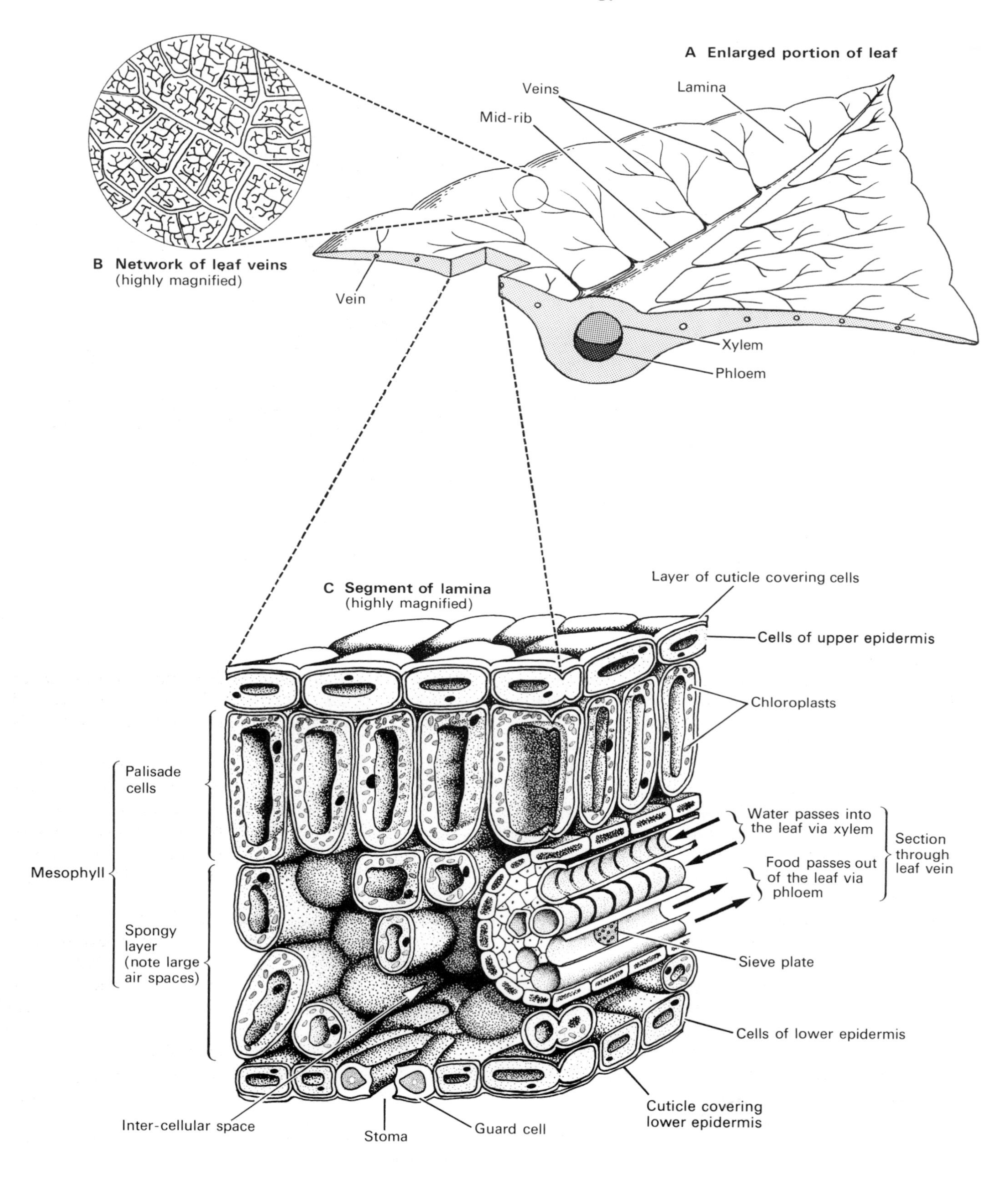

Fig. 5.4 Structure of a leaf

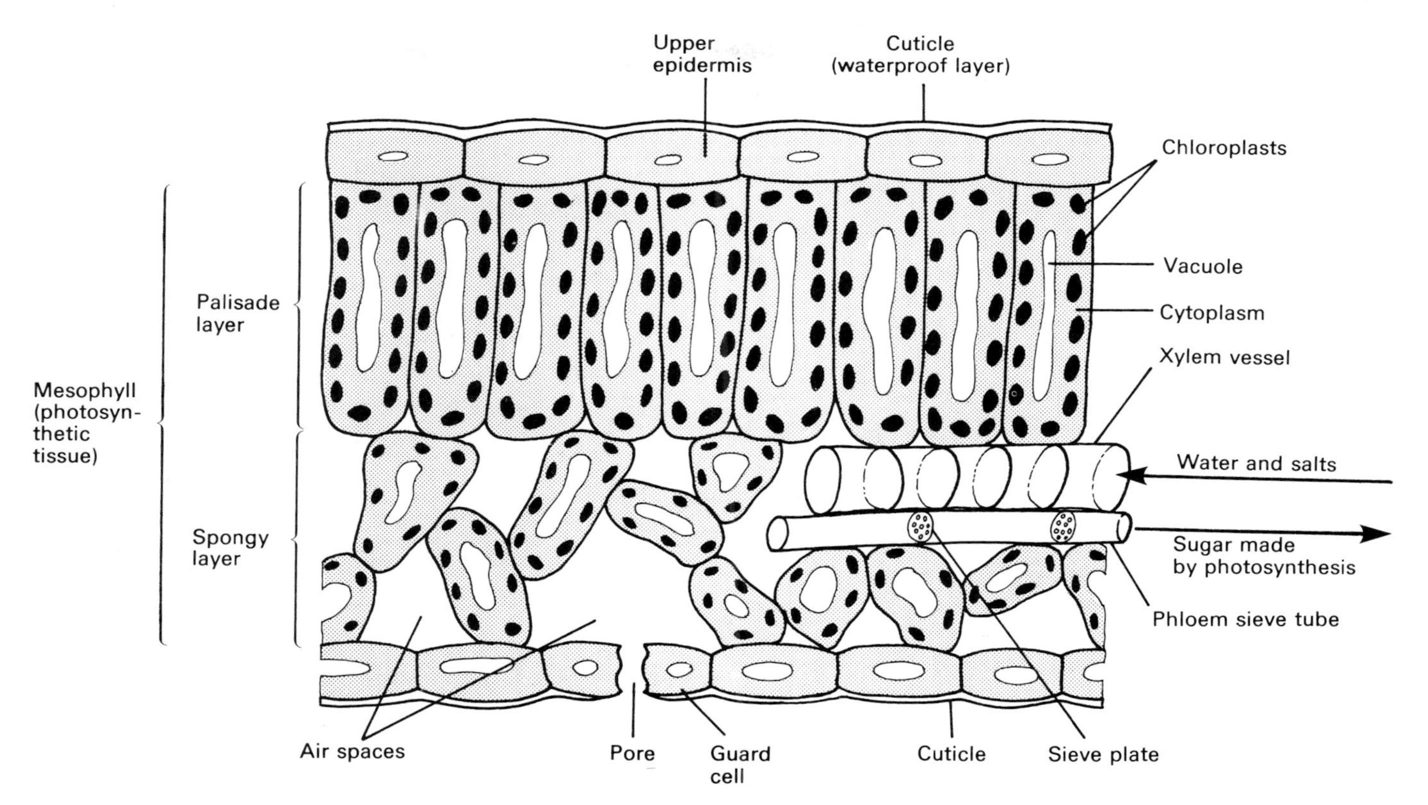

Fig. 5.5 Diagram of the structure of a leaf (simple version of Fig. 5.4C)

Lamina The lamina is the photosynthetic portion of a leaf. It is usually thin and flat and its shape can be a fair guide to the species of plant. Its flat, thin shape allows a large area of chlorophyll to be presented to the light; in fact most leaves are so thin that some light passes right through them. This means that chlorophyll deep inside a leaf is not completely shaded from the light by the outer layers.

In certain plants the petiole can bend so that the lamina is approximately at right angles to the sun's rays throughout the day. Observe a geranium on a window-sill at intervals during the day. Movements of this kind are called tropisms (chapter 11).

Mid-rib and veins Leaf veins consist of many tiny tubes which convey water into the leaf and carry food from it. These tubes enter through the petiole and pass in a thick bundle, called the **mid-rib**, down the leaf centre. Smaller bundles, the **veins**, spread out from the mid-rib to all parts of the leaf giving the net-veined appearance typical of dicotyledons.

Internal structure of a leaf

Figures 5.4 and 5.5 illustrate in detail the internal structure of a typical dicotyledonous leaf.

Epidermis The epidermis is the outermost layer of cells of a plant. This layer is one cell thick, and consists of regularly shaped, often flattened cells which cover the plant like a skin. In leaves there is usually a distinct difference between the upper and lower epidermis. The upper epidermis is often covered by a continuous layer of waxy **cuticle** which protects the plant against disease organisms such as parasitic fungi. The cuticle is also waterproof and to some extent limits the loss of water from the leaf by evaporation.

Plants which live in hot, dry, and windy conditions typically have a very thick cuticle, which gives their leaves a hard and stiff texture, e.g. *Ficus elastica*, the ornamental rubber plant. The lower epidermis is characterized by the presence of pores called **stomata** (singular stoma) at regular intervals over its surface. Each stoma is made up of a pair of crescent-shaped **guard cells** which can change shape, thereby opening and closing the pore (section 7.4). (The terminology which refers to these structures is confusing. The word **stoma** can be used in reference to the pore itself, or to include both pore and guard cells together. In this text 'stoma' refers to both structures.) Stomata are the openings of an extensive system of air spaces between the cells of the leaf. These spaces allow gases to diffuse in and out of the leaf.

In between the upper and lower epidermis of a leaf is the photosynthetic tissue, called the **mesophyll**. Mesophyll cells are the only ones in the leaf which contain chlorophyll, apart from the guard cells of certain species. There are two types of mesophyll:

Palisade mesophyll The palisade layer of a leaf consists of elongated, cylindrical cells which are situated immediately below the upper epidermis. The long axis of each cell is at right angles to the leaf surface. This feature permits light to penetrate deep into the photosynthetic tissue without passing through very many cell walls, so avoiding loss of light by absorption and reflection. Palisade cells contain more chlorophyll than any other type of cell in the leaf. The chlorophyll is contained within organelles called **chloroplasts** which can move within the cells to areas in which illumination is strongest.

Spongy mesophyll Spongy mesophyll consists of irregularly shaped cells with large air spaces between them, and it has the appearance of a sponge when viewed under the microscope. These cells have fewer chloroplasts than palisade, and they receive light at a lower illumination.

Structure of mid-rib and veins There are two types of tube in veins, and each has a different structure and function. **Xylem** is a plant tissue which includes a system of water-conducting tubes, usually called **xylem vessels**, which form a continuous water-transport system all the way from the roots. The walls of xylem vessels are very strong owing to the presence in them of a substance called **lignin** which appears as rings, coils, or continuous layers perforated by tiny holes called **pits** (Fig. 7.2). Xylem in the veins, and especially the mid-rib, provides mechanical support for the softer leaf tissues. In the mid-rib, xylem is situated above much softer tissue. In this position xylem takes a large amount of the tension caused by stretching of the upper leaf tissues when the leaf bends under the force of gravity and wind.

Phloem is a plant tissue which includes a system of tubes which transport food substances from where they are manufactured in the leaf to growing points and food storage areas. Phloem tubes are characterized by the presence of perforated cross-walls called **sieve plates** at intervals along their length. Because of this the tubular components of phloem are called **sieve tubes.**

The leaf in action

To understand the activities within a leaf it is necessary to remember four important points. First, plants respire at *all* times. Second, during daylight hours plants are involved in both respiratory and photosynthetic activities. This is obvious because photosynthesis must by definition be interrupted by darkness. It cannot and need not be continuous because more than sufficient food is made in daylight to last through the night. Respiration, on the other hand, must occur at all times since life depends upon it. Third, photosynthesis is the *reverse* of respiration. Fourth, photosynthesis uses the products of respiration, and respiration uses the products of photosynthesis. The last two facts are illustrated below.

Oxygen + Food —Respiration→ Energy + Water + Carbon dioxide
↑ *The metabolic cycle* ↓
Oxygen + Food ←Photosynthesis— Light energy + Water + Carbon dioxide

With these facts in mind consider what happens inside a leaf during the transition from darkness to light.

In darkness leaves take in oxygen and release carbon dioxide owing to the continuous process of respiration. As dawn breaks photosynthesis begins in the mesophyll cells and, since this is the reverse of respiration, the following sequence of events takes place. The release of carbon dioxide from the leaf slows down and eventually stops because this gas is absorbed in increasing amounts by the mesophyll as a raw material of photosynthesis. Simultaneously, the absorption of oxygen from the air slows down and stops because this gas is produced in increasing amounts by leaf photosynthesis. For a brief moment the movement of gases in and out of the leaves stops altogether. This moment is called the **compensation point**. At this time photosynthetic and respiratory processes are equal and compensate each other. As dawn turns into day a further increase in sunlight intensity results in food production requiring more carbon dioxide than is supplied by respiration alone. Likewise, more oxygen is produced than is used by respiration. Leaves therefore absorb carbon dioxide and produce oxygen for the remainder of the day, until another compensation point is reached as the sun sinks.

During daylight hours sieve tubes of the phloem continually transport sugar made in the leaves to growth and food storage regions in the plant. In bright light, however, plants often produce more sugar than can be removed by the phloem. When this happens excess sugar is temporarily stored in the leaves. Some plants convert this sugar into starch grains which remain in the leaf until night when they are re-converted

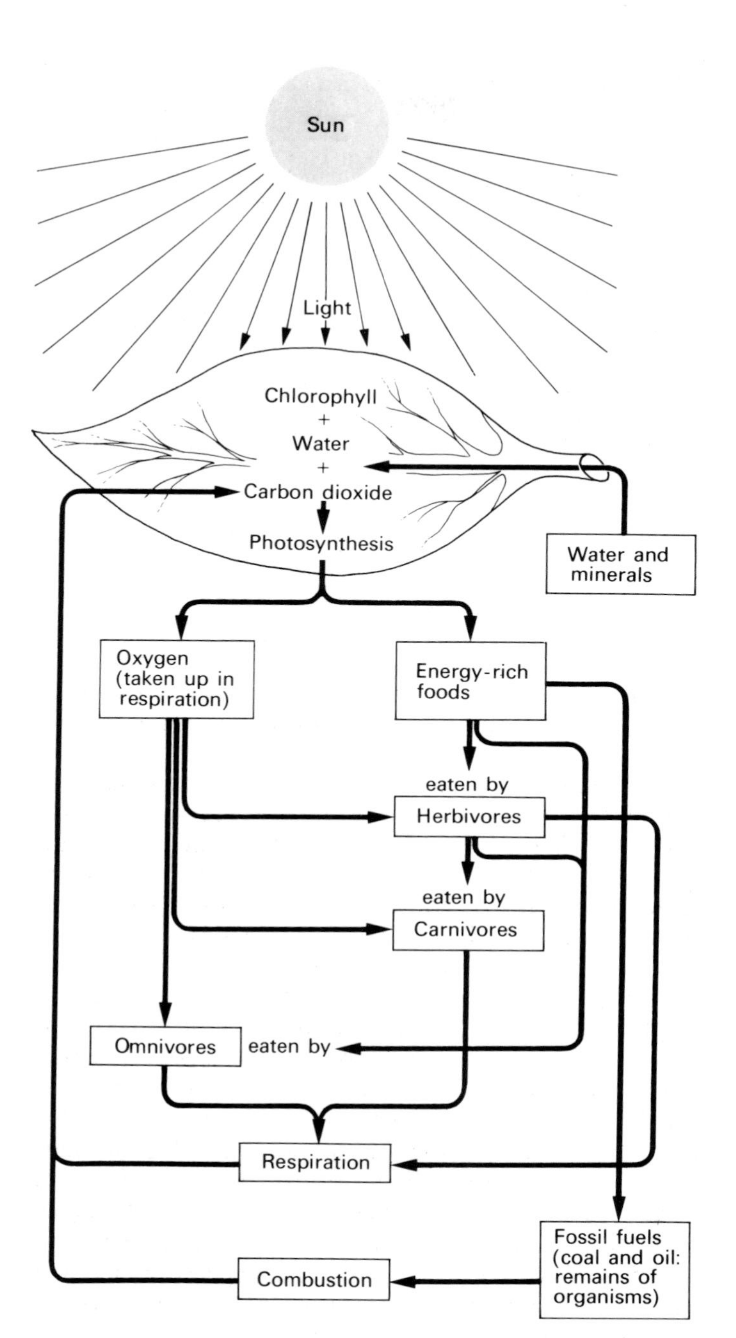

Fig. 5.6 Diagram showing the conversion of light energy by plants into substances which feed other living things and give rise to fossil fuels such as coal and oil

into sugar and transported from the leaf. The presence of starch in a leaf of this type is evidence that photosynthesis is, or has recently been taking place. This fact is applied in experiments described later.

5.4 The importance of photosynthesis

This section examines some of the reasons why photosynthesis is of vital importance to the continuation of life on earth.

Production of oxygen

It has been explained that photosynthesis is the reverse of respiration. This explains why the supply of atmospheric oxygen is not exhausted by processes such as respiration and combustion.

At present the sum total of photosynthetic organisms on earth is sufficient to maintain atmospheric oxygen at a level which can support life and combustion. Should it ever fall below this level animals and many other heterotrophs would suffocate.

Food production

Plants manufacture all their body-building and energy-producing substances from simple raw materials, using the energy of sunlight. Animals, however, are not independent in this way; they live by eating plants and/or each other.

This fundamental difference between animals and plants (or, to be more exact, between heterotrophic and autotrophic organisms) has extremely important consequences. Consider what would happen if the sun were to go out (ignoring the fact that the earth would quickly freeze).

1. Green plants, deprived of sunlight energy, would be the first organisms to die.
2. Herbivores, deprived of plant food, would be the next organisms to die.
3. Carnivores, deprived of herbivores to eat, and omnivores, deprived of herbivores and plants to eat, would be the next to die.
4. Parasites, deprived of host organisms, would be the next to die.
5. Finally, saprotrophs, the organisms which live on the decaying remains of other organisms, would die when all these remains had been used up.

This means that the ability of plants to use the energy of sunlight to manufacture food leads indirectly to the feeding of the whole living world on the products of light, air, water, and minerals from the soil. Figure 5.6 summarizes this principle, and in addition shows that plants are responsible for the presence of fossil fuels in the earth: these are the remains of long-dead organisms. Motor-cars, aeroplanes, home-heating systems, and industrial complexes all make use of sunlight energy which reached the earth and was absorbed by plants millions of years ago.

5.5 Plant mineral requirements

Julius Sachs, mentioned briefly in section 5.1, produced the first clear evidence that plants cannot grow

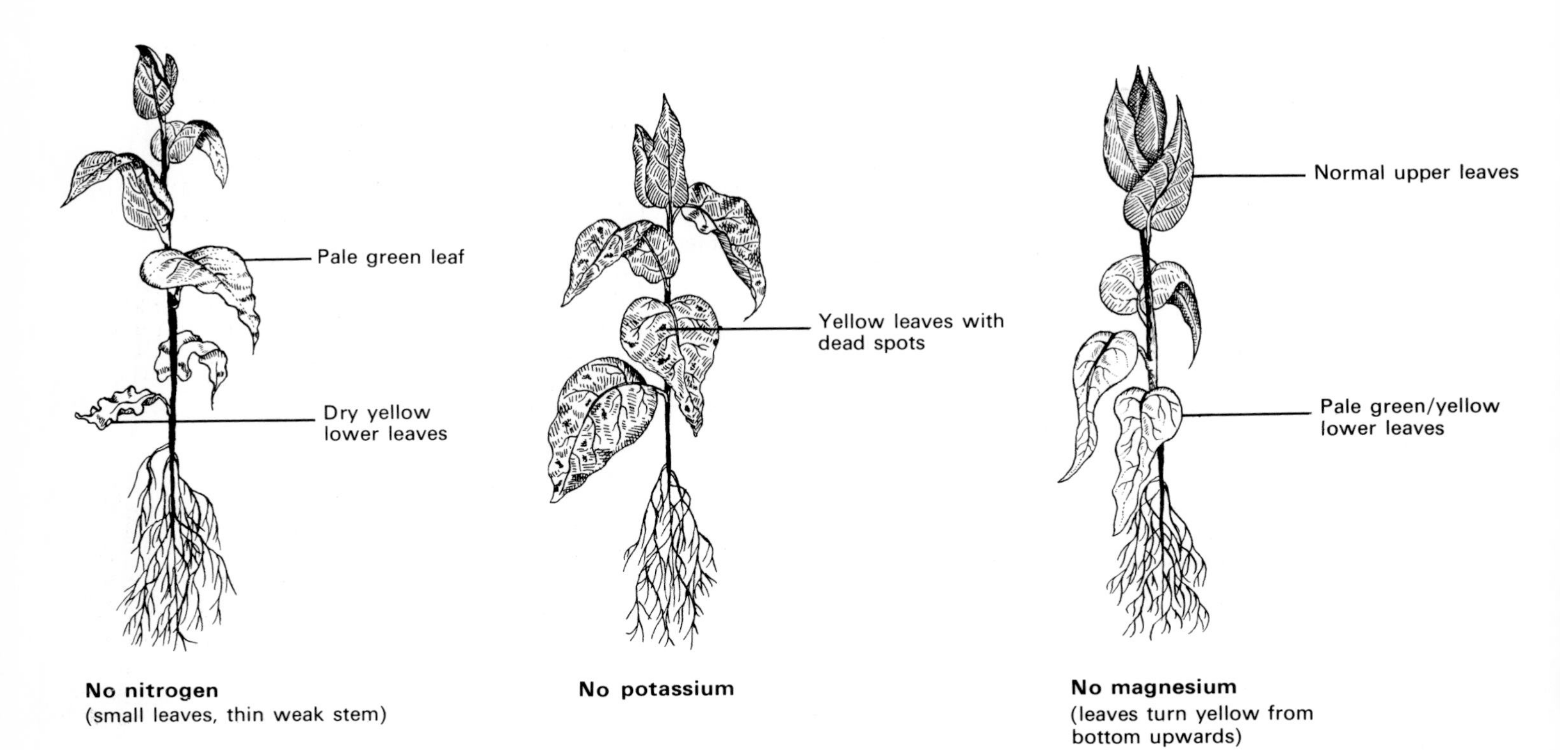

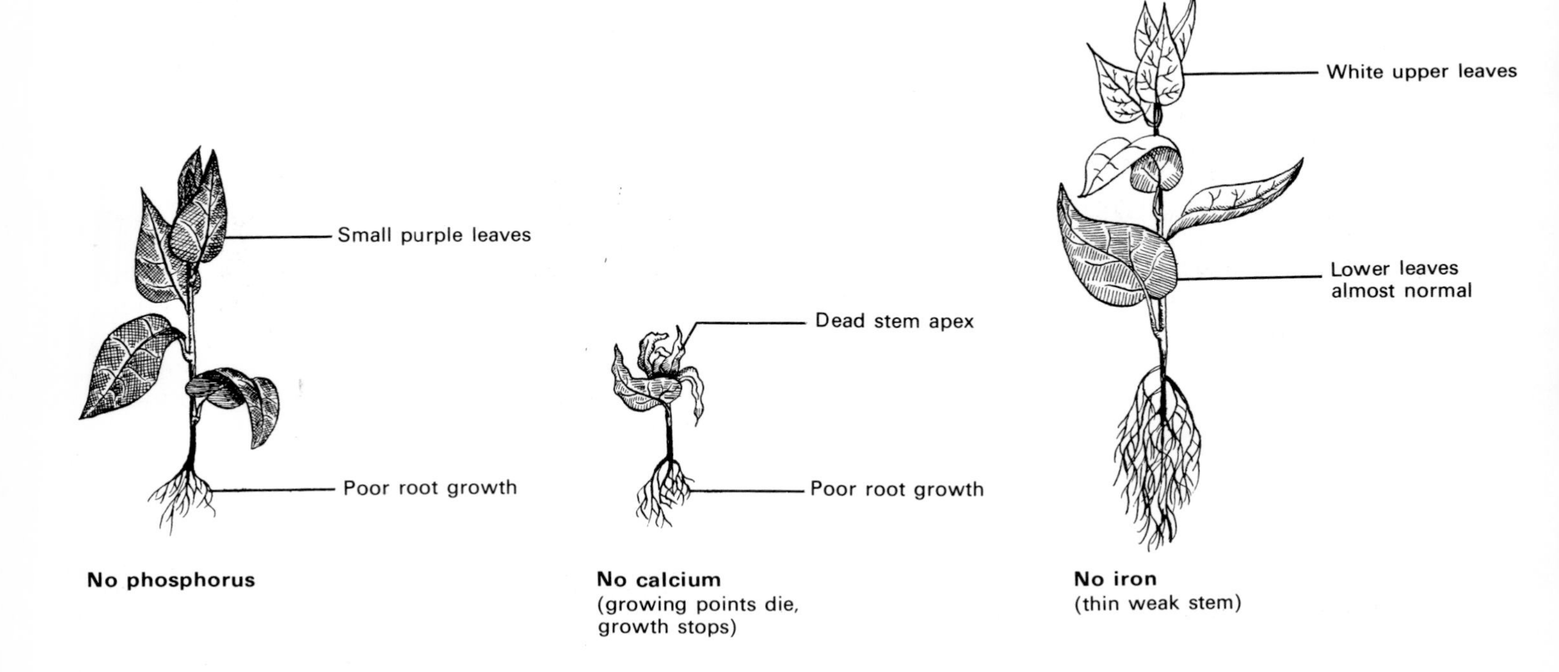

Fig. 5.7 Symptoms of some mineral deficiency diseases

and thrive on air and water alone. He found that in addition to these substances plants require supplies of minerals which they extract, often in minute quantities, from the soil.

Sachs grew some plant seedlings with their roots suspended in distilled water and found that their development was very poor when compared with other seedlings whose roots were in solutions of various minerals. This method is now called **water culture** (or hydroponics when carried out on a large scale) and, along with other techniques, has led to the discovery that plants require at least twelve minerals for healthy growth.

Certain of these minerals are called the **major elements**, because they are required in relatively large quantities. In the water culture method, for example, major elements must be present in several hundred parts per million for healthy plant growth. These are major elements: nitrogen, phosphorus, sulphur, potassium, calcium, and magnesium.

Other minerals are called **trace elements** because they are required in very small amounts. In water culture they need only be present in quantities as low as one part per million. Some of the trace elements are: manganese, copper, zinc, iron, boron, and molybdenum.

Sachs's method can be used in the following manner to find out more about plant mineral requirements. A series of plants are grown in solutions each of which lacks just one of the mineral elements necessary for plant growth. These plants are compared with control plants grown in solutions containing all the necessary elements. By this method it is possible to discover exactly what happens to plants when they lack a particular mineral. These defects in a plant's growth are called **mineral deficiency symptoms**, and they are of great value in agriculture since they indicate the type of treatment needed to improve mineral-deficient soils. Figure 5.7 shows an imaginary plant suffering from typical deficiency symptoms. Minerals are used by plants mainly in the manufacture of proteins.

Verification and inquiry exercises

A *Testing a leaf for starch*

The presence of starch in a leaf is an indication that photosynthesis is or has recently been taking place. Starch can be detected by the method illustrated in Figure 5.8.

B *To verify that plants require light, carbon dioxide, and chlorophyll for photosynthesis*

Carry out the following experiments either on one plant with variegated leaves, e.g. *Tradescantia*, *Coleus*, or *Pelargonium*, or on separate plants. The plant must first be 'destarched' by keeping it in the dark for forty-eight hours, and then a leaf tested to verify that it is starch-free (test A above).

1. *Light* Partly cover one or more leaves with light-proof paper (Fig. 5.9A), while still attached to the plant. After the plant has been in the light for one day, detach the covered leaves and test them for starch. Only those parts exposed to light will assume a blue-black colour (Fig. 5.9A1).

2. *Carbon dioxide* The plant must first be 'destarched' by keeping it in the dark for forty-eight hours, and then a leaf tested to verify that it is starch-free (test A above). Insert a leafy stem into a flask as shown in Figure 5.9B. The potassium hydroxide in the flask will absorb all carbon dioxide which diffuses through the cotton wool

Fig. 5.8 How to test a leaf for starch

1 Take a leaf from the experimental plant and boil it for 30 seconds to kill it and make its cells more permeable to iodine.

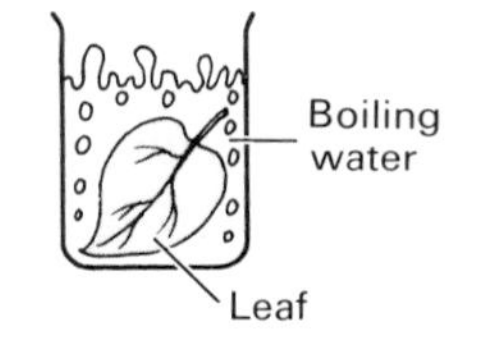

2 Boil the leaf in alcohol (using a hot water bath to avoid risk of fire) to remove chlorophyll and make the starch/iodine reaction easier to see.

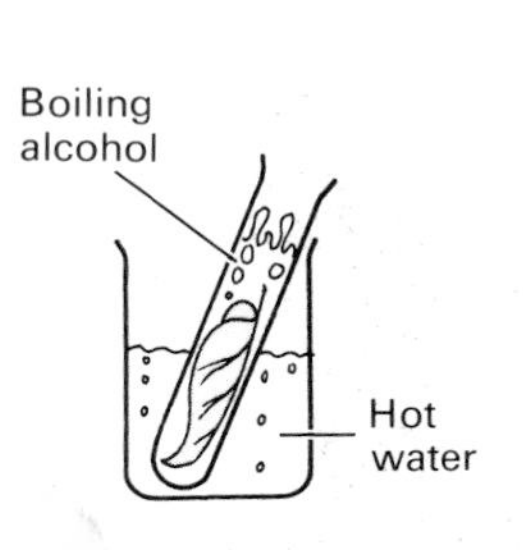

3 Soften the leaf by dipping it in boiling water again.

4 Spread the leaf on a white tile and drop iodine on it. A blue-black colour indicates that starch is present.

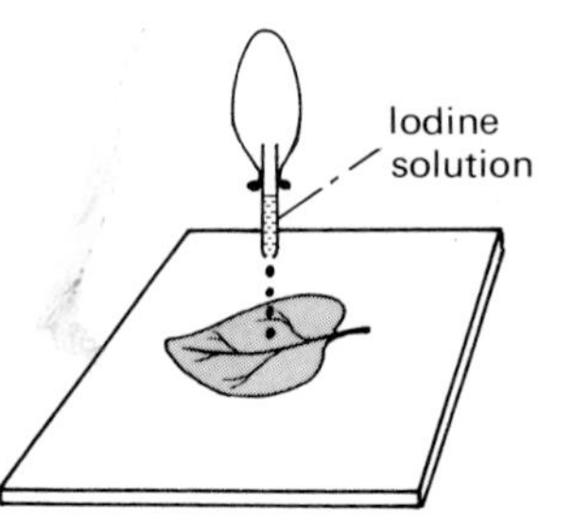

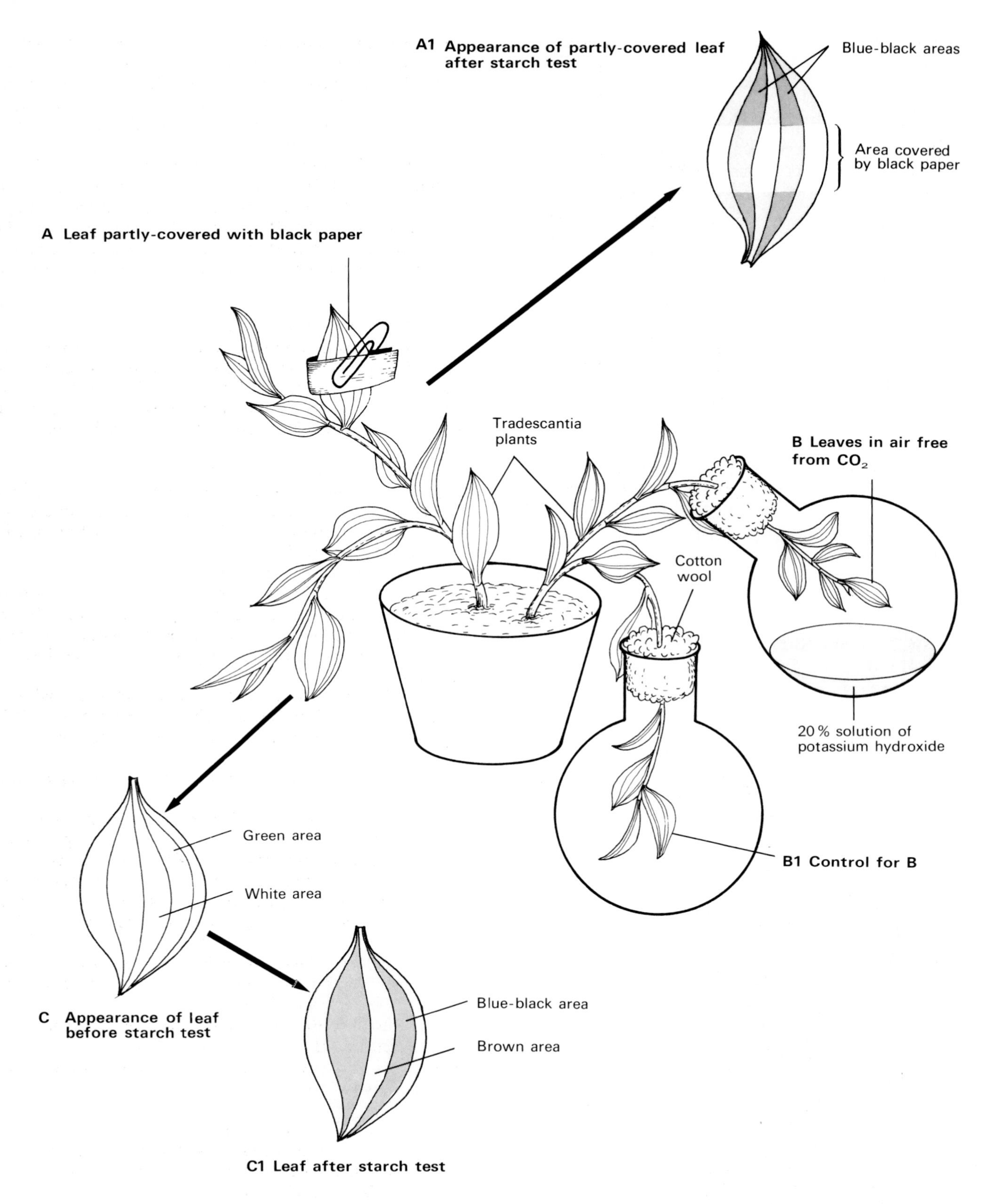

Fig. 5.9 Experiments with *Tradescantia* to verify that photosynthesis requires light, carbon dioxide, and chlorophyll

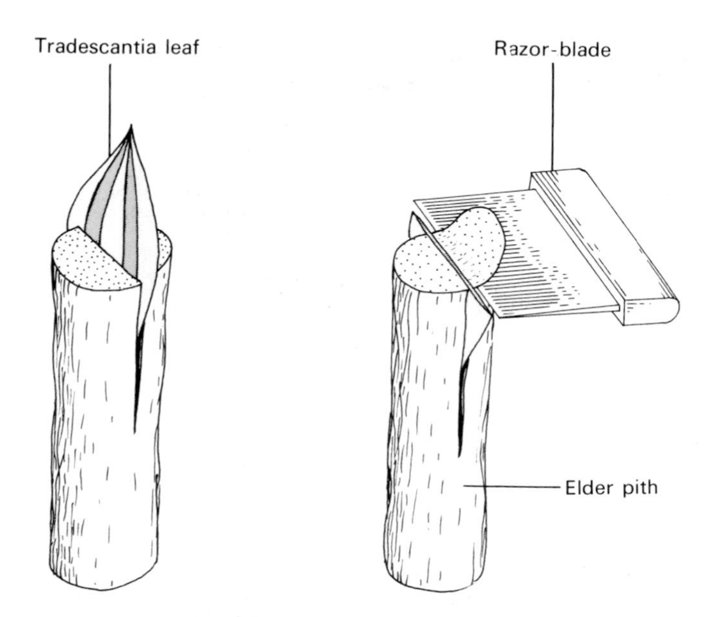

Fig. 5.10 How to cut a thin cross-section of a leaf

Fig. 5.11 Verification of oxygen production in photosynthesis

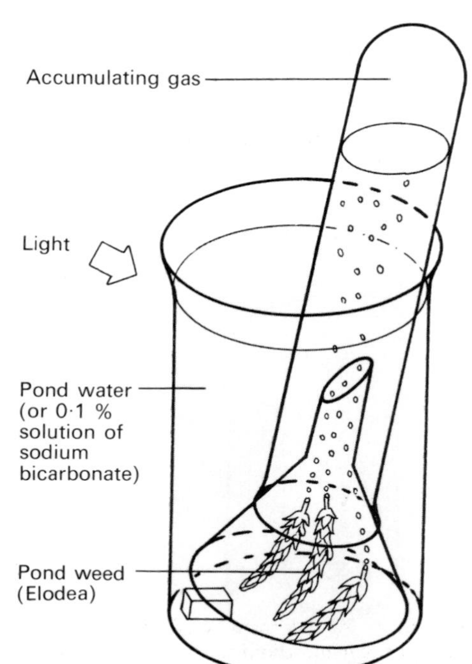

plug. Set up a control as shown in Figure 5.9B1. After the plant has been illuminated for one day test the experimental and control leaves for starch. Only control leaves will contain starch. Why is cotton wool used to plug the flasks instead of an airtight rubber bung? Why is the control experiment necessary?

3. *Chlorophyll* *Tradescantia* leaves usually have green areas and chlorophyll-free areas in alternate stripes. Draw the pattern of green areas of one leaf while it is still attached to the plant (Fig. 5.9C). After the plant has been illuminated for one day detach the leaf and test it for starch. Only the green areas should contain starch (Fig. 5.9C1).

C *To verify that mesophyll cells make starch*

Starch grains may be seen in the mesophyll cells of an iodine-stained leaf by cutting thin cross-sections of leaf tissue using the method illustrated in Figure 5.10. Mount the sections in 25% glycerine solution on a microscope slide and observe them under high magnification.

D *An investigation of leaf structure*

1. Cut thin cross-sections of various leaves by the method shown in Figure 5.10, mount them on a slide in 25% glycerine solution, and observe under low and high magnification. Use Figures 5.4 and 5.5 as a guide to the identification of structures observed.

2. By roughly tearing a leaf into pieces it is possible to expose a small area of lower epidermis. Try this with privet or rhubarb leaves. Mount small pieces in water and look for stomata under low magnification. Use Figure 7.7 as a guide.

3. In late autumn look for leaf 'skeletons', which are formed when all but the leaf veins have rotted away. These show the extent and intricacy of the leaf vein network.

E *To verify the production of oxygen during photosynthesis*

Using the apparatus illustrated in Figure 5.11 demonstrate that a pond weed such as *Elodea* produces a colourless gas when illuminated. After the gas has accumulated for a day or so test it with a glowing wood splint. The splint will re-light, proving that the gas contains a higher proportion of oxygen than normal atmospheric air.

F *An investigation into the influence on the rate of photosynthesis of variations in light intensity, temperature, and carbon dioxide concentration*

These experiments involve bubble counting. That is, the number of bubbles given off by a sprig of *Elodea* over a given period is used as a rough indication of photosynthetic rate.

1. *Influence of temperature*

a) Set up at least four sets of apparatus as illustrated in Figure 5.12, ensuring that the *Elodea* in each is about the same size and of equal quality.

b) Each set of apparatus should be brought to a different temperature: 0°C (using ice and salt if necessary); 10°C; 20°C; and 30°C. Check throughout the experiment that these temperatures are maintained.

c) The plants should be allowed to adjust to each temperature for at least five minutes in front of a bench lamp which is at a distance, tested prior to the experiment, known to produce a steady flow of bubbles from a plant at 20°C.

d) Record the number of bubbles produced per minute over a five minute period at each temperature. Graph the results.

e) Do variations in temperature affect photosynthetic rate? If so, suggest why. What are the temperature limits within which the rate is highest?

2. *Influence of light intensity*

a) The same sprig of *Elodea* can be used throughout this exercise. Set it up in the apparatus illustrated in Figure 5.11 and keep the water bath at a constant 20°C. Cover half of the water bath with black paper so that the *Elodea* receives light only from a bench lamp.

b) Begin the experiment with the apparatus 80 cm from the lamp. Allow it to adjust to these conditions for five minutes before a five minute bubble count is made. Move the apparatus to 40 cm from the lamp, wait one minute and then count bubbles for five minutes. Repeat this procedure with the apparatus at 20 cm and then 10 cm from the lamp.

c) Using this method, light intensity is increased four times each time the apparatus is moved closer to the light. Thus, if 80 cm is given a relative light intensity of one, the relative intensity of the other distances will be as follows:

Distance from light	80 cm	40 cm	20 cm	10 cm
Relative light intensity	1	4	16	64

d) Plot a graph of your results with relative light intensity on one axis and rate of bubbling on the other.

e) Do variations in light intensity influence photosynthetic rate? If so, is there a point at which further increase in light intensity produces no further increase in photosynthesis? Under these conditions what factor could be preventing further increase?

3. *Influence of carbon dioxide concentration*

a) Set up four sets of apparatus as illustrated in Figure 5.12 each with a fixed temperature of 20°C and at a distance from a bench lamp which produces a steady flow of bubbles.

b) Each plant should be in a tube with a different concentration of sodium bicarbonate solution: e.g. 0.1%, 0.3%, 0.5%, plus a control in distilled water.

c) After five minutes of adjustment, make a five minute bubble count and graph the results.

d) Do variations in carbon dioxide concentration affect the rate of photosynthesis? Do these results and results from experiments 1 and 2, suggest methods by which horticultural yields may be increased?

G *Designing controlled experiments*

These exercises provide an opportunity to solve problems by designing controlled experiments.

1. *Materials*

Bromothymol blue indicator; large test-tubes; beakers; sprigs of *Elodea*; carbonated water (e.g. from a soda syphon). All of these materials should be used in conjunction with the apparatus shown in Figure 5.12, after reading the notes below.

2. *The problem*

The problem is to devise experiments which give conclusive evidence that plants: (a) use up carbon dioxide, but (b) only when exposed to light, and (c) that they produce carbon dioxide in the dark.

3. *Points to note*

a) Bromothymol blue is an indicator; that is, its colour depends upon (i.e. indicates) the degree of acidity/alkalinity to which it is exposed. Discover and note its various colours by dropping some into weak acid, distilled water, and weak sodium hydroxide solution.

b) Carbonated water is a solution of carbon dioxide (i.e. a weak solution of carbonic acid). When placed in carbonated water *Elodea* can carry out photosynthesis very rapidly. Verify this by dropping some indicator into carbonated water, and by comparing the production of bubbles from *Elodea* in carbonated and distilled water.

c) Using the materials listed, devise experiments to solve 2 (*a*), (*b*), and (*c*) above, noting that bromothymol blue is not poisonous to plants, and that somewhere in the experiment it is necessary to show that the indicator is not affected by exposure to light.

Fig. 5.12 Apparatus for exercises F and G

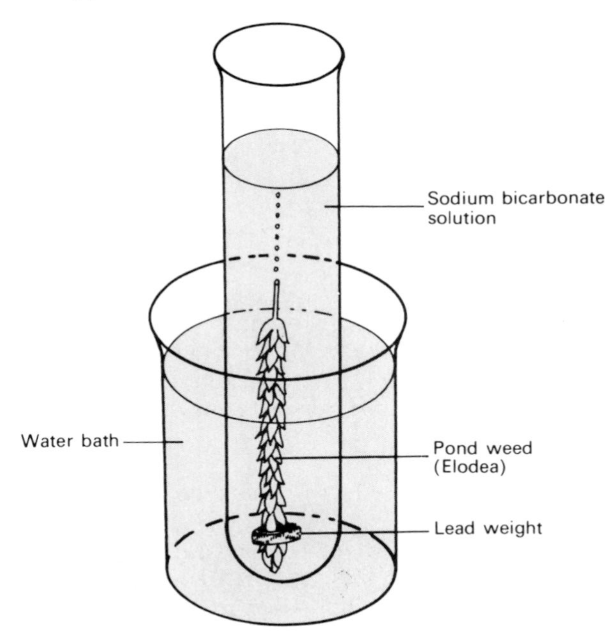

Comprehension test

1. Six bottles were filled with water and prepared as described below. They were sealed to make them completely airtight.

Three bottles were placed in the dark.	*Three bottles were placed in the light.*
No. 1 contained a water-snail and water-weed.	No. 4 contained a snail and water-weed.
No. 2 contained a snail only.	No. 5 contained a snail only.
No. 3 contained water-weed only.	No. 6 contained weed only.

Give reasoned answers to the following questions:

a) In which bottle will the most carbon dioxide be produced?

b) In which bottle will there be the greatest surplus of oxygen?

c) In which bottle will life persist the longest time?

d) In which bottle will life persist the shortest time?

e) In which *two* bottles will life persist for about the same length of time?

2. Look at Figure 5.5, remembering that the upper surfaces of leaves face the light in most plants, and decide how each of the following features of leaves contributes to photosynthetic efficiency:

a) the absence of chloroplasts in the upper epidermis;

b) the fact that palisade cells have more chloroplasts than spongy mesophyll cells;

c) the fact that xylem vessels are closer to palisade cells than phloem sieve-tubes;

d) the presence of stomata in the lower rather than upper epidermis;

e) the presence of air spaces between the cells of a leaf.

3. The graphs below show the changing concentrations of nitrates and green algae found in the upper regions of temperate seas throughout the year.

a) Suggest two reasons why the numbers of algae increase at A.

b) Suggest one reason why algae decrease at B.

c) Suggest why algae numbers decrease in winter.

d) Suggest why nitrate concentration increases in winter.

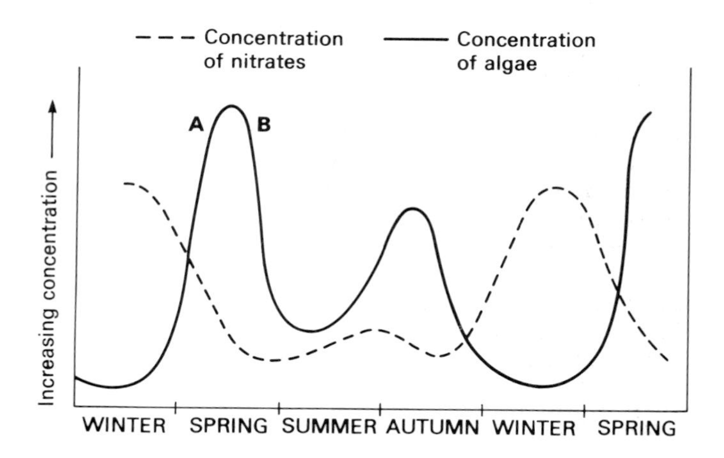

Summary and factual recall test

During photosynthesis plants take in (1) gas which enters through pores called (2) mainly in the (3) epidermis of leaves, and they take in (4) which enters through the (5). The green pigment (6) absorbs (7) energy which drives this process, the traditional chemical equation for which is:

$$(8) + (9) + (10) \xrightarrow{(11)} (12) + (13)$$

Recent work has shown that the oxygen liberated in photosynthesis comes from (14) which is split up into (15) and (16) during the (17) reaction. This knowledge makes it necessary to rewrite the traditional equation as follows:

$$(18) + (19) \longrightarrow (20) + (21) + (22)$$

Chlorophyll is contained within organelles called (23) which are found mostly in the (24) cells of the (25) layer of leaves. (26) vessels transport (27) to these cells from the roots, and phloem (28) tubes transport (29) from the leaves to the plant's (30) and (31)-storage areas.

All life on earth ultimately depends on photosynthesis for two main reasons, which are (32) and (33).

6
Transport in mammals

Living things constantly absorb useful substances like oxygen and food which must then be distributed throughout their bodies. In addition, they produce a continuous stream of waste materials such as carbon dioxide, which must be removed from their bodies before they accumulate to harmful levels.

In microscopic organisms such as *Amoeba* and *Paramecium*, the volume of the body is so small that useful substances can be distributed and waste materials removed by a process called **diffusion**. Oxygen, for example, enters an amoeba and spreads out, i.e. diffuses, in all directions at a rate approximately equal to the rate at which oxygen is consumed in respiration. Similarly, carbon dioxide diffuses out of an amoeba with sufficient speed to prevent it accumulating to harmful levels within the cell. In large multicellular organisms, however, the body volume is so great that diffusion alone is far too slow a process for the adequate distribution of oxygen and food, and the removal of waste. The cells in a multicellular animal relying on diffusion alone would be like people in a tightly packed crowd: those in the middle would not get enough oxygen. But most large organisms do not rely on diffusion for their supply of food and oxygen; they have a transport system of some kind to carry these substances to all the cells in the body.

In the human body, for example, the transport system consists of a pump called the **heart**, which propels a liquid called **blood** around a complex system of tubes called **blood vessels**. As it passes through these blood vessels the blood picks up oxygen from the lungs and transports it to every cell in the body. Blood also picks up waste products such as carbon dioxide from the cells and transports them to organs which remove them from the blood and excrete them from the body.

This chapter describes transport systems in mammals, using the human blood system as an example. It begins by describing the composition of blood, goes on to describe the heart and blood vessels through which blood circulates around the body, and finally summarizes the main functions of the blood.

6.1 Composition of blood

The average person has about 5.5 litres of blood. Although blood is a liquid, about 45 per cent of it is made up of solid particles held in suspension. The remaining 55 per cent is a straw-coloured fluid called **plasma**. The solid matter in blood consists of **red cells**, sometimes called red blood corpuscles or **erythrocytes**; **white cells** or **leucocytes**, which are actually colourless; and tiny particles called **platelets**.

Red cells

Human red cells are tiny bi-concave discs, i.e. concave on both sides (Fig. 6.1A). They measure 0.008 mm in diameter and are 0.002 mm thick, which means that 125 of them side by side would make a row 1 mm long. They are called 'cells' but they have no nucleus. This is one reason why they live for only about four months, after which they are broken down in the spleen and the liver. Some of their component chemicals, notably iron, are re-used to make new red cells. The new red cells are made in the bone marrow, particularly at the ends of the long arm and leg bones, in the ribs, and in the vertebrae. More than two million red cells are destroyed and replaced every second in the human body.

When seen individually under a microscope red cells are not red at all; they are golden yellow. It is only when they are seen massed together in a drop of blood that the red colour becomes apparent. This colour is due to the presence in each cell of a substance called **haemoglobin**.

Haemoglobin is the substance which enables blood to transport large quantities of oxygen from the lungs to the body tissues. As blood flows through vessels in the lungs it meets oxygen which has entered the body from the atmosphere. Haemoglobin in the red cells reacts with this oxygen to form an unstable substance called **oxyhaemoglobin**. When blood leaves the lungs and flows through vessels among the body cells in which there is little oxygen, the oxyhaemoglobin breaks

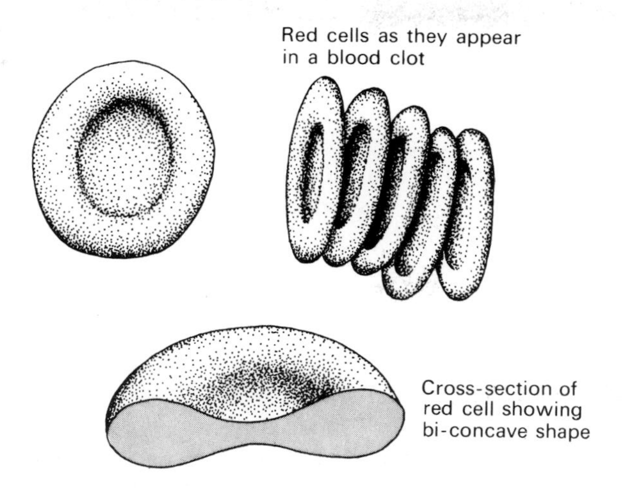

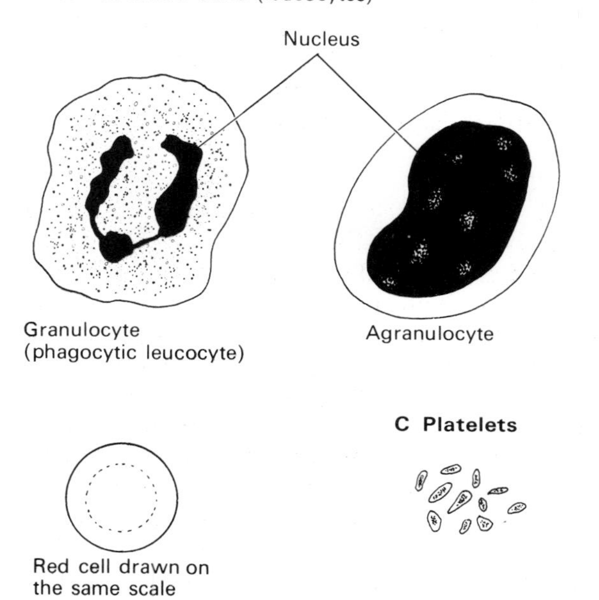

Fig. 6.1 Blood cells

down and releases the oxygen, which diffuses into the cells and is used in respiration.

These reactions, which are summarized below, form the body's oxygen transport system. This is described in more detail in section 6.4.

$$\text{Haemoglobin + Oxygen (Dark red)} \underset{\text{Low oxygen concentration}}{\overset{\text{High oxygen concentration}}{\rightleftharpoons}} \text{Oxyhaemoglobin (Bright red)}$$

There are about 5 million red cells in one cubic millimeter of human blood. This means that there are some 30 million million red cells in an average person's blood.

White cells or leucocytes

Leucocytes are colourless and have a nucleus (Fig. 6.1B). There are only about 8000 of them per cubic millimeter of human blood, which means there are about 75 million in the whole body. They are much larger than red cells, measuring up to 0.02 mm in diameter.

About 75 per cent of leucocytes have large lobed nuclei, and are irregularly shaped, rather like an amoeba. They have tiny granules in their cytoplasm and because of this they are called **granulocytes**. This type of leucocyte is made in the bone marrow but by different cells from those which make red blood cells.

Granulocytes are **phagocytic**, a word which means 'cell eater'. They 'eat' or engulf other cells, particularly bacteria, in rather the same way that amoebas feed (Fig. 4.4). Granulocytes form the body's chief defence mechanism against disease-causing bacteria. They gather in wounds and destroy bacteria before they can enter the body. They are also capable of leaving the blood vessels altogether by squeezing between the cells which make up the vessel walls. In this way granulocytes can reach infected areas anywhere in the body.

Red and white cells of human blood. p is a typical phagocyte; l is a typical lymphocyte; and m is a lymphocyte belonging to a type called monocytes.

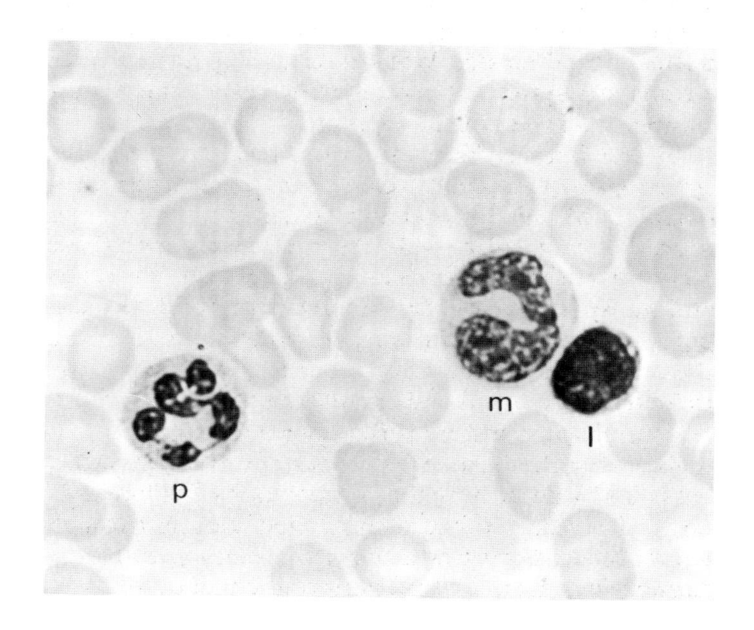

The remaining 25 per cent of leucocytes or white cells are rounded in shape, have very large round nuclei, and a thin layer of cytoplasm without granules. These cells are called **agranulocytes** (Fig. 6.1B). This type of leucocyte does not originate in the bone marrow, but in the lymphatic system, and for this reason they are sometimes called **lymphocytes**. Agranulocytes can move like amoebas but they are usually not phagocytic. The function of agranulocytes is to produce chemicals called **antibodies** which help prevent disease. This is described in chapter 20.

Platelets

Platelets are irregularly shaped objects about 0.003 mm in diameter (Fig. 6.1C). There are about 250 000 platelets per cubic millimeter of blood. They originate in the bone marrow, probably as fragments which become detached from cells. They play an important part in the system which causes blood to clot in wounds (chapter 20).

Plasma

Plasma is the liquid portion of blood. It consists of 92 per cent water and contains many important dissolved substances, including the products of digestion, such as glucose, fatty acids, glycerol, amino acids, vitamins, and minerals; plasma proteins such as albumin, fibrinogen, and antibodies; substances called hormones; and waste materials, such as urea and carbon dioxide.

Owing to the presence of the protein **fibrinogen** plasma can clot: that is, it can turn into a semi-solid jelly. This characteristic is extremely important because it is responsible for sealing off damaged blood vessels in wounds, thereby preventing excessive loss of blood.

Plasma can be forced through blood vessel walls under high pressure, carrying with it food and oxygen from the blood stream. Once it is outside the blood vessel walls plasma forms a liquid called **tissue fluid** which bathes every cell in the body. A large quantity

Fig. 6.2 Simplified version of Harvey's experiment to show that blood flows towards the heart in a vein

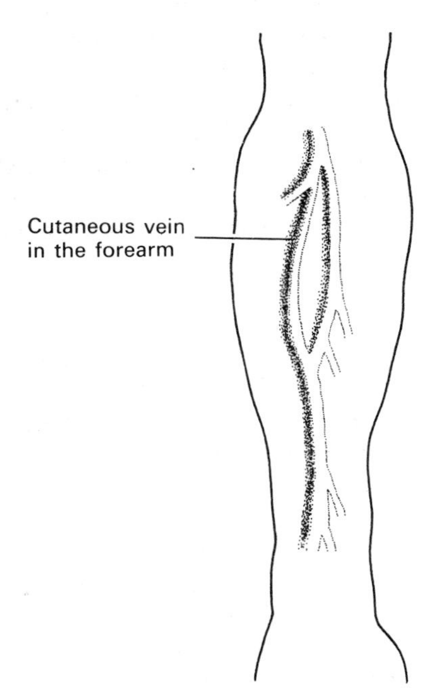

1 Hold the arm downwards for a few seconds. The veins will become prominent.

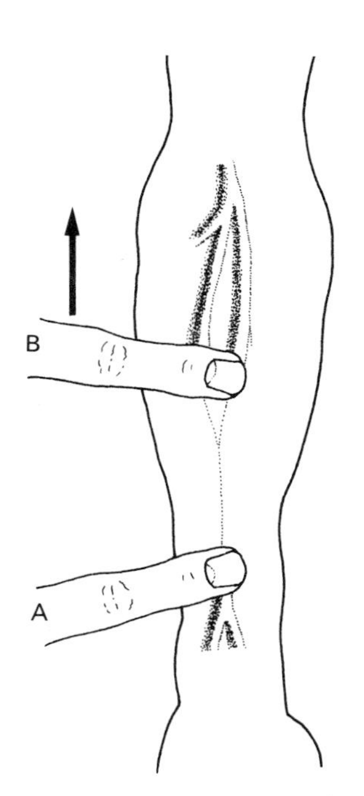

2 Place one finger (A) firmly on the wrist. Sweep blood upwards with another finger (B). Lift finger B from the arm. Blood will not flow back into the empty vein, because it never flows *away* from the heart.

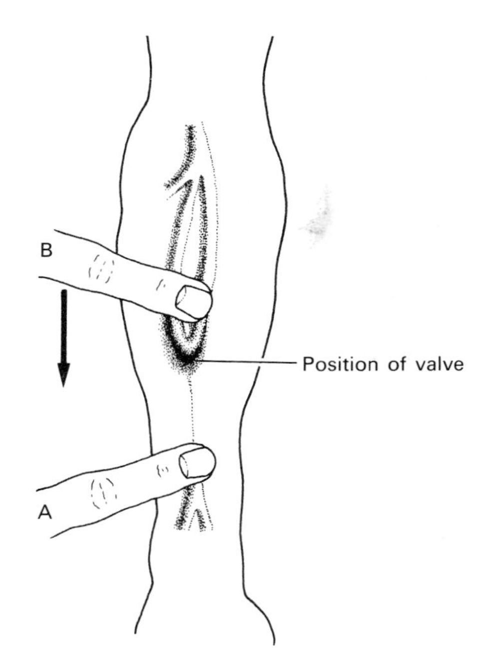

3 Move finger B down the vein. A bulge will appear showing where a valve prevents blood flowing away from the heart. Remove finger A and watch the empty vein fill again.

of plasma in the form of tissue fluid is constantly circulating among body cells supplying food and oxygen, and removing waste products.

It is important to realize that plasma and the tissue fluid derived from it form the environment which keeps body cells alive. In a sense, these fluids are equivalent to the pond and sea water in which unicellular organisms live, which supplies their food and oxygen, and into which they excrete waste. The difference is that the chemical composition and temperature of tissue fluid is controlled within very precise limits by the liver, kidneys, and other organs. As a result, body cells live and grow in a watery environment which in a healthy person is always perfectly suited to their needs.

6.2 **Circulation of blood in mammals**

In the early seventeenth century it was believed that blood flowed from the heart to the body organs and back again via the same blood vessels on the return journey. This is sometimes called the 'tidal theory' because it suggests that blood ebbs and flows like the tide.

In 1628 an English doctor called William Harvey produced experimental evidence suggesting that blood flows away from the heart in one set of vessels which are now called **arteries** and towards the heart in a separate set of vessels now called **veins**. One of Harvey's most famous experiments is illustrated in Figure 6.2. This shows that blood will flow in only one direction in a vein: always towards the heart. Furthermore, it shows that it will only flow one way because valves are present at intervals along the veins. The existence and structure of these valves was already known in Harvey's time (Fig. 6.3), but he was the first to demonstrate their function.

From this and other evidence, Harvey proposed the theory that the heart and blood vessels form a **circulatory system**, rather than a tidal system. He believed that the heart pumps blood along a circular route: outwards from the heart along the arteries, through the body organs, and back to the heart along the veins. Unfortunately, Harvey could not prove that blood circulates in this way because he was unable to find any connection between arteries and veins at their furthest point from the heart.

The connections between arteries and veins were discovered some seventy years later by the Italian biologist Marcello Malpighi. While he was examining the lungs of a frog under a microscope, Malpighi discovered a network of extremely narrow blood vessels

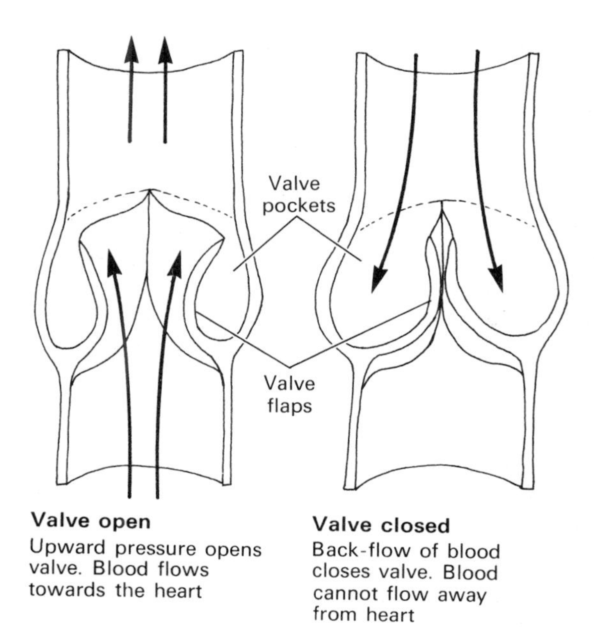

Fig. 6.3 Diagram of how vein valves prevent blood flowing away from the heart

which completed the circuit between the artery leading into the lung and the vein leading out of it. These tiny blood vessels are now called **capillaries**. Harvey did not use a microscope for his work and could never have found capillaries with the unaided eye because they are only 0.005 mm to 0.02 mm in diameter.

To summarize: the circulatory system of mammals consists of the heart, which pumps blood into vessels called arteries, which divide inside body tissues and organs into extremely fine vessels called capillaries, which in turn empty into veins that carry blood back to the heart (Fig. 6.4).

Double circulation in mammals

Figures 6.4 and 6.5 show the double circulatory system in mammals. Their blood circulates through two systems: the **pulmonary circulation** which conveys blood to and from the lungs, and the **systemic circulation** which conveys blood to and from all other parts of the body. These two systems are connected at the heart, but before the connections can be traced it is necessary to understand the structure of the heart (Fig. 6.6).

In mammals the heart consists of four chambers. There are two chambers on each side of the heart and these are completely separated by a central wall. The uppermost chambers on each side of the heart are called **atria** (singular atrium), and they have relatively thin muscular walls. (Some biologists prefer to call these chambers auricles.) Below each atrium is a thick-walled chamber called a **ventricle.** A system of valves on each side of the heart permits blood to flow from the atria into the ventricles, but not in the reverse direction. Traditionally, the two sides of the heart are always described from the animal's point of view. (The right atrium is the atrium on the animal's right-hand side and *not* the viewer's right-hand side.)

Using Figures 6.5 and 6.7 it is possible to trace the double circulation of blood starting from where it enters the right atrium. This blood has come from the systemic circulation: that is, from all the organs except the lungs. On its journey it has lost its supplies of oxygen and so it is called **deoxygenated blood.** The deoxygenated blood flows into the right atrium until it is full. Then the muscles in the atrium walls contract (in time with the left atrium) forcing blood down into the right ventricle, which is relaxed at this stage. A fraction of a second later when it is full the right ventricle contracts (in time with the left ventricle) forcing blood into the pulmonary artery. Blood is prevented from flowing back into the right atrium by the **tricuspid valve** (so called because it consists of three valve flaps). Blood is prevented from flowing back into the right ventricle by pocket-like **semi-lunar valves** located at the point where the pulmonary artery leaves the heart. Blood flows through the pulmonary artery to capillaries in the lungs where it absorbs more oxygen, and after this it is called **oxygenated blood.** This oxygenated blood returns from the lungs through the pulmonary vein, which empties into the left atrium. When full the left atrium contracts forcing blood into the left ventricle, and then the left ventricle contracts forcing blood into the main artery of the body, which is called the **aorta.** Blood is prevented from flowing back into the atrium by the **bicuspid valve,** which has two valve flaps, and it is prevented from flowing back from the aorta into the ventricle by another set of semi-lunar valves. The left ventricle has thicker muscular walls than the right ventricle. This gives the left ventricle the extra muscular power necessary to pump oxygenated blood all around the systemic circulation, which is far more extensive than the pulmonary circulation.

When deoxygenated blood returns via the main veins (the **superior** and **inferior vena cavae**) to the right atrium, the double circuit has been completed.

The heart

The heart's pumping action is driven by **cardiac muscle** in the walls. Cardiac muscle differs from

Fig. 6.4 Diagram of double circulation in a mammal

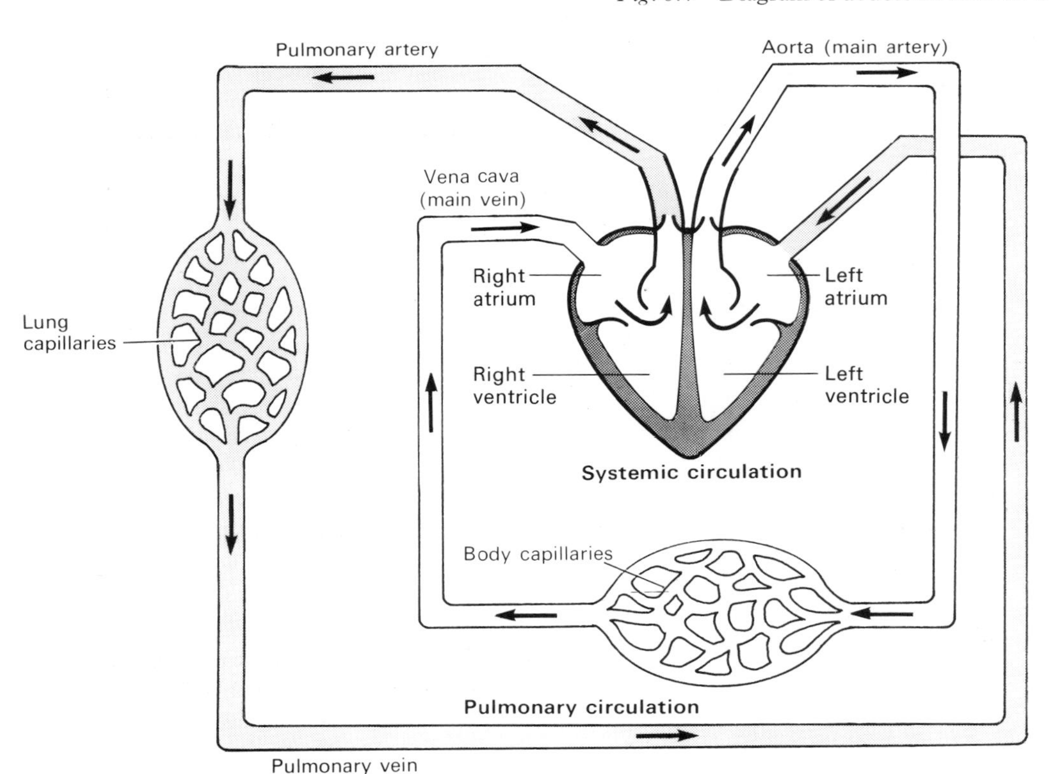

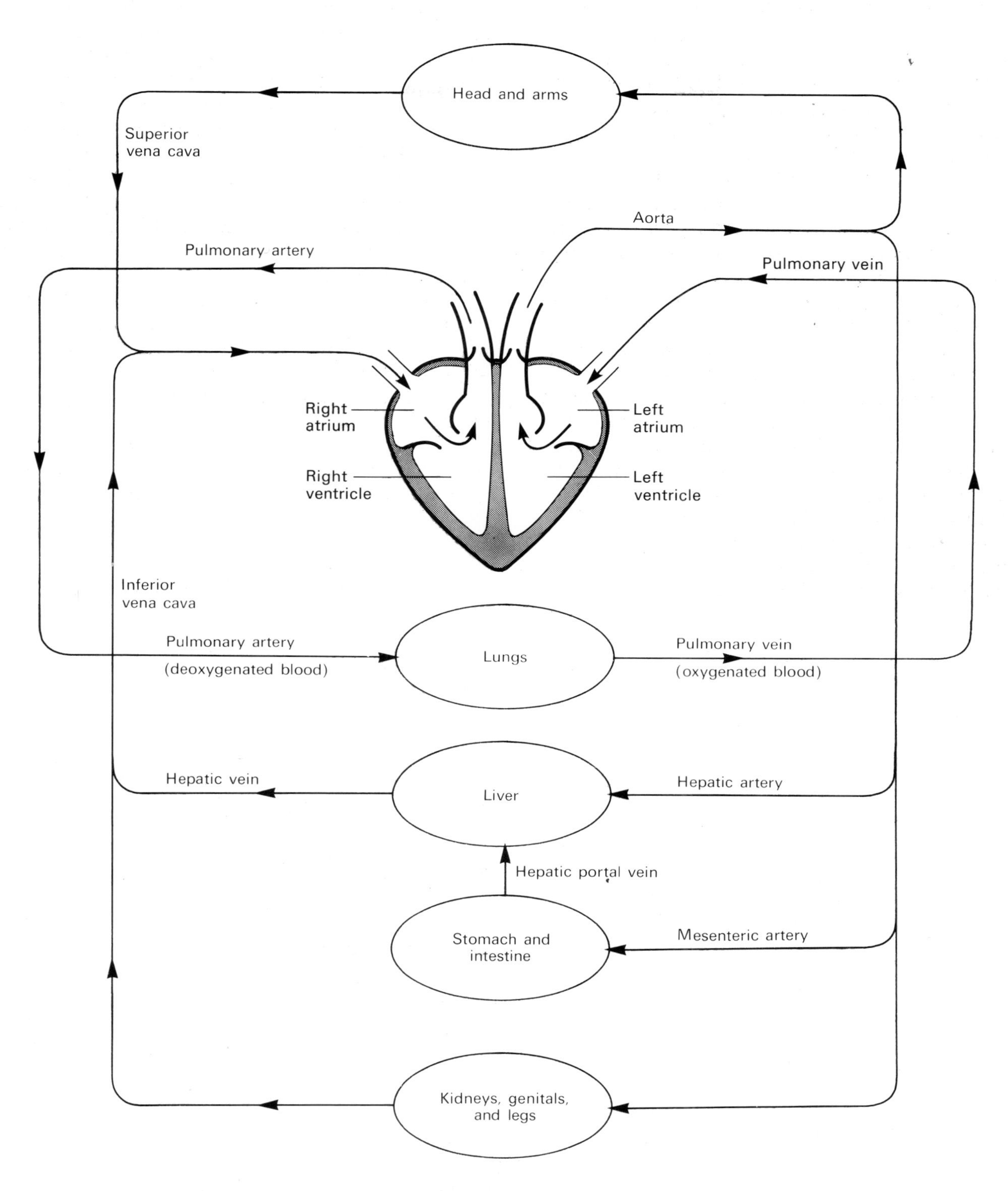

Fig. 6.5 Diagram of the main arteries and veins in a mammal, and the structure of a mammalian heart

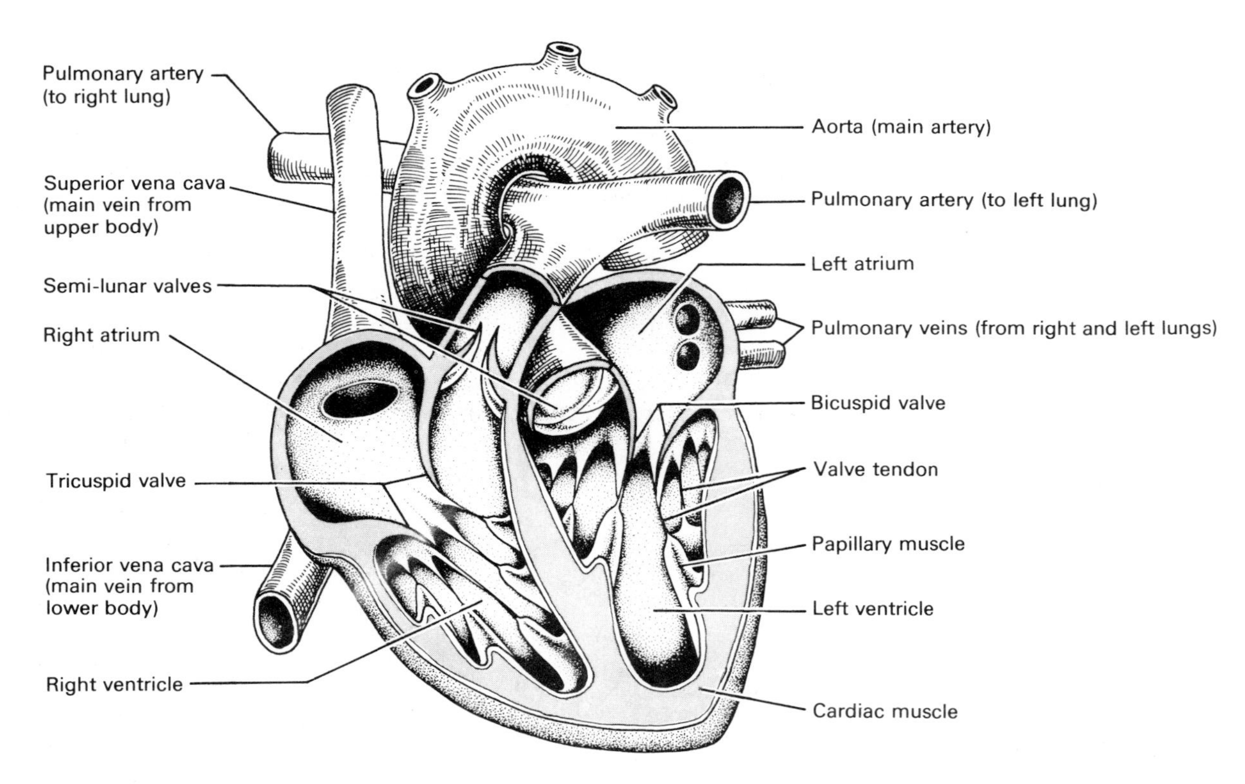

Fig. 6.6A Structure of the mammal heart

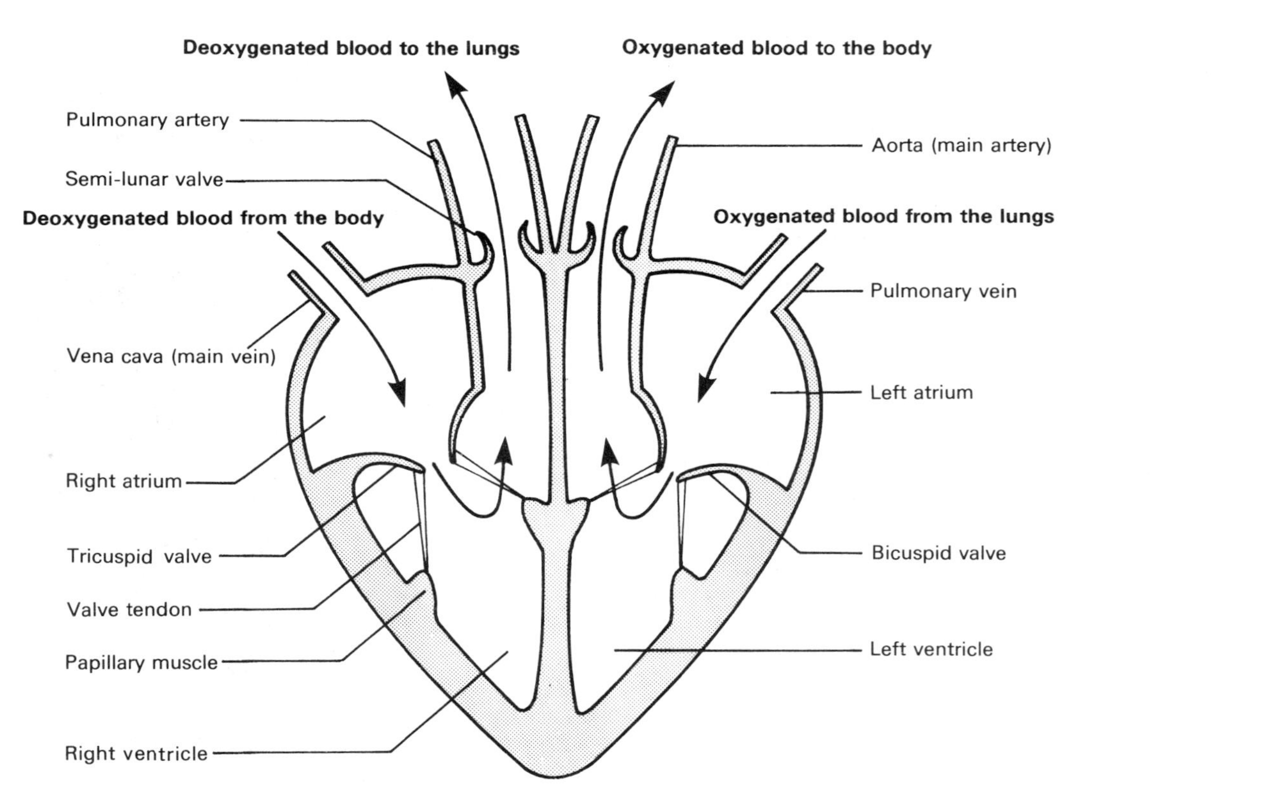

Fig. 6.6B Diagram of the heart showing direction of blood flow

other types of muscle in at least three important ways. First, it is made up of branching muscle fibres connected to each other in the form of a network, and not unbranched parallel fibres as in the muscles which move the arms and legs. This characteristic allows contractions to begin at one point in the heart and spread outwards in all directions. Second, cardiac muscle contracts and relaxes rhythmically in what are called 'beats'. This natural rhythm is generated within the muscle itself and not by impulses from the nervous system. William Harvey saw this in 1623 when he cut strips from a living heart and watched the pieces beat for some time all on their own. Third, cardiac muscle does not fatigue despite continuous rapid contractions over many years. The heart beats about 60 to 70 times a minute in a resting adult human, increasing to 150 or more during strenuous exercise. This adds up to an average of over 100 000 beats a day, in the course of which the heart pumps about 14 000 litres of blood. Compare this performance with the type of muscle attached to the skeleton by opening and closing one hand 70 times in a minute, and noting how tired it becomes.

Blood is driven around the body mainly by the pumping actions of the heart. But the flow of blood is greatly assisted by arteries and veins as well.

Arteries

Artery walls consist of an inner membrane one cell thick surrounded by a heavy layer of interwoven muscle fibres and elastic fibrous material. On the outside is a further layer of fibrous material (Fig. 6.8A).

When ventricles contract blood is forced out of the heart under high pressure, and this pressure stretches the aorta and artery walls outwards (Fig. 6.7A). Later, when the ventricles relax and the semi-lunar valves close, the elastic fibres in the artery walls recoil (like someone letting go of a stretched elastic band). Therefore a powerful force presses inwards on the blood, pushing it away from the heart. It must move in this direction because the semi-lunar valves have closed behind it, preventing it flowing back into the heart (Fig. 6.7B). This sequence of events begins at the heart and moves outwards along the arteries. Every time a ventricle contracts a volume of blood moves through the arteries preceded by a wave of artery wall expansion, and followed by a wave of elastic recoil and muscular contraction which forces the blood onwards. These waves are the 'pulse' which can be felt with the fingers in arteries at the wrist and neck.

Arteries divide to form smaller vessels called **arterioles** which have a thinner muscular/elastic layer. Arterioles divide many times to form a dense

Aorta wall expanded by blood pressure
Semi-lunar valve opened by blood pressure
Bicuspid valve closed by pressure in the ventricle
Left atrium fills with blood
Left ventricle contracts squeezing blood into aorta

A Ventricle contracts (systole)

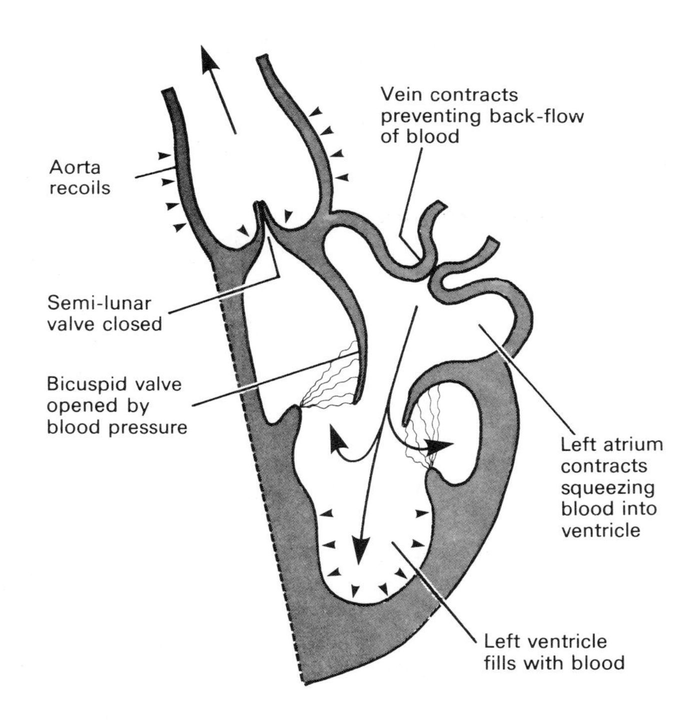

B Ventricle relaxes (diastole)

Fig. 6.7 Diagram of the heart's pumping action. (Only the left side of the heart is shown)

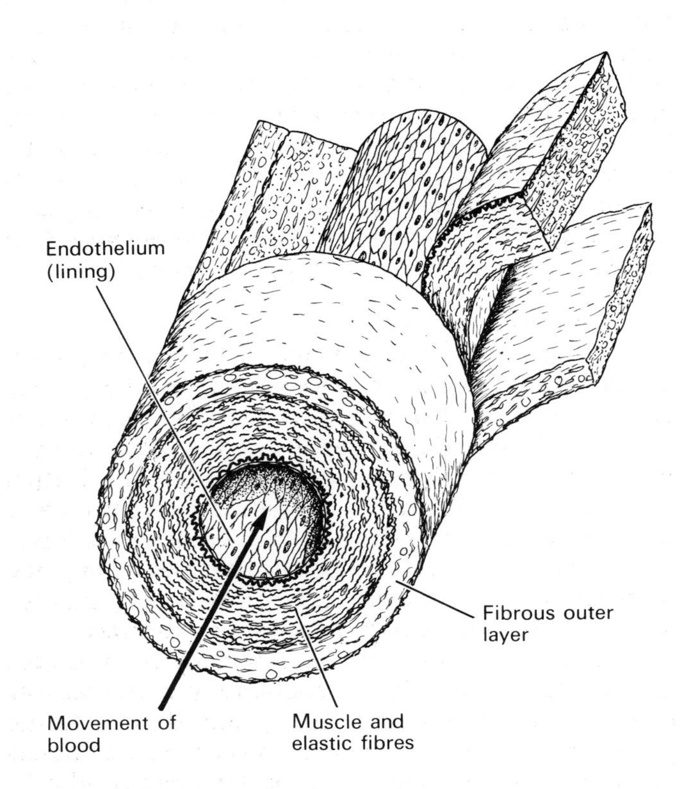

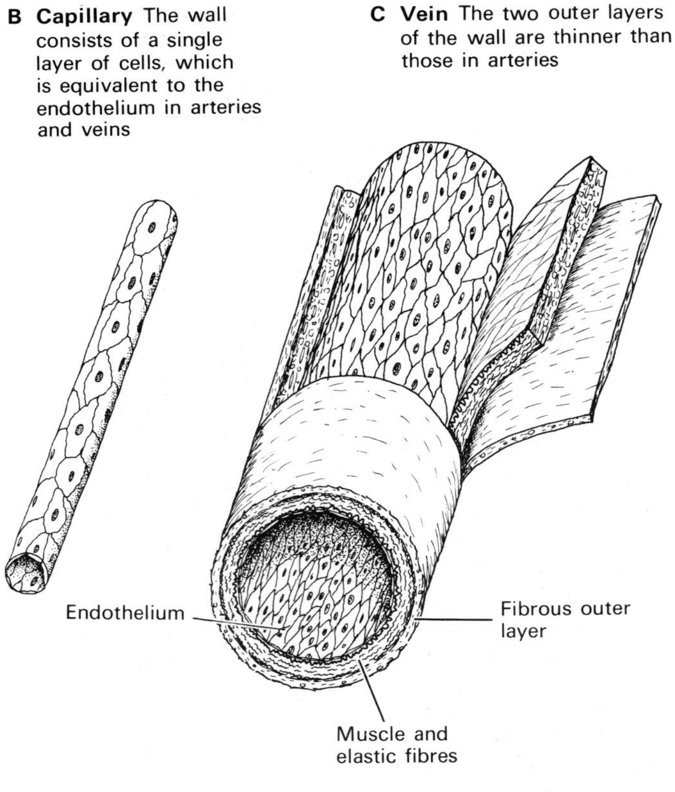

Fig. 6.8 Structure of an artery, capillary, and vein

network of capillaries which have walls only one cell thick (Fig. 6.8B).

Capillaries

The capillary network of the body is so extensive that if the rest of the body were dissolved away an outline of it and all its organs would still be visible owing to the mass of capillaries left behind. In fact, the capillaries are so densely packed together that none of the body's 25 million million cells is more than a fraction of a millimetre from one of these tiny vessels.

Capillary walls are extremely thin, and are permeable to (i.e. let through) water and all dissolved substances except the larger protein molecules. Water and dissolved substances are forced through the capillary walls by the pressure of blood and the resulting tissue fluid passes between the cells of the body. All the blood's functions depend on the permeability of capillary walls. The other parts of the circulatory system simply deliver high pressure oxygenated blood loaded with food to the capillaries, and remove low pressure deoxygenated blood from them.

Capillaries unite to form wider vessels called **venules**, and these eventually unite to form veins.

Veins

Compared with arteries, veins have a thinner muscular elastic layer in their walls, and a wider passageway or **lumen** for the movement of blood (Fig. 6.8C).

Blood in the veins is at low pressure, and its flow towards the heart is assisted in two ways. First, veins contain valves which permit blood to flow only towards the heart (Fig. 6.3). Second, many veins are situated between the large muscles of legs, thighs, arms, etc. Veins are squeezed flat when these muscles contract, which forces blood towards the heart. Muscles are never completely still for long even when a person is asleep, so blood flows along the veins at all times.

6.3 Tissue fluid and the lymphatic system

It has been explained that each tissue and body organ contains a dense network of capillaries. These are usually called **capillary beds** (Fig. 6.9). It has also been explained that a liquid called tissue fluid is forced under pressure through the capillary walls. This process tends to occur at the artery end of a capillary bed, since blood pressure is greatest at this point.

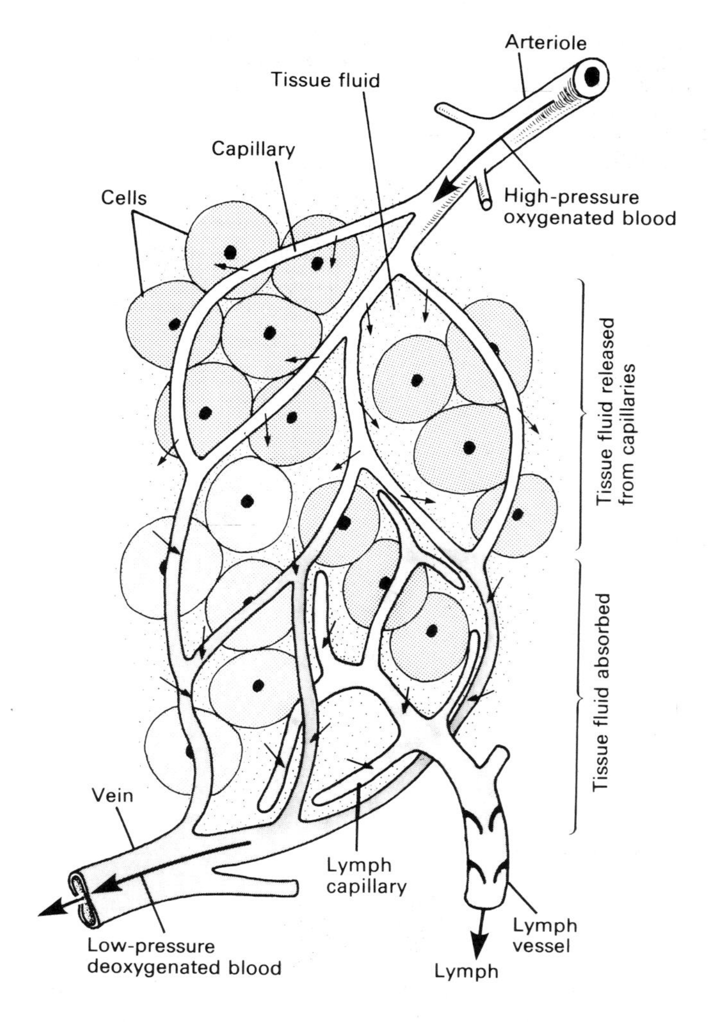

Fig. 6.9 A capillary bed

Tissue fluid

When tissue fluid is being forced out of the capillaries the capillary walls act as a filter holding back red blood cells, most of the white cells, and large protein molecules. The following substances which pass through the capillary walls make up the tissue fluid: water with dissolved oxygen, glucose, fatty acids, glycerol, amino acids, vitamins, minerals, and hormones.

Tissue fluid flows away from the capillaries and passes among the body cells, which extract oxygen and food from it and at the same time release carbon dioxide and other waste materials into it. Meanwhile, blood in the capillaries drains to the other end of the capillary bed and loses most of its pressure. At this stage the blood is a highly concentrated solution of protein molecules because it has lost water and soluble materials at the artery end of the bed. At the same time tissue fluid has become a weak solution, having given up most of its soluble contents to the body cells as it passed among them. Under certain circumstances concentrated solutions absorb water from weaker solutions by a process called **osmosis** (explained in chapter 7), and this is what happens at the vein end of a capillary bed. The blood absorbs some of the tissue fluid by osmosis through the capillary walls. These processes are illustrated in Figure 6.9, and summarized diagrammatically in Figure 6.10. Any tissue fluid not absorbed in this way passes into the lymphatic system.

Lymphatic system

Lymph vessels, usually called **lymphatics**, begin as tiny blind-ended tubes within the capillary beds and are about as numerous as the blood capillaries (Fig. 6.9). Tissue fluid which is not absorbed into the blood stream drains into these lymphatics and is then called **lymph**. The small lymphatics drain into larger ones, which are similar to veins in some ways. They have valves which ensure that lymph flows in only one direction, and they are situated mainly among muscles which squeeze the lymph along as they contract.

At intervals along the lymphatics there are structures called **lymph nodes**. These contain a system of narrow channels through which the lymph drains. Large phagocytes are attached to the walls of these channels and their function is to engulf bacteria and dead cells from the lymph. There are large lymph nodes in the groin, under the arms, and in the neck (Fig. 6.11).

The large lymphatics unite into two main lymphatic ducts which empty their contents into the blood stream at the subclavian veins near the heart (Fig. 6.11). All tissue fluid eventually drains back into the blood.

6.4 The main functions of blood

Transport of oxygen from the lungs

Oxygen is transported from the lungs to the body cells by the red cells of the blood in the form of oxyhaemoglobin. The ability of haemoglobin to carry oxygen in this form depends on the presence in every haemoglobin molecule of four atoms of iron. Each of these four atoms attracts and combines loosely with one molecule of oxygen. There is enough iron in the haemoglobin of one red cell to carry about 1 000 000 000 oxygen molecules.

When red cells reach the lungs they are deoxygenated. They contain no oxygen because they have left it all behind on their travels around the body tissues. But when red cells leave the lungs they are oxygenated, because they have picked up oxygen on their way through the capillary network of the lungs. On average, it takes only 45 seconds for each red cell to

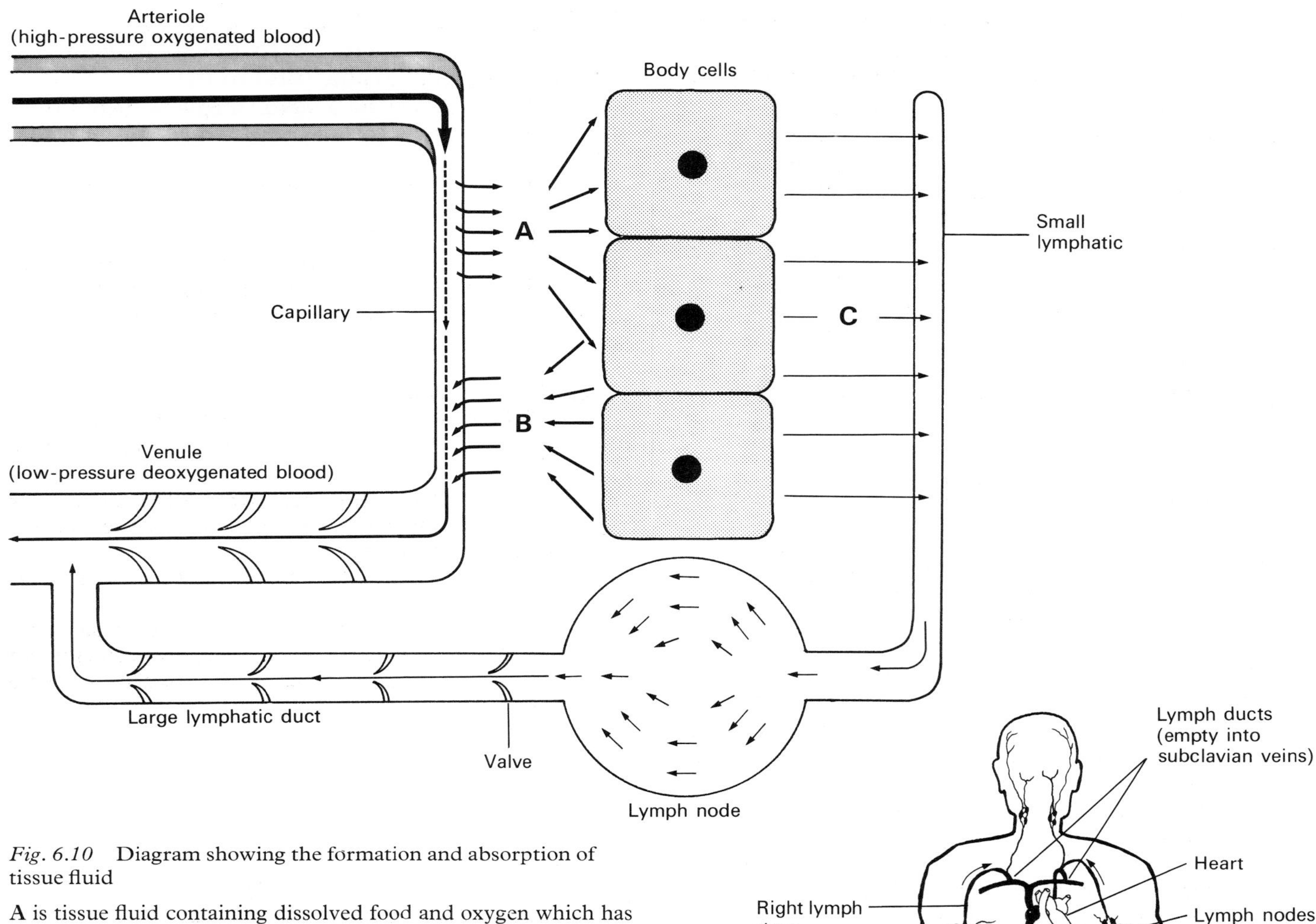

Fig. 6.10 Diagram showing the formation and absorption of tissue fluid

A is tissue fluid containing dissolved food and oxygen which has been forced out of the artery end of a capillary by high blood pressure. **B** is tissue fluid moving, partly by osmosis, into the vein end of a capillary. **C** is tissue fluid moving into the lymphatic system

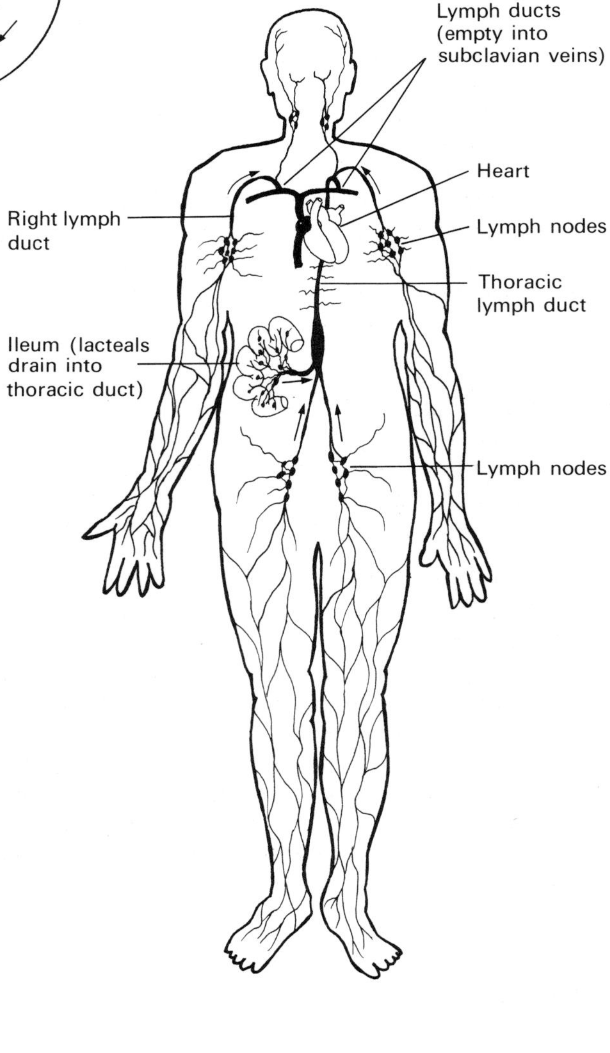

Fig. 6.11 Diagram of the human lymphatic system (greatly simplified)

collect a full load of oxygen from the lungs, deliver it to the body cells, and return for more.

Transport of carbon dioxide to the lungs

Carbon dioxide is produced constantly by cell respiration. It diffuses out of the cells, into tissue fluid, and then into blood passing through capillary beds in the tissues and organs. Once it is in the blood, carbon dioxide is transported to the lungs in one of three ways.

1. About 85 per cent of the carbon dioxide is transported in the form of **sodium bicarbonate** dissolved in the plasma. This substance is produced by a sequence of chemical reactions, some of which

take place inside red cells. During its formation, however, the reactions within the red cells release a by-product which triggers off the breakdown of oxy-haemoglobin, so that oxygen is released from the blood. Consequently, whenever blood enters a region of the body with a high carbon dioxide content, oxygen is automatically released from red cells.

2. About 5 to 10 per cent of carbon dioxide transported by the blood is in the form of a substance called **carbamino-haemoglobin**. This is formed from carbon dioxide which diffuses into red cells where it reacts with a certain amino acid contained in haemoglobin.

3. About 5 per cent of carbon dioxide transported by the blood is in the form of carbonic acid dissolved in the plasma.

When blood loaded with carbon dioxide reaches the lungs, it takes in oxygen and oxyhaemoglobin forms in its red cells. This reaction triggers off the simultaneous breakdown of carbamino-haemoglobin in the red cells and sodium bicarbonate in the plasma, so that carbon dioxide is released from both these substances. Consequently, the absorption of oxygen by red cells automatically releases most of the blood's carbon dioxide. The carbonic acid in the plasma changes into carbon dioxide gas.

Transport of urea to the kidneys

A number of chemical reactions in the liver concerned with the metabolism of protein release poisonous substances as waste products. These poisons are quickly converted into **urea**, which is relatively harmless. Urea travels in the blood stream in solution until it reaches the kidneys, where it is removed and excreted from the body in urine.

Transport of digested food

The soluble products of digestion pass into the blood stream through villi which line the ileum (section 4.5). From here, the soluble food is transported to the liver in the hepatic portal vein (Fig. 6.5). The liver releases food into the blood as the body requires it.

Distribution of heat, and temperature control

Whatever the weather conditions, mammals and birds can maintain a high and relatively constant body temperature. This is possible partly because the blood transfers heat from places where it is produced, such as the muscles, and distributes it fairly evenly throughout the body. In addition, there is a mechanism which helps to control the rate at which heat is lost through the body surface by controlling the amount of blood which flows through capillaries close to the surface of the skin.

Verification and inquiry exercises

A *An investigation of blood cells*

1. *Red cells* Obtain a blood sample by gently stabbing with a sterile lancet the area of skin shown in Fig. 6.12A.

a) Smear the blood on a slide (Fig. 6.12B) and allow it to dry.

b) Place a drop of glycerine and then a cover slip on the smear. Examine the red cells under a microscope.

2. Repeat with blood from a bird and a frog. Note any differences between red cells in these animals and those in mammals.

3. *Stained red and white cells* Obtain a blood smear as described in 1 above.

a) When the smear is *nearly* dry pipette on to it about three drops of Leishman's (or Wright's) stain, and leave for 30 seconds.

b) Pipette an equal number of drops of distilled water on to the stain, and rock the slide to mix the liquids. Leave for 10–15 minutes.

c) Wash off the stain with water. Place a drop of glycerine and a cover slip on the slide.

d) Observe the slide under low and high magnification. Red cells are pink, granulocytes have blue nuclei and pink granules, agranulocytes have blue nuclei and pale blue cytoplasm.

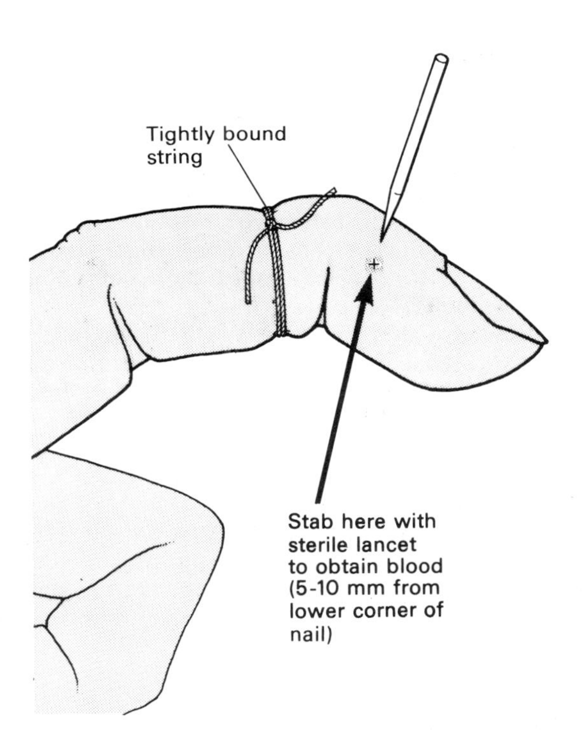

Fig. 6.12A How to obtain a blood sample

Fig. 6.12B How to make a blood smear

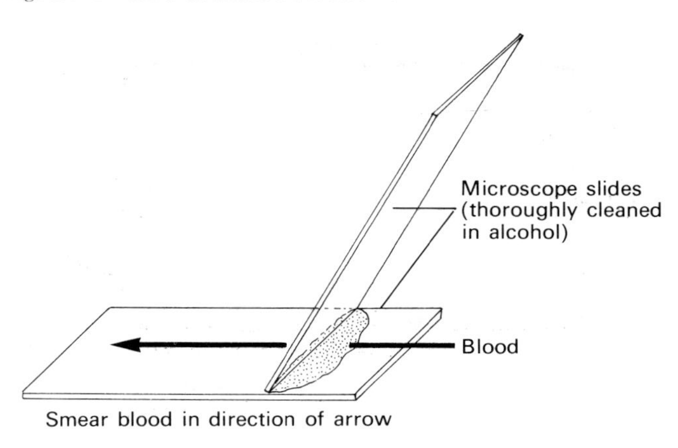

B *An investigation of blood clotting*

Place a drop of blood on a slide using the method given in A above, but do not smear it. Slowly pull the tip of a needle through the blood every 10–15 seconds and note the time taken before the needle draws a fine thread of fibrin from the blood.

C *An investigation of capillary circulation*

Capillary circulation is clearly visible in the web of a fish tail, the web of a frog's foot, or the gills and tail of a tadpole (especially *Xenopus*). Small animals can be observed in a watch-glass of water, or when held flat on a slide by a strip of wet cotton wool. Large fish and frogs should be lightly anaesthetized in weak chloroform water, then wrapped in wet cotton wool and loosely strapped on to a rectangle of thin wood with a circular hole at one end. The web of the fish tail or frog's hind foot should then be spread and pinned out over the hole in the wood, and viewed under a microscope. Wet the web every two minutes. (*Remember*, investigators have a responsibility to act humanely towards animals. A number of fish and frogs must be available so that none are kept under observation for more than 10 minutes.)

1. Observe the movement of red cells through the vessels and distinguish between arteries (arterioles), capillaries, and veins (venules), by comparing the relative sizes of the vessels.

a) Is the rate of blood-flow the same in each type of vessel? If not, which has the fastest flow and which the slowest?

b) Estimate the size of the vessels. A frog's red cells are oval and between 0.02 mm and 0.015 mm in diameter.

2. Cool the specimen with ice or refrigerated water.

a) How does this affect the rate of blood-flow and capillary diameter?

b) What does this result reveal about the possible effects of cold weather upon capillary circulation?

3. Apply one or two drops of the following chemicals, in the order indicated, directly on to the web either of different animals, or the same animal after thoroughly washing away all traces of the previous substance: adrenalin, alcohol, nicotine, lactic acid. Allow two minutes for each to diffuse into the web.

a) How does each chemical affect the diameter of blood vessels and the rate of blood-flow?

b) Why does a drink of alcohol appear to make a person feel warmer in cold weather, and yet contributes to rapid heat loss from his body?

c) Strenuous exercise results in the accumulation of lactic acid in muscles. How will this affect the circulation of blood through the muscles?

d) Why should an athlete refrain from smoking before an event?

D *An investigation of gaseous uptake in blood*

Obtain fresh blood from a slaughter-house and immediately add about 5 cm^3 per litre of 0.1% sodium oxalate to prevent it clotting. Alternatively, use an aqueous solution of crystalline haemoglobin.

1. Place equal amounts of blood or haemoglobin solution in three flasks.

a) Bubble oxygen through one flask and carbon dioxide through another. Note any colour changes compared with the blood in the third flask.

b) Reverse the procedure, i.e. bubble carbon dioxide through the flask which previously received oxygen and oxygen through that previously treated with carbon dioxide. Note any colour changes.

c) Do these experiments help to verify that haemoglobin both absorbs and releases oxygen? If so, explain why.

2. Use a vacuum pump to evacuate air from the flask containing oxyhaemoglobin. Note and explain any colour change.

E *Observing capillaries in human skin*

Soak the area of skin between a finger-nail and first joint in cedarwood oil. This should clear the skin sufficiently to make capillaries visible, in strong light, under low-power magnification.

F *An investigation of pulse rate*

Pulse points occur in the wrist and in the throat (on either side of the wind-pipe). Find these and count the number of pulsations over a 30-second period: when sitting at rest in a chair, and after vigorous exercise such as press-ups. Record the time taken for the pulse to return to its original rate. Account for the changes in rate after exercise.

G *An investigation into the effect of temperature on heart-beat in cold-blooded animals*

Observe the heart-beat of a *Daphnia* (water-flea) by mounting one in water on a cavity slide under a microscope. Count the heart-beats by tapping a pencil point on paper in time with the pulsations for 5 or 10 seconds, then count the pencil dots. Do this several times and find the average. Compare heart-beat rates with the animal in water at various temperatures.

1. Assuming that the effect of temperature on *Daphnia's* heart is typical of most cold-blooded animals, why is it necessary for species of this type either to hibernate during the winter, or survive in the form of fertile eggs?

2. In view of the experimental results, name some advantages of the high, constant body temperature found in mammals and birds.

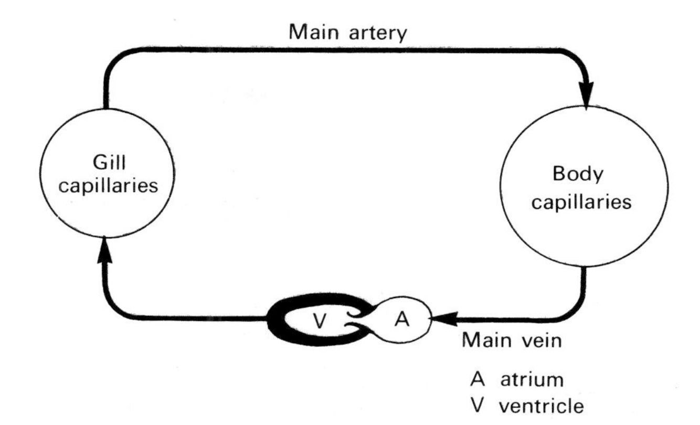

Fig. 6.13 Diagram of single circulation in a fish

Comprehension test

1. People who are born and spend their lives in villages high in the Andes mountains have more red blood cells than people who live at sea level. How is this condition related to atmospheric conditions at high altitudes?

2. Trace the path of a glucose molecule (naming only the larger blood vessels) from intestinal capillaries to the brain where it is respired, and then trace the path of respiratory carbon dioxide from the brain to the lungs.

3. *a)* Why is there five times as much cardiac muscle in the left ventricle as there is in the right ventricle?

b) Both ventricles have an equal internal volume. Why is this necessary?

4. Figure 6.13 is a diagram of the single circulation in a fish.

a) What relative difference in blood pressure will there be between blood reaching body capillaries in a fish and blood reaching the body capillaries in a mammal?

b) What has a mammal's four-chambered heart to do with this pressure difference?

c) In view of the answers to the above, explain the main advantage of double over single circulatory systems.

5. Blood which flows through capillaries in the feet is at low pressure. Explain how it is returned to the heart against the force of gravity.

6. Unlike oxygen, carbon monoxide gas (present in car exhaust fumes and certain household gas supplies) forms a chemically stable compound with iron atoms in haemoglobin.

a) In view of this, explain why it is very dangerous to breathe this gas for more than a few minutes.

b) If a person's blood is damaged by carbon monoxide gas, what two processes may eventually return his red cell supply to normal?

7. Prolonged starvation reduces the amount of protein in the blood. One consequence of this is an increased amount of tissue fluid which tends to gather in the abdomen and lower limbs. What has this to do with B in Figure 6.10?

8. When papillary muscles (Fig. 6.6) contract they pull on the valve tendons and draw the edges of heart valve flaps together. At what point does this event occur in the sequence of atrial and ventricular contractions, and what function does it serve?

9. Lymph drains into the blood-stream through the subclavian veins (Fig. 6.11), which are close to the heart. Why would it be impractical to have lymph draining into an artery near the heart?

Summary and factual recall test

The liquid part of blood is called (1). It consists of 92 per cent (2), and contains dissolved substances such as (3–name twelve). When this liquid is forced through the (4) walls under pressure it forms (5) which bathes all body cells, supplying them with (6) and (7), and removing their (8) products.

Red cells are (9) in shape, are made in the (10), and have no (11). They are red because of a chemical called (12) which contains (13) atoms that combine loosely with (14) forming an unstable substance called (15).

Granulocytes are leucocytes which are (16) in shape, have a (17) nucleus, and contain (18) in their cytoplasm. They are said to be (19) which means they 'eat' cells. Their function is to (20). Agranulocytes have (21) nuclei, and no (22) in their cytoplasm. Most of them produce (23) which help prevent (24). Platelets help blood to (25) in wounds. Blood is pumped by the (26) into vessels called (27) which have (28) muscular walls. In the tissues these vessels branch to form very narrow vessels called (29), the walls of which are (30) to water and dissolved substances. These vessels join together and lead into (31) which have (32) at intervals along them which ensure that blood flows only towards the (33).

The blood of mammals circulates through two systems, the (34) and the (35). Deoxygenated blood fills the (36) at the (37)-hand side of the heart, and oxygenated blood fills a chamber with the same name at the (38)-hand side. Both these chambers contract, forcing blood to open the (39) and (40) valves and pass into the (41). These chambers contract forcing blood from the (42)-hand side to go to the lungs through the (43) artery, and blood from the (44)-hand side to go to the body cells through the (45).

7

Support and transport in plants

The roots of plants absorb water and minerals from the soil. These substances are then transported up the stem to the leaves where they take part in many different processes. Water is combined with carbon dioxide to form sugar during photosynthesis, and minerals are used in the manufacture of proteins and other complex chemicals. Later, these end-products are transported to other parts of the plant where they may be stored, or used in the growth of new tissues.

Substances move from one part of the plant to another through a system of narrow tubes known collectively as **vascular tissue**. Plants which have vascular tissue are known as the **vascular plants**, and these include the Pteridophytes (ferns), the Gymnosperms (mostly conifers), and the Angiosperms (flowering plants).

This chapter describes the structure of vascular tissue in flowering plants; the movement of water and minerals through this tissue from roots to leaves; and the movement of photosynthetic products from the leaves to storage areas and growing points.

7.1 Vascular tissue in flowering plants

There are two types of vascular tissue: **xylem** and **phloem**. In the roots of most flowering plants the xylem tissue is arranged in an X-shaped mass and the phloem is between the 'arms' of the X (Fig. 7.1). In most stems, however, the xylem and phloem are arranged together in compact masses called **vascular bundles** (Fig. 7.2). Usually the xylem is situated in the part of each bundle near the stem centre, with the phloem towards the outer surface of the stem. The arrangement of xylem and phloem in leaves is illustrated in Figures 5.4 and 5.5.

Xylem

Xylem is a tissue composed of tubes called **xylem vessels**, and long pointed **fibre cells**. These are best observed in vertical sections through a plant stem, as illustrated in Figure 7.2.

Xylem vessels develop from cylindrical cells arranged end to end, in which the cytoplasm dies and the cross-walls break down leaving a dead, empty tube (Fig. 2.6A-D). In trees these tubes may be over 30 metres long. Xylem vessels form a transport system through which water and dissolved minerals move from the roots, through the stem, and into the leaves.

Both the vessels and the fibre cells of xylem are very strong owing to the presence in their walls of a substance called **lignin**. In xylem vessels lignin appears either as rings, coils, or layers perforated at intervals by tiny holes called **pits**. The strength of xylem tissue provides a great deal of support for the softer tissues of roots, stems, and leaves against the force of gravity and the pressure exerted by strong winds. In other words, xylem is the hard woody 'skeleton' of plants. (The structure of this skeleton is best observed in trees, whose trunks consist of layers of xylem laid down year by year in concentric rings. The xylem is so strong that in gale force winds trees are more likely to be uprooted than suffer a broken trunk.)

Phloem

Phloem tubes are also formed from cylindrical cells arranged end to end. Unlike xylem vessels, however, the cross-walls of the phloem tubes do not disappear; they develop perforations like a sieve. These cross-walls are called **sieve plates**, and the tubes are called **sieve tubes**. Sieve tubes form a transport system through which photosynthetic products, such as sugar, move from the leaves to the storage areas and growing points of a plant.

Phloem tissue also contains **companion cells**, so called because they grow side by side with sieve tubes and are apparently essential to their transport functions. Unlike xylem vessels and fibre cells, sieve tubes and their companion cells contain living cytoplasm, and have thin delicate walls which are unable to provide support.

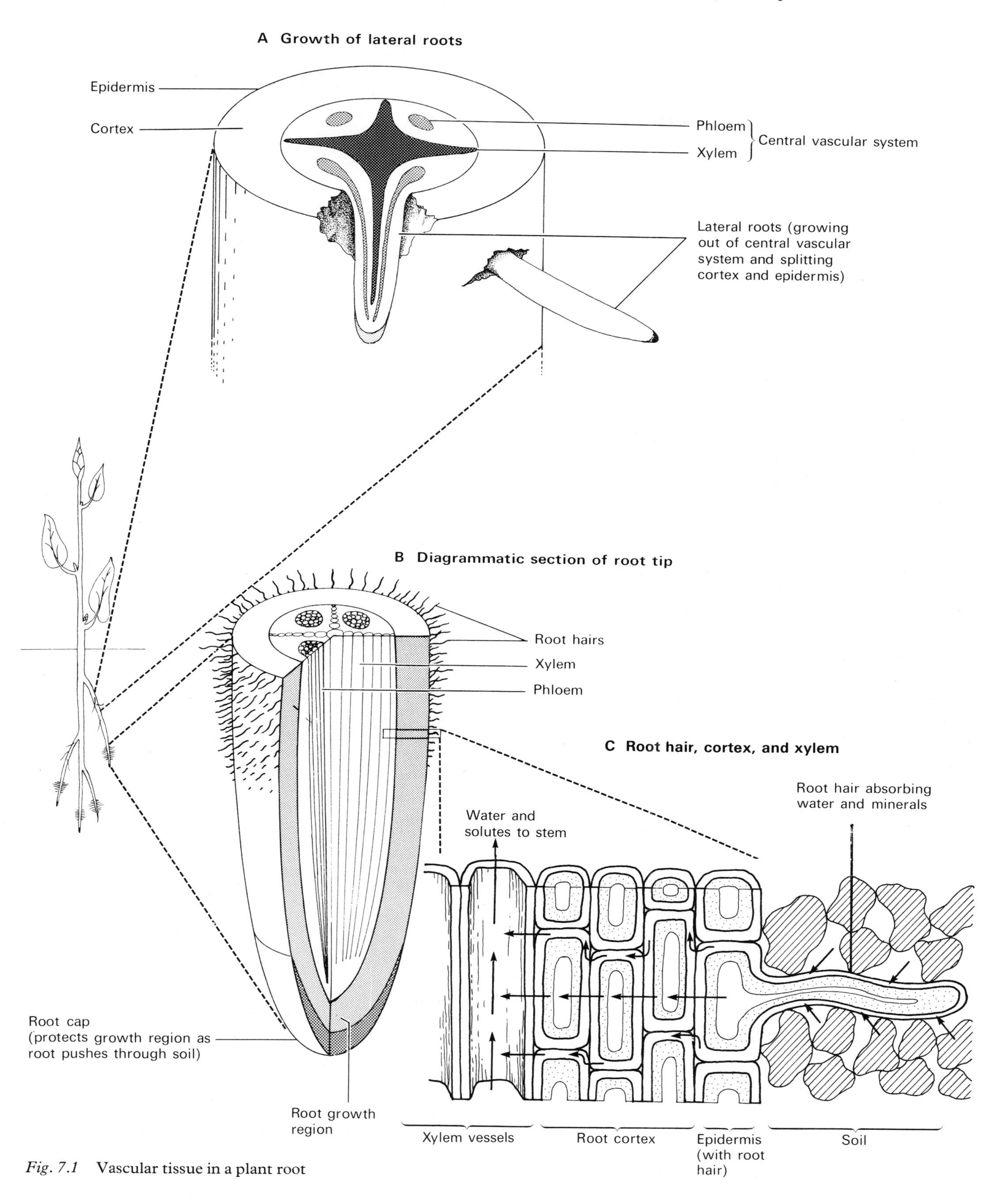

Fig. 7.1 Vascular tissue in a plant root

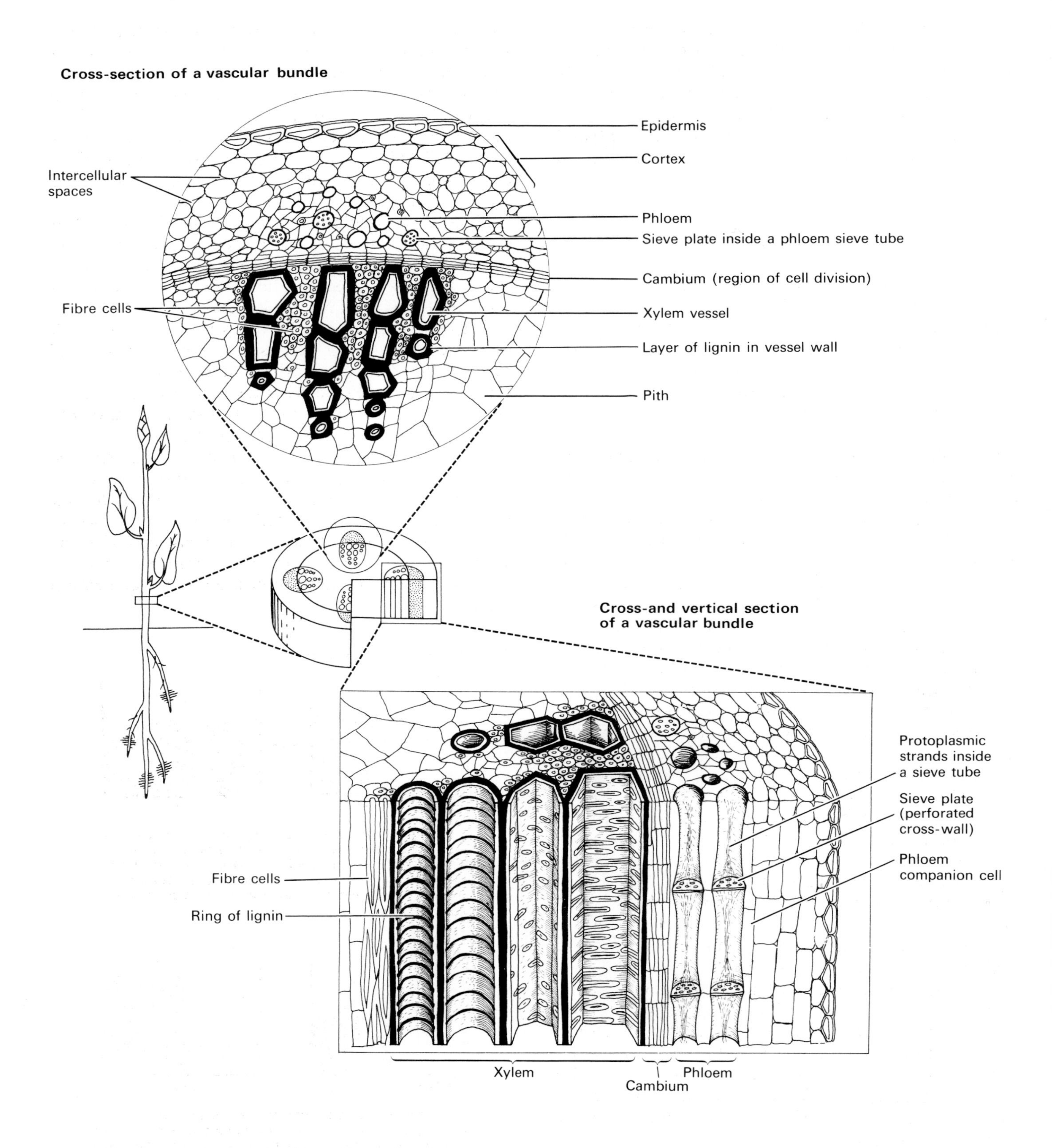

Fig. 7.2 Vascular tissue in a plant stem

In addition to the mechanical support provided by xylem and fibre cells (wood), plants are supported by the water in their tissues. Seedlings, non-woody plants, and soft parts like leaves and flowers would be limp were it not for the fact that their cells are inflated with water which makes them firm, like a tyre inflated with air. Water enters cells by a process called osmosis.

7.2 Osmosis and diffusion

Osmosis is a special kind of diffusion, so it is first necessary to explain what diffusion is.

Diffusion is the movement of molecules from where they are concentrated to where they are *less* concentrated. In technical terms, molecules diffuse down **concentration gradients** and the 'steeper' the gradient the faster diffusion occurs.

If you let a drop of ink fall gently into a cup of water, ink molecules will diffuse outwards from the blob of ink where they are concentrated into regions where they are less concentrated. At the same time water molecules will diffuse from pure surrounding water where they are concentrated into the ink blob where they are less concentrated. Diffusion will continue until the ink and water molecules are evenly dispersed throughout the cup.

Diffusion occurs for two reasons. First, there is a great deal of empty space between the molecules of all substances. This space is greatest in gases, much less in liquids, and least of all in solids. Second, all molecules are in a state of constant random movement so that they collide and intermingle all the time.

Think once more about a drop of ink in a cup of water. Both the ink and the water molecules are in constant motion which causes them to intermingle. The ink molecules move into the empty spaces between the water molecules and the water molecules move between the ink molecules.

In solids diffusion occurs extremely slowly, if at all, because the molecules are packed tightly together. Liquids and gases diffuse freely because their molecules are widely spaced.

Consider next the difference between *equal* volumes of pure water and a strong sugar solution. The water will have more water molecules (i.e. a high concentration) than the sugar solution because, in the sugar solution, sugar molecules take up some of the space. Similarly, a weak sugar solution will have a higher concentration of water molecules than a strong sugar solution.

With this in mind, think what will happen in the situation illustrated in Figure 7.3A. Here, a strong sugar solution is separated from a weak sugar solution by a membrane through which both water and sugar molecules can pass. Sugar will diffuse from the strong solution to the weak solution until it is uniformly distributed on both sides of the membrane. Likewise, water will diffuse from the weak to the strong solution until it is uniformly distributed on both sides.

Next consider what must happen in Figure 7.3B. A strong sugar solution is here separated from a weak sugar solution by a membrane that will allow water but *not* sugar to pass through. Such a membrane is said to be **semi-permeable**. Under these circumstances the sugar molecules cannot diffuse from high to low concentration because they are 'imprisoned' behind the semi-permeable membrane. Only the water molecules can move: they will diffuse from the weak to the strong solution until they are uniformly distributed on both sides of the membrane.

Water movements of this type are called osmosis. *Osmosis can be defined as: the diffusion of water molecules through a semi-permeable membrane from a weak to a strong solution.* (Although generally used to describe diffusion of water, osmosis can also be applied to the diffusion of any solvent across a semi-permeable membrane in response to a concentration gradient.)

Fig. 7.3 Diffusion and osmosis

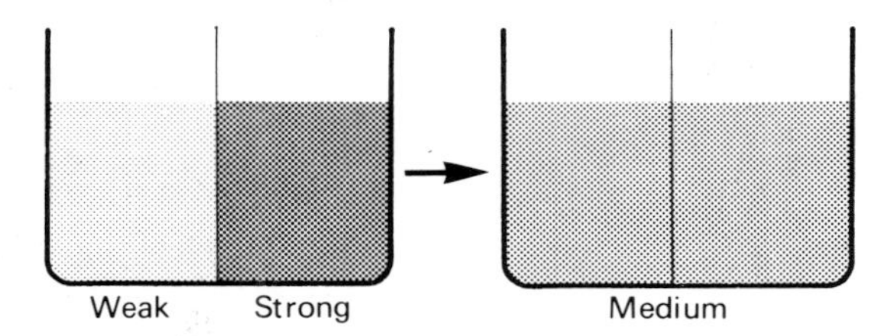

A Diffusion If a weak sugar solution is separated from a strong sugar solution by a membrane permeable to sugar and water, sugar will diffuse from the strong to the weak solution, and water will diffuse from the weak to the strong solution, until both solutions are of equal strength.

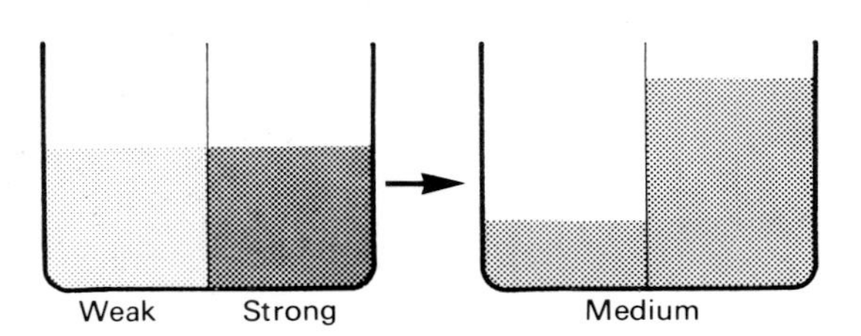

B Osmosis If a weak sugar solution is separated from a strong sugar solution by a membrane permeable only to water, water will diffuse from the weak to the strong solution until both solutions are of equal strength.

Theoretical explanation of osmosis

Figure 7.4 shows how to demonstrate osmosis with an **osmometer**, made with a commercially available semi-permeable membrane called Visking tubing (exercise B). This Figure illustrates the generally accepted theory that such membranes have microscopic pores through which small molecules like water can pass but larger molecules like sugar cannot. Thus, in an osmometer, Visking tubing acts as a molecular net, holding back large molecules like sugar but allowing small molecules like water to pass through.

It is believed that osmosis occurs in living cells because the cell membrane (described in section 2.1) is semi-permeable, having microscopic pores like those in Visking tubing. In fully-formed plant cells there is a second important semi-permeable membrane: a membrane called the **tonoplast** which lines the cell vacuole (Fig. 7.5). Consequently, water entering plant cells by osmosis passes through the cell membrane, the cytoplasm, and finally the tonoplast before entering the vacuole. It is customary, however, to treat the cytoplasm and the two membranes on either side of it as if they were a single semi-permeable membrane. When water reaches the vacuole it mixes with a liquid called **cell sap**.

Osmotic potential

The osmic potential of a solution can be defined as: a measure of the pressure with which water molecules could diffuse out of the solution if it were separated from another solution by a semi-permeable membrane.

Osmotic potential depends upon the amount of dissolved material (solute) in a solution. The greater the concentration of solute, the *lower* the osmotic potential of the solution. Pure water has the highest possible osmotic potential, and so it follows that addition of solutes lowers this potential.

In osmosis, therefore, water molecules will diffuse across a semi-permeable membrane from regions of high osmotic potential to regions of lower osmotic potential.

7.3 Osmosis in plants

Plant cells behave like osmometers: they absorb or lose water by osmosis depending upon the concentration of solutes in their sap. Cells with a low osmotic potential (high concentration of solutes) will absorb water from cells with a higher osmotic potential, and also from liquid which bathes all the cells of a plant, provided this liquid has a higher osmotic potential. The source of this liquid is water which plant roots take up from the soil.

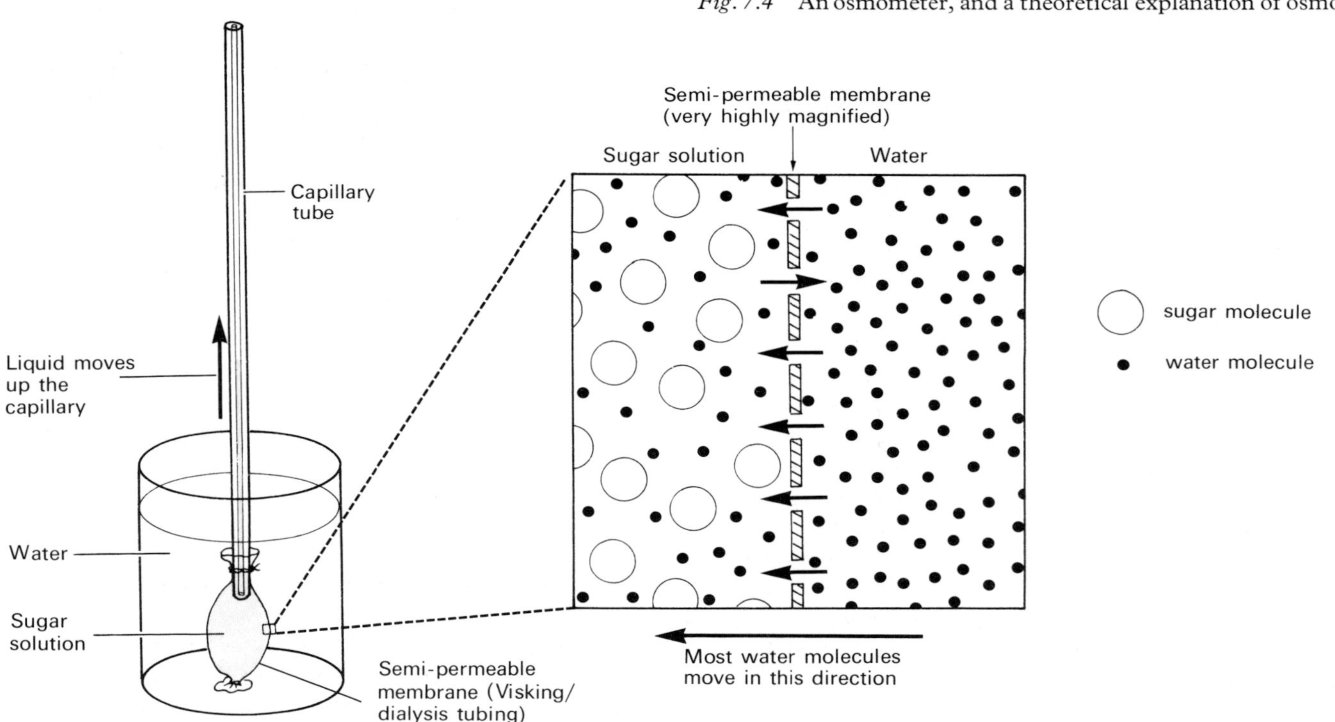

Fig. 7.4 An osmometer, and a theoretical explanation of osmosis

As cells take in water by osmosis they undergo several changes. The stream of water entering a cell causes it to swell, and this causes the cellulose cell wall to stretch slightly. When this wall can stretch no further it becomes taut and firm.

Wall pressure and turgor pressure
To be more exact, a fully stretched cell wall is exerting a restraining inward force called **wall pressure** on cell contents. This wall pressure is balanced by an equal but oppositely-directed force called **turgor pressure** in the cell contents, and when the cell wall can stretch no further the cell is said to be **fully turgid** (Fig. 7.5A).

As turgor pressure builds up it squeezes water molecules out of the cell and, when the cell is fully turgid, water leaves it under the influence of turgor pressure at the same rate as water enters it by osmosis.

Turgidity and support
Most of the time plant cells take in water not from each other but from the liquid which fills the spaces between cells and penetrates the tiny cavities between the cellulose fibres which make up cell walls. Since most cells are bathed in this liquid, and since it almost always has a higher osmotic potential than cell sap, it follows that the bulk of a plant will consist of fully turgid cells.

Turgid cells make a plant firm, maintain its shape, and allow it to function efficiently. All young seedlings, herbaceous (non-woody) plants, and plant structures like leaves and flowers, depend entirely upon the turgidity of their cells for support.

Plasmolysis
If a cell is surrounded by a liquid with a lower osmotic potential than its own sap, the cell loses water by osmosis. Turgor pressure drops to zero as water is drawn from the cell. The vacuole and cytoplasm contract, which usually pulls the cell membrane away from the cell wall (Fig. 7.5B).

The effects of water loss are called **plasmolysis**, and cells in this state are said to be **plasmolysed**. Plasmolysed cells are deflated and soft owing to lack of turgor pressure. The technical name for this condition is **flaccid**. Flaccid cells give no support to a plant and wilting occurs; that is, the leaves, flowers, and other non-woody tissues droop and become limp.

Plasmolysis occurs very rarely in nature. It can happen if land is flooded with sea water because this leads to a sudden steep decline in the osmotic potential of water surrounding roots. This, in turn,

A Cell becoming turgid as it takes in water by osmosis

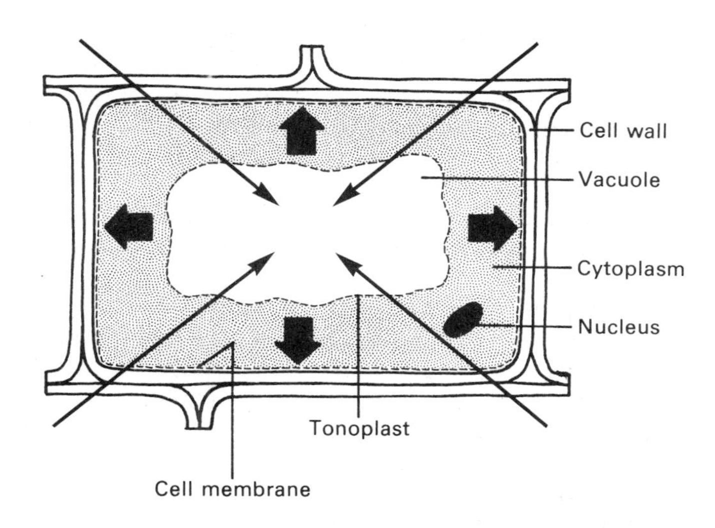

B Cell becoming plasmolysed as it loses water by osmosis

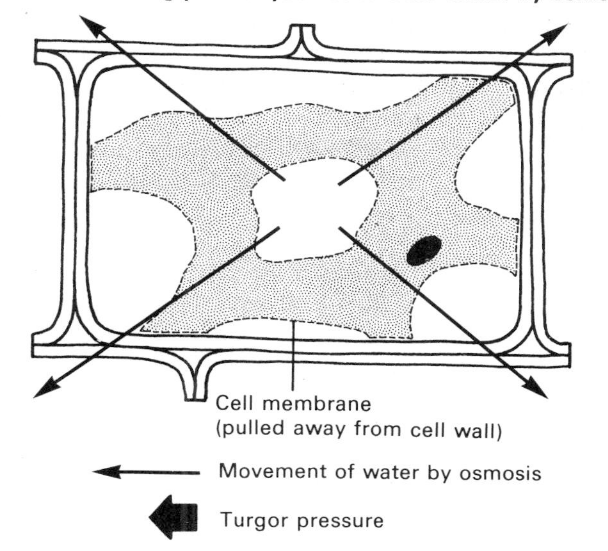

Fig. 7.5 Turgidity and plasmolysis. (Pores shown in the semi-permeable membrane are not drawn to scale.)

leads to loss of water by osmosis from the plants. Glasswort and other sea shore plants do not suffer in this way because their cell sap has a lower osmotic potential than sea water.

7.4 Flow of water through plants

Water flows in a continuous stream through a plant. It enters through the roots and flows up the root and stem xylem vessels to the leaves, where it evaporates from the surface of spongy mesophyll cells and diffuses out of pores in the leaves called stomata (Figs. 5.4 and 7.6).

Evaporation of water from leaves generates the main force which moves water through a plant. The technical name for evaporation of water from leaves is **transpiration**, and the flow of water through a plant which this causes is called the **transpiration stream**.

Transpiration and the transpiration stream

Water lost from a leaf by transpiration is replaced by more water which flows from leaf xylem vessels (veins). Water flows from leaf xylem vessels to the surface of mesophyll cells by two routes.

Most of the water flows directly through and around the porous cellulose walls of the mesophyll cells (Fig. 7.6). But a small amount of water is thought to flow by osmosis from cell to cell. Loss of water by evaporation from cells near stomata gives these mesophyll cells a lower osmotic potential than cells next to xylem vessels, where water is plentiful. In other words, a **gradient of osmotic potential** is created which is highest near the xylem and lowest in cells near stomata. Movement of water down this gradient is very slow because of the great resistance caused by the many cell membranes and layers of cytoplasm through which it must pass.

Flow of water up the xylem

Transpiration causes water to be continually removed from leaf xylem vessels. The result is a lower water pressure at these points than in root xylem vessels. This pressure difference causes water (and dissolved minerals) to be sucked up from the roots to the leaves, like lemonade being sucked up a drinking straw.

Flow of water from soil to root xylem

Water sucked up xylem vessels from the roots to the stem and leaves is replaced by water which flows from the soil through and around the porous cellulose walls of cells in the root cortex (Fig. 7.6). This flow of water continues as long as a plant transpires.

In spring, however, before leaf buds open, there is little or no transpiration. At this time water flows across the root to the xylem by osmosis. When there is no transpiration, mineral salts accumulate in root xylem. At certain concentrations these salts give xylem sap a lower osmotic potential than adjacent cells in the root cortex. As these cells lose water by osmosis to the xylem sap they develop a lower osmotic potential than root hairs and other root epidermal cells, since these are bathed in soil water. Thus, a gradient of osmotic potential is created which is highest in cells nearest the soil and lowest in cells next to the root xylem. Water flows from cell to cell by osmosis from the high to the low osmotic potential and then into the root xylem.

Root pressure

Water entering root xylem by osmosis causes a build-up of pressure in the xylem vessels. This is called **root pressure**, and is responsible for forcing water, and minerals, up into the stem in early spring before the transpiration stream is established.

Root hairs

These are extremely narrow tubes which grow out from individual cells on the root surface (Fig. 7.1). They develop in a narrow zone of the root close to the root tips. There are millions of root hairs on even quite small plants, and they give the plant a far greater surface area over which it can absorb materials from the soil than if the root surface were smooth.

In addition, root hairs stick to the soil particles between which they grow, holding the soil firmly in place around the root and thereby helping to anchor the plant in the ground.

Individual root hairs usually function for a few days and are then replaced by new ones which grow nearer the root tip as it grows through the soil. In this way the root hair zone is always in contact with new regions of soil.

7.5 Absorption of minerals by roots

It was once thought that mineral salts were absorbed in solution along with water which plants take in through their roots. But there is now considerable evidence that water and mineral salt absorption are independent processes.

Plants absorb minerals from the soil against concentration gradients; that is, until their cell sap contains a far greater concentration of them than is present in surrounding soil water. This suggests that neither osmosis nor simple diffusion can be responsible for mineral absorption since, by definition, these two processes could do no more than produce the same concentration of minerals in plants as there is in soil water.

The actual mechanism of mineral absorption is very complex. All that need be said here is that evidence suggests a mechanism which requires the expenditure of energy on the part of plants, and for this reason mineral absorption is described as an **active transport** mechanism.

For instance, it has been shown that the rate of mineral absorption varies according to the rate at which energy is released by respiration in plants. This is demonstrated by growing plants in solutions of minerals. If oxygen is bubbled through the solutions mineral absorption increases, but it slows down when the oxygen is turned off, and almost stops when chemicals which inhibit respiration are added to the solution.

7.6 Transpiration rate

Transpiration rate depends upon the same factors that govern evaporation rate. Although these factors are listed separately below it must be remembered that they affect transpiration simultaneously.

Humidity

In general, transpiration only occurs when there is a lower humidity level (concentration of water vapour) in the atmosphere than exists in the air spaces inside the leaves. Transpiration stops when the atmosphere is saturated with water vapour, and resumes when the air becomes drier. Anything which produces a change either in the humidity of the atmosphere or of the air spaces in leaves will alter transpiration rate.

Temperature

A rise in air temperature affects transpiration rate in two ways. First, it increases the capacity of air to absorb water from leaves. Second, it warms the water inside leaves making it evaporate more quickly. Direct sunlight has the same effect since it warms leaves to a higher temperature than the atmosphere. Transpiration is therefore generally faster on warm sunny days than on cold dull ones.

Wind

Air movements carry away water vapour from leaves and this prevents air around them becoming saturated with water vapour. Consequently, depending upon temperature and humidity, transpiration is faster on a windy day than in still air.

The best conditions for a high rate of transpiration are the same as those needed for drying laundry on a line: a warm, dry, sunny, windy day.

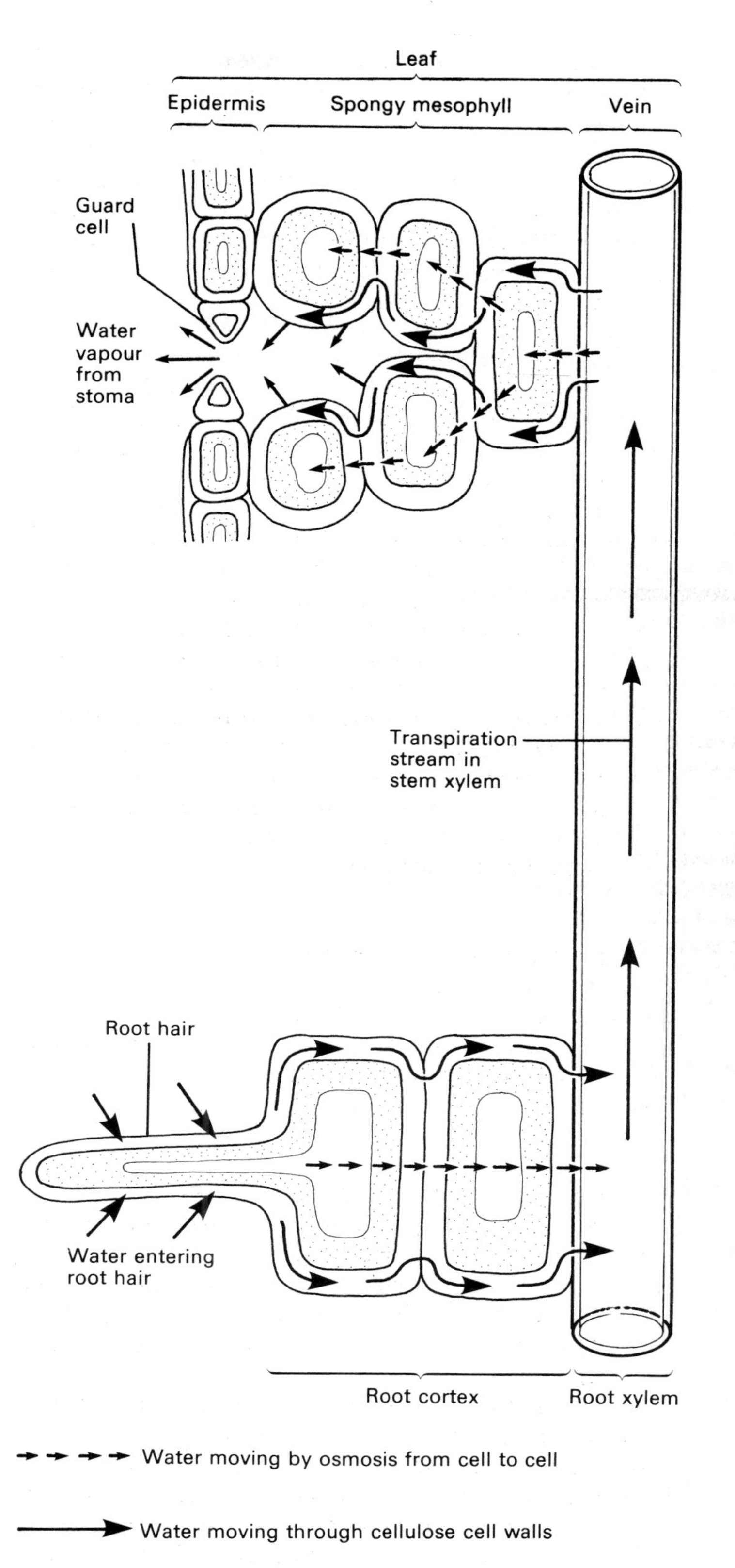

Fig. 7.6 Diagram summarizing water absorption, transpiration, and the transpiration stream

The importance of transpiration

Transpiration is important for at least three reasons. First, it results in the transport of water and minerals from the soil to the leaves where they form the raw materials of photosynthesis and other types of food manufacture. Second, it ensures that the walls of spongy mesophyll cells are kept moist, which is essential for the efficient absorption of the carbon dioxide needed for photosynthesis. Third, evaporation of water from a leaf has a cooling effect which helps prevent hot direct sunlight from damaging delicate cells.

On the other hand, transpiration brings far more water to the leaves than is needed for photosynthesis and, during drought, loss of water by transpiration may result in wilting which can kill a plant.

Apart from plants with special adaptations which prevent excess water loss in dry environments (Fig. 7.12 and comprehension test 3), it seems that most plants suffer from the unfortunate coincidence that most of those features which result in maximum photosynthetic efficiency (thin, flat leaf shape, large area of moist cell walls, many stomata, etc.) are also responsible for continuous rapid loss of water by evaporation from the leaf in warm, dry, windy conditions. However, there is evidence to suggest that two mechanisms protect the plant as it begins to wilt. First, the stomata close, which limits water loss to evaporation through the cuticle of the leaves and stem. Second, various chemical reactions increase the ability of roots to absorb water from the soil. In prolonged dry conditions, however, neither of these mechanisms can prevent the ultimate death of a plant.

7.7 Stomata

Stomata consist of two **guard cells** surrounding a central pore. Stomata are found all over the aerial parts of a plant, even on petals, anthers, and ovaries. In most flowering plants, however, they are concentrated on the under-surface of leaves (Fig. 7.7).

Generally speaking, stomata open in the light and close in the dark. It is almost certain that these movements are the result of variations in guard cell turgidity. It is known that each guard cell has a thick and relatively inelastic wall bordering the pore, and a thin elastic outer wall (Fig. 7.8). Experiments show that in the light guard cells take up water and become turgid, which causes their thin outer walls to bulge outwards, and the thicker inner walls to bend along with them. This gives the guard cells a curved banana-like shape, and opens the pore. In the dark the guard cells lose water, become flaccid, and the pore closes.

7.8 Translocation

Translocation is a term generally used to describe the transport of substances throughout the plant, but in particular it describes the movement of sugar and other manufactured materials from leaves to storage areas and growing points. Sugar, for example, moves from the mature leaves, where it is made during photosynthesis, upwards through phloem sieve tubes in the stem to developing leaves, flowers, and fruits. It also moves downwards into the growing root tips. Consequently, there are regions of stem in which food moves

Fig. 7.7 Stomata

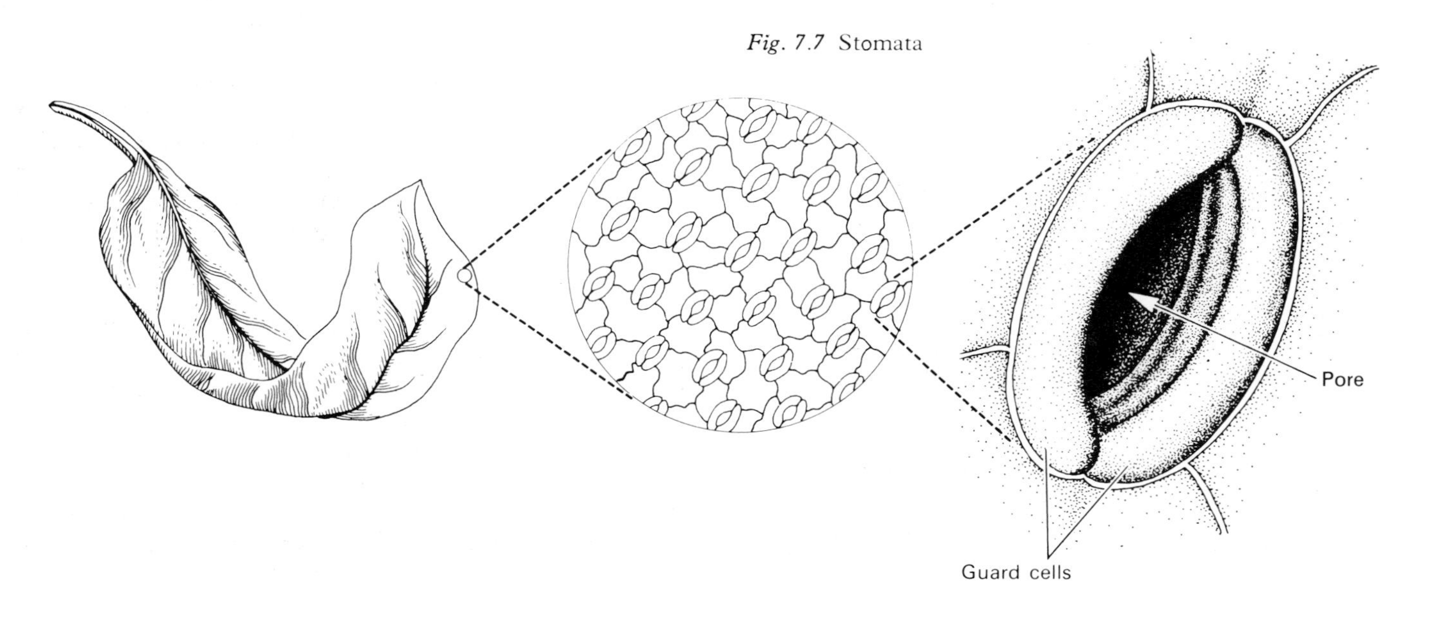

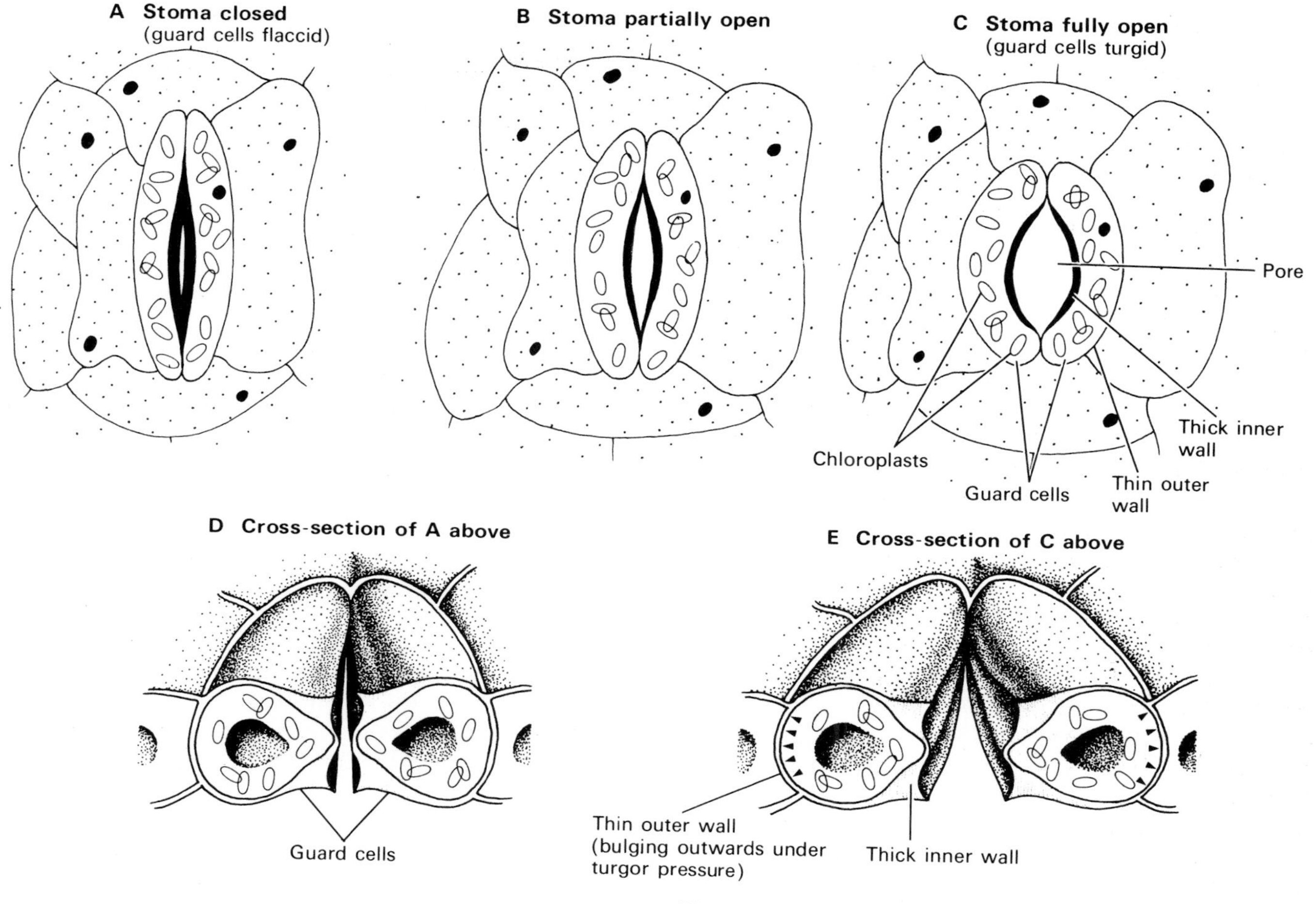

Fig. 7.8 How stomata open and close

Fig. 7.9 Girdling of a woody stem

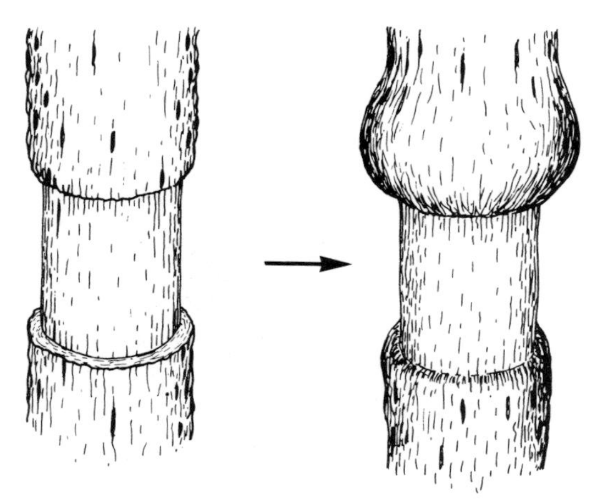

up and down at the same time through adjacent, and possibly even the same sieve tubes.

Evidence of food transport through phloem

Food substances such as sugar are necessary for the development of new tissue at the growing points of a plant. If these foods are transported through sieve tubes, then damage to or removal of the phloem will affect growth. The following experiments demonstrate this in various ways.

Girdling experiments A stem is girdled by completely removing a ring of bark and phloem from one or more places on the stem, leaving the xylem vessels unharmed (Fig. 7.9A).

1. When a girdle is cut around a stem just above ground level, root growth is suppressed and tissues immediately above the girdle swell, owing to the accumulation in them of food which would otherwise have gone to the root (Fig. 7.9B).

2. When a girdle is cut between the mature leaves of a plant and developing flowers, fruits, or leaves, further growth of these organs is suppressed.

Heat and cold treatments

1. The translocation of food from a leaf stops when its petiole is scalded with steam. Similarly, the girdling of a stem with steam gives the same result as the above-mentioned girdling experiments where tissue is removed.

2. Translocation of food from a leaf is considerably reduced by subjecting the petiole to a temperature of 1 °C.

It is argued that heat and cold treatments adversely affect the living sieve tubes but not the dead xylem vessels. The results suggest that food passes through the sieve tubes since, in these experiments, there is no sign of it by-passing the girdle via xylem vessels.

Use of radio-active isotopes It is possible to follow the translocation of sugar during photosynthesis by supplying a plant with carbon dioxide which contains radio-active carbon, i.e. $^{14}CO_2$. This procedure shows that sieve tubes are the major if not the only path for translocation of sugar.

Verification and inquiry exercises

A *To verify that diffusion can take place against the force of gravity*

1. Prepare a small quantity of agar jelly mixed with phenolphthalein indicator.

2. Fill a test-tube with the agar/indicator mixture, and when it has set invert the tube and fix it in a clamp so that its mouth is immersed in a beaker of dilute sodium hydroxide. Leave the apparatus for 1 or 2 days.

3. Phenolphthalein turns purple in alkaline conditions. Use this information to explain the result.

B *To verify that osmosis takes place through a semi-permeable membrane*

1. Prepare the osmometer apparatus shown in Figure 7.4 using a length of Visking/dialysis tubing as the semi-permeable membrane.

a) Cut a length of tubing–about 10 cm–and tie a knot in it at one end.

b) Fill the tube with strong sugar solution and tie it to a length of capillary tubing as illustrated.

c) Insert the tube in a beaker of water and observe the movement of sugar solution up the capillary.

C *An investigation of osmotic potential in potato cells*

1. Dissolve 324 g of sucrose (cane sugar) in 500 cm³ of water, then make up to 1 litre. This gives a 'molar' solution, i.e. 1 gram-molecule (mole) per litre, which is written M/1.

2. Set aside 20 cm³ of the M/1 solution, then dilute the remainder with water to produce 20 cm³ each of M/2, M/4, M/8, and M/16 solutions.

3. The osmotic potential of M/1 sucrose is 2450 kPa at 20 °C. Calculate the osmotic potential of each of the solutions prepared in 2. above.

4. Use a cork borer to obtain twelve potato cores from large potatoes. The cores should be as long as possible and cut to the same length (to the nearest millimetre).

5. Place a pair of cores in M/1 sucrose, a pair in each of the solutions prepared in 2. above, and another pair in distilled water.

6. Measure the length of the cores as accurately as possible every 15 minutes for 2 hours and again after 24 hours. Calculate the average length for each pair on these occasions.

7. *a*) Which pair becomes *more* turgid?
b) Which pair becomes flaccid?
c) Which pair does not change in length?

How can this pair of cores be used to tell you the osmotic potential of potato cell sap?

D *To observe plasmolysis in living cells*

Strip a small piece of coloured epidermis from a rhubarb petiole (leaf stalk) and mount it in M/1 sucrose solution on a microscope slide. The plasmolysis of cell contents may be clearly observed under low magnification owing to the movement of coloured cytoplasm in the cells.

Use Figure 7.5 as a guide. Restoration of turgor may be seen by washing the plasmolysed cells in water.

E *An investigation of root and stem structure*

1. *Root hairs*

Germinate some broad bean seeds by trapping them against the sides of a jar with a cylinder of blotting paper filled with damp sand. Examine the seedlings for root hairs.

2. *Section cutting*

a) Put a bean seedling from 1 above into a beaker with its root immersed in water coloured red with eosin dye. After the dye becomes visible in the petioles and leaf veins, cut very thin slices of root and shoot using the method shown in Figure 5.10. Examine these slices under a microscope looking for the structures shown in Figures 7.1 and 7.2, noting that xylem vessels are stained red with eosin.

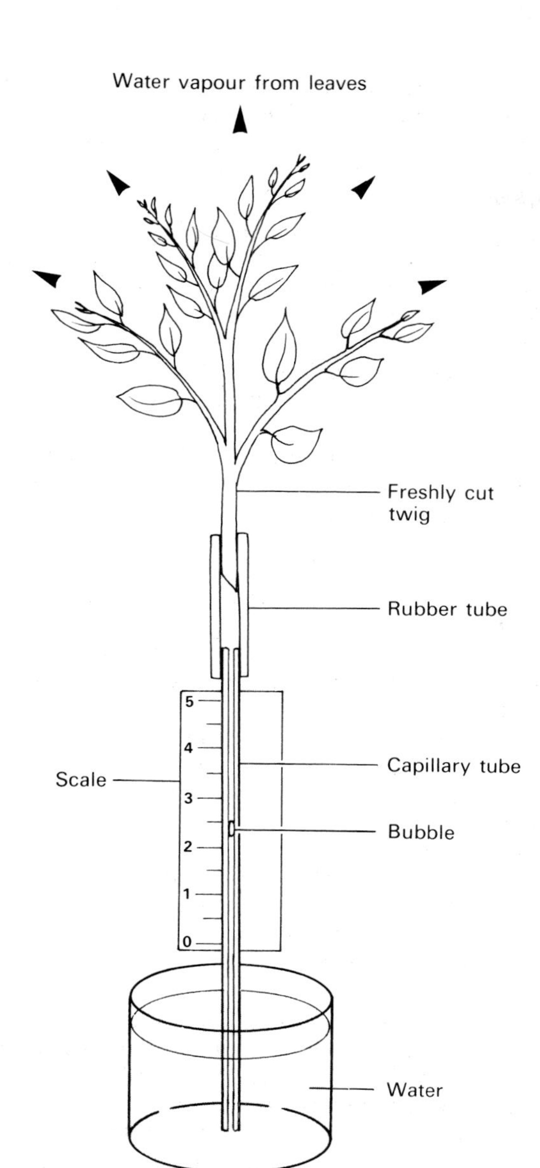

Fig. 7.10 A potometer (exercise F)

b) Cut sections of various plant tissues using the same method. Place the sections on a slide in a freshly made mixture of equal parts of phloroglucinol and dilute hydrochloric acid. This liquid will stain the xylem vessels bright red.

3. *Tissue maceration*

Warm small pieces of root, stem, and leaf separately in 10% potassium hydroxide solution for 5–10 minutes (handle this liquid carefully; it is very caustic). This procedure will separate the cells, and is known as maceration.

a) Put the macerated tissue in water on a slide and tease the cells from each other with mounted needles.

b) Stain the tissues as in 2(*b*) above.

c) Examine the shapes of all types of cell and vessel. Try to relate structure to function, using Figures 7.1 and 7.2 as a guide.

F *An investigation of transpiration rate*

1. Prepare the potometer apparatus shown in Figure 7.10. Be careful to keep the twig stem in water from the moment it is cut from the plant. This is necessary to prevent air entering its xylem vessels.

2. Allow a bubble to enter the capillary by:

a) removing the apparatus from the beaker of water;

b) pressing the rubber tube until a drop of water is squeezed from the capillary;

c) easing pressure on the rubber tube slightly to draw a little air into the capillary;

d) finally, replacing the apparatus in water. It is now ready for use.

3. Measure the time taken for a bubble to move along a specific length of scale when the potometer is:

a) indoors in the dark, and then in the light;

b) outdoors under different weather conditions;

c) next to an electric fan heater blowing hot, and then cool air (note that some unexpected results may be achieved);

d) arranged so that the leaves are covered with a plastic bag.

Between each of these operations return the bubble to the base of the capillary by: squeezing the rubber tube until the bubble is removed from the capillary; easing pressure on the rubber tube to draw water back into the capillary; and then repeating 2(*a*)–(*d*) above.

4. Explain how the results illustrate the explanation of transpiration given in the text.

G *An investigation of stomata*

1. *Stomatal structure*

Obtain a small piece of lower epidermis by roughly tearing leaves into pieces (e.g. privet, *Tradescantia*, or rhubarb). Mount the pieces in water on a slide under a cover slip. Observe under high magnification using Figure 7.7 as a guide.

2. *Epidermal 'prints'*

a) Smear a thin layer of latex rubber (e.g. 'Copydex' adhesive) on the lower epidermis of a leaf and allow it to dry completely.

b) Press a strip of clear self-adhesive tape (e.g. 'Sellotape') firmly on to the dry latex. When the tape is peeled off the leaf the latex comes with it, bearing a perfect imprint of the leaf surface structure including stomata.

c) Press the tape on to a slide and observe the imprints under low and high magnification.

d) Make prints of several different leaves, and of different regions on the same plants. Estimate the number of stomata per square millimetre in each region examined.

3. *Stomatal behaviour*

a) Fill two dishes with water and on each dish float a plant leaf with its lower side uppermost.

b) Place one dish under a bench lamp and the other in the dark for 30 minutes, then take a strip of lower epidermis from each and plunge them into absolute alcohol. Alternatively, obtain a latex 'print' from each using method 2 above.

c) Look at 25 different stomata on each strip and estimate what percentage of stomata are open and what percentage are closed.

d) What do the results reveal about the effect of light on stomatal behaviour?

H *An investigation of water loss through stomata*

1. *Cobalt chloride method* Cobalt chloride is blue when dry, and turns pink in the presence of water.

a) Use a strip of 'Sellotape' to stick a small piece of dry cobalt chloride paper to the upper and lower surface of a leaf which is still attached to a plant.

b) How do the results help to confirm that more water is lost through the lower epidermis than the upper epidemis of a leaf?

c) What control is necessary in this experiment?

2. *Vaselined leaves* Vaseline smeared on the surface of a leaf seals the stomata and prevents water loss. Devise a controlled experiment using this technique to show that water is lost through only one particular leaf surface in most angiosperm leaves. Clues:

a) At least four leaves are required.

b) Detach them from a plant, leave one without vaseline, and treat the remaining three with vaseline in different ways.

c) Hang the leaves by their petioles from a length of cotton and observe the rate at which they dry and shrivel over 1 or 2 days.

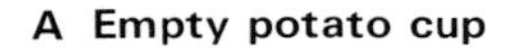

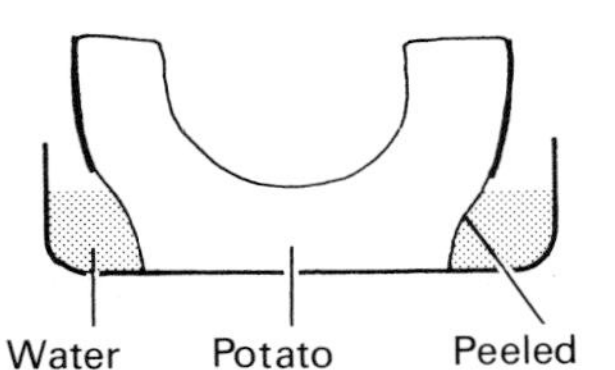

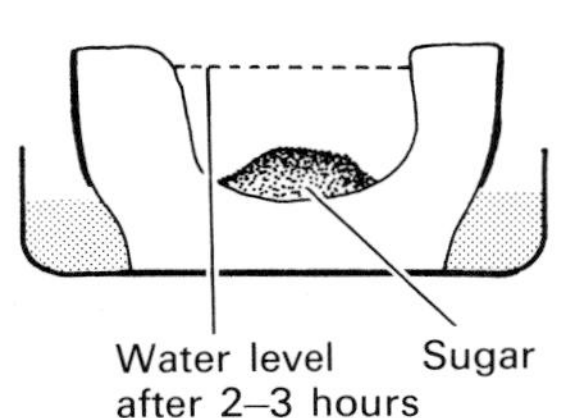

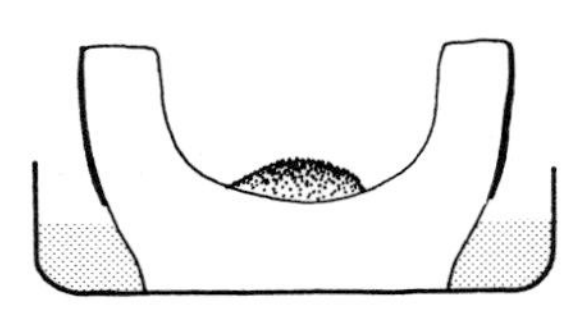

Fig. 7.11 How to demonstrate osmosis using potato cups (see comprehension test 2)

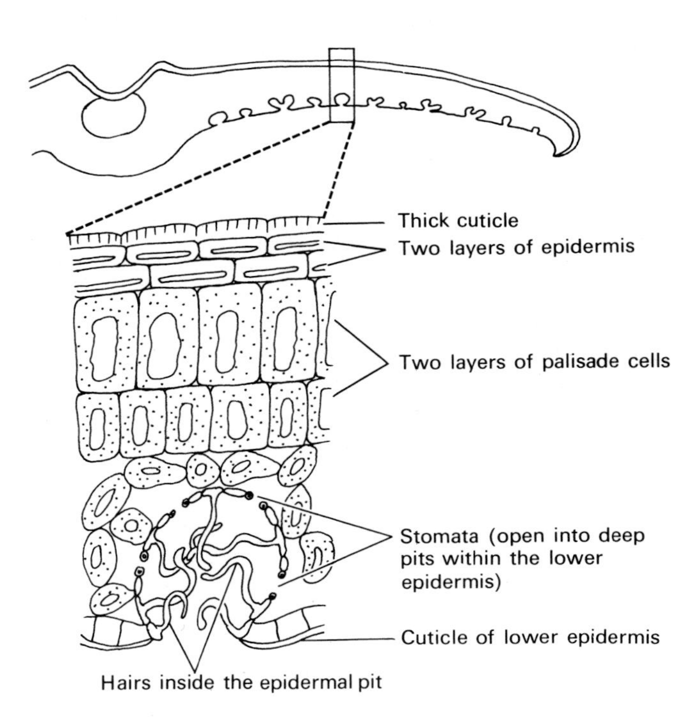

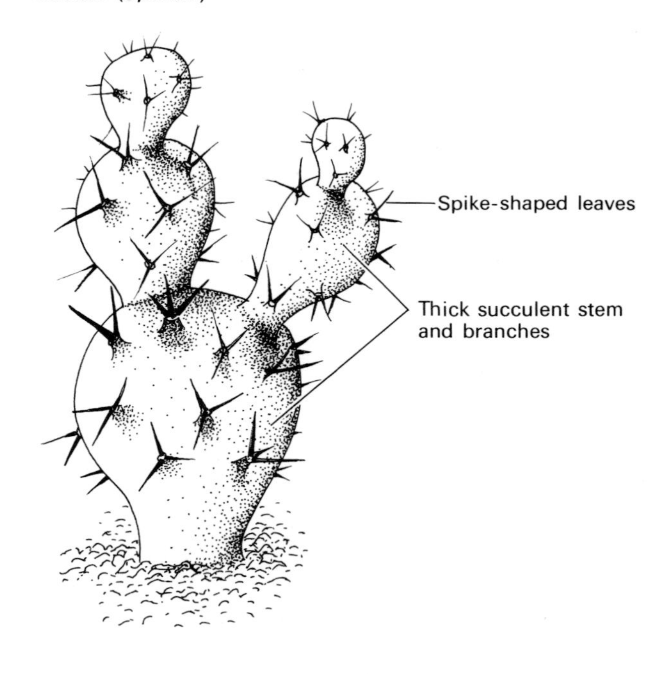

Fig. 7.12 Examples of adaptations which enable plants to survive in dry climates (comprehension test 3)

Comprehension test

1. Most garden plants wilt if watered with strong salt solution. Explain the part which osmosis plays in this result.

2. Carry out the osmosis experiment described in Figure 7.11.

a) Explain in detail why water accumulates in the hollowed out portion of potato B.

b) Explain why water does not accumulate in the hollowed out portions of potatoes A and C.

c) Why is potato A necessary in this experiment?

3. Study Figure 7.12. This illustrates two plants with adaptations which enable them to survive in conditions where water is usually in short supply.

a) List at least three ways (there are more) in which the *Oleander* leaf is adapted to restrict water loss by evaporation through its cuticle and transpiration through its stomata.

b) What is the advantage of a two-layered palisade in *Oleander*? (Clue: what effect will the thick cuticle and two-layered epidermis have on light penetration?)

c) In the course of evolution, the leaves of *Opuntia* have become reduced to non-photosynthetic spines. How does this help to restrict water loss by transpiration?

d) Since *Opuntia* has no leaves, what part of the plant carries out photosynthesis?

e) What other characteristics illustrated in the figure enable *Opuntia* to survive in dry conditions?

4. Study the apparatus illustrated in Figure 7.13 and, if possible, set it up and observe it in operation. (*Note:* This experiment should be carried out in a fume cupboard.)

a) To what part of a plant is the atmometer's porous porcelain equivalent?

b) To what part of a plant is the glass tube equivalent?

c) Describe the various atmospheric conditions which affect the rate at which mercury rises up the glass tube, and explain why these same conditions should also affect the rate at which water rises up plant xylem vessels.

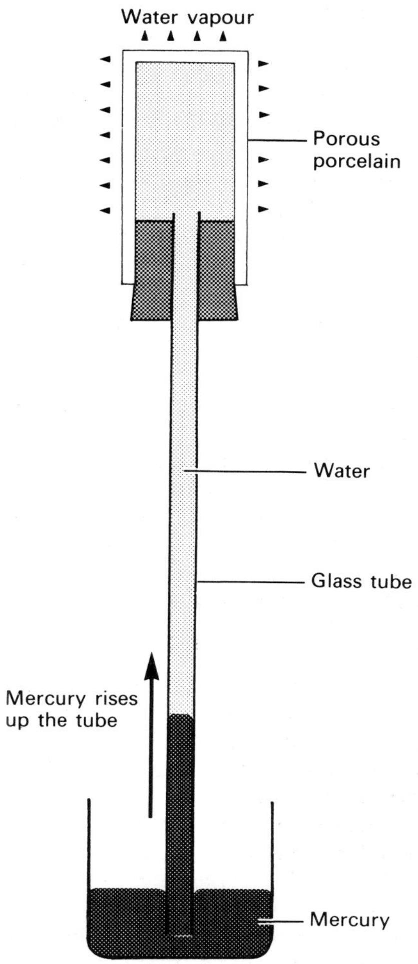

Fig. 7.13 An atmometer (comprehension test 4)

Water evaporates from the surface of a porous porcelain cylinder. This causes mercury to rise up the glass tube

Summary and factual recall test

Xylem (1) form a transport system in which (2) and (3) move from the roots to the (4). Phloem (5) tubes transport substances such as (6) from the (7), where they are made, to (8) and (9).

Osmosis is the (10) of (11) molecules through a (12) membrane from a (13) to a (14) solution. A cell taking in water by osmosis inflates and presses outwards against its (15) cell wall with a force called (16) pressure. This force is opposed by (17) pressure, and when the two forces are equal the cell is fully (18). Osmosis stops at this stage because (19). The following are supported by the turgidity of their cells (20—list four examples).

A cell surrounded by liquid with a lower osmotic potential than its sap will (21) water by osmosis. The cell membrane will pull away from the cell wall, a condition called (22).

Transpiration is the loss of (23) by (24) from the (25) mesophyll of leaves. Water vapour diffuses into the atmosphere through (26) which occur on the (27)-surface of most leaves. These consist of two (28) cells surrounding a pore. Transpiration rate varies according to (29—list three factors). Transpiration is useful because (30—give three reasons), but can harm plants in conditions of (31).

Water lost from the leaves is replaced by water from (32) vessels, which produces a flow of water from root to leaves called the (33) stream.

8 Respiration

The word respiration is derived from the Latin *respirare* which means to breathe. At first this term referred to the breathing movements which cause air to be drawn into and pushed out of the human lungs, but now, when defined with strict accuracy, respiration means something entirely different.

The modern definition of respiration is: the processes which lead to, and include, the chemical breakdown of materials to provide energy for life. These processes occur inside the living cells of every type of organism.

To avoid unnecessary confusion, it is strongly recommended that respiration be used only in this modern sense, so that it is clearly distinguished from the mechanism of breathing, which is concerned with the absorption of oxygen from the air. Breathing and related mechanisms are described in chapter 9. This chapter offers a brief introduction to the ways in which energy is released by the chemical breakdown of body materials.

8.1 The release of energy

The energy for life is released during respiration from substances known loosely as 'food'. There are many different foods, and they are taken into the body in many different ways, but in the majority of organisms all foods are converted into glucose sugar before they are used as a source of energy. For the sake of simplicity, the following descriptions refer to the respiration of glucose.

In most organisms energy is released by a process called **aerobic respiration**, which requires a continuous supply of oxygen molecules obtained from the air or water surrounding the organism. In certain circumstances, however, energy can be released without the use of oxygen molecules. This is known as **anaerobic respiration**. These two different but related types of respiration are described in the following sections.

Aerobic respiration

The aerobic respiration of glucose is summarized by the following chemical equation:

$$\underbrace{C_6H_{12}O_6}_{\text{glucose}} + 6O_2 \longrightarrow 6CO_2 + 6H_2O + 2898\text{ kJ of energy}$$

Aerobic respiration releases all the available energy within each glucose molecule; that is, it produces the same amount of energy that is released when glucose is burnt in oxygen gas.

The chemical equation above gives the false impression that respiration involves only one chemical reaction, because it shows only the raw materials and end-products of respiration. The whole process involves a sequence of some fifty separate reactions, each catalysed by a different enzyme. The result is a controlled release of energy which is far more useful to the organism than a sudden explosive burst of energy.

Look again at the equation, and see what happens to all the hydrogen atoms contained within a glucose molecule. Eventually these atoms combine with oxygen atoms to form water. In fact, the bulk of respiratory energy becomes available to the organism as hydrogen atoms are removed from glucose during respiration. This process is catalysed mainly by **dehydrogenase enzymes**. In other words, the oxygen which an aerobic organism has absorbed combines with hydrogen atoms from glucose or other foods to produce water, which may be excreted from the body.

Anaerobic respiration

Anaerobic respiration differs from aerobic respiration in three important ways. First, anaerobic reactions break down glucose in the absence of oxygen. Second, anaerobic reactions do not completely break down glucose into carbon dioxide and water but into intermediate substances such as lactic acid or alcohol. Third, anaerobic respiration releases far less energy than aerobic respiration, because glucose is not completely broken down.

During periods of maximum effort the heart pumps blood at 34 litres per minute, and this delivers oxygen to the muscles at 4 litres per minute. But this is insufficient to meet the oxygen requirements of the muscles. They change over to anaerobic respiration and produce lactic acid instead of carbon dioxide and water, and begin to incur an oxygen debt

Organisms which respire anaerobically are called **anaerobes**. Certain bacteria are complete anaerobes. They live permanently in conditions where no oxygen exists and rely entirely upon anaerobic respiration for energy. Some of these bacteria are actually poisoned by oxygen, even in small quantities. Many organisms are partial anaerobes, in which case their cells are capable of carrying out both types of respiration, either separately or at the same time.

Anaerobic respiration in micro-organisms Micro-organisms such as yeast and certain bacteria obtain most of their energy by a form of anaerobic respiration called **fermentation**. Typical products of fermentation are alcohol (ethanol) which is formed by yeast; and citric, oxalic, and butyric acids which are formed by certain bacteria. These chemicals are of great commercial value, and their production, with the help of micro-organisms, is now a major industry.

Many types of yeast are used in alcoholic fermentation. The equation for the fermentation of glucose is as follows:

$$C_6H_{12}O_6 \longrightarrow \underbrace{2C_2H_5OH}_{\text{ethanol}} + 2CO_2 + 210\,\text{kJ of energy}$$

Compare this with the equation for aerobic respiration of glucose, and note two things. First, water molecules are not released in fermentation. This is because the reactions do not involve the removal of hydrogen atoms from glucose and their subsequent combination with oxygen. Second, very little energy is released in fermentation compared with aerobic respiration.

Yeast is not completely anaerobic, and in the production of alcoholic drinks aerobic conditions are maintained for some time so that yeast cells can carry out both types of respiration. In these conditions they grow rapidly and reproduce.

The type of alcoholic drink produced by fermentation depends largely upon the source of the sugar solution used. Fermentation of apple juice produces cider, grape juice produces wine, and malt extract from germinating barley produces beer. Distillation of certain fermentation products gives rise to much stronger alcoholic solutions called spirits. Brandy is a spirit produced by distilling wine.

The equation for anaerobic respiration shows that carbon dioxide is a product of alcoholic fermentation. In the making of bread, bakers' dough 'rises' because the yeast mixed into it produces carbon dioxide gas which fills the dough with bubbles as it escapes.

Anaerobic respiration in plants Green plants can respire anaerobically for short periods. The time limit for this is determined by the rate at which alcohol accumulates in their tissues, since this substance is poisonous in high concentrations.

The ability of plants to live as temporary anaerobes allows them to survive in conditions where animals would quickly die of suffocation. When flooding occurs, for example, plants can survive for several days completely immersed in water, and for several weeks in waterlogged, airless soil. Anaerobic respiration is also necessary in the initial stages of germination, when the plant embryo is completely enclosed within an air-tight seed coat.

Anaerobic respiration in vertebrate muscle
Breathing rate and heart beat increase during exercise but there is a limit to the speed at which they can

deliver oxygen to muscles and this, in turn, limits the speed at which aerobic respiration can supply energy to muscles.

However, muscles can respire without oxygen (anaerobically) for a short time. During strenuous exercise, when oxygen supplies are no longer sufficient to meet energy demands, the rate of anaerobic respiration increases rapidly in muscles and this allows them to work faster than if they relied on aerobic respiration alone. Under these circumstances muscles obtain most of their energy anaerobically, but this type of respiration produces lactic acid instead of carbon dioxide and water:

$$C_6H_{12}O_6 \longrightarrow \underbrace{2CH_3CH(OH)COOH}_{\text{lactic acid}} + 150\,\text{kJ of energy}$$

Lactic acid accumulates in muscles eventually preventing further contraction.

The body of a trained athlete can tolerate up to 127 g of lactic acid before it prevents further effort, and for every 10 g of acid the body must absorb 1.7 litres of oxygen in order to break the acid down into carbon dioxide and water. This oxygen requirement is called the **oxygen debt**, because the body has 'spent' energy in excess of oxygen absorption, and now requires oxygen to 'pay' the debt.

The oxygen is used to break down one-sixth of the lactic acid into carbon dioxide and water, which releases enough energy to convert the remaining five-sixths back into glucose.

Research into the chemistry of respiration has led to the discovery that anaerobic and aerobic respiration are closely linked. In fact, respiration in the majority of organisms consists of both anaerobic and aerobic reactions, which occur in the following order. First, glucose takes part in a short sequence of anaerobic reactions (i.e. requiring no oxygen). This stage of respiration produces little energy and involves the breakdown of sugar into an intermediate substance such as lactic acid or alcohol, depending on the organism. The second stage of respiration involves a longer sequence of aerobic reactions (i.e. requiring oxygen) in which the intermediate substance is broken down into carbon dioxide and water with the release of far more energy than is produced by the first stage.

In the absence of oxygen, certain organisms can temporarily suspend the second (oxygen-consuming) stage of respiration in all or part of their bodies, and exist for a short time by the first (anaerobic) stage alone.

8.2 The utilization of respiratory energy

Cells do not use energy as soon as it is released from respiration: the energy is used to build up a temporary energy store, which takes the form of a chemical called **adenosine triphosphate**, or ATP for short.

Molecules of ATP may be thought of as 'go-betweens' because they transfer energy from the chemical reactions which release it to the muscles or other tissues of the body which make use of it. There are two important advantages in having ATP as a go-between. First, ATP molecules release their energy the instant it is required without having to go through fifty different reactions, each controlled by a separate enzyme, which happens in respiration. Second, ATP releases precisely controlled amounts of energy, because each ATP molecule has a specific energy value. The following greatly simplified account of ATP and Figure 8.1 should help to make these points clear.

ATP is formed during respiration from a related substance called **adenosine diphosphate**, or ADP. It takes about 34 kJ of energy to transform one mole of ADP into one mole of ATP, and exactly this amount of energy is released for use in metabolism when one mole of ATP is broken down once more to ADP.

There are four main advantages to the ADP/ATP system. First, ATP takes up some energy which would otherwise have been lost as heat during the breakdown of glucose by respiratory enzymes. Second, energy is released from ATP the instant it is required without the body having to go through the fifty different reactions of respiration. This is important when sudden bursts of energy are required. Third, ATP delivers energy in precise amounts (34 kJ per mole of ATP). Fourth, energy can be transferred from ATP to other substances without any loss, which converts relatively inert substances into highly reactive ones. This is vital in the synthesis of complex chemicals out of simpler raw materials.

Verification and inquiry exercises

A *To verify that respiration produces heat*

1. *Heat production during germination*

a) Obtain sufficient seeds (e.g. wheat, barley, or small peas) to fill two vacuum flasks, and soak them in water for 12 hours.

b) Divide the seeds into two equal portions. Boil half of them for two minutes to kill them. Wash both portions in 10% formalin to prevent the growth of bacteria and fungi.

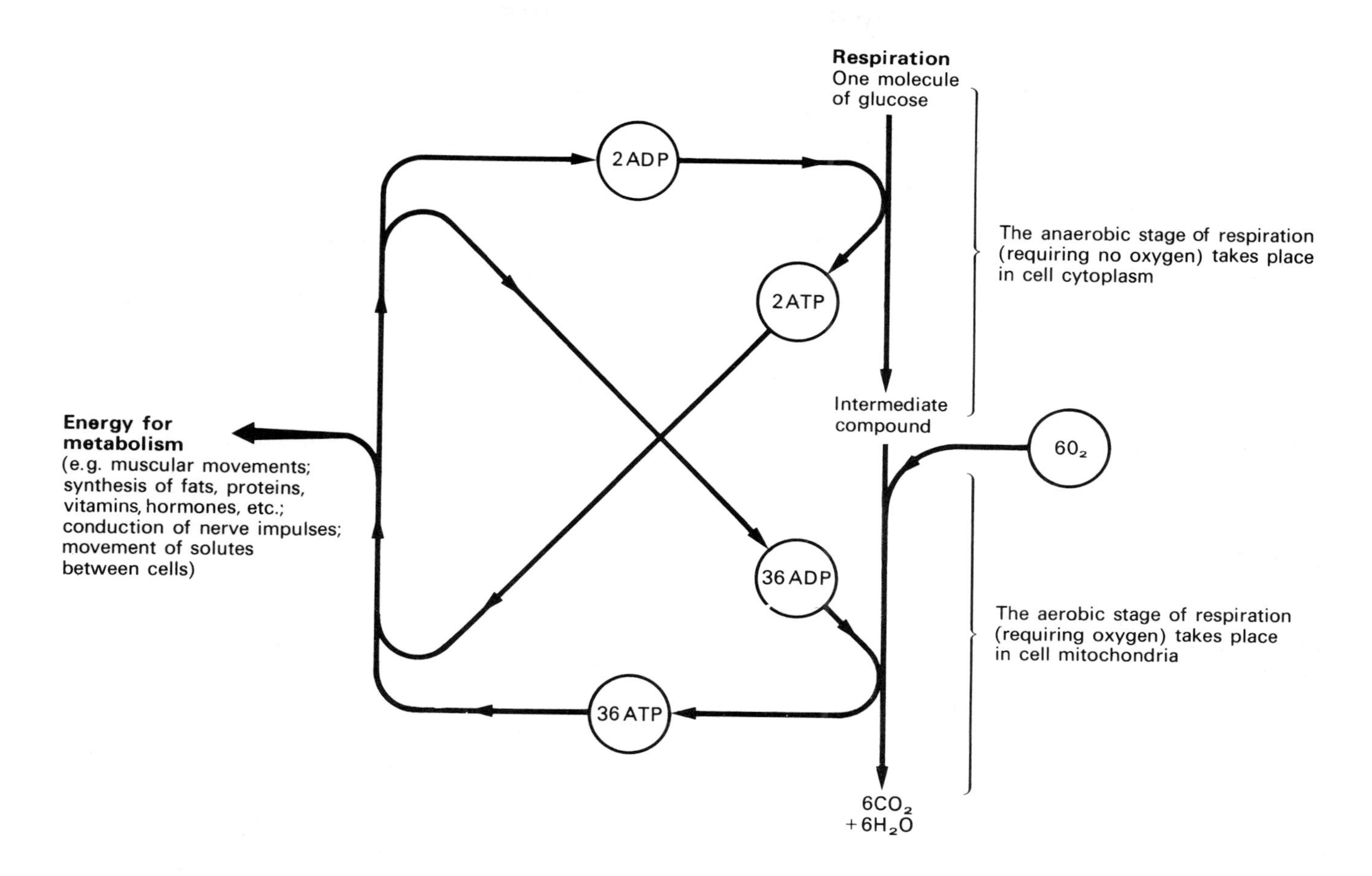

Fig. 8.1 Diagram summarizing the transfer of energy via ATP from respiration to metabolism

c) Place the boiled seeds in one vacuum flask and the live seeds in the other. Plug the mouth of each flask firmly with cotton wool, then insert a thermometer through each plug deep into the flask. Finally, mount the flasks upside-down in clamps.

d) Record any temperature changes over two or three days.

e) Why are boiled seeds included in the experiment?

f) Why are the flasks inverted, and why is cotton wool preferable to a rubber bung? (Clue: heat rises and carbon dioxide is heavier than air.)

g) Why does the temperature rise in one flask and not in the other? What is the source of this heat?

h) Why would a growth of fungi on the seeds interfere with experimental results?

2. *Heat production by animals*

a) Obtain sufficient maggots (fly larvae) to fill two vacuum flasks. (Maggots are obtainable from shops which sell bait for fishermen.)

b) Kill half of them by boiling, then proceed as in 1(*b*) above omitting the formalin treatment and inversion of the flasks.

c) Results should be obtainable after 5 or 10 minutes.

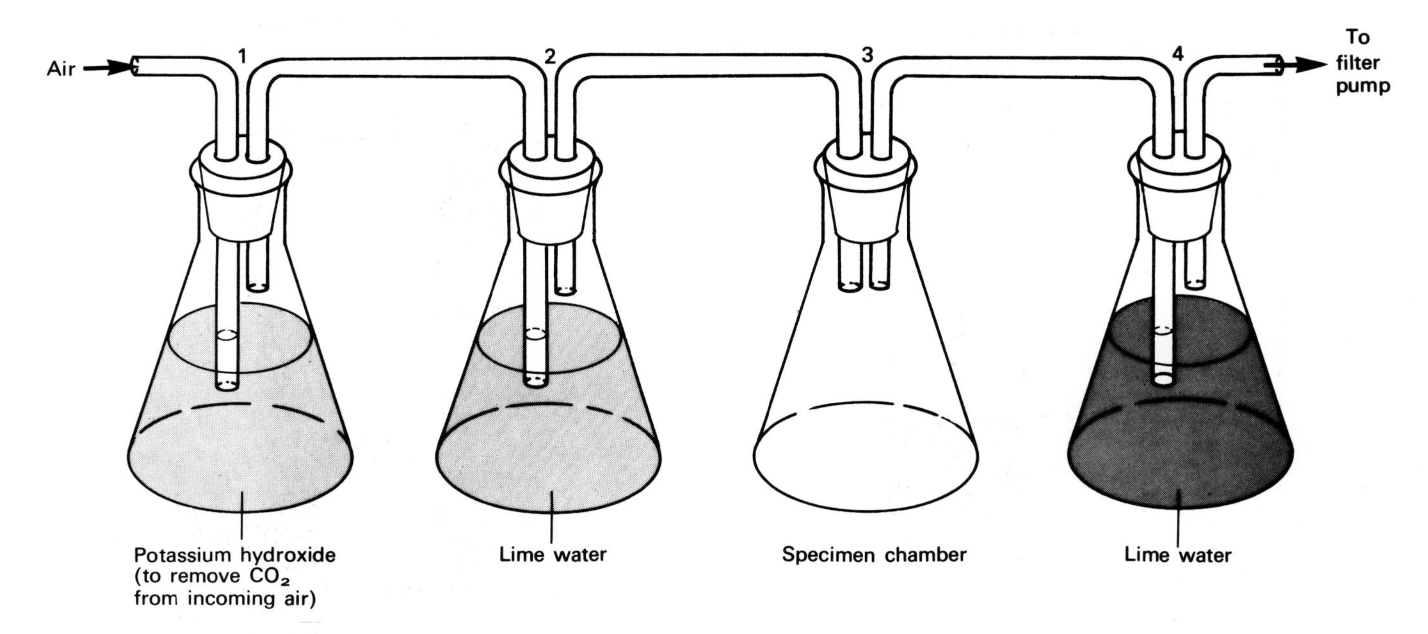

Fig. 8.2 To verify that carbon dioxide is produced during respiration (exercise B)

B *To verify that carbon dioxide is produced during respiration*

1. *To verify that plants and land animals produce carbon dioxide*

a) Set up the apparatus shown in Figure 8.2.

b) Almost any small organisms can be placed in the specimen chamber: wood-lice, maggots, earthworms, germinating seeds, mushrooms or other large fungi, and small plants. (If plants are used place the apparatus in the dark.)

c) Draw air through the apparatus, noting the milky precipitate which develops in flask 4, owing to the effects of carbon dioxide gas on lime water.

d) Why is flask 2 necessary?

2. *Simplified alternatives to B1 above*

a) Prepare the apparatus in Figure 8.3 for use with small animals. Use an identical apparatus without a specimen as a control.

b) Prepare the apparatus in Figure 8.4 to find out whether all parts of a plant produce carbon dioxide. Wrap pieces of root, stem, leaves, and flowers separately in gauze, and hang them inside the apparatus in the manner illustrated. When testing green parts of plants place the flask in a dark cupboard. Prepare an identical flask without a specimen as a control.

C *To verify that organisms consume oxygen during respiration*

1. Prepare the apparatus in Figure 8.5. Almost any

Fig. 8.3 Simplified version of apparatus in Figure 8.2 (exercise B2 (a))

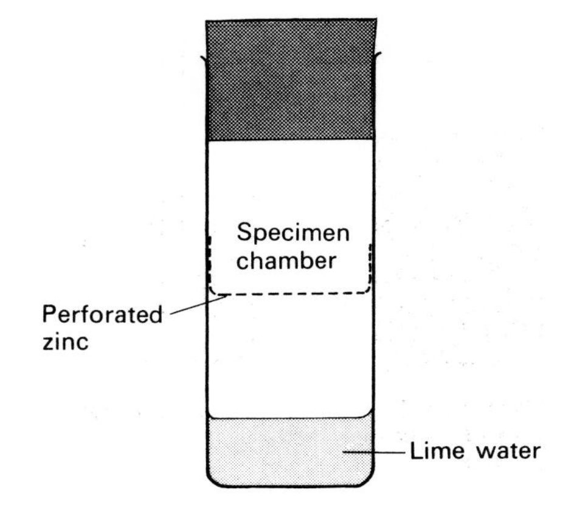

Fig. 8.4 To verify that plants produce carbon dioxide (exercise B2 (b))

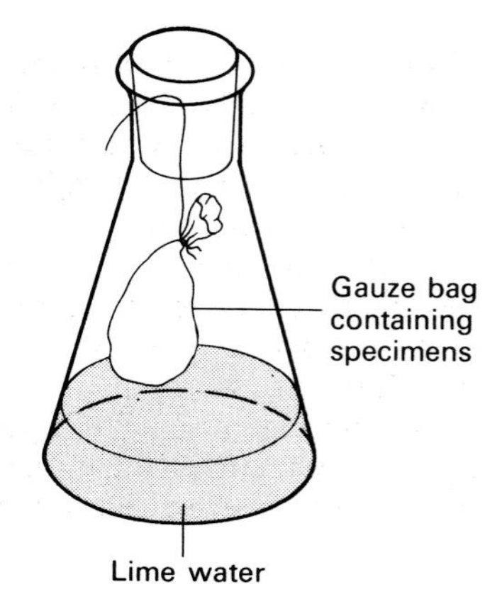

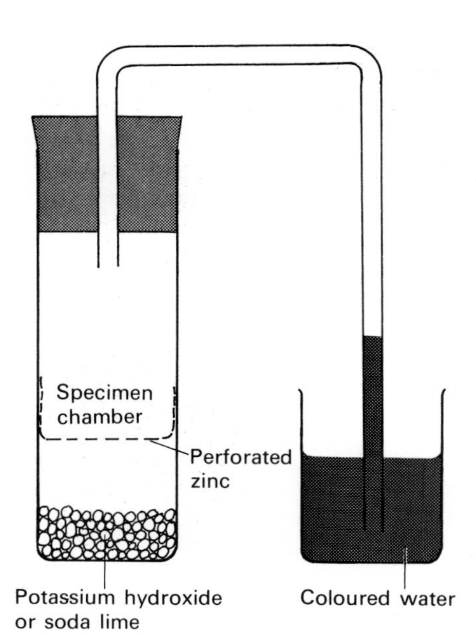

Fig. 8.5 To verify that organisms consume oxygen (exercise C)

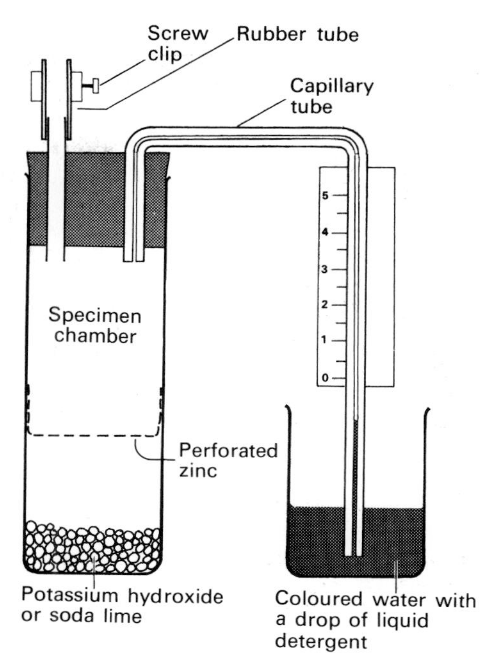

Fig. 8.6 To measure respiratory rate (exercise D)

small organism can be put into the flask (see B1 (b) above). In addition, try a thick suspension of yeast cells in sugar solution. Spread the suspension over crumpled filter paper in the flask.

a) Organisms take in oxygen and produce carbon dioxide at about the same rate. Carbon dioxide is absorbed by the potash (or soda lime) therefore oxygen consumption by the organism reduces air pressure inside the flask, which draws coloured water up the capillary.

b) Does this experiment actually *prove* that oxygen is consumed by the specimens?

c) How would the result differ if potash were omitted from the flask?

D *An investigation into the effect of temperature changes on respiratory rate*

1. Prepare the respirometer apparatus shown in Figure 8.6. This is a development of that shown in Figure 8.5, and works on the same principle (see C1(a) above).

a) Place different kinds of organisms (one type at a time) in the apparatus (see B1(b) above), and leave it, with the screw clip open, in a water bath at 20°C for 5 minutes. After this time close the screw clip and measure the time it takes for fluid to be drawn up a specific length of capillary tube. Repeat this procedure at the same temperature two or three times to obtain an average result. At the end of this stage, remove the rubber bung to let fresh air reach the organisms.

b) Repeat (a) above at 0°C, 10°C, 30°C, and 40°C, using the same organisms, and allowing them 10 minutes to adjust to each new temperature before closing the screw clip.

c) Graph the results, and if the capillary bore is known calculate the volume of oxygen consumed per hour per gram of the organism's body weight, at each temperature.

d) How do these results relate to findings from exercise C in chapter 6, concerning the effects of temperature on blood flow?

Fig. 8.7 To verify anaerobic respiration in yeast (exercise E1)

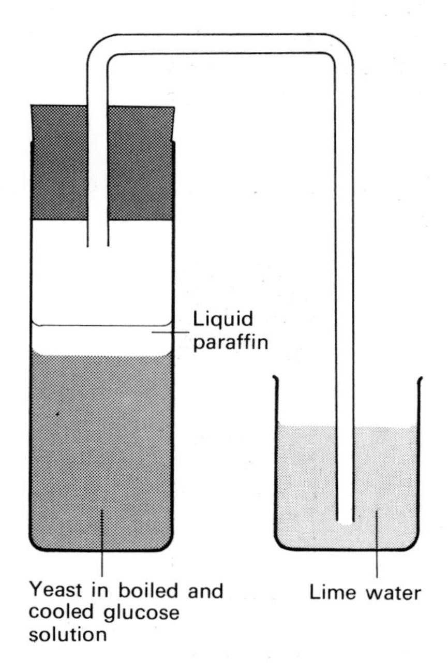

e) Why does temperature affect oxygen consumption in cold-blooded organisms? Is it likely to have the same effect on mammals? Explain your answer.

E *To verify that yeast and plant tissues respire under anaerobic conditions*

1. *Anaerobic respiration (fermentation) in yeast*

a) Prepare the apparatus shown in Figure 8.7, using 5% glucose solution which is boiled and cooled before adding a small amount of dried yeast. Finally, cover the yeast suspension with a layer of liquid paraffin.

b) How does boiling the glucose and then covering it and the yeast with oil produce, and maintain, anaerobic conditions?

c) Design an appropriate control for this experiment.

d) A larger scale version of this experiment (without the liquid paraffin) will produce enough alcohol for distillation.

Fig. 8.8 To verify anaerobic respiration in green peas (exercise E2)

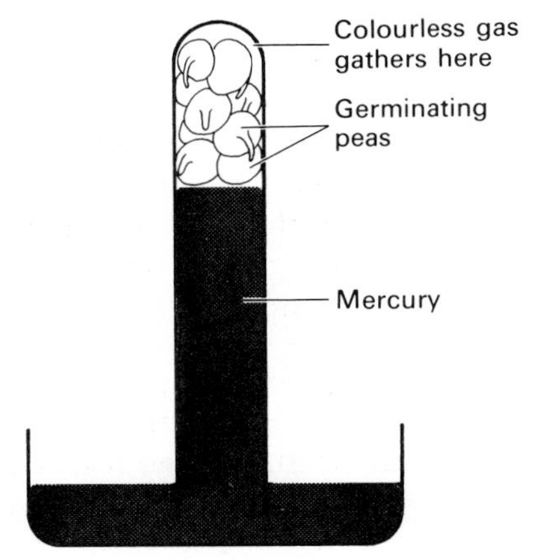

2. *Anaerobic respiration in germinating seeds*

a) Soak 4 or 5 peas in water for 12 hours, put them in a test-tube, then fill the tube to its rim with mercury.

b) Hold a piece of card tightly over the tube mouth, then invert the tube into a deep dish of mercury. Remove the card once the tube mouth is submerged, then clamp the tube to hold it in a vertical position (Fig. 8.8). (*Note:* This experiment should be placed in a fume cupboard or under a large bell-jar.)

c) Over a period of 2 or 3 days a colourless gas appears at the top of the tube. When the tube is half full of this gas test it with lime water.

d) Design an appropriate control experiment.

F *An investigation of energy release from ATP*

1. Obtain ampoules of ATP solution; fresh pork from a butcher or other muscle tissue from a recently killed animal; and a 1% solution of glucose.

a) Use forceps to strip muscle fibres from the meat. The fibres should be as long and thin as possible and separated from each other as far as is practicable. Arrange one or two fibres in straight lines on a clean microscope slide using a glass rod, and measure their length to the nearest mm.

b) Place a few drops of distilled water on the fibres and measure their length after 30 seconds. Tilt the slide to drain off the liquid.

c) Repeat this procedure using first 1% glucose solution, and then ATP solution.

d) Calculate the percentage contraction observed after the application of each substance.

e) Does this result *prove* that ATP causes contraction in living intact muscle tissue within an animal?

f) Explain why different results are obtained when using glucose and ATP separately on the above dead muscle fibres. Is there likely to be this difference when glucose is supplied, via the blood-stream, to living muscles?

Summary and factual recall test

Respiration is the chemical (1) of materials to provide (2) for (3). Aerobic respiration requires a supply of (4) and results in the complete breakdown of glucose into (5) and (6) with the release of (7). Anaerobic respiration requires no (8), produces far less (9), and breaks down glucose into intermediate substances such as (10) and (11).

Micro-organisms such as (12) and (13) obtain energy by fermentation. Commercially useful products of fermentation are (14–name three). Bakers' dough rises because (15).

Green plants can respire anaerobically for some time until (16) accumulates in their tissues. This is (17) to them and prevents further respiration. Vertebrate muscle can respire anaerobically until (18) prevents further (19). This condition is known as an oxygen (20).

ATP can be described as a 'go-between' substance because (21).

9

Breathing and gaseous exchange

Organisms which respire aerobically must absorb oxygen into their bodies. At the same time they must remove from their bodies the carbon dioxide gas which is a waste product of respiration. In other words they must 'exchange gases' with the air or water around them. This process is called **gaseous exchange**.

Gaseous exchange takes place over the whole body surface in microscopic organisms such as *Amoeba* and *Paramecium*, and also in some larger animals such as *Hydra* and jellyfish. But in most higher animals, such as insects and vertebrates, gaseous exchange takes place at a specialized region of the body called a **respiratory surface**, which is often part of an elaborate **respiratory organ** such as fish gills or human lungs. In addition, there is usually a mechanism which ensures that the respiratory surface is well ventilated: that is, it receives a steady flow of air or water. Ventilation mechanisms, such as human breathing movements, increase the rate of gaseous exchange by continually removing carbon dioxide as it emerges through the respiratory surface, and by renewing supplies of oxygen as fast as it is absorbed.

This chapter describes the respiratory organs and ventilation mechanisms of mammals (using the human lungs as an example), fish, and insects.

9.1 Respiratory organs of mammals

The lungs of a mammal, together with the heart and major blood vessels, are situated in the **thoracic cavity**, or thorax. The walls of the thorax are strengthened by the ribs, and its floor consists of a sheet of muscle called the **diaphragm**. A system of passageways leads from the mouth and nostrils into the lungs (Figs. 9.1 and 9.2).

Structure of the respiratory organs

The nasal passages Air entering through the nostrils is drawn into the nasal passages where it is warmed to body temperature and humidified by moisture which evaporates from the warm nasal membranes lining the walls of these passages. The membranes covering the roof of the nasal passages contain the organs responsible for the sense of smell. The walls and base of the nasal passages are lined with a 'carpet' of microscopic hair-like structures called **cilia** (Fig. 9.3). Between the cilia are **goblet cells** which produce a sticky fluid called **mucus**. Dust and germs inhaled from the atmosphere are trapped in the mucus and are carried by the rhythmic beating of the cilia towards the back of the mouth where they are swallowed. This mechanism helps to prevent germs and dirt from entering the lungs. Inhaled particles are trapped and passed out of the body through the digestive system. By the time air reaches the lungs it is relatively dust-free, germ-free, warm, and moist.

Air is drawn out of the nasal passages into a channel called the **pharynx** at the back of the mouth. From here, air is drawn into the **trachea**, or wind-pipe. Food in the mouth is prevented from entering the trachea during swallowing by a mechanism described on page 47. Food accidentally entering the trachea touches a sensitive area and sets off a coughing reflex, which clears the trachea of any obstruction.

The larynx The larynx, or voice-box, is a cavity at the top of the trachea which contains the vocal cords. The vocal cords are two folds of membrane situated on opposite sides of the larynx. They are attached to muscles which vary the tension in the cords and the distance between them. When the muscles relax the cords are separated and slack so that air passes soundlessly between them; but when the muscles contract the cords become taut and close together so that air causes them to vibrate. This produces sound. The pitch of the sound varies according to the tension in the cords and the distance between them.

The trachea The trachea is a tube running from the pharynx (back of the mouth) to the lungs. It is

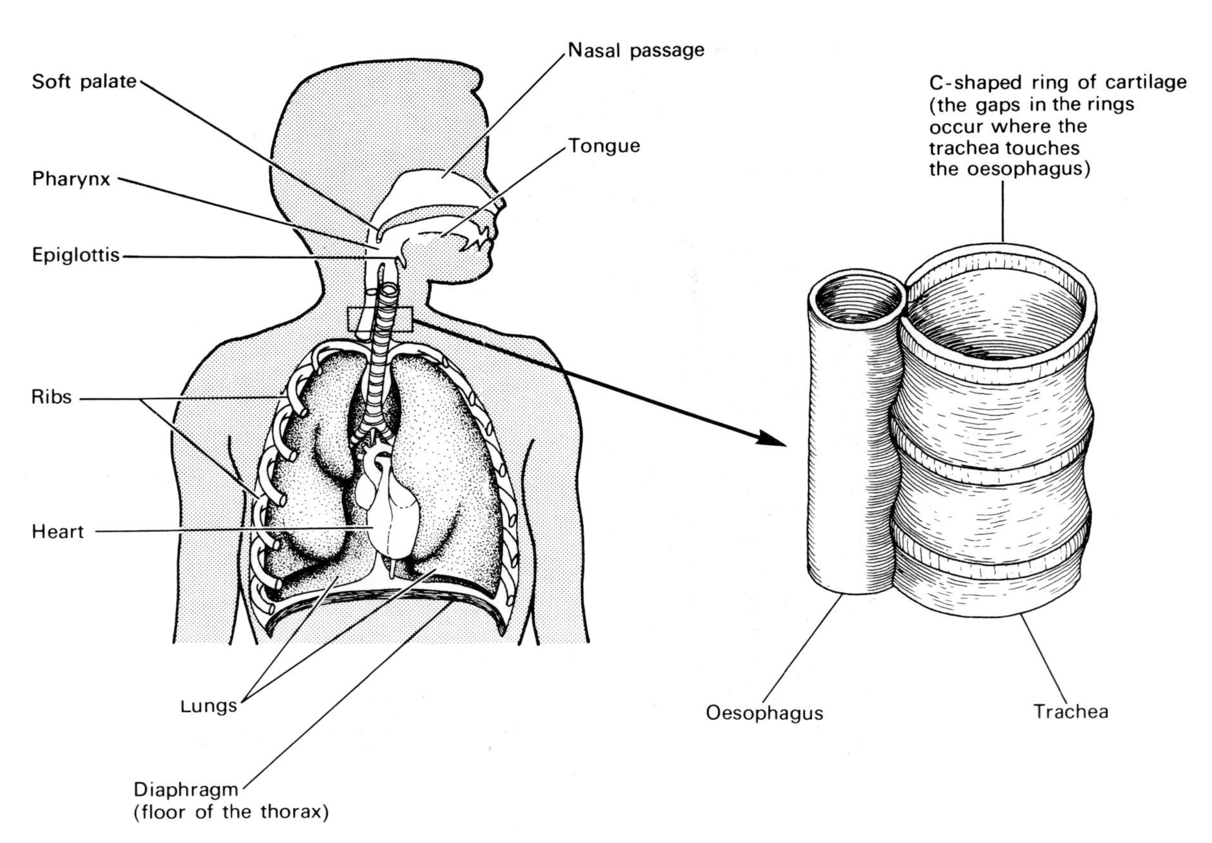

Fig. 9.1 The thoracic cavity

held permanently open by C-shaped rings of cartilage (gristle) in its walls (Fig. 9.1). This feature prevents the trachea from being blocked by 'kinks' every time the neck bends during head movements. The cartilage rings also keep the trachea open when it develops a low internal air pressure during every intake of breath.

Cilia and goblet cells extend from the nasal passages for some distance into the trachea. They create an upward flow of mucus which removes dust and germs as explained above.

At its lower end, the trachea divides and subdivides to form the bronchial 'tree', which is made up of millions of **bronchial tubes**.

The bronchial tree The main trunk of the bronchial tree is the trachea. This divides into two branches, the **bronchi** (singular bronchus), one leading to each lung. Inside the lungs each bronchus divides again and again to form a mass of very fine branches called **bronchioles.** Like the trachea, bronchi have C-shaped rings of cartilage in their walls, but these are only present to the point where the bronchi enter the lungs. From then onwards the rings are replaced by irregularly shaped plates of cartilage which perform the same function of keeping the tubes permanently open. Cartilage support of the bronchioles ceases altogether if they are less than 1 mm in diameter.

The bronchial tree terminates in air passages called **respiratory bronchioles**, which are about 0.5 mm in diameter. These branch into many short tubes of equal diameter called **alveolar ducts**, which end in tiny hollow bags called **air sacs** (Fig. 9.2B). The air sacs have many bubble-like pockets in their walls called **alveoli** (singular alveolus). The alveoli are the respiratory surface of a mammal.

Alveoli There are about 300 million alveoli in one set of human lungs. Each alveolus is about 0.2 mm in diameter, and has walls made of membrane only 0.001 mm thick. It has been estimated that if the alveoli in both lungs could be spread out flat they would cover a surface area of 90 m^2. This is about the area of a singles tennis court.

This arrangement of bubble-like alveoli gives the lungs an appearance and texture similar to sponge rubber. It also gives the lungs a far greater internal surface area than if they consisted of two smooth-walled bags, like balloons.

The whole outer surface of each alveolus is covered

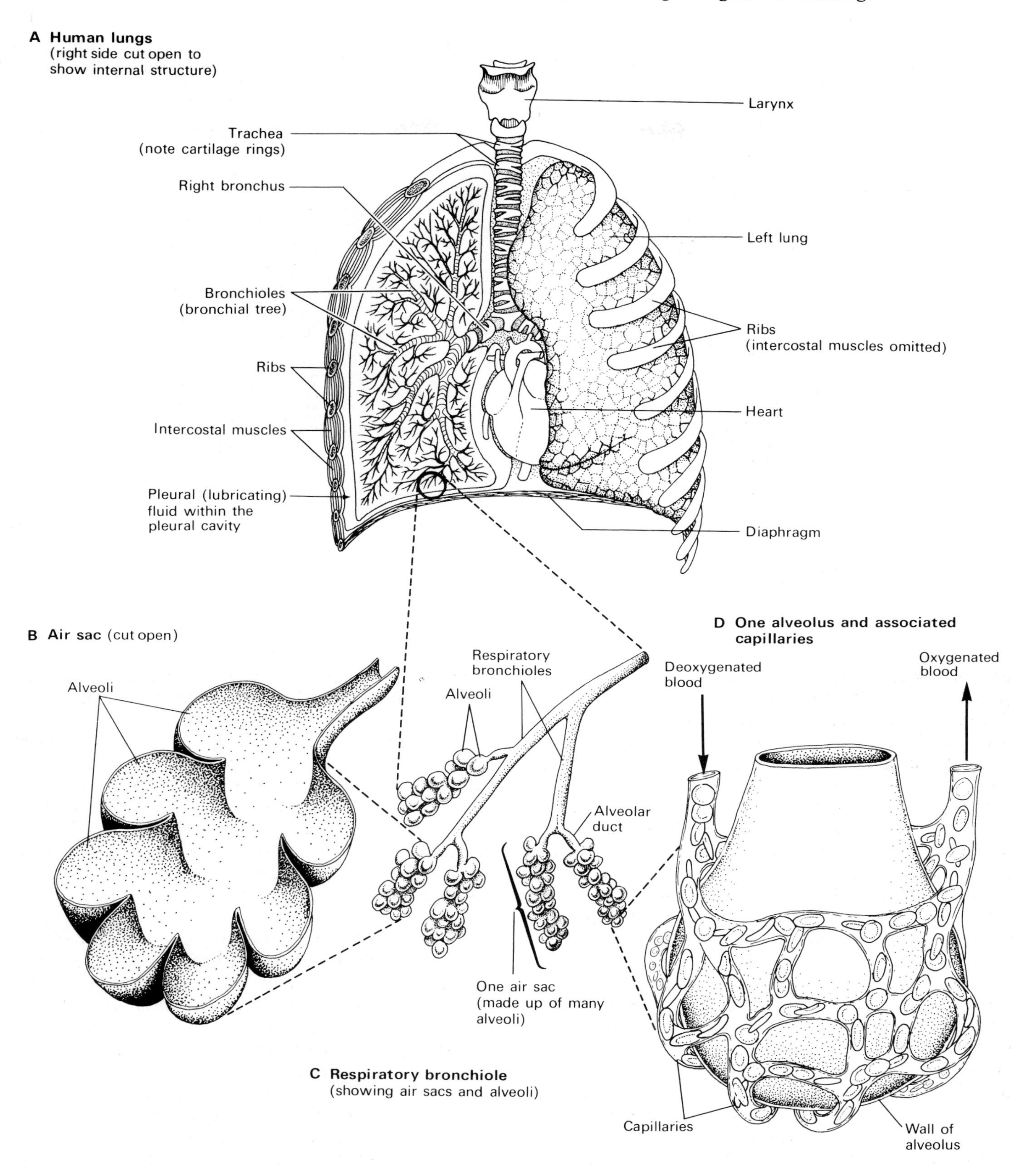

Fig. 9.2 Structure of human lungs

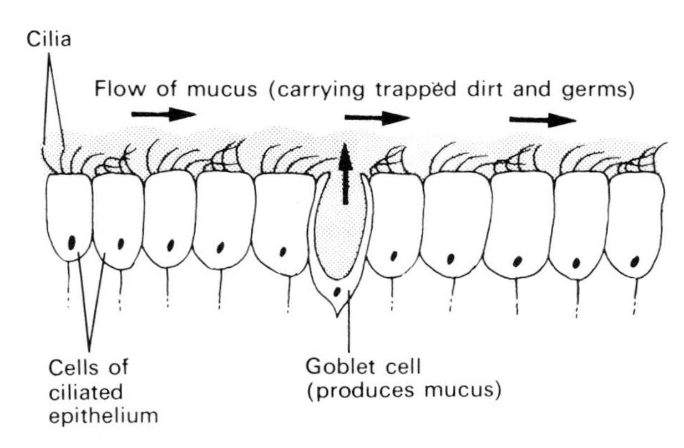

Fig. 9.3 Cross-section of the ciliated cells and goblet cells which line the air passages (highly magnified)

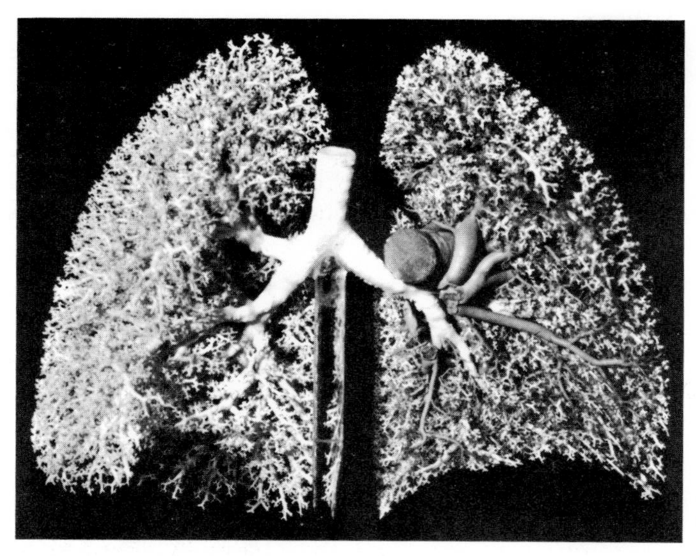

A latex cast of a human lung. The lungs were filled with latex rubber and all the tissues were digested away with enzymes

Part of the previous photograph enlarged. Identify the trachea, bronchus, and bronchioles

by a dense network of capillary blood vessels (Fig. 9.2D). All of these capillaries originate from the pulmonary artery, and eventually drain into the pulmonary vein (Fig. 6.4). If all the capillaries which make up the network covering the entire 90 m^2 area of the lungs could be joined end to end, they would reach almost from London to New York, a distance of about 5000 km.

Blood flowing through this immense network of capillaries absorbs oxygen which diffuses through the alveoli walls from inside the lungs. At the same time the blood releases carbon dioxide which diffuses in the opposite direction into the alveoli. This two-way diffusion takes place through the alveoli and capillary walls which, in humans, have a combined thickness of only 0.005 mm. This is the distance which separates air from blood in the lungs, and across which gaseous exchange takes place.

Absorption of oxygen

In chapter 6 it was explained that blood entering the lungs is deoxygenated, because the haemoglobin in its red cells has given up all its oxygen to the body tissues.

The internal diameter of the lung capillaries is actually smaller than the diameter of the red cells which pass through them. The red cells are therefore squeezed out of shape as they are forced through the lungs by blood pressure (Fig. 9.2D), and the speed at which they move is considerably reduced by the resulting friction. This increases the rate of oxygen absorption in two ways. First, as the red cells squeeze through the narrow capillaries they expose more surface area to the capillary walls through which oxygen is diffusing, and thereby absorb more oxygen. Second, their slow rate of progress increases the time available for oxygen to diffuse into them and combine with haemoglobin.

The continuous removal of oxygen as fast as it diffuses into the lung capillaries, and the continuous arrival of oxygen in the alveoli owing to breathing movements, means that there is always a higher concentration of oxygen molecules in the alveoli than in the blood. This difference causes oxygen to diffuse from the alveoli into the lung capillaries since (as explained in chapter 7) diffusion continues so long as the molecules concerned are unequally distributed.

Release of carbon dioxide

It was explained in chapter 6 that blood entering the lungs is charged with carbon dioxide which it has absorbed from body tissues. This diffuses out of the blood into the alveoli for the same reason that oxygen diffuses in the opposite direction: the continuous arrival of carbon dioxide from the tissues, and its continuous removal from the alveoli by breathing movements, means that there is always a higher concentration of carbon dioxide in the blood than in the alveoli.

Figure 9.4 summarizes the process of gaseous exchange in the lungs and shows that it also occurs, in the reverse order, between body tissues and the blood in the capillaries which serve them.

Ventilation of the lungs

The lungs are ventilated by muscular movements of the thorax wall which alter the volume of the thoracic cavity. When this volume is increased air is drawn into the lungs and they inflate, and when the volume is decreased air is pushed out and the lungs deflate. Before explaining how these volume changes are brought about it is necessary to describe the thorax in more detail.

Look at Figures 9.2 and 9.5 and note that the lungs hang down inside the thorax where they are surrounded by a very narrow space called the **pleural cavity**. This cavity is lined with a shiny, slippery skin called the **pleural membrane**, which produces an oily substance called **pleural fluid**. The pleural fluid acts as a lubricant which greatly reduces friction as the lungs rub against the thorax wall during breathing movements.

The pleural cavity is completely air-tight and contains a partial vacuum: its internal pressure is always less than the atmospheric pressure outside the body. On the other hand, the lungs are open to the atmosphere through the trachea and so there is always a higher pressure in the lungs than in the pleural cavity which surrounds them. This pressure difference is extremely important for two reasons. First, the higher pressure in the lungs than in the pleural cavity around them stretches the thin elastic alveoli walls so that the

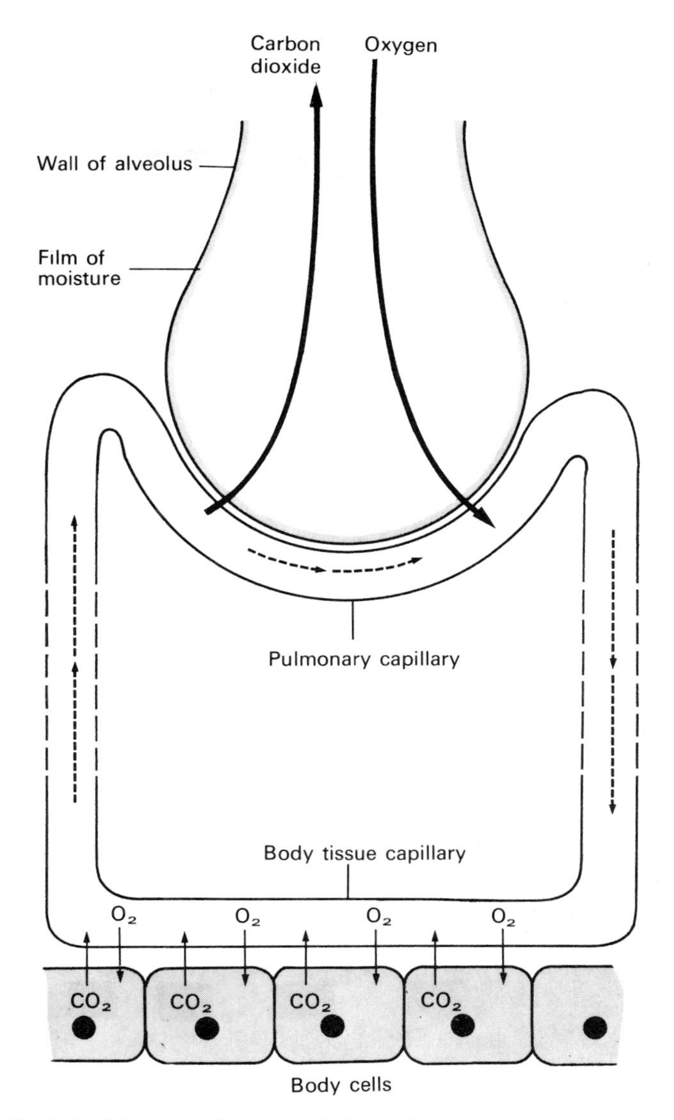

Fig. 9.4 Diagram of gaseous exchange between an alveolus and the blood, and between blood and body cells

lungs as a whole almost fill the thorax. Second, since this pressure difference is maintained during breathing movements, when the thoracic cavity increases in volume the lungs inflate to fill the extra space available.

The muscles which bring about these volume changes are the diaphragm and the intercostal muscles. The diaphragm is a dome-shaped sheet of muscle which forms the floor of the thorax, and the intercostal muscles consist of muscle fibres which cross the gap between each rib (Figs. 9.2 and 9.6).

Inspiration, or breathing in, results from an increase in the thoracic cavity volume brought about by the simultaneous contraction of the diaphragm and intercostal muscles (Figs. 9.5A, 9.5B, and 9.6).

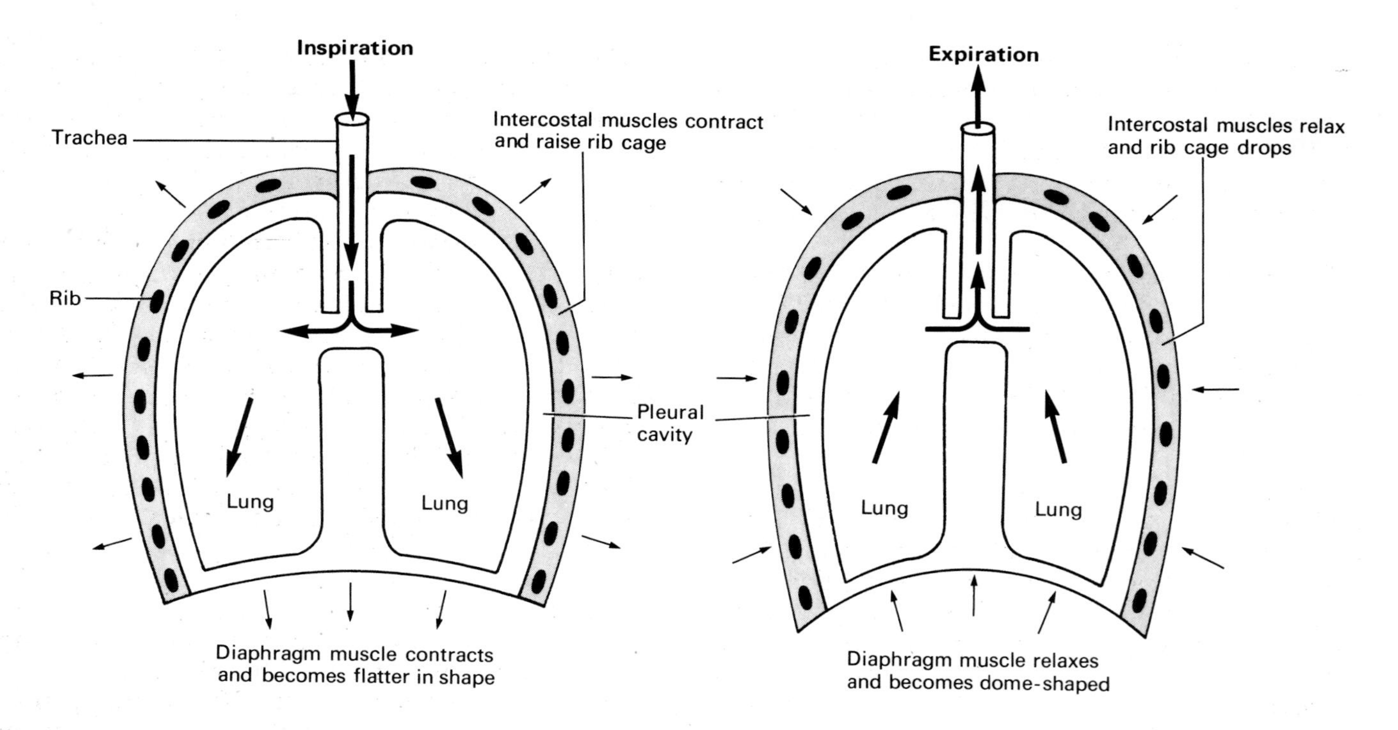

Fig. 9.5A Ventilation of the lungs

1. *Diaphragm movements* Immediately before inspiration the diaphragm is dome-shaped and its muscle is relaxed. Inspiration takes place when the diaphragm muscle contracts, making it flatter in shape. This contraction results in a downward movement of the central region of the diaphragm, and this increases the volume of the thoracic cavity.

2. *Rib movements* At the same time, contraction occurs in the external intercostal muscles between each rib. This raises the rib cage and increases its diameter and volume. Observe this movement by folding the arms across the chest and taking a deep breath. The arms move upwards and outwards along with the rib cage. Note from Figure 9.6 that this movement results from each rib pivoting at the point where it joins the backbone and sternum.

The increase in thoracic cavity volume which results from both these movements is automatically followed by an increase in lung volume. This temporarily reduces air pressure inside the lungs, and so air rushes into them from the atmosphere through the air passages.

Expiration, or breathing out, occurs when the diaphragm and external intercostal muscles relax. The rib cage drops under its own weight and the diaphragm returns to its original dome shape. Both of these movements reduce thoracic cavity volume and the lungs return to their original size, which pushes air out of the lungs. Air can be forced out of the lungs by contracting *internal* intercostal muscles (Fig. 9.6).

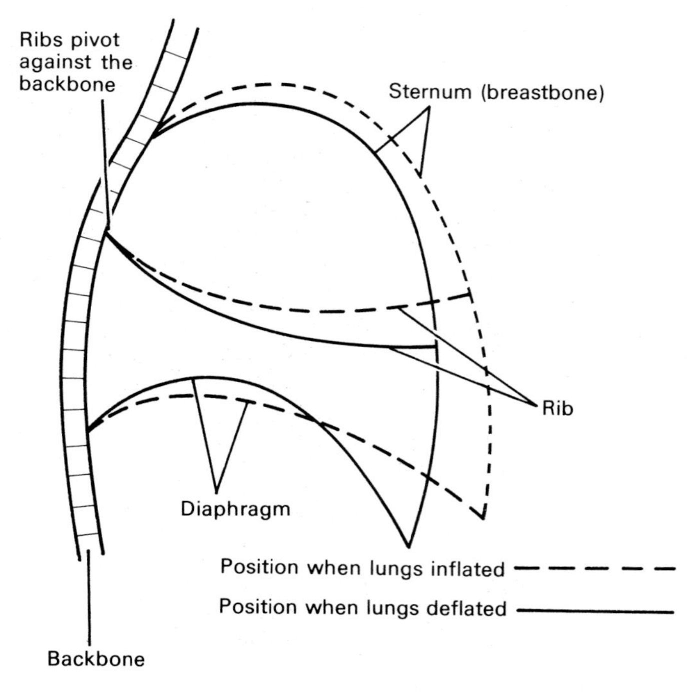

Fig. 9.5B Side view of the rib cage showing rib movements during breathing

Control of breathing rate

A relaxed human adult breathes about 16–18 times per minute, using the diaphragm alone. During exercise this rate automatically rises, and the intercostal muscles come into operation so that the volume of each breath increases, as well as the breathing rate.

The rate and depth of breathing are partly under conscious control. In fact, they can be controlled with great precision, when playing a wind instrument such as a flute for example, or when singing. At other times, breathing is controlled by an unconscious (reflex) mechanism, which operates in the following way.

The walls of certain arteries contain nerve endings which are sensitive to changes in the levels of carbon dioxide and oxygen in the blood. During exercise there is a rise in the level of carbon dioxide in the blood owing to the increased production of this gas by the muscles. The sensory nerve endings detect this rise and send impulses to the medulla of the brain, which automatically increases the rate at which the diaphragm and intercostal muscles contract. As a result of this carbon dioxide does not accumulate in the body and extra oxygen is supplied to the muscles.

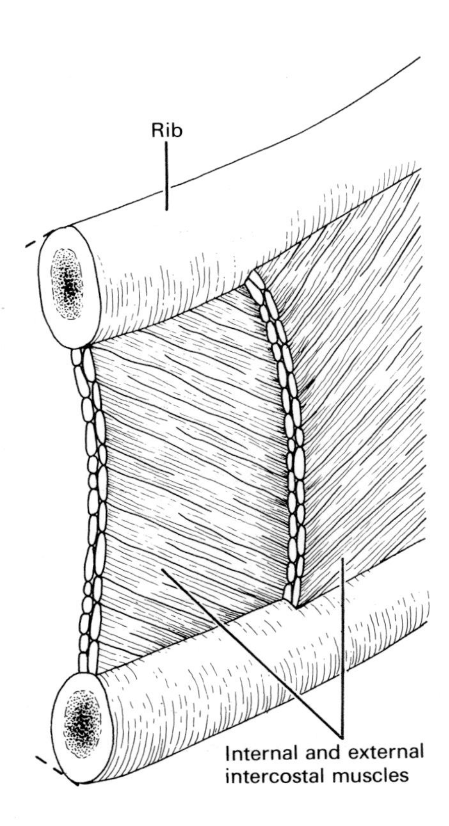

Fig. 9.6 External and internal intercostal muscles

Experiments show that these nerve endings are relatively insensitive to reduced levels of oxygen in the blood. The oxygen level in inspired air can be artificially reduced by 50 per cent (if the carbon dioxide level is held constant) before breathing rate alters. The body does not have to be equally sensitive to both gases since carbon dioxide variations are almost invariably related to corresponding changes in oxygen requirements. However, under conditions of low air pressure which occur, for example, at the top of a high mountain, the rate at which oxygen can be absorbed by the body is greatly reduced. In these circumstances breathing rate increases in response to a low oxygen level rather than a high carbon dioxide level in the blood.

Lung capacity

The total average capacity of adult human lungs is about 5 litres. But in normal breathing only 500 cm^3 of air is breathed in and out. This is called **tidal air**. After a normal tidal inspiration a further 1500 cm^3 of air can be drawn into the lungs. This is called **complemental air**. After a normal tidal expiration a further 1500 cm^3 of air can be forced out of the lungs. This is called **supplemental air**. After a forced expiration a further 1500 cm^3 of air remains in the lungs since the thorax cannot be completely collapsed. This is called **residual air**. Residual air is not stagnant since inspired air mixes with it at each breath.

9.2 Respiratory organs of fish

Fish absorb dissolved oxygen from water by means of their gills. In most fish there are four gills on each side of the body, and in bony fish each set is situated behind a large plate called the **operculum**.

Structure of a gill

Each gill is supported by a long curved bone, the **gill bar**, to which the gill filaments are attached (Fig. 9.7). The filaments consist of many long narrow structures called **gill lamellae** which are situated horizontally one on top of the other. This arrangement allows water to pass between the gills during ventilation movements. The surface of each gill lamella is folded in a way which increases its surface area (Fig. 9.7). Within each of these folds there is a dense network of capillaries which originate from larger vessels situated alongside the gill bar. Gaseous exchange occurs as water is passed over the capillaries in the following way.

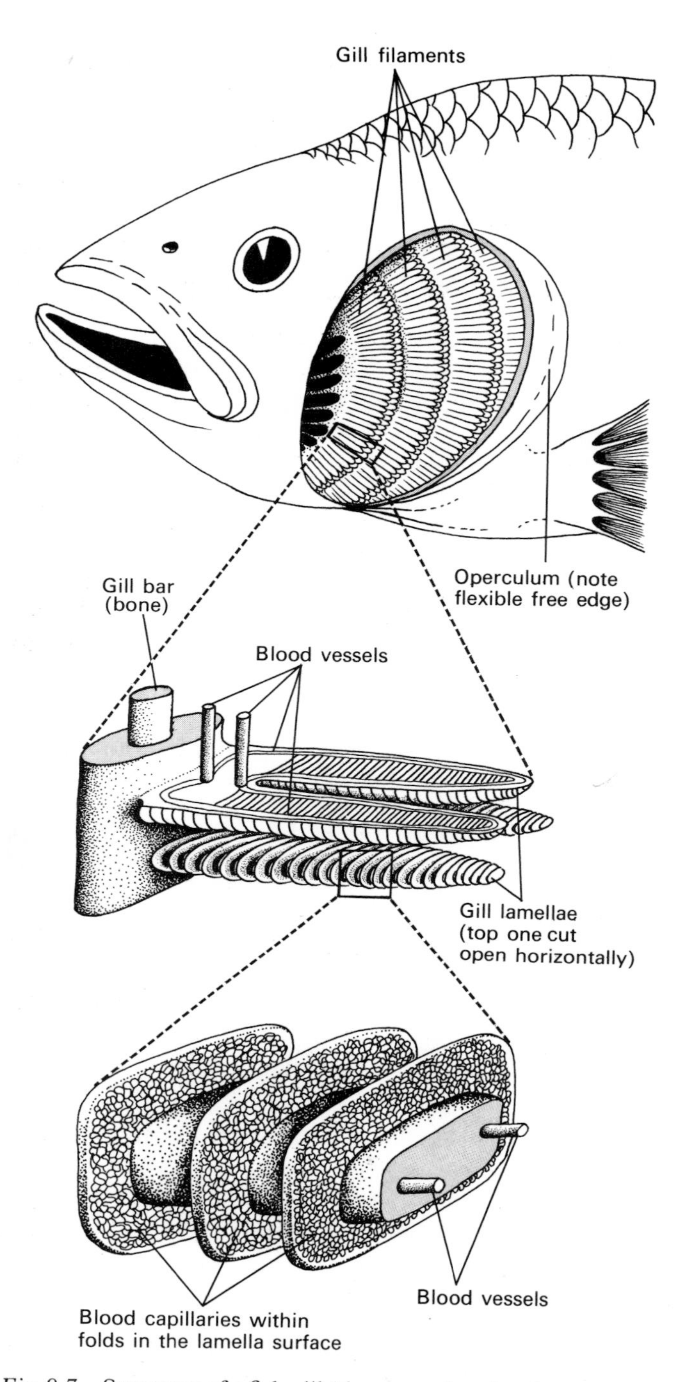

Fig. 9.7 Structure of a fish gill (the operculum has been cut away to expose the gills)

Gill ventilation

In most fish gill ventilation is achieved by a sequence of movements involving the mouth, the floor of the mouth cavity, and the opercula. Follow this sequence using Figure 9.8 as a guide.

1. Muscular contractions lower the floor of the mouth cavity which reduces pressure in this region so that water flows in through the open mouth.

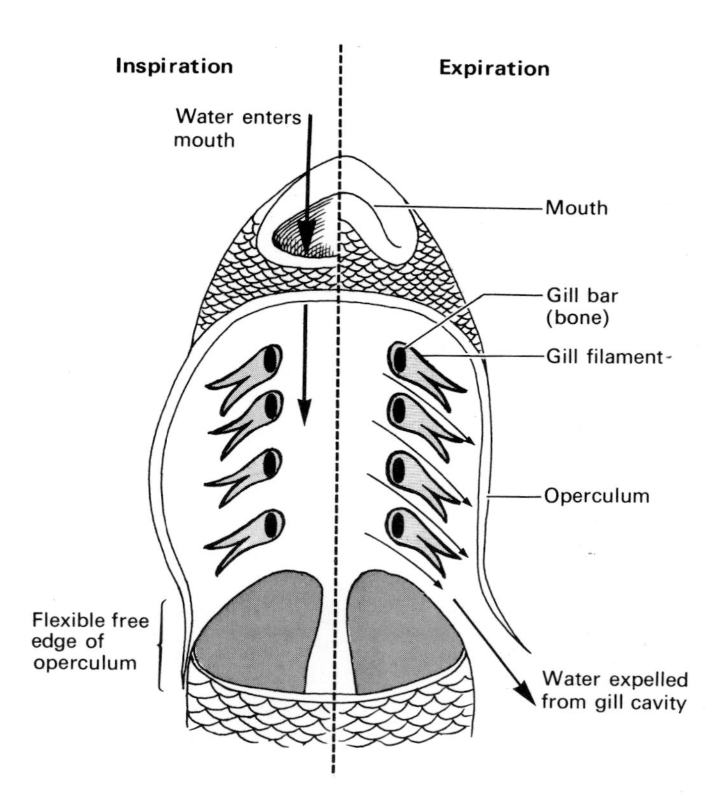

Fig. 9.8 Ventilation of fish gills (seen from below)

2. Muscles in each operculum cause them to bulge outwards. This creates reduced pressure in the gill region so that water flows from the mouth cavity through to the gills. During this time the flexible free edge of each operculum is pressed tightly against the side of the fish by the higher external water pressure. The opercula therefore act as valves which ensure that water enters only through the mouth.

3. Next, the mouth closes and muscles raise the floor of the mouth cavity. This pumps the remaining water from the mouth towards the gills.

4. Muscles then squeeze the operculum walls inwards so that pressure around the gills becomes higher than external pressure. This pressure lifts the flexible free edge of each operculum letting water flow between the gill lamellae and out of the fish.

This sequence occurs in such a way that water flows almost continuously through the gill lamellae.

9.3 Respiratory organs of insects

Oxygen and carbon dioxide are not transported by blood in insects. Unlike the animals described above, the respiratory organs of insects transport these gases to and from body tissues. These respiratory organs consist of a network of tubes called the **tracheal system** which permeate all body tissues somewhat

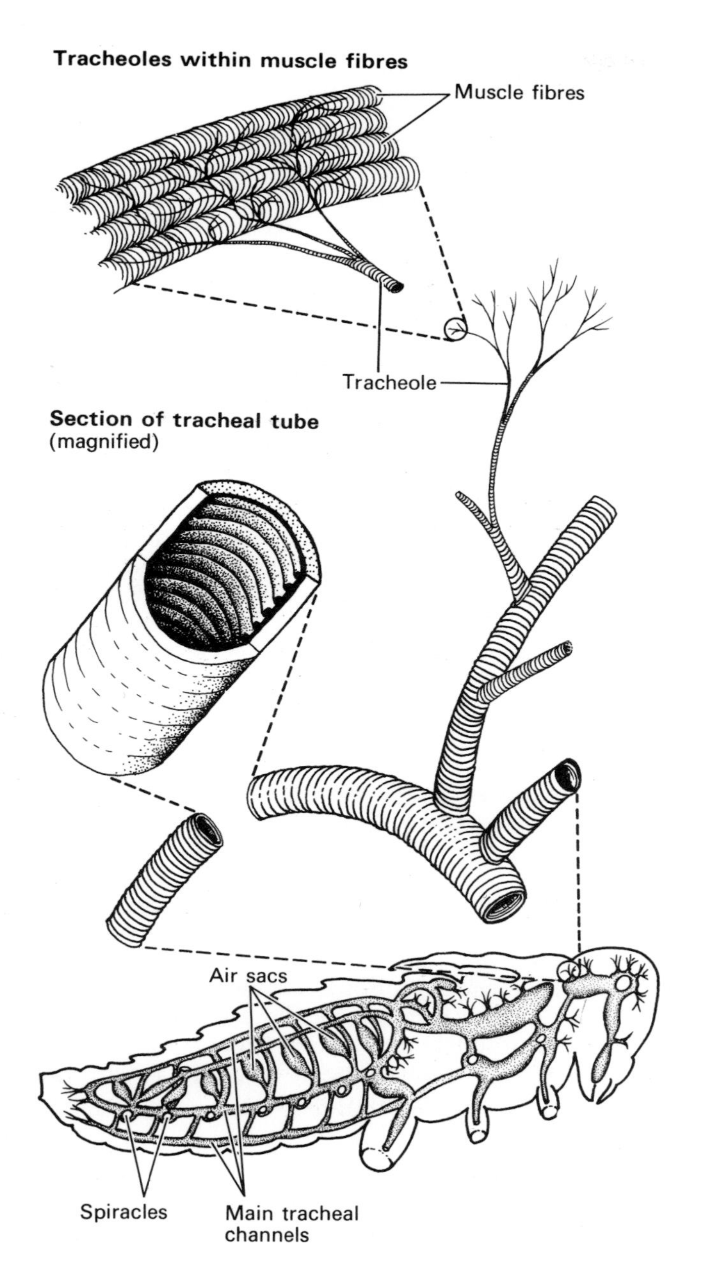

Fig. 9.9 Tracheal system of a locust

like blood vessels, but they are filled with air and not liquid.

Structure of the tracheal system

The tracheal system opens on to the body surface by pores called **spiracles**. There is usually one pair of spiracles on each body segment of an insect, but some species have only one or two spiracles on the whole body. Each spiracle generally has a muscular valve which, under certain circumstances, contracts and seals the opening so that gases and water vapour cannot pass through. (Spiracle valves are an important part of an insect's water conservation system since they help prevent the insect from drying up through loss of water vapour out of the tracheal tubes.)

Each spiracle leads into a **tracheal tube** between 0.5 mm and 0.1 mm in diameter (Fig. 9.9). These tubes are strengthened by spiral folds of cuticle in their walls. This helps to prevent them from becoming flattened by internal body pressure, but they can be extended and contracted, like an accordion.

In most insects the tracheal tubes from each spiracle are linked by large channels running lengthwise along each side of the animal. As these tubes pass deeper into the insect's body they divide and sub-divide many times and eventually form tracheal 'capillaries' or **tracheoles**, which vary from about 0.005 mm to 0.002 mm in diameter. Tracheoles possess spiral supportive thickenings in their walls which can be seen under very high magnification. They penetrate deep into the tissues, passing between individual cells and muscle fibres. Gaseous exchange takes place through the walls of the narrowest tubes.

The tracheal system of a bed bug. The tracheal tubes have been injected with a chemical which stains them black

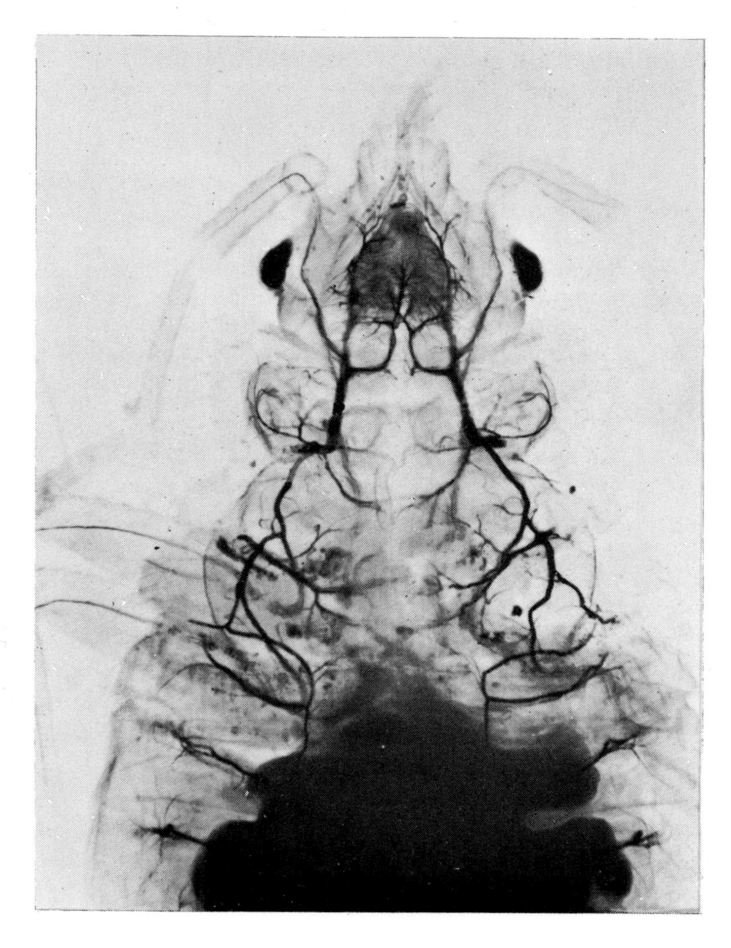

Ventilation of the tracheal system

Ventilation is achieved by movements of the abdomen walls which alternately squeeze air out of the tracheal system and then draw it in again. In bees and wasps, for example, muscles move the abdominal segments in and out lengthwise, like the opening and closing of a telescope. But in cockroaches, locusts, and beetles, muscles squeeze the abdomen laterally.

These pumping movements are often co-ordinated with movements of spiracle valves, so that air is not merely pumped in and out of the same pores, but is made to enter through spiracles near the head of the insect, pass down the main channels of the tracheal system, and out of spiracles at the rear.

In many insects, such as flies, beetles, and locusts, there are regions where the tracheal tubes are expanded into thin-walled **air sacs** which have no supporting spiral thickenings in their walls. These air sacs increase the volume of tidal air which is pumped in and out of the animal. Some of the air sacs are alternately compressed and expanded by changes of pressure in surrounding body fluids during the ventilation movements described above. Others are situated between layers of flight muscle and so are compressed and expanded whenever the insect flaps its wings. This system automatically increases the rate of tracheal ventilation and gaseous exchange during flight.

Verification and inquiry exercises

A *An investigation of the mammalian respiratory system*

1. *Gross structure*

Obtain the lungs of an animal from a butcher or slaughter-house.

a) Identify and examine the trachea and bronchi.

b) Slice open one lung and look for bronchioles.

c) Note and account for the bright red colour of lungs and their spongy texture.

2. *Lung capacity*

a) Squeeze all the air out of a large plastic bag, take a deep breath, and then exhale as much air as possible into the bag.

b) Seal the bag, without letting air in or out, by tying a knot at its mouth.

c) Measure the approximate volume of the bag by packing it into a large measuring cylinder, or other previously calibrated glass vessel.

3. *Analysis of inhaled and exhaled air*

a) Prepare the apparatus illustrated in Figure 9.10.

b) Measure the time taken for lime water to turn milky when breathing exhaled air down tube A; and again when sucking inhaled air through tube B (using fresh lime water for the second part of the experiment).

c) Explain why one procedure takes longer to turn the lime water milky than the other. What exactly do the results prove?

4. *Breathing rate*

a) Count the number of breaths per minute of a person at rest when he is unaware that he is being observed, and then when he is aware.

b) Count the breathing rate after one minute of vigorous exercise.

c) Account for any differences in the three rates.

B *An investigation of respiration in fish*

1. Obtain a whiting or herring.

a) Observe the structure of its operculum.

b) Dissect out one gill, remove some of its lamellae, and observe their structure under a microscope.

2. Observe the respiratory movements of a live fish (e.g. goldfish) in a bowl of water at room temperature. Count the number of respiratory movements it makes in one minute.

a) Reduce the temperature of the water by 10°C, by *slowly* adding ice. Leave the animal for 5 minutes at the new temperature, then measure its breathing rate by counting the respiratory movements.

b) Raise the temperature of the fish's water to 10°C above room temperature by *slowly* adding warm water (stop if the fish shows signs of distress). Measure its breathing rate again after it has had 5 minutes to acclimatize.

c) Account for any differences in the three breathing rates.

3. Prepare the following liquids: cold boiled water; water which has had air bubbled through it for 10 minutes; solutions of sodium bicarbonate at 0.25% and 5% (17.5g of sodium bicarbonate in 1 litre of water = 1% solution).

a) Place a goldfish in the first liquid for 5 minutes; then count its breathing movements.

Fig. 9.10 Apparatus for analysing inspired and expired air (exercise A3)

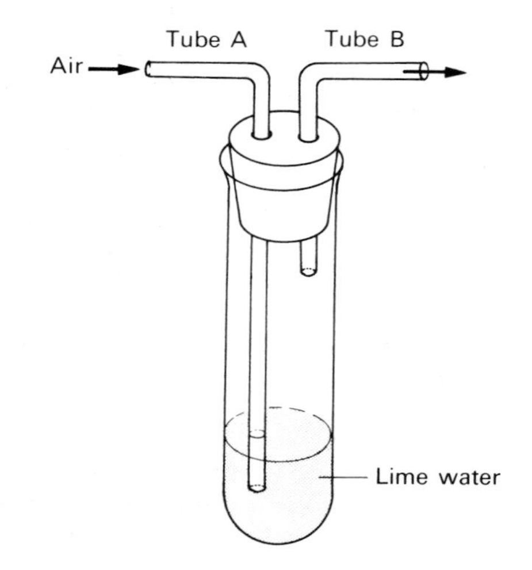

b) Place the same fish in the other liquids in turn (allowing 5 minutes for acclimatization each time) and note the number of breathing movements per minute. All the liquids must be at the same temperature.

c) Account for any differences in breathing rate.

Comprehension test

1. What conclusions may be drawn from the following table?

	Unbreathed air %	*Breathed air from a sleeping man* %	*Breathed air from a running man* %
Nitrogen	78	78	78
Inert gases	1	1	1
Oxygen	21	17	12
Carbon dioxide	Trace	4	9
Water vapour	Variable	Saturated	Saturated

2. Explain the following experimental results:

a) An animal was placed in an apparatus which ensured that it re-breathed the same air continuously. Its breathing rate increased.

b) An animal was placed in the same apparatus but all carbon dioxide was removed from its expired air before it was breathed again. Its breathing rate remained normal for some time and then suddenly increased.

c) An animal was placed in a chamber containing air which was maintained at a constant 35 per cent oxygen level. The air pressure in the chamber was then lowered by one third. The animal experienced difficulty in breathing.

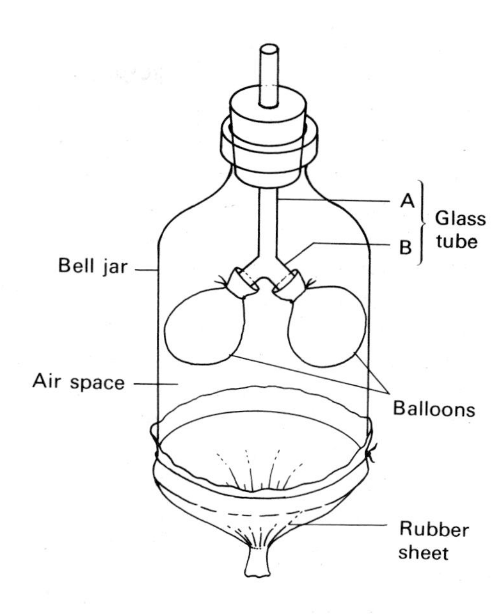

Fig. 9.11 Model of the lungs and thorax (comprehension test 5)

3. Study the apparatus illustrated in Figure 9.11, and construct it if possible. When the rubber sheet is pulled downwards the balloons inflate; when it is raised the balloons deflate.

a) To what parts of a mammal's body are the *labelled* parts of the apparatus equivalent?

b) How is each labelled part of the apparatus similar to, and yet different from, the part of a mammal's body to which it is equivalent?

c) How does the method of inflating the balloons resemble the mechanism of lung inflation? How does it differ?

d) Which mechanism of lung inflation is not illustrated by this apparatus?

Summary and factual recall test

In mammals the dome-shaped (1) muscle contracts and becomes (2) in shape. At the same time the (3) contract making the ribs pivot where they join the (4) and (5), so that the rib (6) lifts upwards and outwards. All these movements (7) the volume of the (8) cavity, and this automatically expands the lungs because (9). This temporarily reduces air (10) in the lungs and they fill with air.

(11) and (12) in inhaled air are trapped in a layer of (13) produced by (14) cells. This substance is moved by the beating movement of (15) to the back of the (16) where it is (17).

The bronchial tree consists of the (18) or wind-pipe which divides to form two (19) which further sub-divide into (20). These end in millions of bubble-like (21), which are surrounded on the outside by a network of (22). This is where (23) exchange takes place.

Red cells are (24) in diameter than lung capillaries, and this increases the rate of oxygen absorption in two ways: (25) and (26).

Fish gills are situated behind plates called the (27). Each gill is supported by a bone called the (28) and consists of a filament made up of (29) covered by a network of (30). During gill ventilation the opercula act as (31) which ensure that water enters only through the (32).

In insects (33) and (34) are transported to and from the tissues by tubes called the (35) system, and not by the (36). This system opens through pores called (37) which lead into extremely narrow tubes called (38), in which gaseous (39) takes place.

10

Homeostasis: the maintenance of a constant internal environment

In biology the word 'environment' means the conditions in which organisms live. It was explained in chapter 6 that every cell in the human body lives in a liquid called tissue fluid which supplies it with food and oxygen, and removes waste products. Because of this, tissue fluid can be described as the **internal environment** of the body.

This liquid environment in which body cells live can be compared with the pond water in which an amoeba lives. In a sense, tissue fluid is the body's own 'private pond', but it differs from an amoeba's pond in several important ways.

The temperature of pond water depends on the time of day, the weather conditions, and the season of the year, whereas tissue fluid in the human body is kept at a fairly constant 37°C regardless of conditions outside the body. The amount of oxygen and carbon dioxide in a pond varies according to the number of water plants which are carrying out photosynthesis, and the number of respiring animals. But in tissue fluid the level of these gases remains constant most of the time, unless the body is involved in some form of exercise. If this happens the oxygen and carbon dioxide levels are automatically adjusted, and return to normal when the exercise stops. The amount and variety of dissolved substances in pond water depends mostly on the nature of the local soil and rock, whereas in the human body certain mechanisms control the chemical composition of tissue fluid so that it normally contains a particular range of chemicals in precisely controlled amounts. Pond water may be poisoned by chemicals from, for example, a local industrial area and these may kill all the amoebas. If poisons enter the human body, however, they can often be neutralized or removed before harm is done, unless they are absorbed in large doses.

To summarize, the main difference between an amoeba's pond water environment and a body cell's environment of tissue fluid is that the body cell's environment is kept fairly constant, while the amoeba's environment varies considerably.

There are several mechanisms in the human body which work non-stop to stabilize the composition and temperature of tissue fluid at levels which are as near perfect as possible for the health, growth, and efficient functioning of body cells. The technical term for all these processes is **homeostasis**, which can be defined as: 'the maintenance of a constant internal environment'.

10.1 The meaning of homeostasis

Homeostasis involves constant adjustments to the contents and temperature of the blood. This also controls the contents and temperature of tissue fluid, which is formed from liquid squeezed through capillary walls from the blood-stream (section 6.1).

The organs which carry out these adjustments act rather like a boy trying to balance a stick on the end of one finger. The stick is always falling sideways in one direction or another but it does not fall to the ground if the boy moves his finger quickly enough in the right direction to bring the stick back to an upright position. The boy and the organs of homeostasis can never relax: the finger must be moved constantly, and the organs of homeostasis must work continuously, in an effort to achieve perfect balance.

There is another similarity between stick-balancing and homeostasis: they are both controlled by **feed-back** mechanisms. The boy keeps his stick upright by attending to messages from his eyes and other sense organs such as touch receptors in his fingertips. The sense organs feed back information to his brain and from here messages are sent to muscles in his arm and hand which move accordingly. The results of these movements are detected by the same sense organs and this information is again fed back to the brain, which decides on subsequent responses. There is a continuous sequence of processes involving sense organs which detect change and feed back information: the brain decides upon the necessary action; muscles respond and produce further change; and

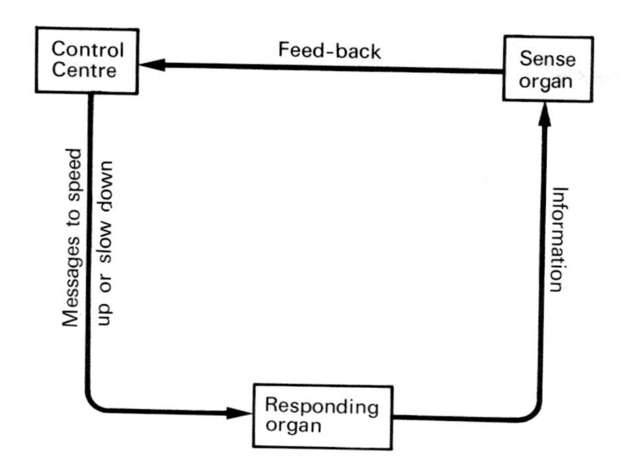

Fig. 10.1 The three parts of a typical feed-back mechanism

then more information is fed back to the brain where more decisions are taken (Fig. 10.1).

A sense organ, a control centre in the brain, and responding organs of many types are involved in most homeostatic feed-back processes. An example is the mechanism which maintains a constant body temperature. There are microscopic sense organs in the walls of blood vessels in a certain region of the brain. These sense organs respond to changes in the temperature of the blood. If the temperature rises, for example, this information passes to the temperature control centre of the brain and from here messages are sent to structures such as the sweat glands in the skin which cause the body to lose heat and so cool down. The rate of cooling is detected by the same sense organs and this information is similarly fed back to the brain so that the process can be stopped when body temperature is back to normal.

There is an important difference between homeostatic feed-back mechanisms and that used by the boy to balance a stick: in homeostasis feed-back is largely an unconscious activity–that is, the person is unaware that it is taking place.

The ability to control the internal environment of the body is best developed in man, but is also found in most mammals, and in birds. In these animals there are organs which keep the following features of blood and tissue fluid at a fairly constant level: temperature; dissolved substances such as carbon dioxide, oxygen, food, urea, and various poisonous substances; and osmotic pressure. The organs which regulate these features are the lungs, liver, skin, and kidneys, and they are therefore said to have homeostatic functions. Control of carbon dioxide and oxygen levels by the lungs has been described in section 9.1. The rest of this chapter describes the part played by the liver, skin, and kidneys in maintaining a constant internal environment in the human body.

10.2 The liver

Before reading this section check the liver's position in the human body from Figure 4.15, and then its position in the circulatory system from Figure 6.5. The homeostatic functions of the liver are illustrated in Figure 10.2.

The liver of an adult man weighs about 6 kg, and is situated immediately below the diaphragm, where it extends from one side of the body to the other. It is dark red in colour owing to the large number of blood vessels which it contains. The liver is the largest gland in the body. It is an enormous chemical 'factory' which makes and releases into the body many useful substances. Obviously a chemical factory needs a large supply of raw materials, and the liver obtains its supplies from blood in the hepatic portal vein. In fact, the liver receives through this vein practically all of the digested food absorbed by the intestine.

The liver's main homeostatic function is to regulate the amount of food which reaches the blood and tissue fluid. It does this mainly by absorbing and storing the food which it receives, and then releasing it into the circulatory system at a rate which depends upon the body's current needs. This, and some additional functions, are described under the following headings.

Regulation of blood sugar

The liver, together with a set of glands in the pancreas, control with great accuracy the amount of glucose sugar in the blood. It is very important that glucose is maintained at a certain constant level, first because this sugar is the body's main source of energy, and second because even slight changes in glucose concentration alter the blood's osmotic pressure and therefore alter the rate at which water moves in and out of body cells by osmosis.

In man, the level of glucose in most arteries is normally 85 mg/100 cm^3. After a heavy meal of carbohydrates it may temporarily rise to 180 mg/100 cm^3. It never falls much below normal except during prolonged starvation.

Whenever there is an increase in the blood's glucose level glands in the pancreas produce a substance called **insulin**. Insulin stimulates the liver cells to extract glucose from the blood. At first the liver converts this glucose into glycogen and stores it in its cells, but the liver can hold only 100 g of glycogen. When this limit is reached any remaining excess glucose in the blood is converted into fat and transferred to more permanent storage areas under the skin and around various body organs.

When there is less glucose in the blood than normal the pancreas slows down insulin production. This

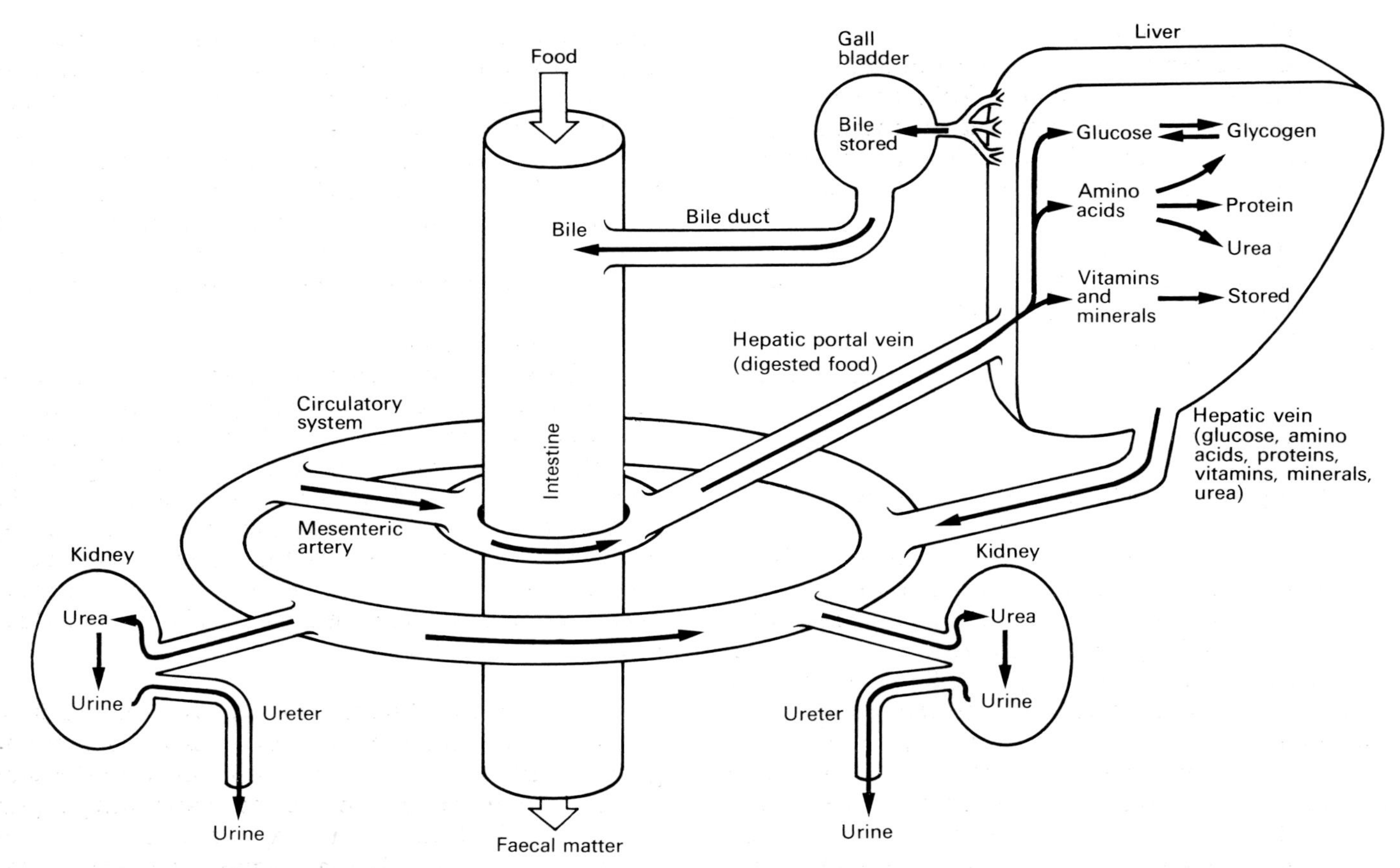

Fig. 10.2 Diagrammatic summary of the functions of the liver and kidneys

causes the liver to convert its stored glycogen into glucose, which then passes into the blood. When all the glycogen has been used up in this way stored fat is converted into glucose, and if after prolonged starvation there is no more fat in the body, protein is converted into glucose. In this way the liver keeps the body supplied with food for as long as possible when food is not available elsewhere.

Regulation of amino acids and proteins

The body can store only very small amounts of amino acids and proteins. When a meal supplies more of these than the body can use the liver gets rid of them by a process called **deamination**.

Deamination is the removal from each amino acid molecule of the part which contains nitrogen: that is, the **amino group** (which has the chemical formula NH_2). These amino groups would automatically change into ammonia (NH_3) which is very poisonous, were it not for the action of liver cells which immediately convert them into urea, which is far less poisonous. Urea then passes from the liver into the blood and is eventually removed from the body by the kidneys. The remaining part of each amino acid molecule, which contains no nitrogen, is either converted by the liver into glucose and respired, or stored.

Storage of vitamins and minerals

The liver stores vitamins A, D, and B_{12}, together with minerals such as iron, copper, and potassium, until they are required by the body.

Purification of the blood

Many poisonous substances are produced by metabolism and by disease-causing organisms. Other poisons, including certain drugs and alcohol, are deliberately taken into the body. The liver can make most poisons harmless, after which they can be removed from the body by the kidneys.

Production of fibrinogen

The liver manufactures an important blood protein called fibrinogen which is vital to the clotting of blood in wounds. In this way the liver indirectly helps to preserve the body's supplies of blood and tissue fluid.

Production of heat energy

The liver produces a great deal of heat as a by-product of the thousands of chemical reactions which take place within its cells. This heat warms the blood as it passes through the liver, which in turn warms body tissues as the blood circulates around the body.

Excretion of bile pigments

The liver excretes (removes from the body) waste substances called bile pigments which are produced during the breakdown of old 'worn-out' red blood cells in the spleen. Bile pigments are excreted into the intestine along with bile.

10.3 Temperature control and the skin

Mammals and birds have a more or less constant body temperature. That is, their body temperature remains about the same despite variations in the temperature of their surroundings. For this reason, mammals and birds are often called 'warm blooded', while all other animals which cannot maintain a constant temperature are called 'cold blooded'. Unfortunately these phrases are very misleading, because there are occasions when so-called warm-blooded animals are cooler than their surroundings, and cold-blooded animals are often warmer than their surroundings. As always, it is better to use accurate technical terms.

Birds and mammals are called **homoiothermic**, which means they have mechanisms of homeostasis that keep their body temperature at a constant level despite changes in their surroundings. All other animals are called **poikilothermic**, which means they have no temperature control mechanism and so their temperature is approximately the same as that of their surroundings.

The temperature of a poikilothermic animal depends upon many factors, such as:

1. the amount of heat produced in its body by metabolism;
2. the heat which it receives by radiation from the sun and nearby warm objects, like rocks; and which it loses from its body by radiation;
3. the heat which it receives, and loses, by conduction to and from objects touching its body;
4. the heat which it receives from and loses to the air;
5. the heat which it loses by evaporation of water from its skin.

Homoiothermic animals are also affected by these factors but with the difference that these animals have some control over the effect which these factors have upon body temperature. A homoiothermic animal has several mechanisms which work non-stop to balance heat production in its body against heat lost through its skin. This balance is achieved by a temperature control centre in the brain attached to sense organs which are very sensitive to temperature changes in the blood. Whenever such a change occurs this control centre adjusts many different processes concerned with heat production and heat loss, so that a balance is restored and body temperature returns to normal.

Control of over-heating

The body makes adjustments which prevent over-heating under at least two different circumstances: first, whenever conditions outside the body are near to, or hotter than, normal body temperature (37°C in humans); second, whenever there is an increase in heat production by the body which may occur during vigorous exercise, or when the body is fighting a disease. Some of the adjustments which help prevent over-heating in humans are described below and illustrated in Figure 10.3.

1. *Sweating* Sweating is the production of a watery fluid containing dissolved salt from sweat glands in the skin. As sweat evaporates from the skin it has a cooling effect because the evaporating liquid carries away body heat. The evaporation of sweat is an extremely efficient cooling mechanism, and in climates where air temperature approaches body temperature sweating is the only mechanism which can effectively cool the body.

The rate at which sweat evaporates from the body, and therefore its effectiveness in cooling the body, depends on two things: humidity (i.e. the amount of water vapour in the air), and air movements (e.g. winds or fans). Sweat evaporates and cools the body very rapidly in hot, dry, windy conditions. In climates of this type people can tolerate temperatures near to that of their own bodies, and can even take part in vigorous physical exercise without much discomfort. But in hot, humid conditions, especially in still air, sweat evaporates and cools the body very slowly. In climates of this type temperatures near to that of the body can be intolerable, and physical exercise will generate heat that may not be removed from the body. In this case the body temperature rises. The critical body temperature seems to be 41°C, above which sudden collapse and unconsciousness are likely and death may result.

The rapid sweating which results from strenuous exercise in hot climates may cause the loss of up to 30 litres of water per day and 30 g of salt. Loss of water at this rate soon causes the blood to become thick and concentrated so that it no longer circulates

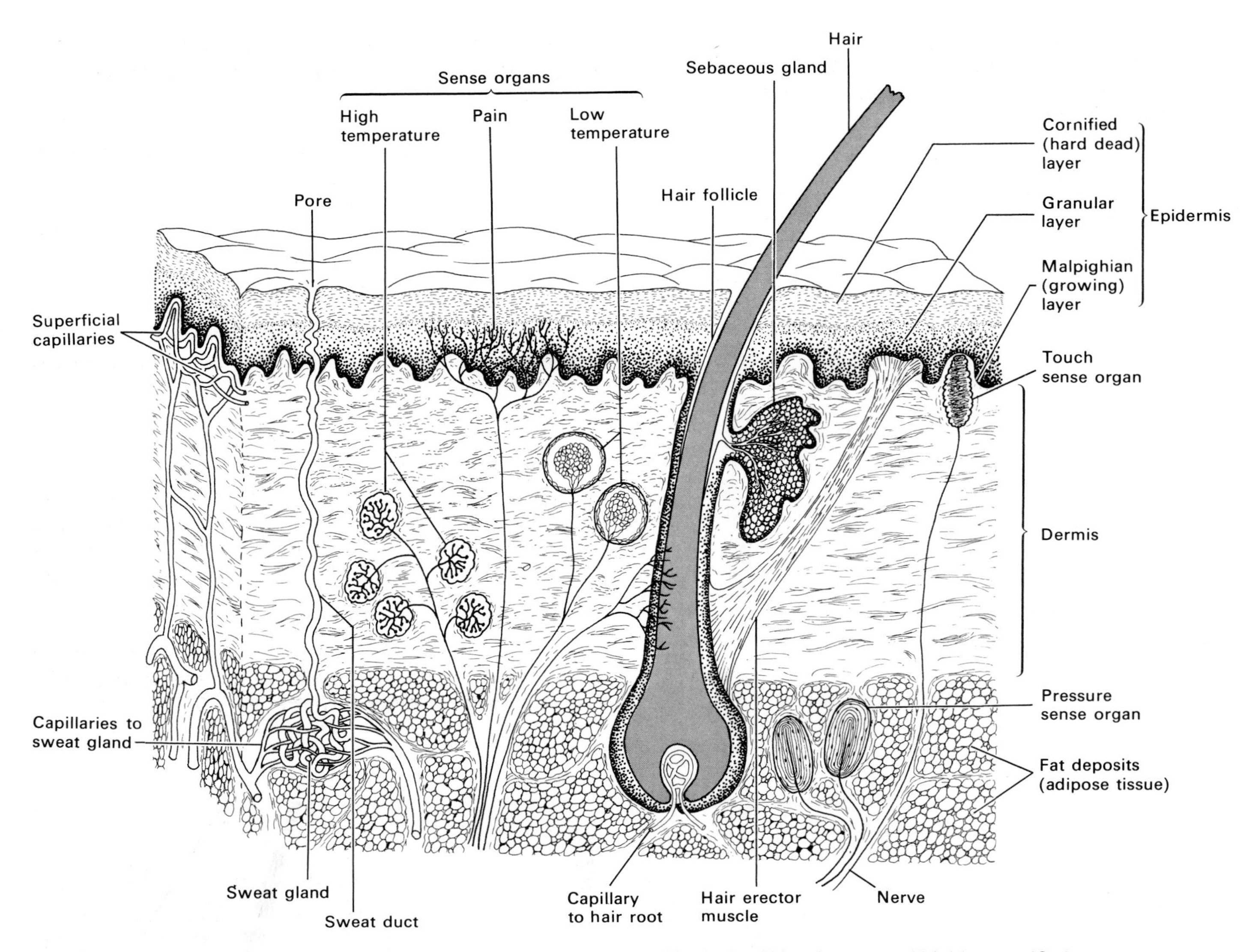

Fig. 10.3 Skin of a mammal highly magnified

properly. Loss of salt causes muscle pains (heat cramp). In hot climates a person must not only drink a lot of fluids; he must also increase the amount of salt in his diet. This is why salt tablets are manufactured for use in the tropics.

When a person is forced to work for long periods in a hot climate his ability to sweat may suddenly fail altogether. This is called **heat stroke**. If the victim is not taken to a cool place immediately his temperature will rise uncontrollably with possibly fatal results.

2. *Panting* Animals of the dog family have sweat glands only in the pads on their paws, therefore evaporation and cooling by sweating is very limited. Their main method of losing heat is to pant rapidly with the tongue hanging out. This causes evaporation from the mouth and lungs and cools the body.

3. *Vasodilation* Vasodilation is the expansion of blood vessels; that is, an increase in their diameter so that more blood flows through them. Whenever the body gets too hot vasodilation occurs in the dense network of capillaries which lie just beneath the epidermis of the skin, i.e. in the **superficial capillaries** shown in Figure 10.3. These capillaries open up and let a large volume of over-heated blood flow very close to the body surface. Here the blood rapidly loses heat by radiation through the skin, which cools the body. This is why a person's skin becomes a flushed pink colour and feels hot to the touch when he is over-heated.

4. *Relaxation of hair erector muscles* The hair erector muscles are shown in Figure 10.3. Whenever heat must be lost from the body, hair erector muscles relax and so the hairs lie more or less flat against the

skin. In this position they offer the least possible obstruction to heat loss by radiation and convection. The reasons for this will be clearer after reading what happens to hair in cold weather.

5. *Feeding and food storage* In hot weather the rate of metabolism is gradually reduced to a lower level, and this in turn reduces appetite for food. Consequently in hot climates there is a lower heat energy output from the body which means there is less heat to get rid of. In addition, the bodies of animals which live in the tropics tend to have stores of fat located in places where they do not insulate the body against heat loss. Camels for example have fat stored in their humps.

Control of over-cooling

There are similarities between the way in which homoiothermic animals keep warm in cold weather, and the way in which people keep their houses warm in winter. A house can be kept warm by turning on a central heating system, by fitting double glazing, and by laying down felt or other insulating material above the ceilings. More heat is produced inside the house, while at the same time the loss of heat from the house is reduced to a minimum. This is exactly what happens in the bodies of mammals and birds when there is a danger of their body temperature dropping below normal.

1. *Increased heat production* The 'central heating system' in a homoiothermic animal is the heat generated by metabolism mainly in the liver and muscles, and in particular the heat which comes from the breakdown of food by respiration in these organs. Vigorous exercise warms the body, because it increases the rate of respiration and heat production in the muscles. However, when a person tries to rest in cold conditions his muscles begin jerky or rhythmic movements against his will. This is called shivering, and it is a mechanism which helps to keep the body warm by automatically causing the muscles to generate heat whenever necessary.

In addition, there is a general increase in the rate of metabolism during cold weather which brings about an increased appetite for food. This further increases heat output, and helps maintain a constant body temperature.

2. *Reduction of heat loss* The equivalent of double glazing and roof insulation in mammals is hair, and the layer of fat beneath the skin.

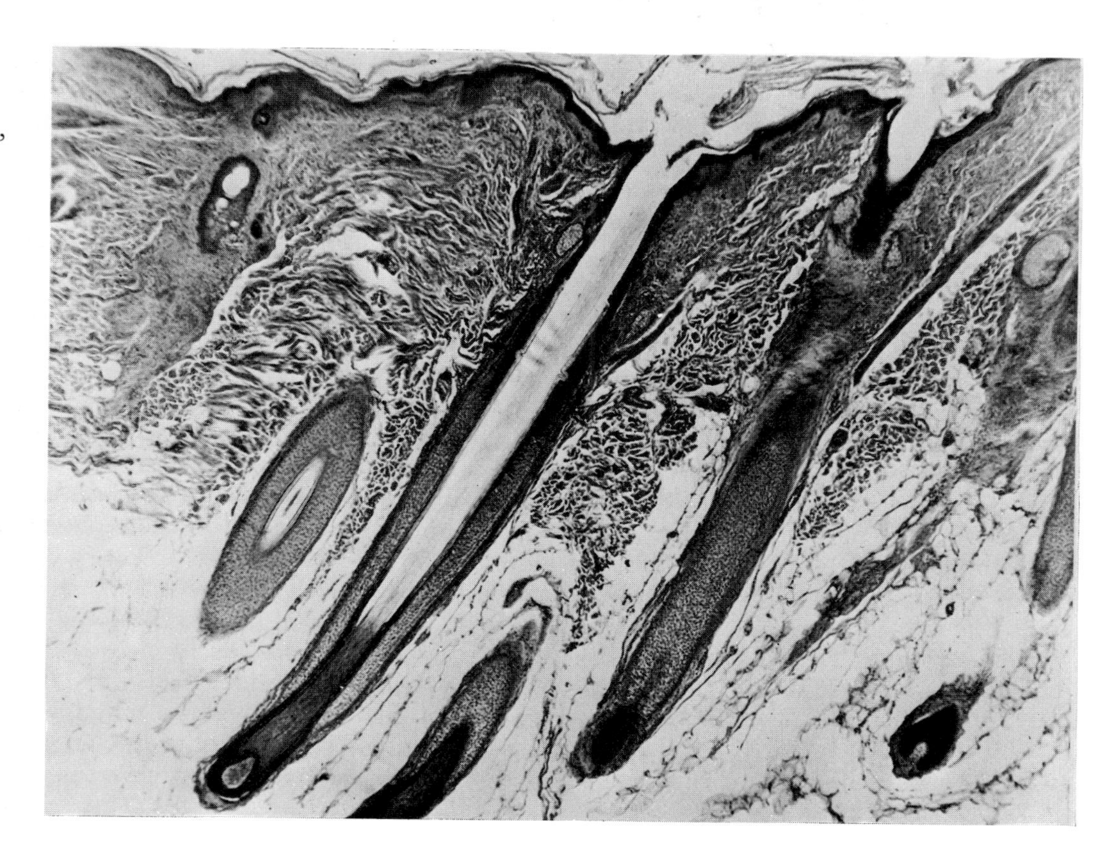

A vertical section through a human scalp. Using Figure 10.3 as a guide identify the epidermis, dermis, hair roots, and hair follicles *a*) In cold weather the hair erector muscles contract, which raises the hairs shafts to an almost vertical position: i.e. the hairs 'stand on end'. This helps prevent heat loss in two ways. First, the upright hairs prevent cold winds from reaching the skin where they would rob the body of heat. Second, the upright hairs cause a layer of still air to develop around the body. This air is slowly warmed by body heat, and helps to insulate the body against heat loss, since air is a very poor conductor of heat (i.e. heat passes very slowly into it from the body). This mechanism does not work very well in humans owing to their lack of body hair. In fact, without clothes the human body can maintain its characteristic 37°C only at outside temperatures no lower than 27°C. It is only by putting on clothing that humans are able to survive in colder conditions. Putting on a warm winter overcoat is man's equivalent of the contraction of the hair erector muscles in other mammals. Birds achieve a similar effect by means of muscles which make their feathers fluff out.

b) Animals which live in cold climates, such as seals and polar bears, have a very thick layer of fat beneath the skin called **adipose tissue** (Fig. 10.3). This fat is quite effective as a layer of insulation and so helps prevent heat loss from the body. It is also a store of food.

c) The sweat glands cease to operate in cold weather. This reduces heat loss by evaporation, but a small amount of water still evaporates through the epidermis from moist underlying tissues.

d) In cold weather, **vasoconstriction** occurs in the skin's superficial capillaries. The capillaries become smaller in diameter, which restricts blood flow near the body surface and so reduces heat loss by radiation through the skin to a minimum. This explains why the skin looks pale in cold weather.

10.4 Excretion and the kidneys

Excretion is the removal from the body of waste substances which are produced by metabolism, and substances of which the body has an excess. In other words, excretion is the removal of unwanted substances from the body. In order to understand where these unwanted substances come from it is necessary to follow what happens to digested food once it gets inside the body cells.

Digested food takes part in the many reactions of metabolism. All these reactions are useful to the body: some produce energy, some build new body tissues, and some produce useful chemicals such as enzymes and hormones. At the same time, most of these reactions give rise to by-products which are useless or even harmful to the body. These substances are called **excretory products**, and are removed from the body by **excretory organs**.

Respiration, for example, produces energy which is useful to the body, but it also produces carbon dioxide and water. The carbon dioxide is not only useless; it can be harmful because it quickly changes into carbonic acid which can damage body cells if allowed to accumulate in large quantities. Consequently carbon dioxide is an excretory product and is removed from the body by the lungs, which are excretory organs. On the other hand, the water produced by respiration is not normally useless or harmful. Indeed, if the body happens to be short of water it uses some of that produced by respiration. Nevertheless, this water is useless to the body when it is surplus to requirements, and can be harmful if allowed to accumulate to the point where it dilutes blood and tissue fluid. Under these circumstances water becomes an excretory product and is removed from the body.

Some by-products of metabolism are useless when they are produced, but are not excreted because they can be turned into useful substances. An example is lactic acid, which is formed in the muscles of vertebrate animals (e.g. man) during vigorous exercise. Lactic acid is useless until converted back into glucose, when it is used in respiration, or stored as glycogen.

From the above it is clear that the body can 'distinguish' between useful and useless substances and can excrete the latter. It is important, however, to distinguish between excretion and two other processes with which it may be confused: defecation and secretion. Figure 10.4 illustrates the differences. **Defecation** is the removal from the body of indigestible substances: that is, the passage of faecal matter through the anus. Defecation differs from excretion in that faecal matter is not produced by metabolism and is not therefore a true excretory product. **Secretion** is the production by metabolism of *useful* substances such as enzymes and hormones.

The most poisonous of all the waste by-products of metabolism is ammonia. Ammonia is formed during the breakdown of excess amino acids in the liver. Ammonia is very soluble and kills cells if its concentration in the blood rises above 1 part in 25 000. The liver immediately converts ammonia into a relatively harmless substance called **urea**, which is released into the blood. The kidneys extract urea from the blood and excrete it from the body as part of a liquid called **urine**. Ammonia and urea contain nitrogen, therefore removal of urea from the body by the kidneys is known as **nitrogenous excretion**. The kidneys are part of a set of organs known as the **urinary system**.

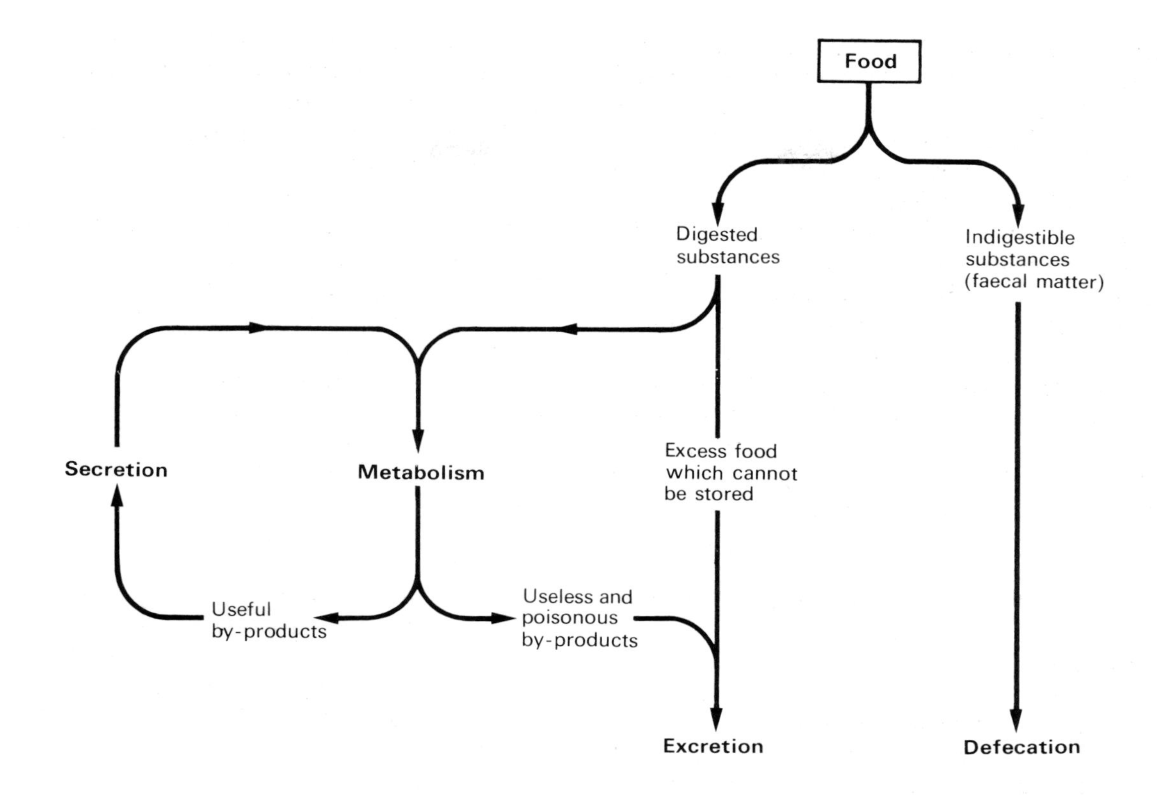

Fig. 10.4 The differences between excretion, defecation, and secretion

The urinary system

The parts of the urinary system and their position in the human body are shown in Figure 10.5.

Human kidneys are about 12 cm long by 7 cm wide, and are bean-shaped. A thin tube, the **ureter**, comes out of the concave side of each kidney and extends downwards to a single large bag called the **bladder**. The bladder has only one exit, a tube called the **urethra**, which leads to the body surface. The bladder end of the urethra is normally held closed by means of a ring of muscle (a sphincter), which controls the release of urine from the bladder.

Urine drains continuously out of the kidneys into the ureters where it is forced downwards into the bladder by wave-like contractions of the ureter walls. The bladder stretches and expands in volume as it fills with urine, and when it is nearly full the stretching stimulates sensory nerve endings in its walls so that nerve impulses are sent to the brain. This is how a person knows when his bladder must be emptied. The

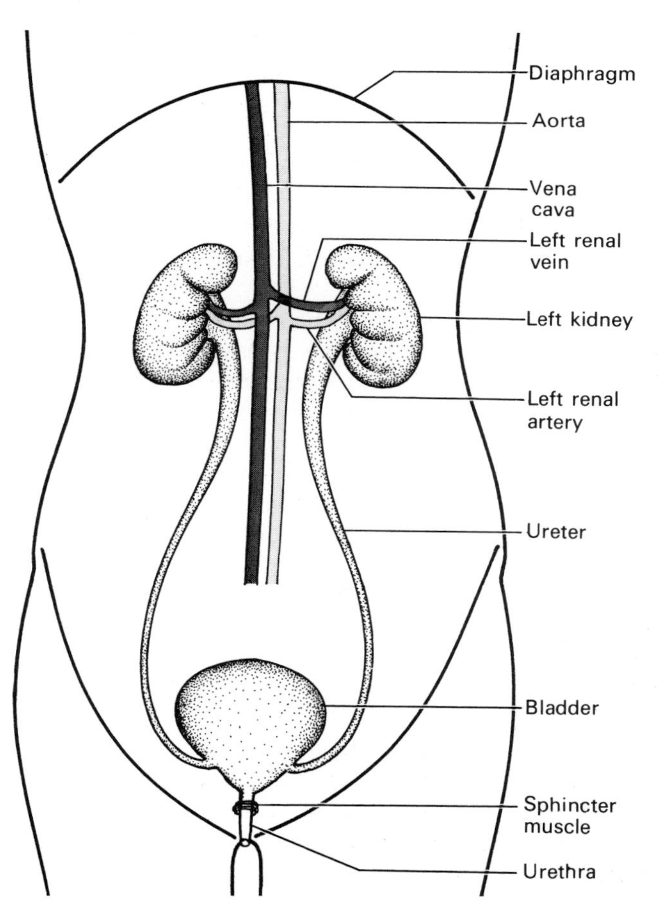

Fig. 10.5 The human urinary system (diagrammatic)

sphincter muscle around the urethra is then voluntarily (i.e. consciously) relaxed to let urine drain from the bladder, through the urethra, and out of the body. This is called **urination**.

Kidney structure

A human kidney contains about 160 km of blood vessels, and more than a million separate lengths of extremely fine tube known as kidney tubules, which have a combined length of about 60 km. The arrangement of blood vessels and kidney tubules is best seen by studying, at higher and higher magnification, a kidney which has been cut in half. This is what Figures 10.6 and 10.7 represent.

A kidney has a dark-coloured outer zone called the **cortex**, and a paler-coloured inner zone called the **medulla**. The medulla consists of several cone-shaped areas called the **pyramids**. Urine drains continuously from the tips of the pyramids into funnel-shaped spaces formed by the top of the ureter.

Kidney tubules In humans, each kidney contains about one million kidney tubules, each of which is about 3 cm in length. Figure 10.6 shows that they begin in the cortex of the kidney, where each one is expanded into a round, cup-shaped object called a **Bowman's capsule**, which is about 0.2 mm in diameter. Each Bowman's capsule almost entirely encloses a ball of finely divided and inter-twined blood capillaries known as a **glomerulus** (plural glomeruli). Glomeruli were discovered by Marcello Malpighi, a seventeenth-century biologist who said that they looked like a coil of worms.

Each kidney tubule emerges from the Bowman's capsule on the side opposite to the glomerulus and, after a complicated series of coils and loops (Fig. 10.7), it joins a wider **collecting duct**. Collecting ducts collect urine from the kidney tubules and transport it straight through the medulla of the kidney to the tips of the pyramids.

Blood supply to the kidney tubules Each kidney receives oxygenated blood at high pressure through a renal artery direct from the main aorta. Inside each kidney the renal artery divides into arterioles which convey blood to glomeruli capillaries with only a small loss in pressure. The blood vessel which leaves each glomerulus branches into a capillary network around the coiled and looped portion of each kidney tubule, before eventually joining the renal vein.

Formation of urine

Urine is formed in two stages. First, blood is filtered into the kidney tubules to form a clear fluid (i.e. a filtrate) which contains the waste substance urea and many useful substances which the body cannot afford to lose. Second, the useful substances are reabsorbed from the filtrate back into the blood leaving only urea and other substances in the kidney tubules which are of no use to the body. The technical words for these two stages are **filtration**, and **reabsorption**.

Filtration In the kidney the liquid to be filtered is the blood. The blood is filtered by the cup-shaped Bowman's capsules, of which there are more than a million in humans. The blood is filtered through two layers of living membrane: the capillary wall of each glomerulus and the inner wall of each Bowman's capsule. Blood is filtered as it passes through these membranes on its way to the kidney tubule. In humans these membranes have a total surface area of one square metre.

However, there are two important points about filtration in the kidney which should be noted. First, blood is forced through the two membranes at a very high pressure. Second, not all the blood is filtered. In fact, the remaining unfiltered blood has an important function to perform after leaving the glomeruli on its way to the renal vein.

The filtering process in kidneys is very similar to the formation of tissue fluid described in section 6.1. Both processes consist of blood being forced at high pressure through capillary walls, with the result that blood cells and large protein molecules are filtered out (i.e. held back) leaving a clear liquid. In kidneys, this liquid is called **glomerular filtrate**. It enters the cavity of the Bowman's capsules and then drains into the kidney tubules.

A high blood pressure in the glomeruli is essential to the filtering process, and this pressure is achieved in three ways. First, blood entering the glomeruli is already at high pressure because the renal artery takes blood from the main aorta at a point close to the heart. Second, this pressure is further increased by the pressure build-up which results from the fact that the vessel which leaves a glomerulus is narrower than the one which enters it. Third, there is a tiny ring of muscle around the vessel which leaves each glomerulus. These ring muscles automatically alter in diameter to maintain a constant pressure in the glomerular capillaries despite changes in blood pressure in the rest of the body.

In humans, the kidneys filter about 60 litres of blood an hour, and it takes them only five minutes to filter an amount which is equal to the body's entire blood supply. The filtering process produces about 7.5 litres of glomerular filtrate an hour, and this liquid contains not only urea, but many useful substances such as glucose, amino acids, mineral salts, and vitamins, dissolved in a large amount of water. If all of this were excreted the body would lose most of its water and soluble food supplies in a few hours. This does not happen because 99 per cent of the glomerular filtrate is reabsorbed.

Reabsorption Reabsorption occurs as glomerular filtrate passes out of the Bowman's capsules and down the kidney tubules towards the collecting ducts. All along this route cells in the kidney tubule walls extract useful substances from the glomerular filtrate so that by the time it reaches the collecting ducts a liquid called **urine** has been formed. Urine contains only waste excretory substances. Humans produce

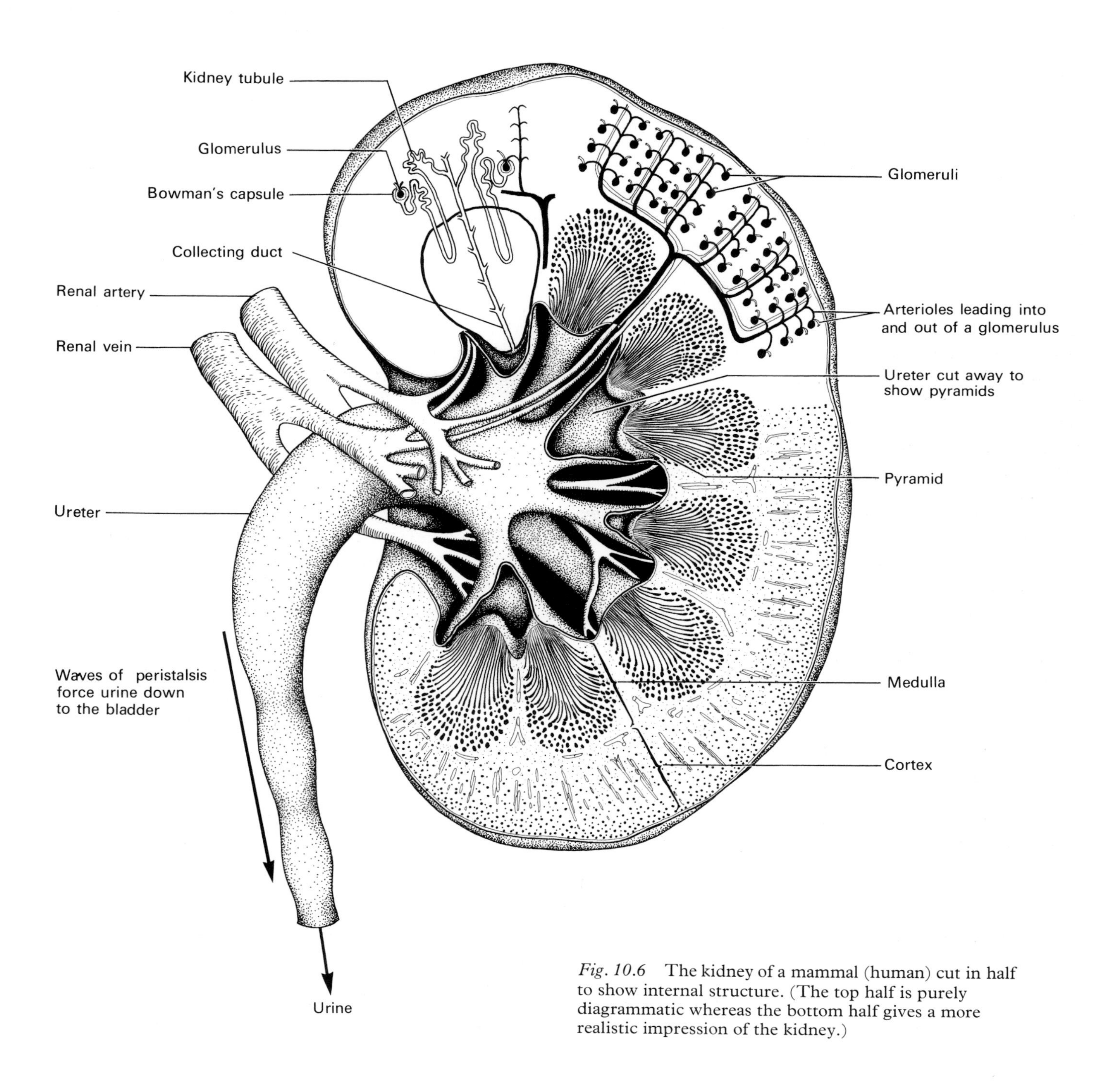

Fig. 10.6 The kidney of a mammal (human) cut in half to show internal structure. (The top half is purely diagrammatic whereas the bottom half gives a more realistic impression of the kidney.)

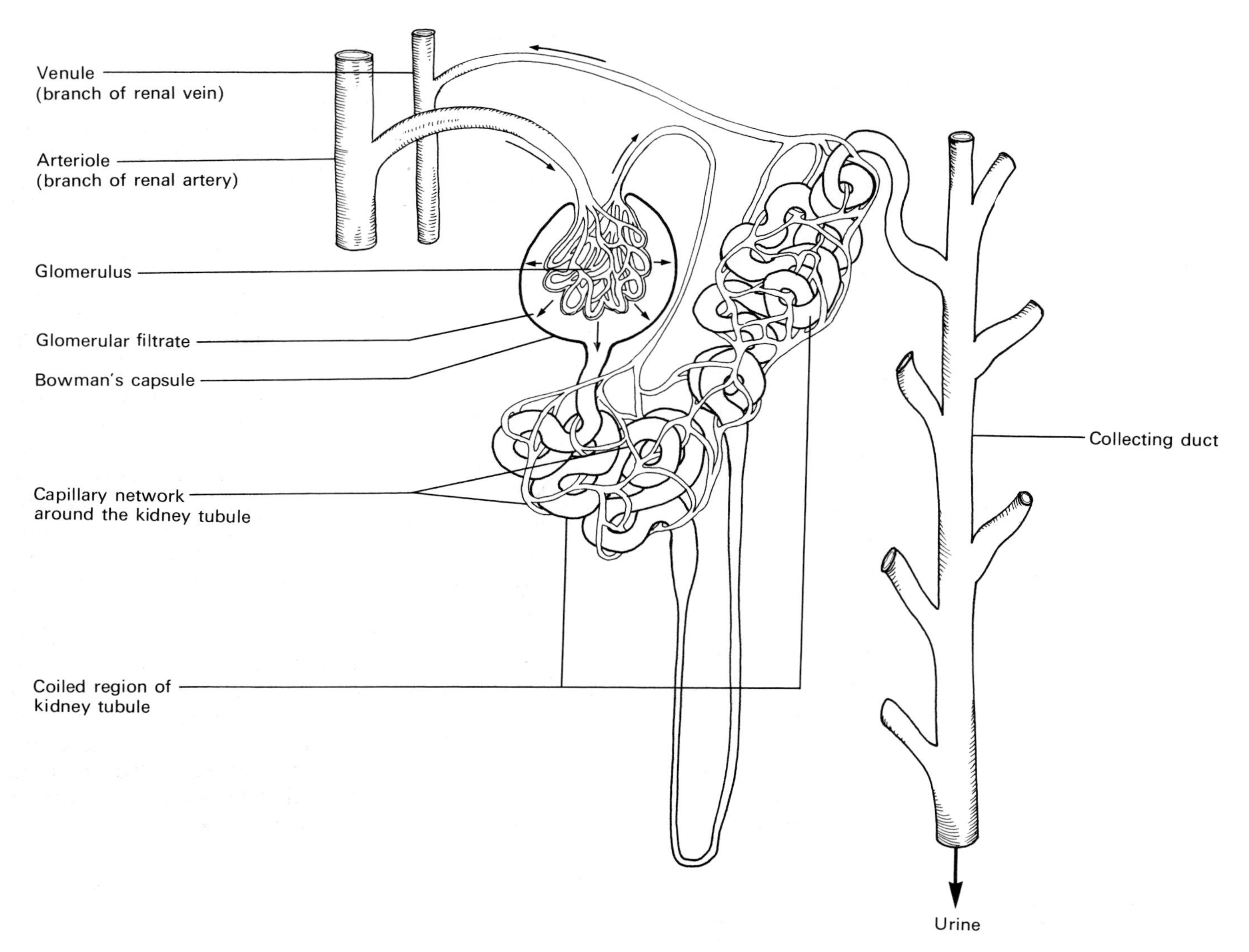

Fig. 10.7 One kidney tubule and its glomerulus. (Fig. 10.8 gives a simplified version)

about 1.5 litres of urine daily depending on the amount of liquid they drink. The full extent of reabsorption is shown in Table 1:

Table 1 Summary of filtration and reabsorption in humans

Substance	*Daily Output*	
	Glomerular Filtrate	*Urine*
Glucose	200 g	Trace
Sodium	600 g	6 g
Potassium	35 g	2 g
Calcium	5 g	0.2 g
Urea	60 g	35 g
Water	180 litres	1.5 litres

The useful reabsorbed materials pass back into the blood through the capillaries which surround the coiled part of each kidney tubule (Fig. 10.7).

Reabsorption is not yet fully understood, but it is known to require a lot of energy. For this reason blood which leaves the kidneys has not only lost most of its waste material but some oxygen and food as well, which have been taken up by the kidneys for respiration.

Reabsorption is not merely a rescue operation to save useful materials in danger of being lost from the body. It is also a means of controlling the level of water and dissolved substances in blood and tissue fluid so that the osmotic pressure of these liquids remains at a constant level. The technical name for the process which controls osmotic pressure is **osmoregulation**.

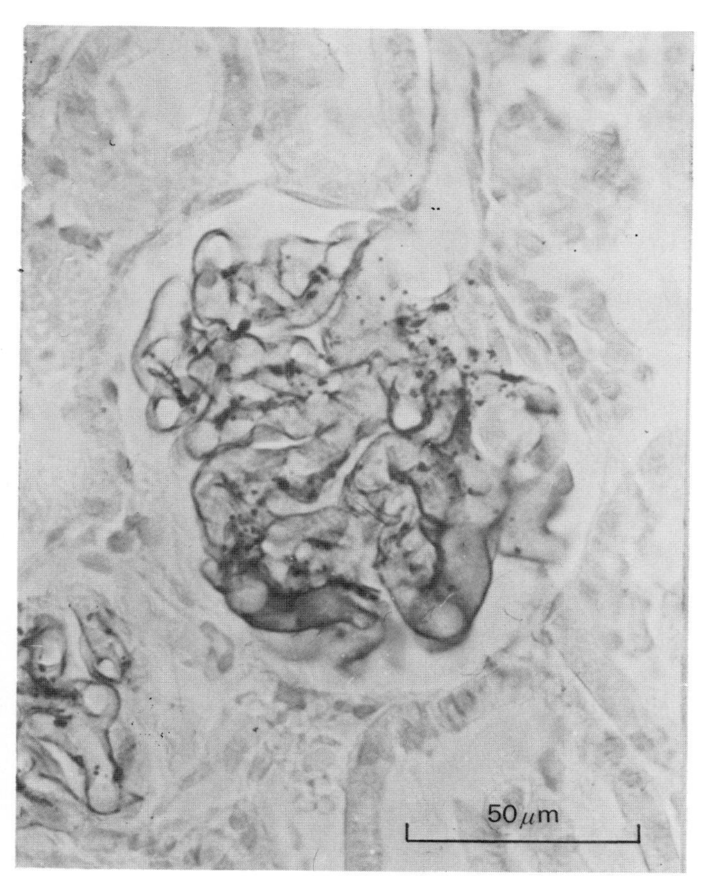

Vertical section through a glomerulus from a rat kidney. The space around the glomerulus is the cavity of the Bowman's capsule

10.5 Osmoregulation

Blood and tissue fluid must be kept at a constant osmotic pressure to avoid unnecessary movements of water in and out of cells by osmosis. If, for example, the osmotic pressure of these liquids is too high, cells lose water by osmosis and the body becomes dehydrated.

In chapter 7 it was explained that the osmotic pressure of a solution depends upon its strength, that is, the amount of dissolved substance which it contains. For instance, the osmotic pressure of a sugar solution depends upon the amount of sugar dissolved in it. If more sugar is added, its osmotic pressure increases. If more water is added its osmotic pressure decreases. Therefore, osmotic pressure can be controlled in two ways: either by altering the amount of dissolved substance in a liquid, or by altering its water content. It is by these methods that kidneys control the osmotic pressure of blood and tissue fluid. Quite simply, they vary the amount of water and dissolved substances which are reabsorbed back into the blood.

Sometimes the osmotic pressure of blood rises because it contains too much dissolved food. This can happen after a very heavy meal, or because of failure in the liver-insulin mechanism described in section 10.2, which controls the level of blood sugar. In these circumstances the kidneys do not reabsorb all of the food material from glomerular filtrate, so that the excess amounts pass from the body in urine.

Fig. 10.8 Diagrammatic summary of nitrogenous excretion

Blood at low pressure on its way to renal vein

Blood at high pressure from renal artery

Glomerulus

Bowman's capsule

Glomerular filtrate (consists of water, glucose, amino acids, vitamins, mineral salts, and urea)

Cells in kidney tubule walls reabsorb some water, and all the glucose, amino acids, vitamins, and mineral salts required by the body

Capillaries pick up reabsorbed substances

Collecting duct (transports urine to tip of pyramid)

Urine on its way to the bladder (consists of water, urea, and some mineral salts)

The osmotic pressure of blood also varies according to its water content. If blood contains too much water, as happens when a lot of liquid is drunk, less water is reabsorbed by the kidney tubules. The result is a large amount of dilute urine. On the other hand, if blood osmotic pressure rises because too little fluid is drunk then the kidney tubules reabsorb a maximum amount of water, and the result is a small quantity of very concentrated urine. In hot weather this same mechanism ensures that more water is available for cooling the body through perspiration.

To summarize: the main function of the kidneys is nitrogenous excretion. They also help conserve the body's water supply so that more is available for perspiration in hot weather. In addition, they make fine adjustments to the blood's dissolved contents and this, among other things, maintains a constant osmotic pressure in the blood and tissue fluid.

Comprehension test

1. In what way should someone with only one kidney adjust his diet?
2. As sweat evaporates, the salt which it contains is left behind on the skin. For this reason, prolonged sweating causes the formation of a progressively stronger salt solution on the skin's surface. Salt solution evaporates more slowly than pure water. In view of these facts, why should a person who lives in the tropics take frequent showers or baths?
3. The thermostat on a central heating system switches the heating on when the building falls below a certain temperature, and automatically switches it off when the desired temperature is reached. How does this system compare with the feed-back mechanism which regulates temperature in the human body?
4. A string vest traps pockets of air between the skin and clothes worn over the vest. Explain why string vests are often worn by arctic explorers?
5. How are the functions of the liver related to the fact that it is an extremely nutritious food?
6. There is a technique whereby an animal's blood pressure can be reduced to half its normal level. Explain why the animal's urine production stops when this is done, whereas most other body functions are relatively unaffected.
7. Explain why an increase in a man's daily intake of protein food causes a corresponding increase in the amount of urea in his urine, whereas this happens to a lesser extent when a child eats more protein.

Summary and factual recall test

Homeostasis is the (1) of a constant (2). It is achieved mainly by (3)-back mechanisms. Homeostasis is best developed in (4), but is found in most (5), and in (6). In these animals there are homeostatic organs which keep the following features of (7) and (8) fluid at a constant level: (9–list seven features).

The main homeostatic functions of the liver include the regulation of (10), (11), and (12) in the blood. The liver helps to achieve homeostasis indirectly by storing vitamins (13–list three), and minerals such as (14–list three). In addition, the liver purifies the blood by (15); helps the (16) of blood in wounds by producing (17); produces (18) energy which is distributed by the blood and so (19) body tissues; and excretes (20) pigments.

A homoiothermic animal is one which (21). Examples are (22–name two). All other animals are said to be (23), which means (24). It is misleading to use the phrases 'warm blooded' and 'cold blooded' because (25).

When the body is over-heated (26) glands in the skin produce a liquid which (27) and so carries away (28). In addition, the (29) capillaries of the skin undergo expansion, or (30), so that a large volume of (31) blood flows (32) to the body surface where it loses heat by (33). Over-cooling is controlled mainly by an involuntary (34) of the muscles which generates extra (35), and by contraction of the hair (36) muscles which reduces heat loss in two ways: (37) and (38).

Excretion is the removal of (39) substances produced by (40), and the removal of substances of which the body has an (41). The kidneys extract a waste substance called (42) from the blood and excrete it in a liquid called (43). This is called nitrogenous excretion because (44). Urine is formed in two stages. Blood is filtered by passing through the walls of capillaries arranged in tiny balls called (45). It then passes into cup-shaped objects called (46). From here the filtrate passes along the kidney (47) where useful substances such as (48–list four) are removed from it and passed back into the blood. This process is called (49).

11
Sensitivity and movement in plants

When a farmer scatters seeds over the soil he does not worry which way up they land. He knows that even if they land on their sides, or upside-down, they will all automatically send their roots down into the soil, and their stems and leaves up towards the light and air. If a potted geranium is placed on a windowsill it is not necessary to worry which way its leaves are facing. The plant will automatically turn its leaves towards the light, and it will also arrange them so that those nearest the light do not overlap those further back. As a result no leaves are in the shade and the plant receives maximum illumination.

Both these examples and a great deal of experimental evidence show that plants respond to certain kinds of stimulation. In biological terms, they respond to a **stimulus**. Plant roots grow downwards in response to the pull of gravity, and they also grow towards water. Plant shoots, on the other hand, respond to gravity in the opposite way by growing upwards, and they respond to light by growing towards it. In short, plants are stimulated by, and can respond to, gravity, water, and light.

The response of a plant to a stimulus is known as a **tropism**, or a **tropic response**. The tropic response to light is called **phototropism**; the response to gravity is called **geotropism**; and the response to water is called **hydrotropism**. This chapter describes tropisms in general with special reference to phototropism and geotropism. A detailed description of hydrotropism is beyond the scope of this book.

11.1 Tropisms

Tropisms are defined as growth movements. This is because the 'movement' involved in the response is produced by a plant's growing points, such as those immediately behind the root and shoot tips. What happens is that the direction of growth alters according to the direction of the stimulus received.

A potted geranium on a windowsill, for instance, detects light coming from only one direction – through the glass. Its stem and leaf stalks respond by growing towards the light, i.e. they bend so that the leaves face the light. But the tropic response is different when a geranium is planted out of doors. Here, it is illuminated from above and it responds by growing straight upwards.

Tropic responses can be either positive or negative. A root, for example, is behaving in a **positively geotropic** manner when it grows downwards in the same direction as the pull of gravity. But shoots are **negatively geotropic** when they grow upwards in the opposite direction from the pull of gravity.

Tropisms are essential to a plant's survival. They enable seedlings to become established with their roots in the soil where they can obtain water and minerals, and their leaves displayed in the air for maximum illumination. Furthermore, tropisms enable plants to alter their pattern of growth at any time throughout their lives to suit changing circumstances. For example, if a plant survives being blown over in a strong wind, new growth curves upwards under the influence of negative geotropism. Similarly, if the water supply from one direction dries up, and the roots can detect water in another direction, new root growth curves in that direction under the influence of positive hydrotropism.

11.2 Phototropism

Phototropism is a growth movement in response to the direction of light; that is, the direction of the growth movement depends on the direction from which the light is coming. Normally, plant stems grow *towards* a source of light, i.e. they are positively phototropic.

Phototropic responses are not normally obvious in plants grown out of doors because they are evenly illuminated from above by the sun and so they grow straight upwards. It is quite possible that the way in which light affects plant growth was first discovered accidentally by farmers who observed some seeds which had grown in the dark, and others which had begun to grow in the shade where they received light from only one direction. If this happened, the farmers would have seen plants which look like those in Figure 11.1. A plant grown out of doors, and evenly illuminated from above, has a short straight vertical stem, and well-developed leaves held horizontally. One grown in the dark has a long thin spindly stem and poorly developed yellow leaves which lack chlorophyll. A plant which is illuminated from only one side has a stem which curves towards the light, and has leaves held at right angles to the light rays.

The responses of plants B and C in Figure 11.1 show how plants are equipped to survive in difficult conditions. If, for example, a seed falls into a shady place the seedling which it produces does not merely die off from lack of light. It uses food reserves stored in the seed to make a long thin stem, and does not waste energy by producing properly developed leaves. Even if it did grow proper leaves they would contain no chlorophyll, because light is necessary for its formation. The light-sensitive stem is able to grow rapidly over or around obstructions towards any source of light until all the food reserves are exhausted. If this race against time is successful and the light is reached before the food reserves run out, the plant produces green leaves and then restores its depleted food reserves by photosynthesis. Phototropism is therefore part of a plant's 'survival kit' in a dangerous world.

Fig. 11.1 The effect of light on the growth of broad bean plants. **A** was grown out of doors; **B** was grown in the dark; **C** was illuminated from one side

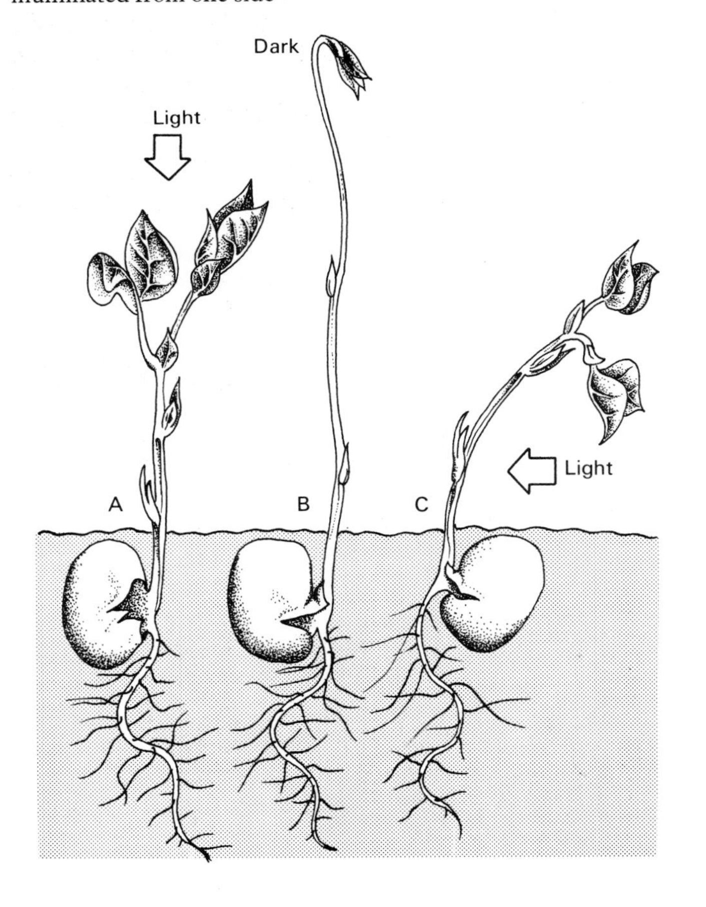

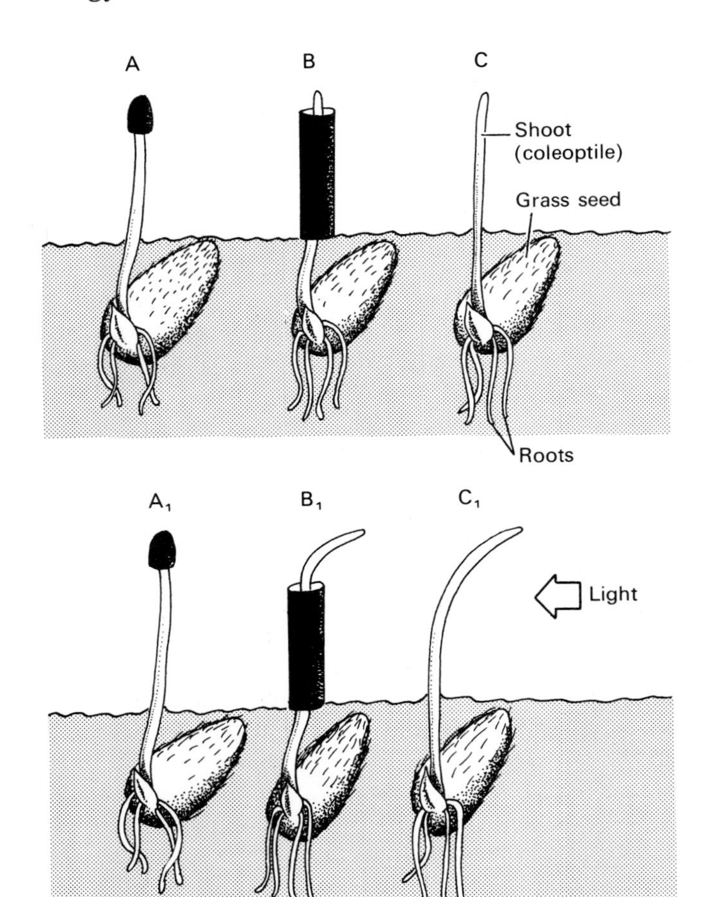

Fig. 11.2 Darwin's experiment to discover which part of a plant responds to light

Early experiments on phototropism

One of the first people to make a scientific study of how plant growth is affected by light was Charles Darwin, the English biologist famous for his study of evolution. Late in the nineteenth century Darwin experimented with a type of grass seedling to discover which part of a plant detects light.

Darwin's method, illustrated in Figure 11.2, was to grow three sets of seedlings. The first set had their shoot tips covered with black paper caps; the second set had everything *except* their shoot tips covered with black paper; and the third set was left to grow uncovered, as a control. Darwin then placed all the seedlings near a window where they received light from one direction only.

The completely uncovered (control) seedlings, and those with their shoot tips uncovered, grew towards the light. The seedlings whose shoot tips were covered with black paper caps grew straight up. Darwin concluded that the shoot tip of a plant is sensitive to light. In addition, he assumed that some 'influence' passes from the tip down the stem causing it to bend.

This experiment is a fine illustration of Darwin's scientific genius. The method is extremely simple and yet it gives clear evidence that shoot tips are sensitive to light. What is more his use of grass seedlings is important because plants in this group germinate (begin to grow) by producing a hollow sheath-like tube known as a **coleoptile**, which encloses and protects the developing leaves until they are free of the soil. Thus, coleoptiles are experimental material which is entirely unencumbered by leaves, side-shoots, and buds. For these reasons Darwin's technique was borrowed and greatly developed early in the twentieth century.

In 1910 it was discovered that an oat coleoptile stops growing when the tip is cut off. But when the severed tip is replaced a few hours later, growth is resumed (Fig. 11.3). This result gave added support to Darwin's theory that the tip of a plant in some way influences the lower part of a shoot. Could this 'influence' be a chemical, or chemicals?

Figure 11.4 illustrates an experiment which suggests that this is true. Growth stops when the tip of a coleoptile is separated from the rest of the plant by a thin sheet of mica (a substance which does not let chemicals pass through). On the other hand, growth does not stop when a thin piece of agar jelly is placed between the tip and the rest of a plant (agar does let chemicals pass through). Subsequent experiments suggested that the mysterious chemical, or chemicals, which influence plant growth can be collected from a shoot tip. This was done by putting the severed tip of a coleoptile on a block of agar for an hour or so, then discarding the tip and placing the agar block on a coleoptile stump (Fig. 11.5). After this the coleoptile resumed normal growth.

These experiments were the beginning of a long line of investigations which eventually led to the discovery

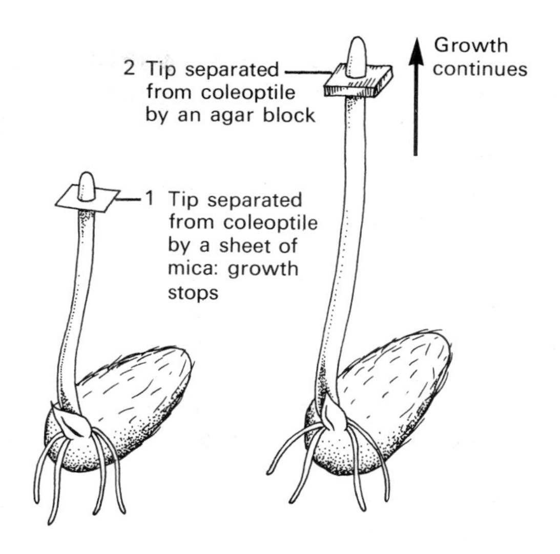

Fig. 11.4 Experiment to show that a growth-promoting substance diffuses from the coleoptile tip

Fig. 11.3 Experiment to demonstrate the growth-promoting properties of a coleoptile tip

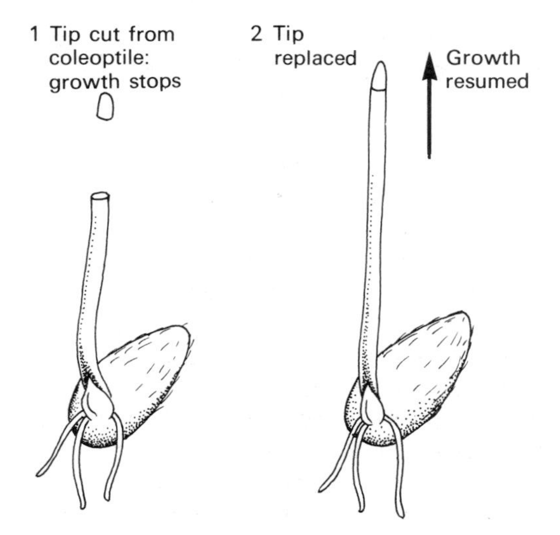

Fig. 11.5 Experiment to show that a growth-promoting substance can be collected from the coleoptile tip

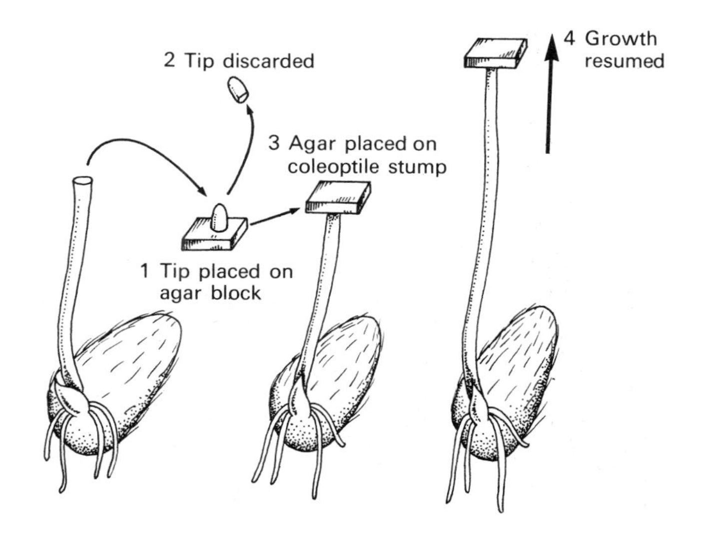

that growth in a coleoptile is controlled by a plant hormone of a type known as **auxin**.

There is now strong evidence that auxin is produced by the region of dividing cells which exist at the tip of a coleoptile, and it is now clear that auxin is produced in the growing tips of most other plants. The auxin passes downwards where it stimulates the growth of cells lower in the shoot. Under the influence of auxin cells grow lengthwise (i.e. they elongate) which results in a corresponding increase in the shoot's length. In other words, auxin promotes growth in plant shoots by increasing the rate of cell elongation. Auxin is the mysterious 'influence' discovered by Darwin more than a hundred years ago.

Auxin and phototropism

What is the relationship between the growth-promoting properties of auxin, and the behaviour of a plant undergoing a phototropic response? A possible answer is illustrated in Figure 11.6. This shows that, under certain conditions, auxin will cause a stem to grow in a curve just as if it were responding to light from one direction.

The tip of a coleoptile is cut off and placed on a block of agar for an hour or so. The tip is then discarded and the block placed on a coleoptile stump a little to one side of centre, i.e. asymmetrically. In time the coleoptile grows in a curve as shown. The most popular explanation of this result is as follows. Auxin diffuses from the block down only one side of the shoot. Since auxin promotes cell growth, there will be a faster rate of growth on this side of the shoot than on the other side which presumably has less auxin

Fig. 11.6 Experiment to show that the growth-promoting substance from a coleoptile tip can cause a growth curvature

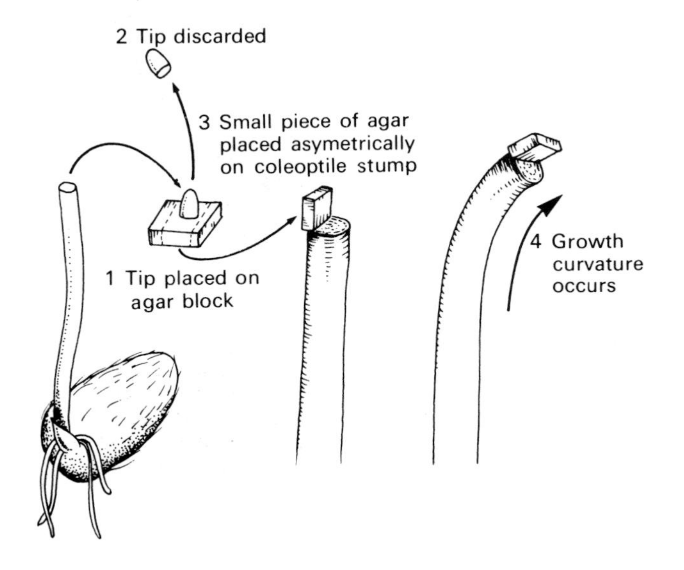

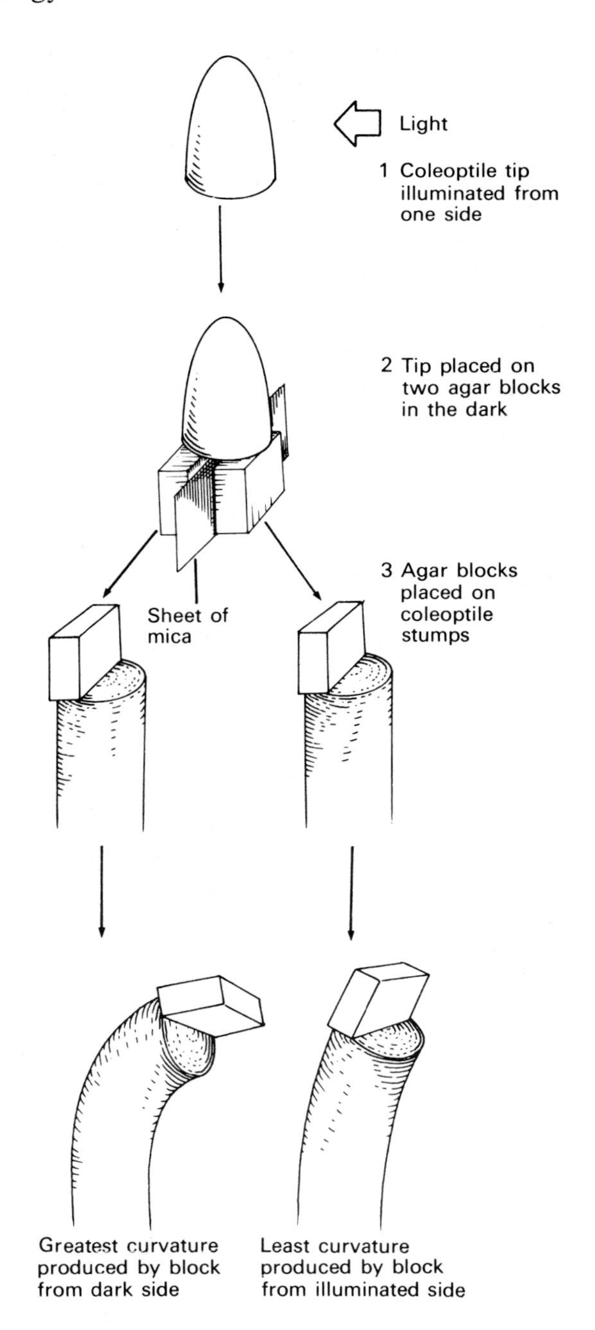

Fig. 11.7 Experiment to show that exposure to light from one direction causes the redistribution of growth-promoting substances in a coleoptile

diffusing through it. If this is so, then growth curvature results from an unequal rate of growth (i.e. cell elongation) under the influence of auxin.

Figure 11.7 illustrates an experiment which suggests that light may bring about a growth curvature by causing more auxin to pass down one side of the stem than the other. The method is to expose a coleoptile tip to light from only one direction. The tip is then placed in the dark on two agar blocks which are separated from each other by a strip of mica. When, still in the dark, these two blocks are placed on coleoptile stumps, the block from the side of the tip which was furthest from the light causes a greater curvature than the block from the side nearest the light. A generally accepted conclusion is that light somehow causes more auxin to gather on the side of the coleoptile tip furthest from the light, i.e. the shaded side. Consequently, more auxin passes into the block under the shaded side of the tip, and it is this block which causes the greatest amount of curvature when placed on a coleoptile stump.

To summarize phototropism so far: it is thought that light affects the direction of growth in most plants by somehow controlling the distribution of auxin in the shoot tip, so that more auxin gathers on the side furthest from the light. Therefore more auxin passes down the shaded side of the shoot where it increases the rate of cell elongation and causes a growth curvature towards the light.

11.3 Geotropism

Geotropism is a growth movement in response to the pull of gravity. When a young bean seedling, for example, is placed in a horizontal position its shoot curves upwards and its root curves downwards. Therefore the root is said to be positively geotropic and the shoot negatively geotropic.

This experiment is more informative if the seedling's root and shoot are marked with lines at 1 mm intervals, as illustrated in Figure 11.8. The spacing of these lines after curvature has taken place shows that the actual bending occurs within only a small region behind the root and shoot tip. There is a way of showing that this region is, in fact, the only part of a plant which can form a growth curvature. The method is simply to turn a seedling from the above experiment through 90° (Drawing 3) which brings the root and shoot once more into a horizontal position. The plant does not respond by straightening out existing growth curvatures, it produces new curves and these are again in the region behind the root and shoot tips. The explanation is that curvatures can form only in these regions because they are the only parts of a plant where the majority of cells are growing by elongation. Elsewhere cells are either dividing, e.g. at the extreme tips of the root and shoot, or are fully grown and often specialized for a particular function.

The seedlings described above seem to be responding to the downward force of gravity, and yet scientific method demands a control experiment to give actual evidence that this is so. A control must somehow be devised which provides a plant with a force equivalent to gravity which pulls equally in all directions rather than just downwards. This is done by using a **clinostat** (Fig. 11.9). A clinostat consists of a cork disc which can be rotated at various speeds by a motor. This machine can be set up as shown to rotate a seedling in a horizontal position. When the speed of rotation is

Fig. 11.8 Experiment to demonstrate geotropism in root and shoot

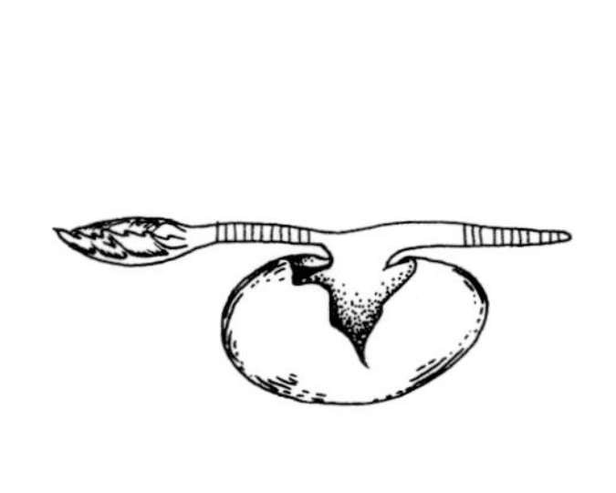

1 Bean seedling placed in a horizontal position in the dark, with markings at 1 mm intervals

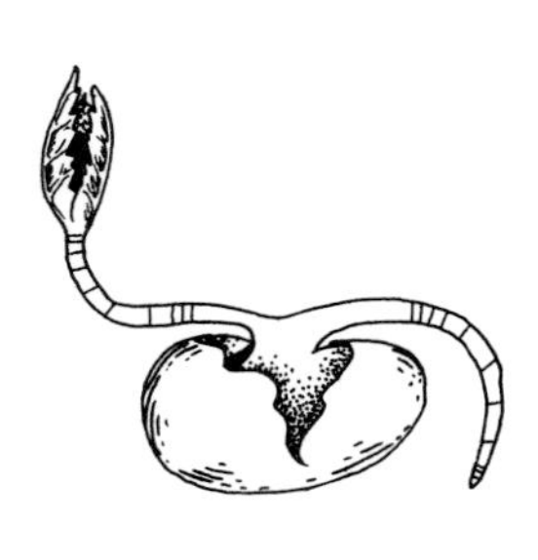

2 The shoot develops an upward (negatively geotropic) curvature, while the root develops a downward (positively geotropic) curvature. The markings indicate where cell elongation has occurred

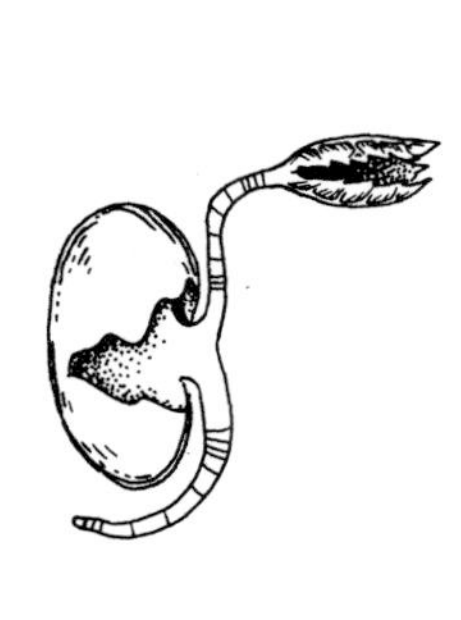

3 Seedling rotated through 90°

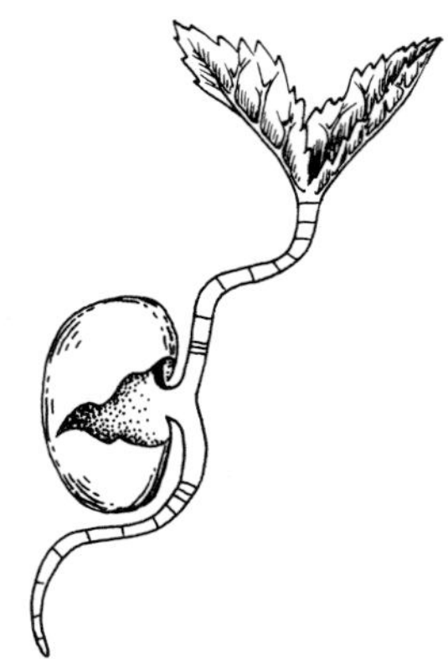

4 Old growth curvatures remain unaltered because they are established (differentiated) tissue. New curvatures develop only behind the root and shoot tips

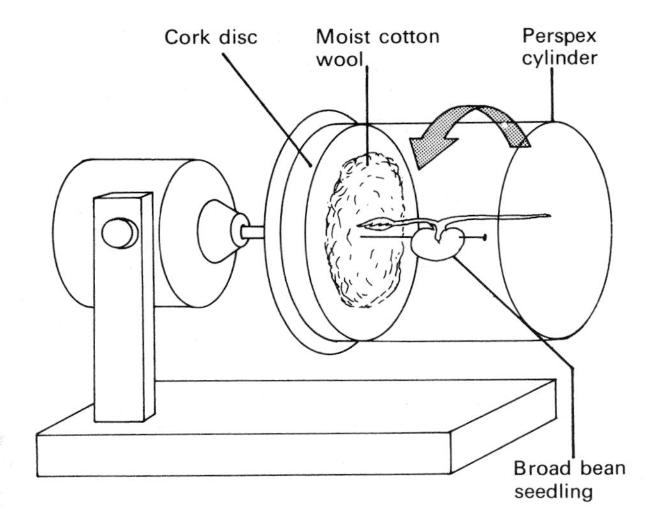

Fig. 11.9 A clinostat in operation

set at about once per hour, the force of gravity acts equally on all parts of the plant. Furthermore, the seedling does not remain in any one position long enough to make a geotropic response. Hence, its shoot and root grow horizontally. At slower speeds of rotation the plant does have time to begin a geotropic response but the direction of the stimulus changes while this is happening. The result is that the root and shoot develop twisted curvatures, like a corkscrew.

There is evidence that auxin plays a part in geotropic responses. One theory is that when a plant is placed in a horizontal position the force of gravity causes auxin to gather in the lower half of the root and shoot (Fig. 11.10). Presumably the auxin accelerates cell elongation in the lower half of the shoot thereby causing an upward curvature. On the other hand, if auxin becomes distributed by gravity as illustrated then it must have an opposite effect on root cells: in order to produce a downward curvature auxin must

Fig. 11.10 Gravity is thought to cause a redistribution of auxin in root and shoot

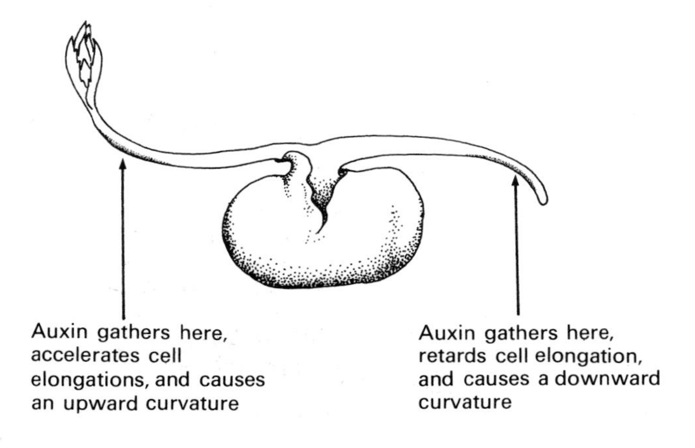

retard cell elongation in the lower half of the root. Recent investigations suggest that this is an over-simplified explanation of geotropism. Apparently auxin is only one of several plant hormones at work in geotropic responses.

Plants are sensitive to other stimuli than light and gravity, and not all responses are too slow to see. Some of the following exercises suggest ways in which other types of response may be investigated.

Verification and inquiry exercises

A *To verify that light affects stem growth*

1. Germinate four sets of mustard seeds in small dishes on moist cotton wool in the dark.

a) Keep one dish of seedlings permanently in the dark.

b) Put another dish of seedlings in a cardboard box with a circular hole about 3 cm in diameter cut in one end. Position the box so that light shines into the box through the hole.

c) Arrange a clinostat so that its disc is in a horizontal position and set it to rotate about once an hour. Place a dish of seedlings on the rotating disc, then put the whole apparatus in a cardboard box with a hole cut in one end, as in (b).

d) Put another dish of seedlings out of doors.

e) After a few days compare the shape of plants in each of the four situations, and explain the results.

f) From the position of the leaves in experiment 1 (b) explain why leaves are sometimes described as 'dia-phototropic'?

B *An investigation of phototropic responses*

1. *An 'obstacle course' for plants*

a) Prepare three cardboard shoe-boxes as shown in Figure 11.11.

b) Put a dish of germinating mustard seeds at the end of one box furthest from the hole.

c) Put a pot of germinating broad beans at the same end of another box.

d) Put a pot containing a sprouting potato at the same end of another box.

e) Compare the ability of each type of plant to reach the light. Why are some more successful than others?

f) Examine and explain the shape of each type of plant at the conclusion of the experiment.

g) Devise other types of 'obstacle course' to test the ability of plants to reach light.

2. *Positive and negative phototropism*

a) Prepare a jar with a cork bung as shown in Figure 11.12 and then cover the whole jar with black paper except for a narrow vertical slit down one side. (Alternatively cover the jar with black paint except for a slit down one side.)

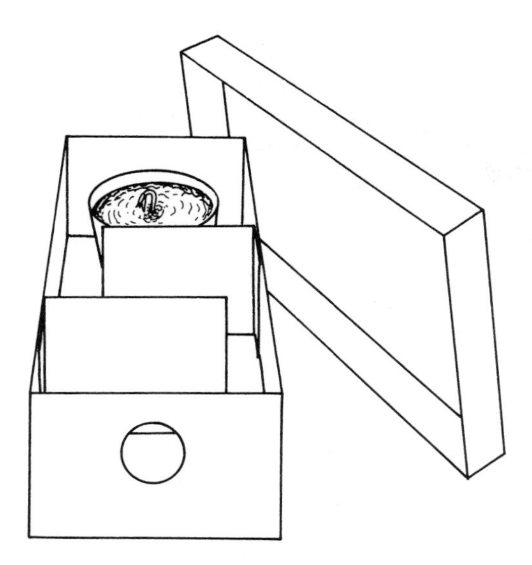

Fig. 11.11 An obstacle course for plants (exercise B1)

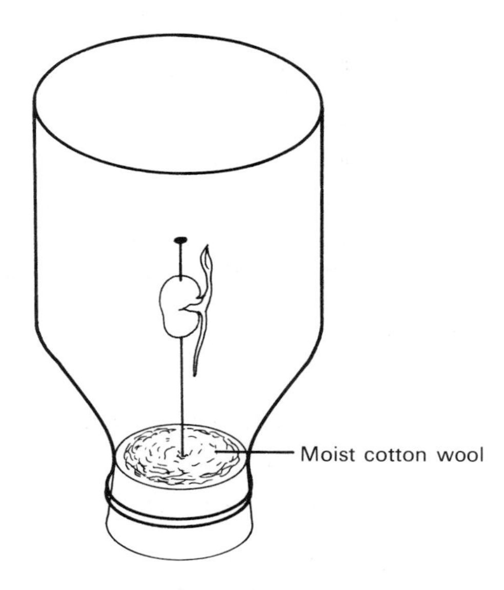

Fig. 11.12 Apparatus to demonstrate positive and negative phototropism (exercise B2)

b) Place the jar in a position where light will shine through the slit.

c) Explain the growth curvatures which occur in the bean root and shoot.

C *To verify that gravity affects the growth of plants*

1. *a*) Prepare a clinostat as shown in Figure 11.9, and place it in the dark.

b) Using different seedlings each time run the clinostat for two or three days at different speeds. Repeat the experiment without rotating the clinostat.

2. *a*) Explain the results obtained in each experiment.

b) Why must the clinostat be placed in the dark?

c) The moist cotton wool provides the seedlings with a uniformly humid atmosphere. Give at least two reasons why this is a necessary part of the experiment.

D *To verify that auxin affects growth curvatures*

1. Germinate three sets of oat or wheat grains in small dishes on moist cotton wool in the dark. Put about five grains in each dish. Obtain some commercially prepared lanolin paste containing the auxin indoleacetic acid (IAA for short), and some plain lanolin. *All* of the following operations must be carried out in red light, after which the plants must be kept in the dark.

a) After about five days, when the coleoptiles are about 2 cm long, select at least two of the straightest plants in each dish and cut down the rest. Do not use any plant in which the leaves have broken through the coleoptile sheath.

b) Leave one set of plants untouched as the control group. Smear a small quantity of warm lanolin with IAA down *one* side of another set of plant coleoptiles. Repeat this operation on the third set of plants using the plain lanolin.

c) Observe the plants again (in red light) after about three hours and explain any growth curvatures which have taken place.

d) Coleoptiles are insensitive to red light. Why is it necessary to observe the plants only in light of this colour?

F *An investigation of other tropic responses*

1. *Hydrotropism* Hydrotropism is growth curvature in response to water.

a) Place some pea seeds at the bottom of a sieve and cover them with moist bulb fibre. Prop up one end of the sieve so that it rests at an angle (Fig. 11.13). Keep the fibre well watered.

b) Note what happens to the pea seedling rootlets when they emerge from the sieve. Which stimulus do they respond to: gravity or water?

c) Why must the sieve be placed at an angle?

2. *Haptotropism (thigmotropism)* Haptotropism is a tropic response in which the stimulus is an object in contact with the plant. Tendrils of pea and vine, and the petioles of nasturtium are particularly sensitive to stimuli of this kind.

Fig. 11.13 Apparatus to demonstrate hydrotropism (exercise F1)

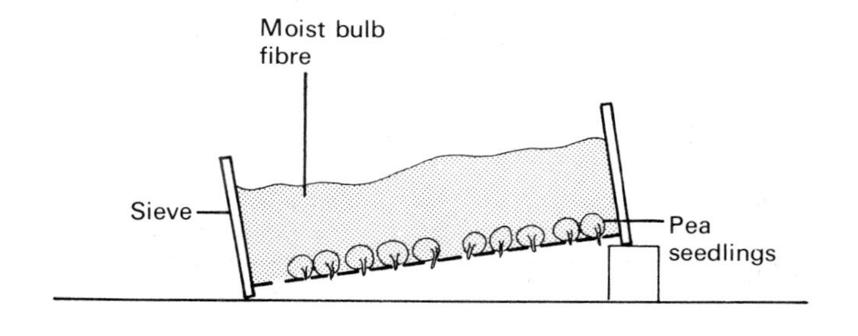

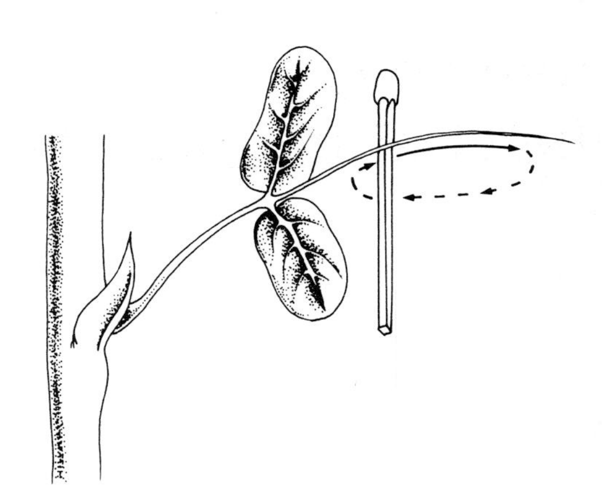

Fig. 11.14 A method of demonstrating haptotropism in a pea leaf tendril (exercise F2)

a) Chose a fairly straight young tendril of a pea plant or a vine. Gently stroke the inside of the curve with a matchstick (Fig. 11.14). Note what happens to the stroked tendril in comparison with others of similar shape which are not stroked.

b) What advantage is haptotropism to these plants?

3. *Nastic movements* Unlike the tropisms already described, nastic movements are responses in which the direction of the movement is not determined by the direction of the stimulus. Certain flowers, for example, open and close in response to changes in light intensity, but the movement of their petals is entirely independent of the direction from which the light shines. Another difference is that most nastic movements are not caused by elongation of cells in the growth regions. They are caused by changes in the turgidity of cells in regions such as the bases of the petals and the bases of certain leaves. These turgor changes cause the cells to inflate or deflate thereby moving the petals or leaves, sometimes within the space of a few seconds.

a) *Sleep movements* Examine one or more of the following plants by day and by night and note any differences in the arrangement of either their leaves or petals: wood sorrel, evening primrose, clover, daisy.

b) *Response to touch* Raise seedlings of *Mimosa pudica* in potting compost in a heated green house or near a window in a warm room. (Seeds of this plant are available from the larger seed suppliers.) The mature leaves consist of many small leaflets. Touch them gently with a finger and observe their remarkably quick response.

c) *Response to chemicals* Insectivorous plants have leaves which are adapted in various ways so that insects are trapped in them and eventually digested. In this way the plants obtain a supply of nitrogen, which is lacking from the soil in which they grow. Obtain seedlings of Venus's flytrap from a seedsman, or search along the banks of a moorland stream for sundew plants. Place small pieces of raw meat on the leaves of these plants and observe their response over a period of about an hour.

Comprehension test

1. Auxin is known to reduce the plasticity of cell walls in a coleoptile. That is, under the influence of auxin the cell walls are more easily stretched, and they do not spring back to their original length (like elastic) when the stretching ceases.

Light from only one side causes auxin to gather on the opposite (shaded) side of a coleoptile.

Cells continue taking in water by osmosis until their walls can stretch no further.

Use these three facts to *explain* the following:

a) the elongation of a coleoptile;

b) the fact that elongation ceases when the tip is removed;

c) the curvature of a coleoptile when illuminated from one side;

d) the fact that once such a curvature has formed it becomes a permanent feature of the plant;

e) the fact that no bending occurs when a coleoptile is illuminated from one side provided its tip is covered by a black paper cap;

f) the fact that shading the tip does not prevent elongation.

2. Why is statement (a) below a more accurate, and more scientific description of tropisms than statement (b)?

a) Shoots bend towards the light because light causes an unequal distribution of auxin in the stem, which in turn causes an unequal rate of cell elongation.

b) Shoots bend towards the light in order to receive more light for photosynthesis.

Summary and factual recall test

Tropisms are defined as (1), because (2). The tropic response to light is called (3). Three sets of cress seedlings were treated as follows: set A was placed in the dark; set B was placed in a box with a slit cut out at one end; set C was evenly illuminated from all sides. The tallest set was (4). Set (5) had yellow leaves. The stems of set (6) were curved. Set (7) was the control group and had (8) leaves. These plants were necessary to show that (9). It is concluded that shoots are (10) phototropic, and that light is required for (11) formation.

Response to gravity is called (12). A bean seedling was placed in a horizontal position and its shoot curved (13), showing a (14) response, while its root curved (15) showing a (16) response.

There is evidence that tropisms are controlled by a hormone called (17). Light, for example, is thought to control the (18) of plant growth by affecting the distribution of this hormone at the shoot (19), so that more of it passes down the shaded side of the shoot where it (20) the rate of cell (21) thereby causing growth curvature (22) the light.

equivalent to three average-sized family cars!

At the same time bone is very light. This is important because less energy is required to move a light skeleton, and less material is required to make it. The large bones, like the thigh bone (femur), are light for their size partly because they have hollow **shafts**, and also because at their ends, or **heads**, the bone is full of holes like a sponge (Fig. 12.3). This hollowness does not weaken a bone very much. In fact bone absorbs most of the stresses to which it is subjected over its surface regions, so a hollow bone does almost as well as a solid one. The hollow region inside a bone is occupied by tissue which manufactures red and white blood cells.

Joints

Joints occur wherever two or more bones touch. In some cases the bones are joined firmly together by fibrous tissue and may even have their edges dovetailed into one another; this is the case with the flattened bones which make up the roof of the skull. Other joints have a pad of flexible gristle (cartilage) bound

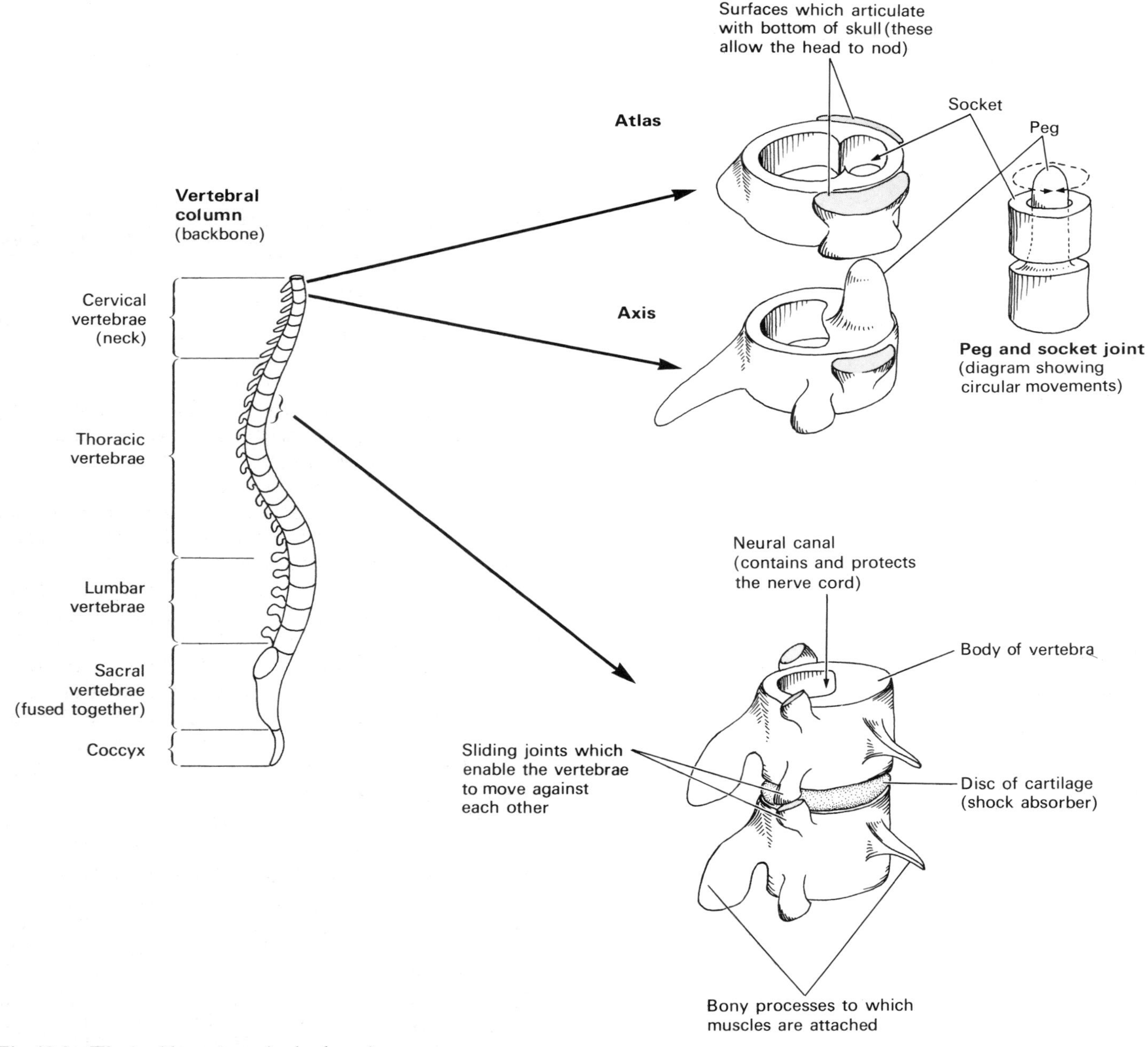

Fig. 12.2 The backbone (vertebral column)

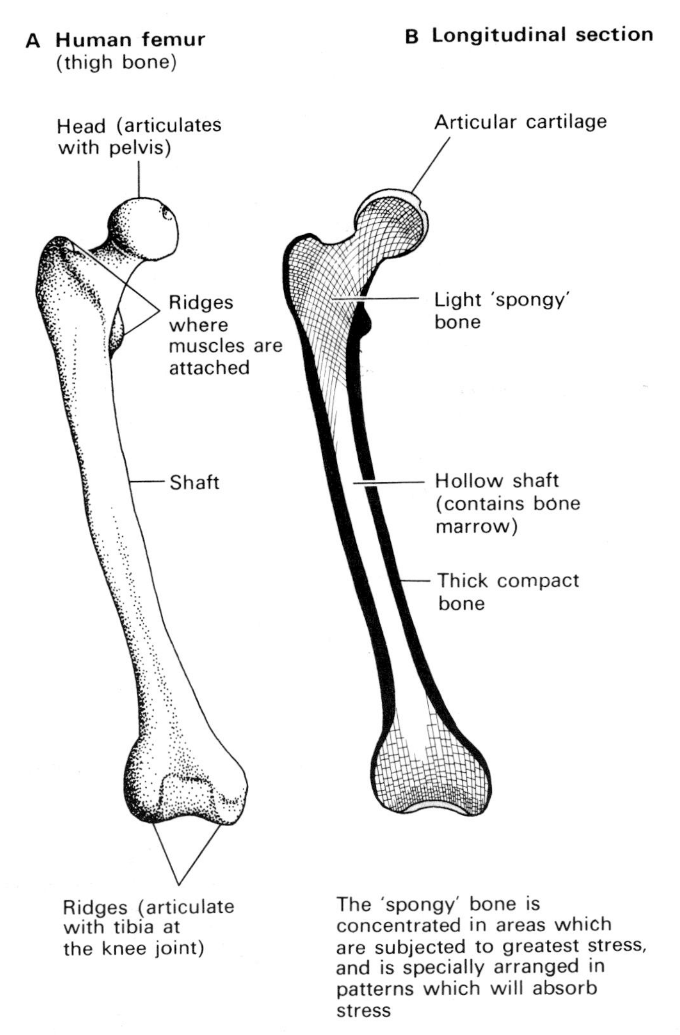

Fig. 12.3 Structure of a bone (human thigh bone)

between the bones, and slight movement is possible, e.g. the places where ribs pivot against the sternum (breastbone) during deep breathing (Fig. 9.5B), and between adjacent vertebrae in the backbone (Fig. 12.2). In the latter case the cartilage pads, or **intervertebral discs**, act as shock absorbers. That is, they absorb shocks and jolts transmitted to them through the limbs during running, jumping, and similar activities. Without these discs, shocks would clatter along the line of vertebrae like the shock waves which pass down a line of shunted railway carriages.

There are also about seventy freely moveable or **synovial joints** in the mammalian skeleton. Synovial joints are so called because the region where one bone rubs against the next is enclosed within a capsule filled with **synovial fluid** (Fig. 12.4A). This fluid lubricates the joint when the bones move. Another feature of synovial joints which reduces friction is the layer of slippery **articular cartilage** which covers the surfaces of the bones that rub together, i.e. the articular surfaces. The whole joint is enclosed within a layer of tough fibres, the **ligament**, which holds the bones firmly in place and yet allows free movement of the joint.

Types of movement at the joints

Imagine two players engaged in a game of tennis. They are using all of their seventy-odd synovial joints and more than six hundred muscles to achieve the agility required to hit and recover the tennis ball. By analysing body movements in such a situation it is possible to learn a great deal about the marvellous flexibility of the human bone and muscle machinery. Study Figure 12.1 in conjunction with the following notes.

Longitudinal section through the upper end of a human femur. Identify the parts using Figure 12.3 as a guide

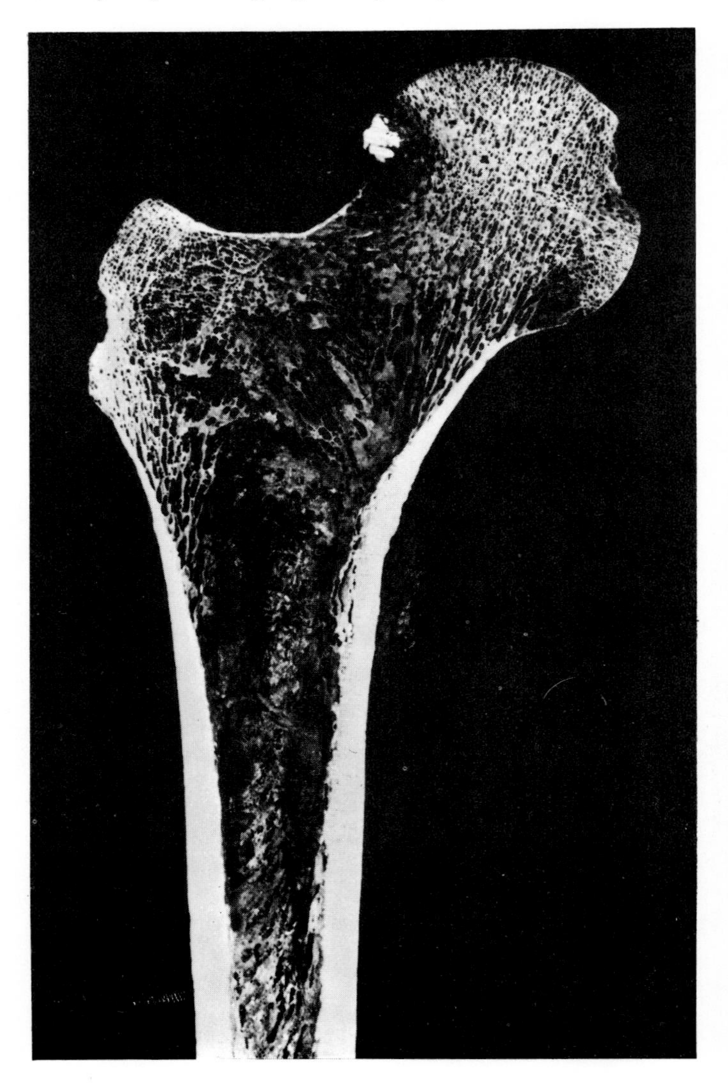

Hinge joints The elbows, knees, and knuckle joints of the fingers are examples of hinge joints. They move, like the hinge of a door, in one plane only.

Ball-and-socket joints The shoulders and hips are examples of ball-and-socket joints. Here the rounded head of one bone fits into a cup-shaped socket in another. These joints give the greatest flexibility of movement of all joints (Fig. 12.4B).

Other types of joint There is an intricate arrangement of small sliding surfaces between adjacent vertebrae (Fig. 12.2). These permit the backbone to bend at the waist and neck in all directions, but particularly forwards, together with a certain amount of twisting. Movements between vertebrae add considerably to the body's flexibility. To verify this point think how restricted one's movements would be if the backbone were a solid rod. The greatest amount of flexibility in the backbone exists where the skull articulates with (i.e. moves against) the topmost vertebrae. Here, the **atlas vertebra** takes the weight of the skull (like the Greek god Atlas, who according to mythology balanced the world on his shoulders). This articulation

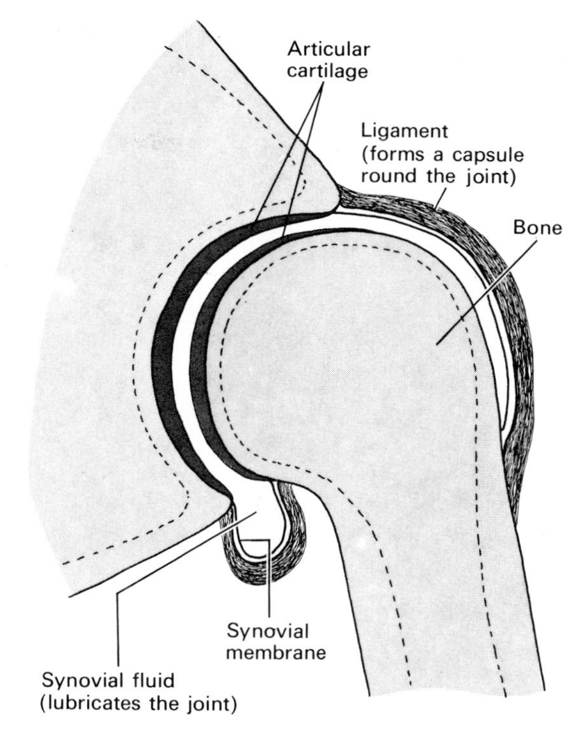

Fig. 12.4A Structure of a synovial joint (human shoulder joint)

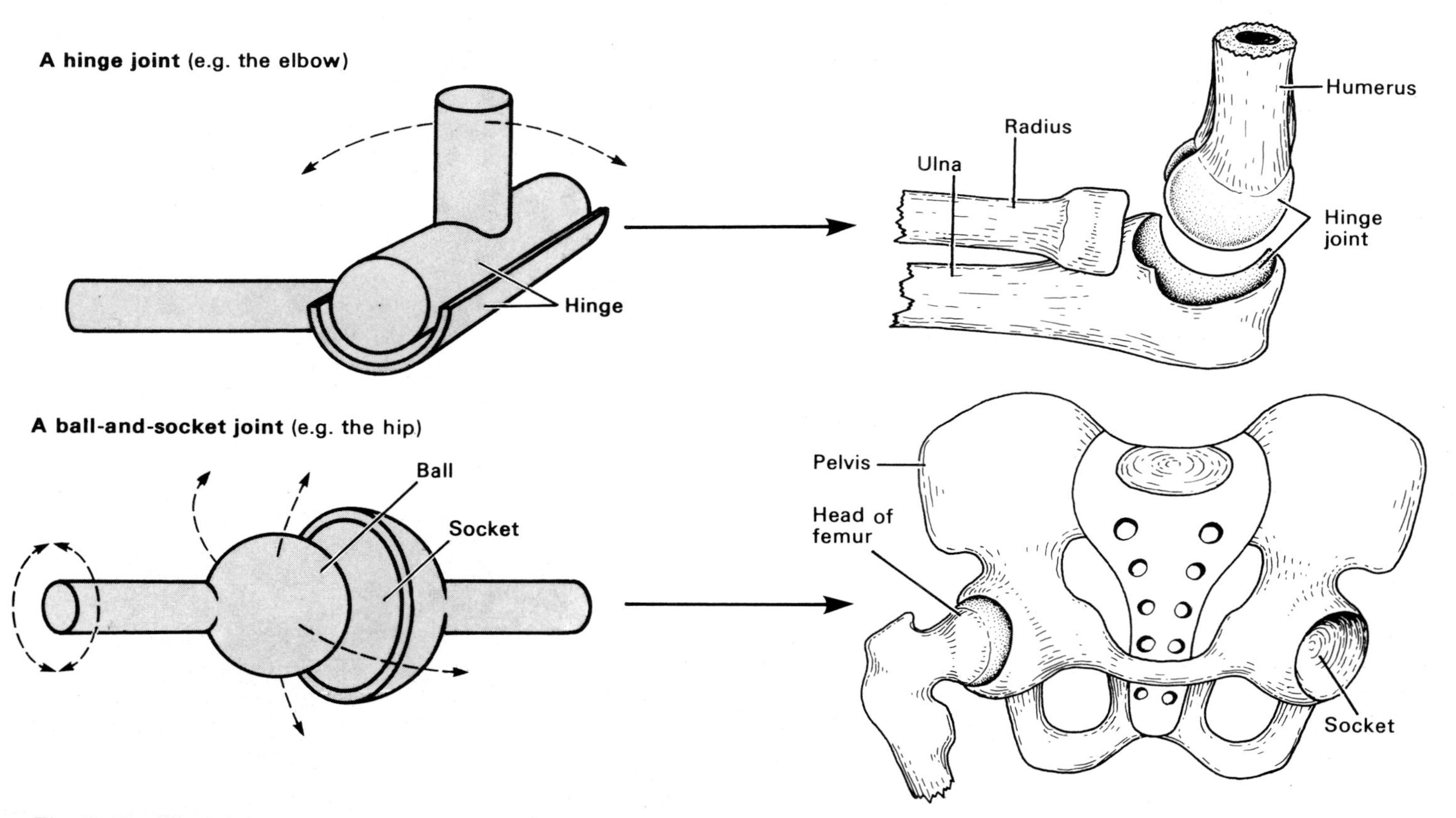

Fig. 12.4B Hinge joints and ball-and-socket joints

Fig. 12.5 Muscles and bones of the elbow joint

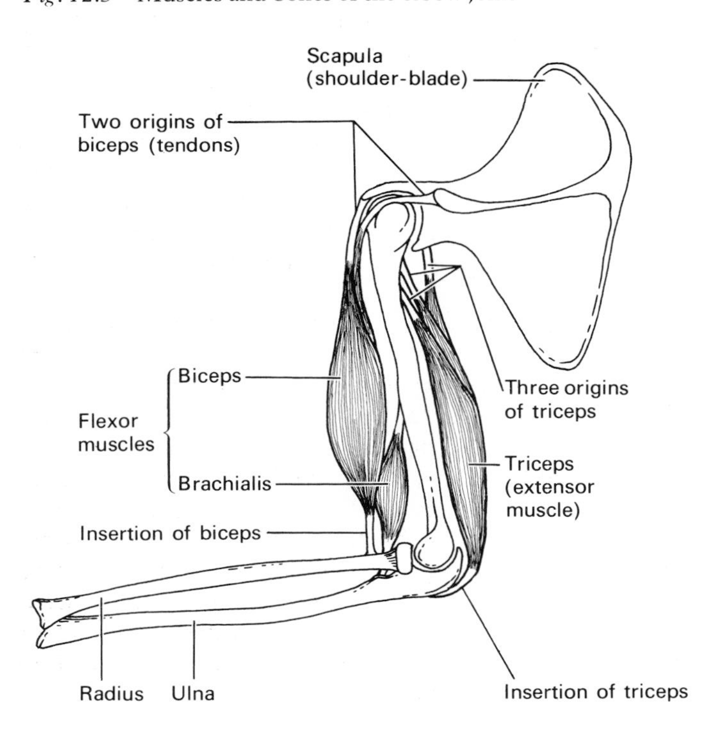

permits nodding movements of the head in all directions. Below the atlas, the **axis vertebra** permits swivelling or pivoting movements of the head, which occur, for example, when a person shakes his head to say no.

Another pivot joint occurs at the head of the radius bone near the elbow (Fig. 12.4B). This permits the forearm to twist, as it does when turning a screwdriver.

Muscles

Muscles make up 40–50 per cent of body weight in most mammals. They are the 'meat' of the body, and consist largely of protein, which is why they are a valuable food.

The muscles which move the body are referred to as **voluntary**, or **skeletal muscles**. They are under conscious control, and are attached to bone at both ends. The point of attachment between a muscle and a bone is called a **tendon** (Fig. 12.5). Tendons consist of very strong inelastic (i.e. non-stretchable) fibres which begin inside the bone and penetrate deep into the muscle tissue, attaching it firmly to the bone. In freshly killed meat tendons appear as glistening silver-grey strands between bone and muscle.

When a muscle contracts it exerts tension between its two points of attachment. One of these points remains fixed. This is the anchorage point, or **origin**, of the muscle. The other end moves as a result of this tension, and is called the **insertion** of the muscle. The origins and insertion of the biceps muscle are illustrated in Figure 12.5.

In mammals, as with all other animals described in this chapter, muscles work opposite each other in **antagonistic systems**. In each of these systems one set of muscles causes the bending of a joint. These are the **flexor** muscles. The **extensor** muscles are those which work in the opposite direction to straighten the joint. Flexor and extensor muscles are illustrated in Figure 12.5.

Sometimes both sets of muscles in an antagonistic system may contract at the same time. This is done to lock the joint at a particular angle. An example of this is the locking of arm muscles to hold something firmly in one particular position so that it doesn't move in any direction.

In locomotion the muscles do not work at random but in a precisely ordered sequence controlled by the nervous system. This sequence generally has to be learned, which is one reason why babies cannot walk at birth.

Even when a person is sitting perfectly still many of his muscles are at work to counteract the force of gravity on the body. These muscles are responsible for body posture, and if they were to stop working suddenly a person would collapse to the floor like a rag doll. The main posture muscles are those which run from the hip bone (pelvis) to the back of the lumbar vertebrae. These are the muscles which may become fatigued and cause backache if the body has to stoop forwards for prolonged periods.

Muscles and leverage

During any kind of movement the bones act as levers, and all lever systems consist of three parts. The **fulcrum** is the point round which the lever pivots. In the skeleton every moveable joint is a fulcrum. All levers are moved by **effort**, which in the body is supplied by muscles. Lastly, all levers carry a **load** of some kind. The levers in the skeleton support the weight of the body, and additional loads such as objects carried in the hands.

A pair of scissors is a kind of lever. The fulcrum is the hinge in the middle of the scissors, the effort is supplied by the hands on the scissor handles, and the load is the resistance offered by whatever is being cut. The fulcrum of a pair of scissors is between the load and the effort, and Figure 12.6A shows a similar type of lever in the human body. The fulcrum is the atlas vertebra, and upon this point the head (load) is

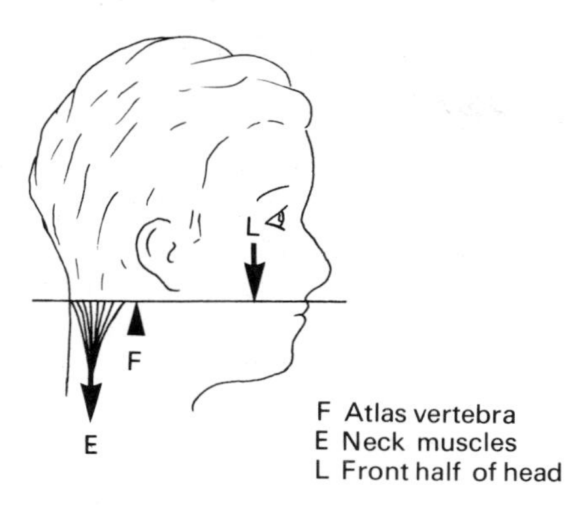

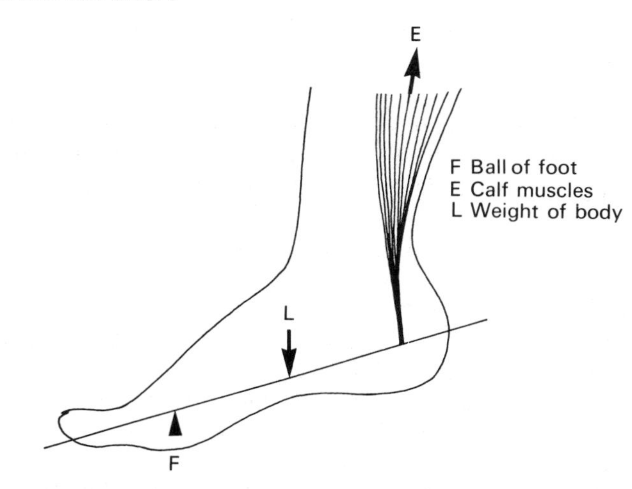

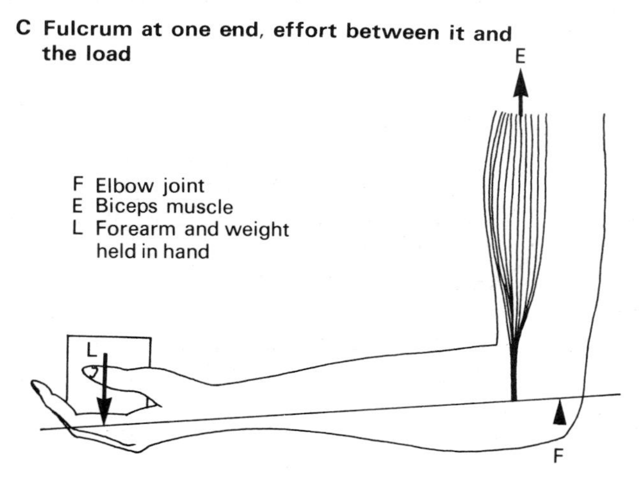

Fig. 12.6 Examples of leverage in the human body

moved by the neck muscles (effort). Another example of this type of lever is the trunk of the body (load), balanced on the lumbar vertebrae (fulcrum), and moved by the muscles of the back (effort).

A wheelbarrow is another example of a lever system. The fulcrum (wheel) is at one end of the lever, and the load (objects carried in the barrow) is between the fulcrum and the effort (force needed to lift the barrow by its handles). Figure 12.6B shows this kind of lever system at work in the skeleton when a person stands on tiptoe.

Yet another type of lever is one in which the effort is applied between the fulcrum and the load. Sugar tongs are an everyday example. Figure 12.6C shows that the elbow joint is a similar example in the human body. A little thought is necessary to see that the biceps muscle is situated in a position where it must produce a great deal of effort to lift a comparatively small load. Even so, a very small movement of this muscle produces a much larger movement at the hand end of the lever. Hence, the design of this lever system produces large fast movements, rather than small powerful ones which is the case with the other two types of lever described above.

12.2 Birds

Birds, and all other flying animals, remain airborne only as long as they are lifted by forces at least equal to their body weight. To put it another way, these lifting forces–simply called **lift**–are necessary to overcome the force of gravity and allow the animals to remain in the air.

In birds, lift results partly from the flapping movements of their wings. These movements also drive the birds forwards through the air. However, the most remarkable thing about birds is that their wings can generate lift whether they are flapped up and down or not. In fact many birds can remain airborne for long periods with hardly any wing movements, provided there is a current of air over their wings. Air currents are easily generated either by gliding downwards at an angle through the air with wings outstretched, or by heading into wind currents. Gulls, terns, swifts, and albatrosses spend most of their lives flying in this way.

The secret behind this kind of flight lies in the shape of a bird's wing as seen in cross-section. This shape is called an **aerofoil** (Fig. 12.7), and a study of its special properties explains not only how birds can defy the pull of gravity, but also how man has conquered the air with aeroplanes.

Fig. 12.7 How an aerofoil generates lift

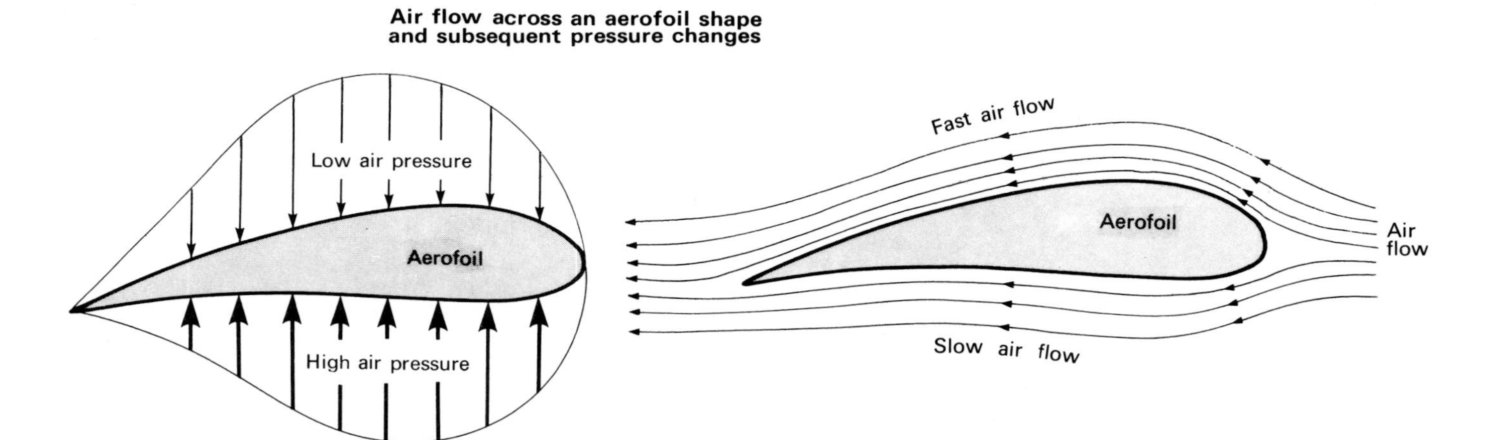

How an aerofoil generates lift

The shape of an aerofoil is such that air flowing across its upper surface moves faster than air flowing across its under surface. The importance of this is that fast-moving air always has a *lower* pressure than slow-moving air. Consequently, fast-moving air above an aerofoil creates a zone of low *downward* pressure, while slow-moving air below an aerofoil creates a zone of high *upward* pressure (Fig. 12.7). The downward pressure is less than the upward pressure and the difference between these two oppositely directed forces is a smaller upward force: the lift.

The power of this lifting force increases as the bird's air speed increases. The same principle is used by humans to send huge aircraft high into the atmosphere.

The next thing to consider is the lifting and propulsive forces that are generated by wings as they are moved up and down in active flapping flight.

Flapping flight

To understand what happens when wings are flapped up and down it is necessary to know a little about the skeleton of a bird. Figure 12.9 shows that a bird's wing is similar in structure to a human arm. The 'forearm' part of a wing is covered with **secondary feathers** arranged in such a fashion that they give it the aerofoil shape described above. The 'hand' part of a wing is covered with **primary feathers**. The whole wing is moved up and down by elevator and depressor muscles shown in Figure 12.8. In addition there are muscles which enable the 'hand' to be moved independently in a circular twisting movement at the 'wrist' joint.

Downstroke The depressor muscles pull the wing downwards and forwards. During this stroke the leading edge of the wing is lower than the trailing edge and so the wing does not merely press down on the air to give lift; it pushes the air backwards to drive the bird forwards. Before the wing tip has reached the lowest point of the downstroke the 'forearm' begins to rise so that the wing bends at its 'wrist' joint (Fig. 12.10). Therefore the 'hand' part of the wing begins its upstroke slightly after the rest of the wing.

Fig. 12.8 Flight muscles of a bird

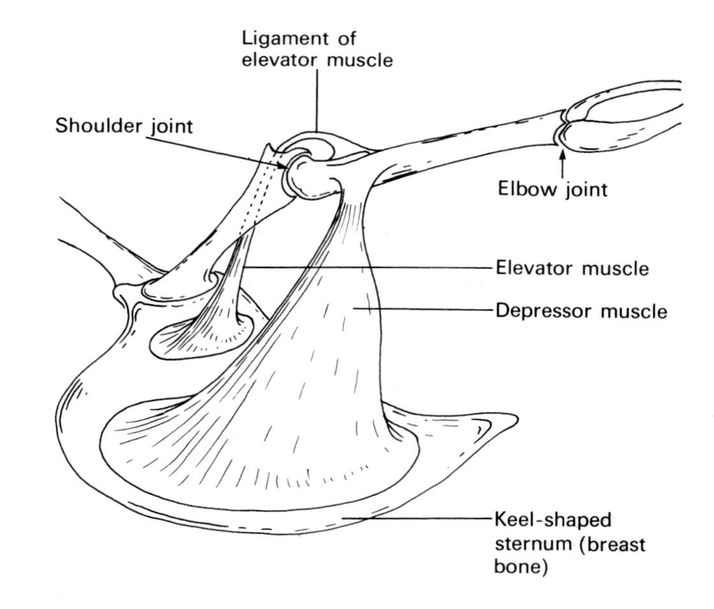

Upstroke At the bottom of the downstroke the wing twists so that its leading edge is now higher than its trailing edge. This causes an instantaneous build-up of pressure under the wing which flips it quickly upwards and backwards ready for the next downstroke.

Figure 12.11 shows a mechanism, present in some birds, which greatly reduces wing resistance during the upstroke, thereby increasing its speed. The feathers swivel like the slats of a venetian blind. The feathers are 'closed' on the downstroke for maximum resistance

Fig. 12.9 Diagram to compare the structure of a bird's wing with a human arm

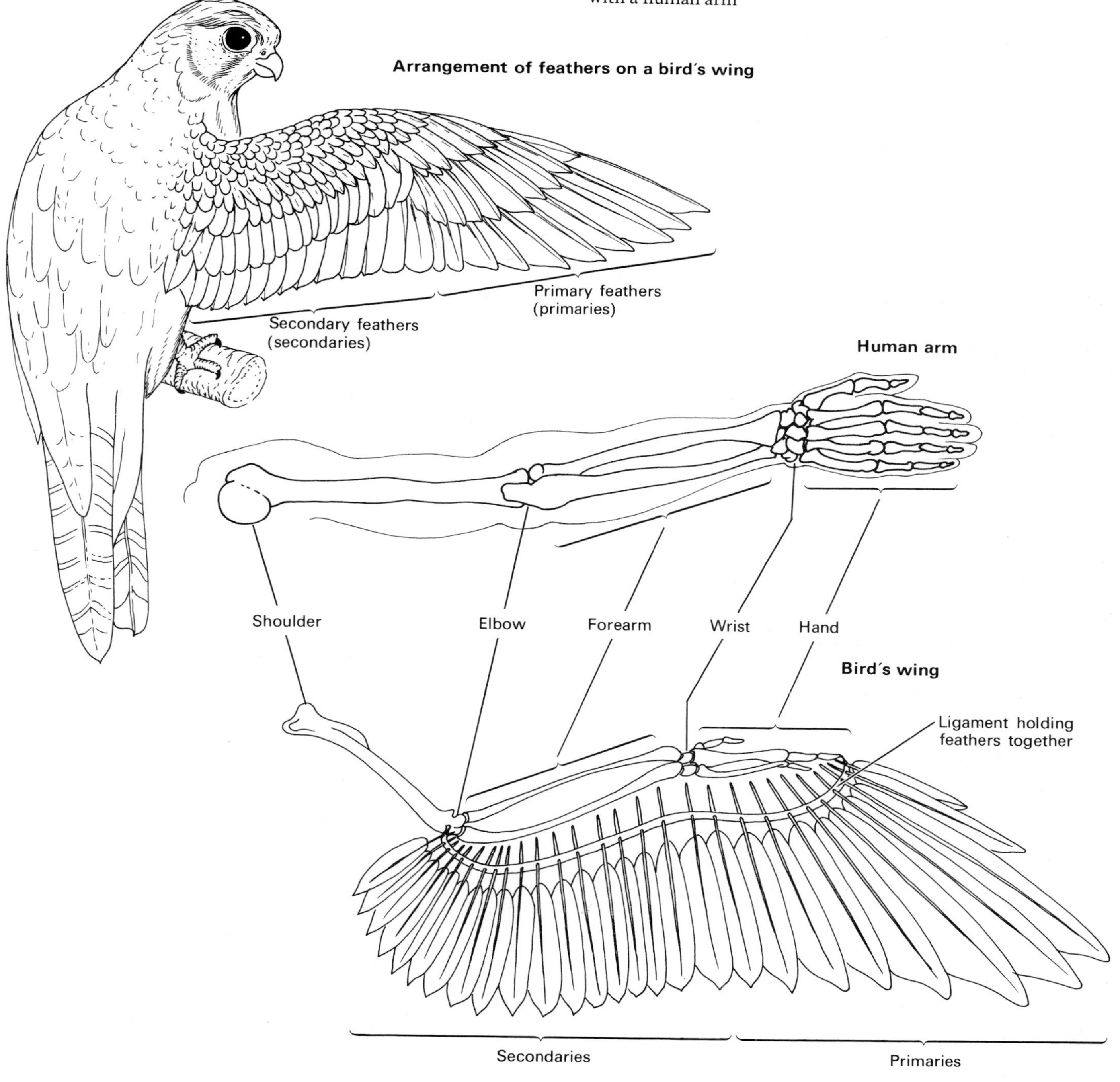

and lift, whereas they are 'open' on the upstroke which lets air rush through the wing as it slips rapidly upwards.

To summarize, the wing tip is moved through an elliptical pathway in the air. The 'power' stroke is the down-beat. This provides both lift and forward propulsion. During each complete movement the 'hand' part of the wing twirls round like a propeller driving the bird upwards and forwards. At the same time the aerofoil shape of the 'forearm' provides additional lift to help overcome the force of gravity.

12.3 **Fish**

The density of water is almost the same as the density of the bodies of fish and many other aquatic creatures. Consequently these creatures are practically weightless when in water. This fact is extremely important because it means that, unlike birds, most swimming creatures do not have to expend much energy in overcoming the force of gravity; they can use most of their energy in moving through the water. In the majority of fish, however, there is an organ called the **swim**

A Sequence of wing positions during upstroke and downstroke

B Halfway through the upstroke
Wings are bent at the wrists and the leading edge is higher than the trailing edge

C Halfway through the downstroke
Wings are outstretched and the leading edge is lower than the trailing edge

Fig. 12.10 Wing movements during flapping flight

Fig. 12.11 Changing angles of the wing and primary feathers during one wing beat

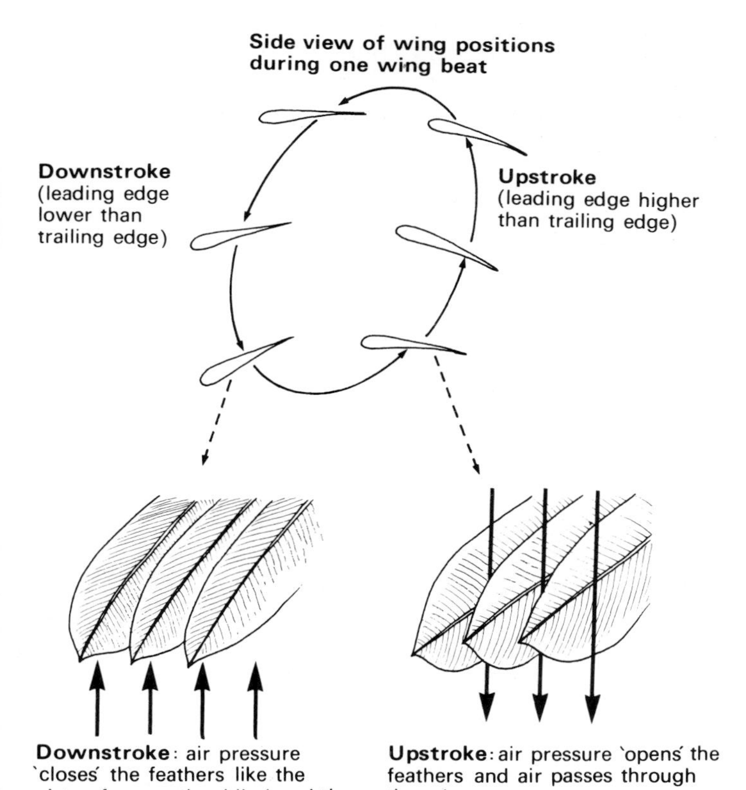

bladder which enables them to become completely weightless. Therefore they need waste no energy at all in overcoming gravity (Fig. 12.12).

Functions of the swim bladder

A swim bladder is a gas-filled space inside the fish's body. The amount of gas in this space can be increased, which makes the fish less dense, or it can be decreased, making the fish more dense. In this way a fish alters its density until it matches the density of water. The fish is then completely weightless and floats effortlessly. Adjustments to the amount of gas in a swim bladder are made in two ways.

The simplest method, found in herrings and eels, is by means of a tube from the swim bladder which is attached to the gut near the mouth. Air is gulped through the mouth and into the swim bladder, or released from the swim bladder through this tube and out of the mouth until weightlessness is achieved. Other fish, including trout, perch, and cod, have no such tube. In these fish gas is either released into, or removed from, the swim bladder by blood vessels which surround it.

The amount of air in the swim bladder must be altered during diving and surfacing. If a fish dives from the surface to a depth of 10 metres the pressure of water against its body approximately doubles. In these circumstances, if it were not for an automatic inflation of the swim bladder it would be compressed by the increased water pressure and the fish would become more dense than water. When the fish moves to the surface the swim bladder must be deflated by the same amount. Fish of the shark family have no swim bladder and so are heavier than water. These fish begin to sink whenever their swimming movements stop.

Streamlining

Water is eight hundred times denser than air and therefore sets up more resistance to movement (which is why it is harder to wade through water than run on dry land). Fish, however, have a streamlined shape which passes smoothly through water with a minimum of resistance. As a rule, the more active the fish the more streamlined its shape and therefore the lower its resistance to movement through water. In other words the more perfect the streamlining the greater the speed achieved for a given expenditure of energy. But what provides the force which drives fish through water?

Swimming movements

With rare exceptions the force which moves a fish comes from blocks of muscle situated around the fish and on either side of the backbone (Fig. 12.13). These muscle blocks are clearly visible when the skin is stripped from the side of a fish. They are usually arranged in a zig-zag pattern, one behind the other, from head to tail.

Fig. 12.12 Position of the swim bladder in a fish, and the arrangement of fins

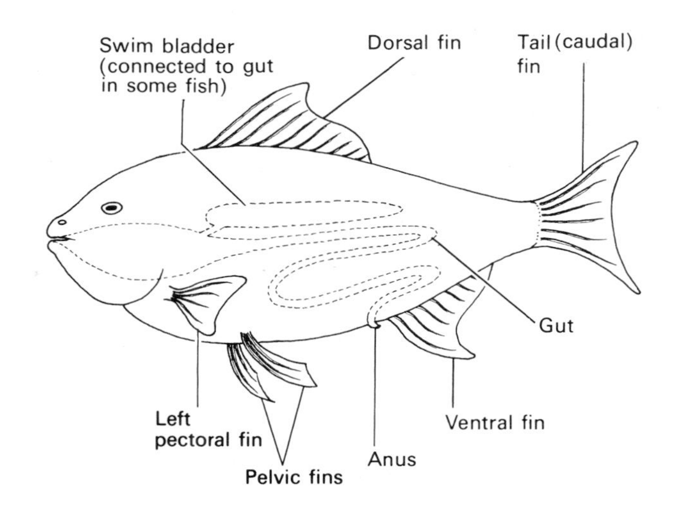

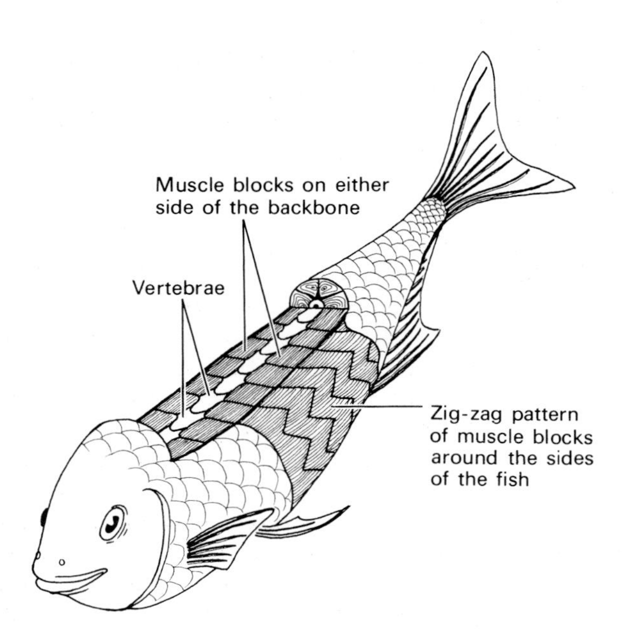

Fig. 12.13 Relationship between swimming muscles and the backbone

These muscles form a large proportion of a fish's body weight, which is why fish is a good source of food. In tunny fish, for instance (Fig. 12.14A), which are capable of very high swimming speeds, the swimming muscles make up three-quarters of the total body weight.

A fish's backbone is made up of many small bones called **vertebrae**, which are fixed together to make a long flexible rod. When muscle blocks on one side of the animal contract, the backbone bends. At the same time, muscle blocks immediately opposite on the other side of the backbone are relaxed and slightly stretched (Fig. 12.14C).

When a fish is swimming, muscle blocks contract and relax alternately on each side of the backbone in a sequence which runs from head to tail. This has the effect of bending the backbone into a continuous series of waves which move rapidly down the length of the animal (Fig. 12.14B). These waves are quite pronounced and sinuous in long thin fish like conger eels. In more compact fish like the tunny, the only part which really bends is the narrow tail stalk. All the muscular power in this type of fish is directed towards bending this stalk, which then moves the tail rapidly and powerfully from side to side.

Fig. 12.14 Fish body shapes and swimming movements

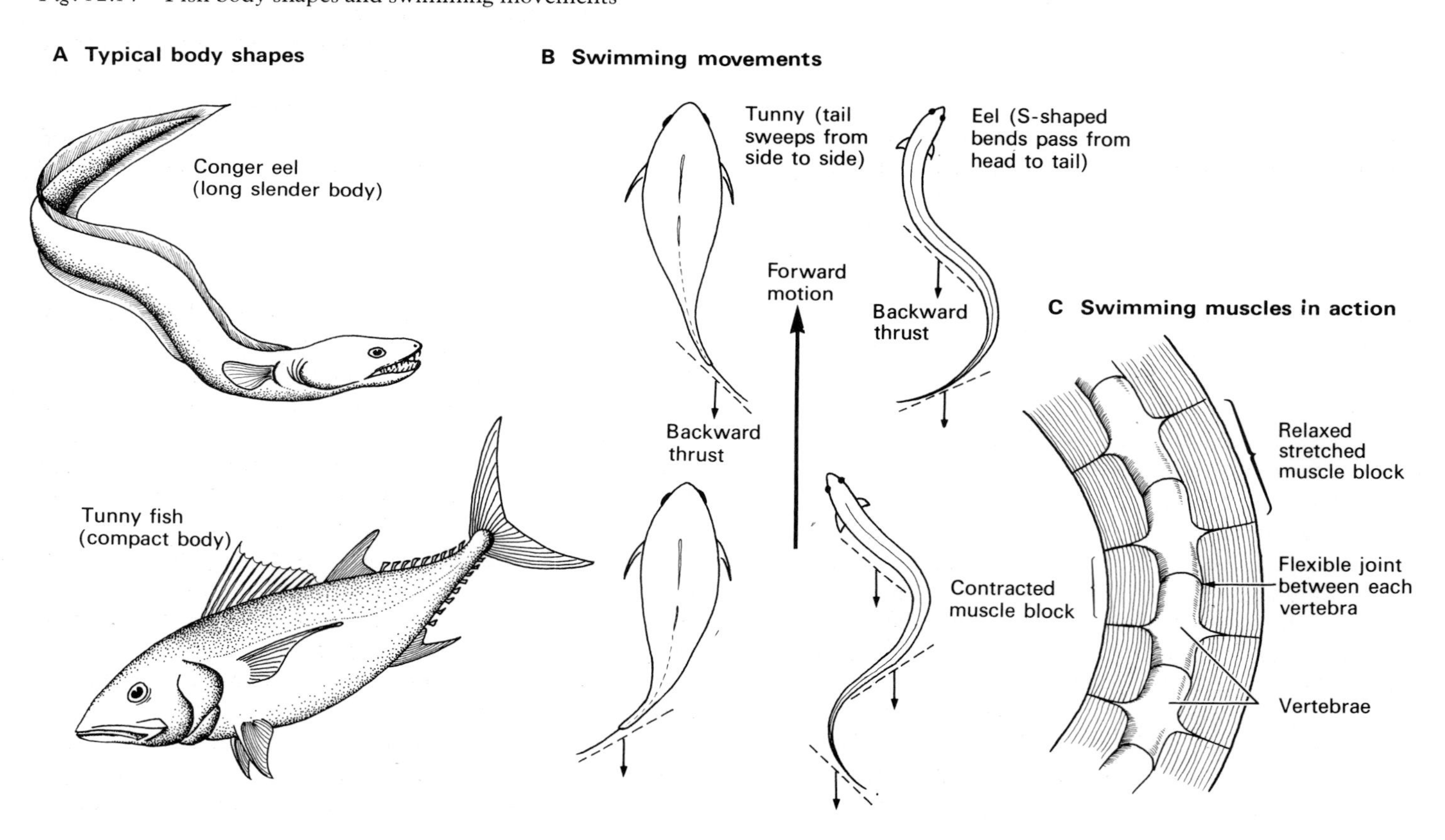

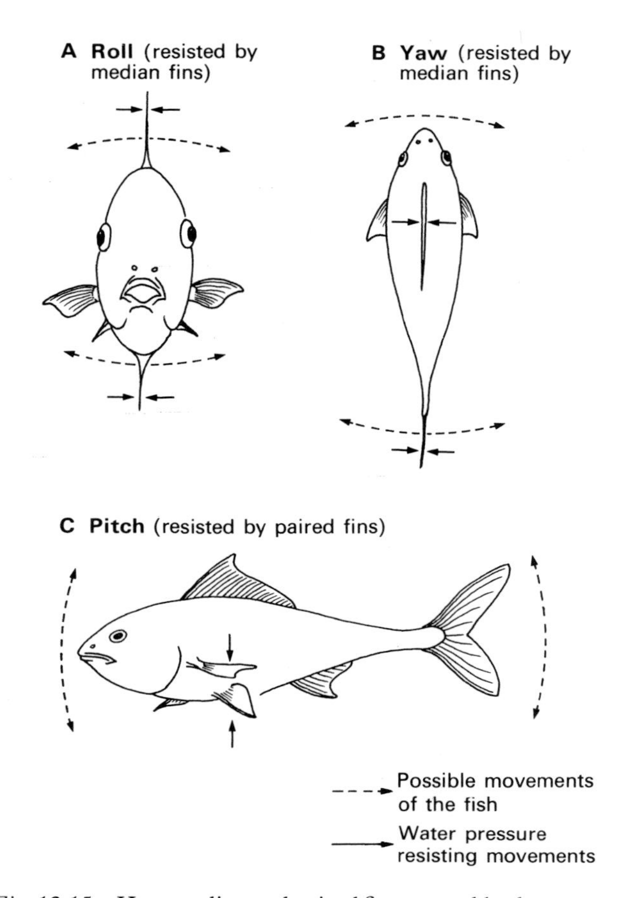

Fig. 12.15 How median and paired fins control body movements

Look at Figure 12.14B, and try to imagine the undulating body of an eel in slow motion. In fish of this type the forward thrust which pushes them through the water comes from all those areas of the body which face backwards (marked with a broken line on the drawings). At any one moment there are several areas facing backwards, and they move quickly down the animal from head to tail pushing against the water all the way. Consequently, the amount of forward thrust produced by eel-like fishes depends on the speed of the wave-like movements which pass down their bodies and the surface area which faces backwards and presses against the water.

In more compact tunny-like fish there is little wave motion along the body. Up to 80 per cent of forward thrust comes from the sideways movements of the tail fin in fish of this shape. The tail generates a forward thrust each time it is swept sideways towards the fish's line of movement (Fig. 12.14B). At this time the tail surface is directed backwards, and therefore it pushes against the water. The amount of forward thrust it gives depends upon the speed of the tail's sideways movement and the extent of its surface area.

To summarize, most fish are propelled through the water by wave-like undulations of their bodies, and sideways movements of their tail fins, both of which push against the water and thrust the fish forwards. What then are the functions of all the other fins possessed by a fish?

Functions of median and paired fins

These fins are necessary for stability, and for precisely controlled movements through the water. Look at Figure 12.12 again: it shows the pattern of fins on a typical fish. Now look at Figure 12.15. The **median fins** (i.e. dorsal and ventral fins) act rather like the flight feathers on an arrow. They help to prevent the fish veering to one side from its line of movement. They do this by increasing the body's vertical surface area, a feature which reduces **rolling** and **yawing** movements to a minimum (Fig. 12.15A and B). The **paired fins** (i.e. pectoral and pelvic fins) have many functions. They help to check **pitching** movements (Fig. 12.15C), but at the same time fish use them to make controlled upward or downward movements. This is done by altering the angle at which the paired fins are held in the water. Fish also use their pectoral fins as brakes to stop forward movement. This is done by spreading the fins vertically and at right angles to the body. When only one pectoral fin is used in this fashion it acts as a pivot round which the fish can make a sharp turn.

Weightless fish, i.e. those with a swim bladder, have freedom which enables them to manoeuvre in almost any direction. Their intricate, co-ordinated patterns of fin and body movement are easily studied by observing aquarium fish, especially the many small tropical varieties.

12.4 Insects

It has already been explained that insects have a tough, non-living exoskeleton made mainly of chitin. The chitin overlies the living cells which produced it, and the chitin itself is covered by a thin layer of wax. The wax makes the insect's body waterproof, and it also reduces loss of water by evaporation from the body. Together, both the chitin and the wax are known as the **cuticle** of the insect.

This layer of dead cuticle presents a problem when the animal grows, because it soon becomes too small for its owner. This problem is overcome by moulting, or **ecdysis**, which is the periodic shedding of the cuticle and its replacement by a new, larger one. During ecdysis the living cells beneath the cuticle produce a substance which dissolves the innermost part of the chitin so that the cuticle becomes very thin and no longer attached to the body surface. Then a new cuticle

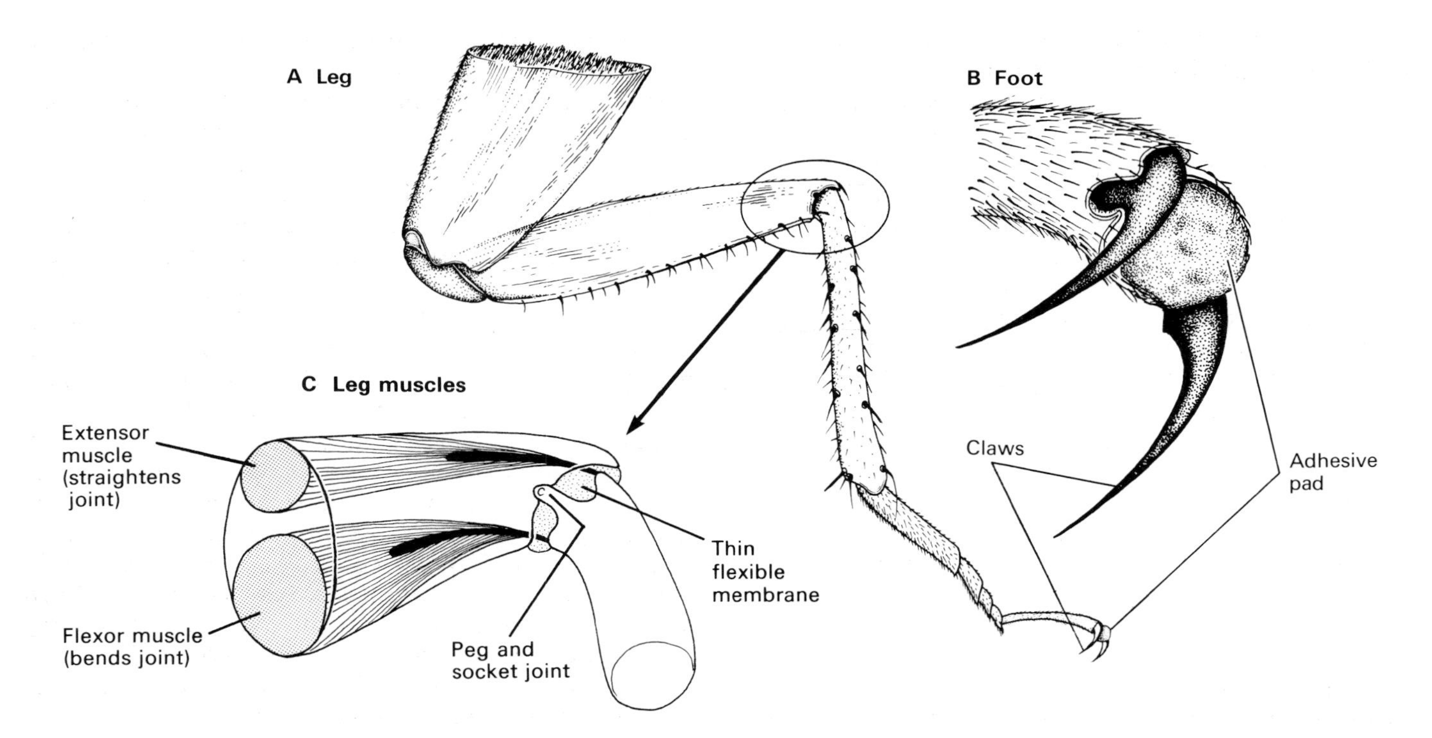

Fig. 12.16 Structure of an insect's leg

is formed beneath the old one. At first the new cuticle is soft and wrinkled, but it slowly stretches as the insect inflates itself by taking in air. This splits open the old cuticle and the insect crawls out of it already clothed in its new larger covering, which soon hardens when exposed to the air.

In places the cuticle extends into the body forming **processes** to which the muscles are attached. The processes and muscles of an insect's leg joint are illustrated in Figure 12.16C.

Walking

An insect's legs consist of a number of stiff, hollow tubes joined together by soft flexible membranes. Figure 12.16 shows how the legs are moved by muscles attached to the inner walls of these tubes on either side of the flexible joints. **Flexor muscles** are those which bend the joints, and **extensor muscles** are those which straighten them. Flexor and extensor muscles together form an antagonistic system.

Insects have six legs and all of them are attached to the thorax, or middle section, of their bodies. The sequence in which each leg is moved in relation to the other five varies according to the type of insect and the speed at which it is moving. Generally, however, three legs are on the ground supporting the body while the other three are raised and moving forward to a new position. The three supporting legs are usually the first and third on one side and the middle leg on the opposite side, like a tripod.

Fig. 12.17 Diagram of insect flight muscles

A The direct flight muscles typical of large insects
(e.g. butterflies and dragon-flies)

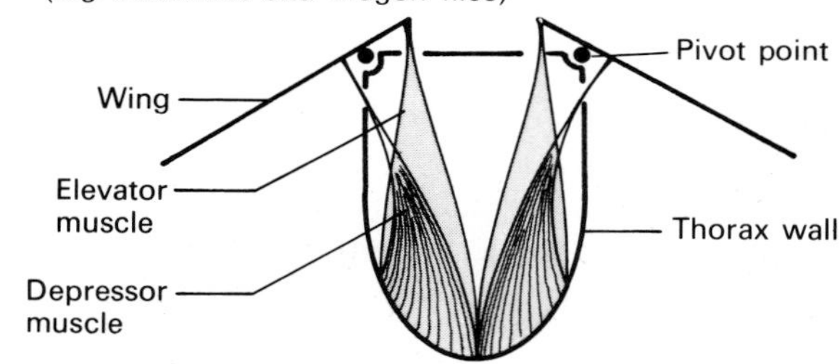

B The indirect flight muscles typical of small insects
(e.g. house-flies and gnats. See Figure 12·18)

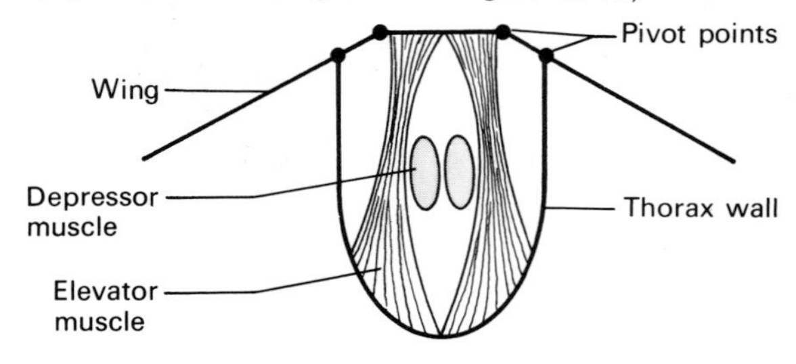

At the end of a typical insect's leg are two claws with a pad of minute hairs between them (Fig. 12.16). The claws are extremely sharp and are moved by muscles. The claws are capable of obtaining a grip on even the slightest lump or groove on an apparently smooth surface. On mirror smooth surfaces, like clean glass or the shiny surface of leaves, the hairy pads are used.

The tiny hairs which form each pad are in fact hollow tubes, which exude a liquid substance from a gland at their base. This liquid helps to stick the pad firmly to any smooth surface. But this is not all. The tips of the hairs are shaped in such a way that the leg may be slid forwards over the surface with ease, but the hairs grip firmly when there is danger of it sliding backwards. This equipment enables a fly to perform remarkable acrobatics like 'skating' up a window pane, and walking upside-down on a ceiling. The pad detaches easily from a surface when the insect flies away.

Flight

When an insect flaps its wings up and down a stream of air is directed downwards and backwards. This has the effect of lifting the insect and driving it forwards through the air.

Insects with large wings, like dragonflies, butterflies, and moths, flap them at a comparatively slow rate: on average about 30–70 times per second. These large-winged insects usually have muscles attached directly to their wing bases (Fig. 12.17A). Muscles arranged in this way are called **direct flight muscles**. One set, the **elevator muscles**, raise the wings and another set, the **depressor muscles**, lower the wings. During flight, nerve impulses are sent to the elevator and depressor muscles in turn, which keeps the wings beating up and down at the correct speed.

Insects with smaller wings must flap them much faster to remain airborne. House-flies, for instance, flap their wings up to two hundred times per second, while tiny gnats and midges vibrate their wings more than a thousand times per second. However, it is impossible for an insect's nervous system to deliver individual impulses to the flight muscles at anything like this rate, and the wing movements of small insects are controlled by a mechanism which differs from that of larger insects in at least two ways.

Fig. 12.18 Indirect flight muscles, and wing movements

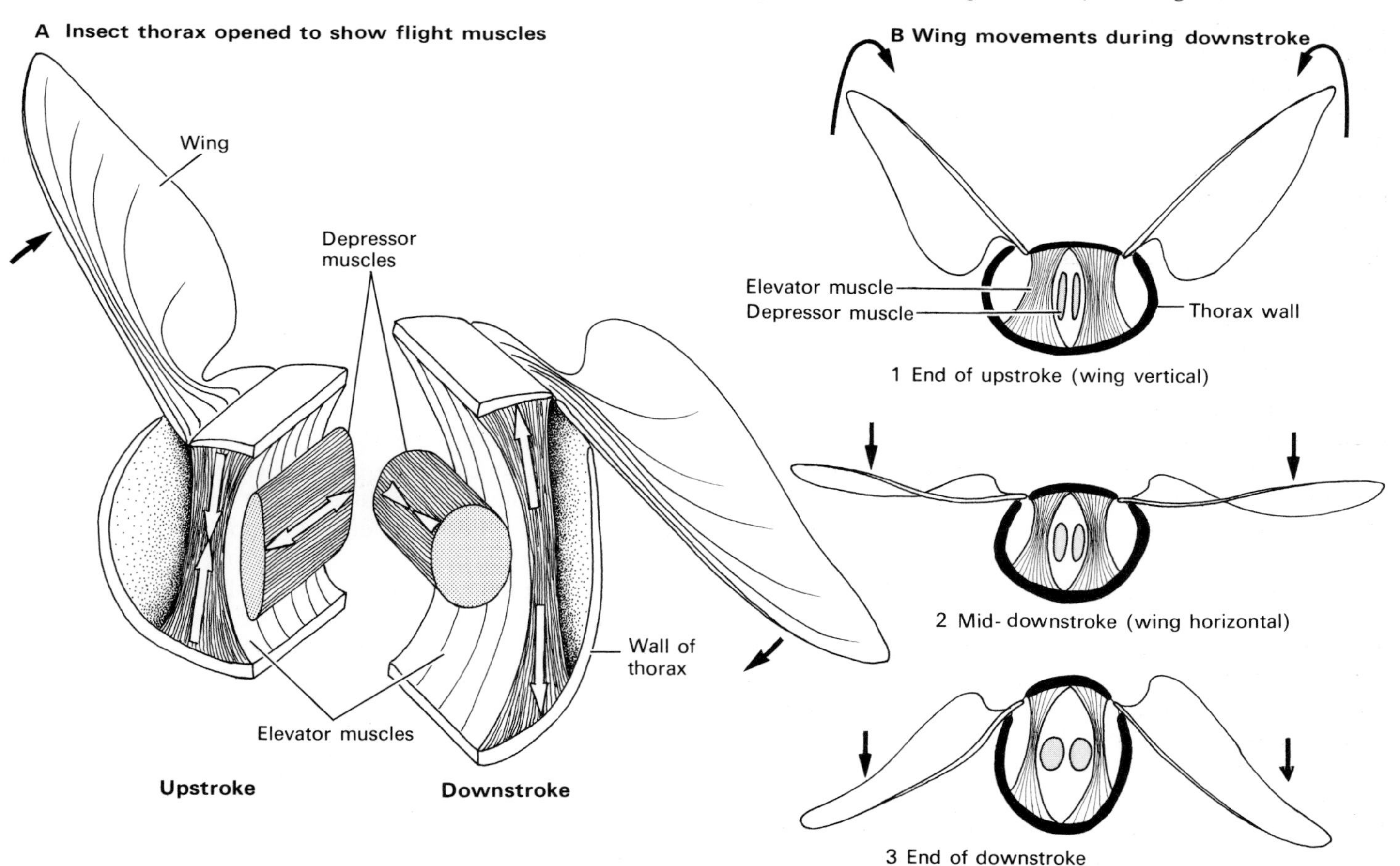

First, in small insects the flight muscles are not attached directly to the wing bases, but to the walls and roof of the thorax (Figs. 12.17B and 12.18). Hence, they are called **indirect flight** muscles.

To raise the wings the elevator muscles contract and pull against the roof of the thorax, moving it downwards. This levers the wings upwards. To lower the wings, the depressor muscles contract and pull in a horizontal direction compressing the thorax *lengthwise*. This raises pressure in the thorax to such an extent that its roof is thrust upwards, which levers the wings downwards (Fig. 12.18A). Then the elevator muscles contract again, and this in turn causes an increase in fluid pressure in the thorax which expands it to its original length and stretches the depressor muscles. In simple terms, whenever one set of muscles contract they not only move the wings, they also increase fluid pressure in the thorax which stretches the opposing set of muscles.

Second, there is the 'click' mechanism which has to do with the structure of the thorax wall. As the elevator muscles begin to contract and raise the wings, these muscles encounter a certain amount of resistance from the walls of the thorax. That is, the walls resist the bending which must take place if the wings are to be raised (compare the shape of these walls in the two drawings of Figure 12.18A). However, at a certain point this resistance suddenly disappears and the wings 'click' upwards at tremendous speed. This movement rapidly stretches the depressor muscles to their fullest extent, which stimulates them into an instantaneous contraction. This in turn stretches and stimulates the elevator muscles and the cycle of events is repeated over and over again with immense rapidity. The flight muscles are in fact stimulating each other, and it has been found that they continue to do this automatically with only an occasional nerve impulse.

Insect wings do not simply move straight up and down. The wing tips move through a figure-of-eight pattern in the air. The wings twist through different angles as they move through this pattern. Look at Figure 12.18B. Drawing 1 shows that during each upstroke the wings twist into a vertical angle and so they slice quickly upwards through the air with a minimum of resistance. On the downstroke (2 and 3) the wings twist into a horizontal position and are slapped downwards flat-on to the air. It is this downstroke which lifts the insect and depending on the direction of the stroke drives the animal forwards through the air, keeps it hovering in one place, or even moves it backwards. But insects are capable of even greater aerobatic feats. By twisting one wing more than the other they can dart sideways or diagonally in any direction. These movements are extremely swift and enable insects to escape predators—and fly swatters.

A Cross-section of an earthworm

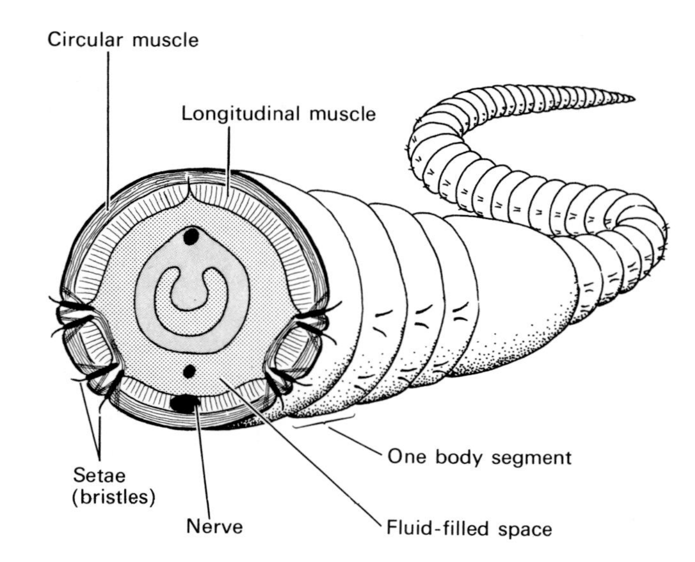

B One seta

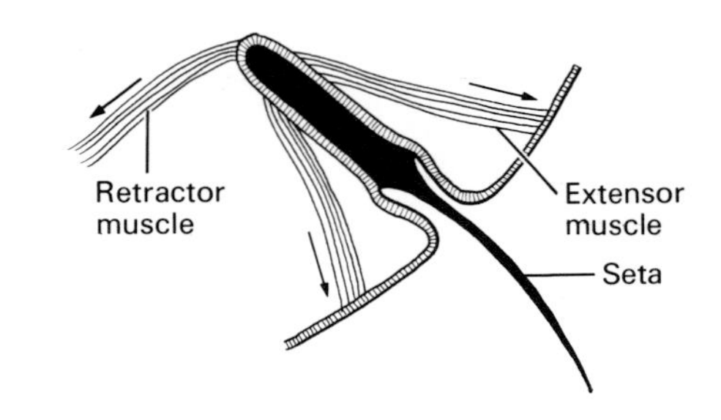

Fig. 12.19 The muscles and locomotory organs of an earthworm

12.5 Earthworms

An earthworm is an example of an animal which is supported entirely by the fluid in its cells and body spaces, i.e. by its hydrostatic skeleton (Fig. 12.19A). There is no hard skeleton whatsoever.

Earthworms move by using muscles to change the shape of their bodies (Fig. 12.19). They move forwards by making their streamlined front ends longer and thinner while their rear ends are anchored firmly in the soil, and then they anchor their front ends and draw up their rear ends by making them shorter and fatter. Figure 12.20 illustrates how this is done.

Circular muscles have fibres which run round the worm's body. When these contract they squeeze inwards against the fluid inside the worm, causing pressure which stretches it and makes it longer and thinner. The fibres of **longitudinal muscles** run along the worm from 'head' to 'tail'. When these contract they make the body shorter, and they also squeeze the body fluids causing the body to become fatter.

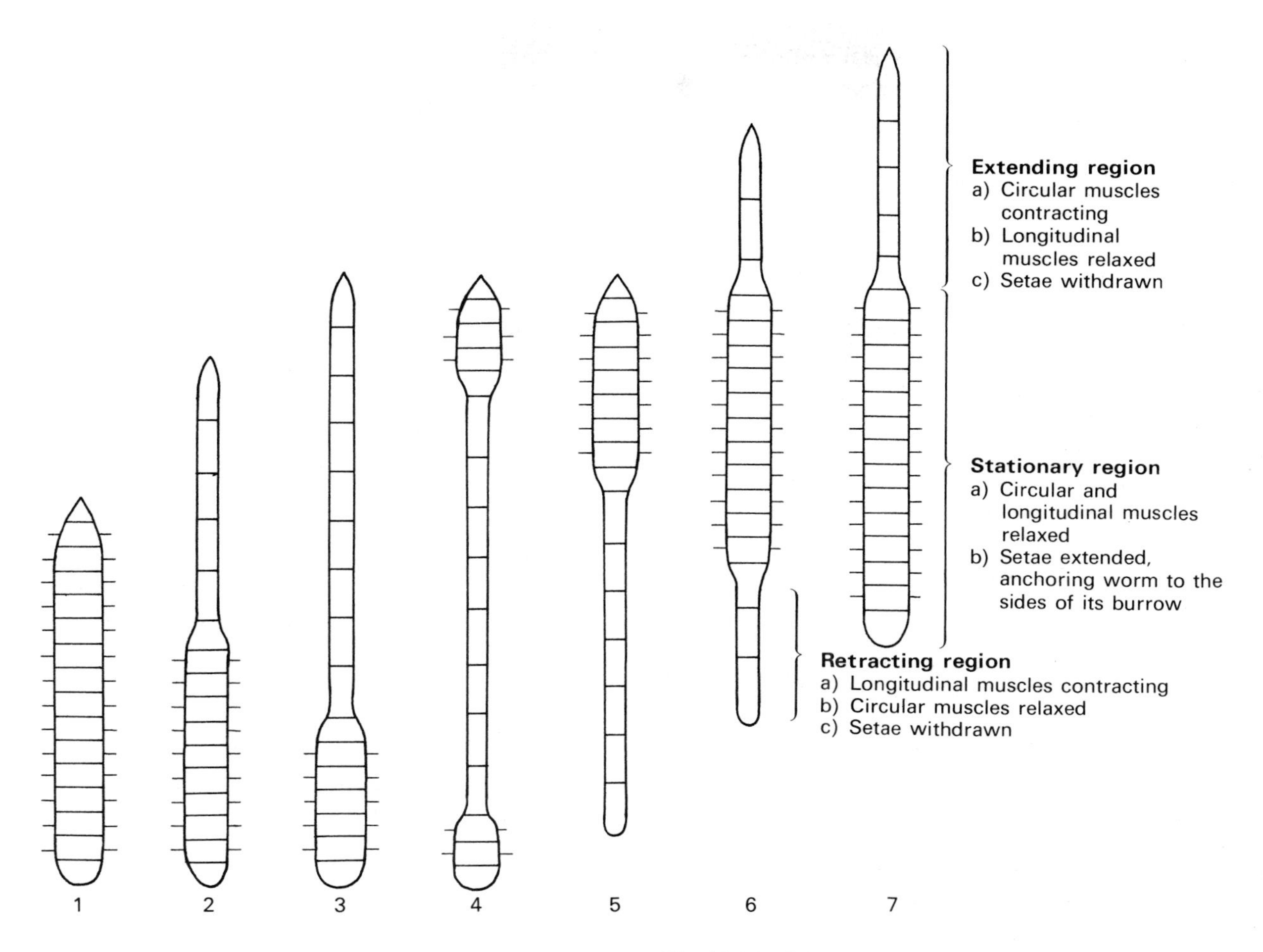

Fig. 12.20 Diagram of how an earthworm moves

In addition to these muscles there are four pairs of bristles, or **setae**, on all but the first and last segments of a worm's body. These setae are operated by another system of antagonistic muscles. One set, called the **extensor muscles**, push the setae outwards so that they protrude from the skin. Another set, the **retractor muscles**, draw the setae in again.

Figure 12.20 illustrates (diagrammatically) the sequence of events as a worm moves through the soil. Drawing 1 shows the animal when it is at rest. Note the shape of its body and the fact that all its setae are protruding. 2 and 3 show what happens when the circular muscles begin to contract at the front end of the worm. This contraction progresses quickly backwards towards the 'tail' of the worm. The circular muscles squeeze inwards on the worm's body, greatly increasing the fluid pressure within it. This pressure stretches the relaxed longitudinal muscles, and makes the animal's front end much longer and thinner in diameter. At the same time, all the setae of this region are pulled in. Since the worm's rear end is firmly anchored by its setae, the front end is forced forwards through the soil. Its thin pointed shape, smooth slimy surface, and lack of protruding setae all contribute to its easy passage through the soil.

4 and 5 show what happens at the next stage. The longitudinal muscles now begin to contract at the front end of the worm, and this contraction progresses quickly backwards. This again increases pressure in the body fluid at the animal's front end, but this time the pressure stretches the relaxing circular muscles. The front end therefore becomes fatter; its setae are protruded; and it becomes firmly anchored in the soil. Once the front end is anchored the animal's rear end is pulled forwards (5). Before this movement is completed the front end is again thrust forwards through the soil (6 and 7).

To summarize: worms move by successive waves of contraction first in the circular muscles, and then in the longitudinal muscles. These waves of body contraction and expansion follow each other down the animal from front to rear.

Verification and inquiry exercises

A *An investigation of exoskeletons and movement*

1. Obtain a live crab from the sea-shore. Keep it in shallow sea-water in an aquarium, and provide it with a pile of stones under which it can hide. It will feed on small pieces of fish or meat, but do not leave uneaten food in the tank; it will decay and foul the water.

a) Watch the animal's movements, noting the range of movements at each leg joint.

b) Always return these animals to the sea-shore when the study is finished.

2. Obtain a dead crab or lobster.

a) Examine all its leg joints, noting the range of movements which they possess.

b) Detach the legs, open the 'body' from the top by sawing through the skeleton with a sharp knife, and remove all the internal organs. Boil the skeleton in several changes of water until only the hard parts remain. Study the *internal* structure of the skeleton and try to explain its functions. Is it strictly accurate to describe a crab's skeleton as 'external'?

3. Obtain a wide variety of insect specimens (or pictures), e.g. water-beetle, 'mole' beetle, grasshopper, praying mantis etc.

a) How do the various leg structures suit the insects' particular ways of life?

B *An investigation of endoskeletons by the transparency method*

It is possible to study complete intact skeletons of small animals by making their soft tissues transparent, and staining their bones deep purple. The method is lengthy, but the results are very informative and extremely beautiful.

1. *Fish*

a) Skin the fish and remove its digestive organs and liver.

b) Soak it in 10% formalin for 1 day.

c) Leave for 24 hours in 95% ethyl alcohol.

d) Wash in running water for an hour.

e) Place in 4% potassium hydroxide solution for 4 days.

f) Wash for 30 minutes in 50% hydrogen peroxide.

g) Place in 4% potassium hydroxide again for one hour.

h) Mix together: 2.5 cm^3 glacial acetic acid, 5 cm^3 of glycerine and 30 cm^3 of chloral hydrate. Use this liquid to make a saturated solution of alizarin stain. Before use, add this solution to 500 cm^3 of 4% potassium hydroxide.

i) Stain the fish in this liquid for 3 days.

j) Soak it in 4% potassium hydroxide again until the stain has been removed from all soft tissues, leaving the skeleton purple (this will take about 2 days).

k) Soak the fish for one day in each of the following: 25%, 50%, 75%, and 100% glycerine solutions. These will remove water from the fish, and make its soft tissues more transparent.

l) Finally, preserve the fish in pure glycerine with a few thymol crystals.

2. *Amphibia and small mammals*

a) Skin the animal, remove its digestive system and liver, then soak it for 24 hours in 70% alcohol with a few crystals of iodine.

b) Soak in pure alcohol for another 24 hours.

c) Soak for 30 minutes in each of the following: 75% alcohol, 50% alcohol, then 25% alcohol, and lastly water.

Fig. 12.21 How to verify the aerofoil principle (exercise D)

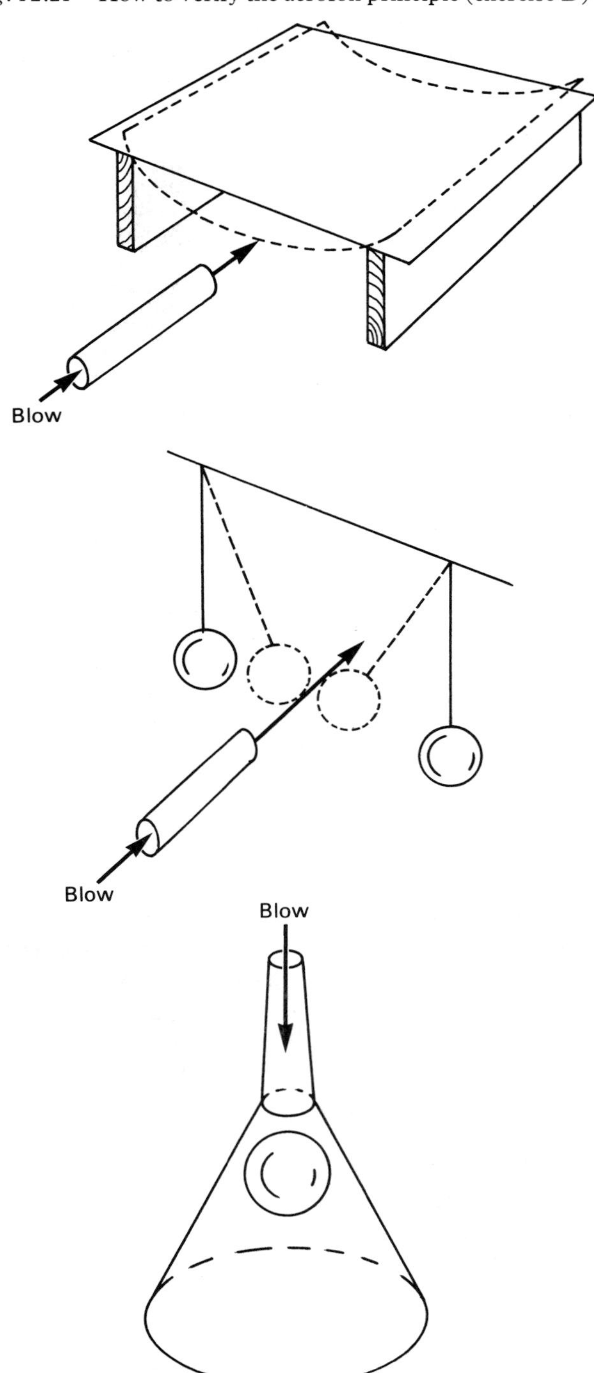

d) Mix yellow alizarin in pure alcohol to make a $\frac{1}{2}$% solution. Mix this with 1% aqueous solution of potassium hydroxide to make a deep purple colour. Stain the animal for 12 hours in this liquid.

e) Put the animal in 1% potassium hydroxide until no more dye colour comes away. This may take 6 to 12 weeks, but must be watched carefully because the animal may disintegrate.

f) Mix equal parts of 1% potassium hydroxide with 880 ammonia (take care, the fumes of this liquid are extremely unpleasant). Brief immersion in this liquid will remove any brown colour from the specimen.

g) Transfer to equal parts of 1% potassium hydroxide and glycerine, and finally to pure glycerine.

C *To verify the aerofoil principle: as speed of air movement increases so its pressure decreases*

1. Lay a piece of paper across two blocks of wood or other supports (Fig. 12.21). Blow a stream of air from a tube *under* the paper and note that the paper bends downwards. The explanation is that blowing under the paper creates a fast-flowing stream of air which has a *lower* pressure than the still air above the paper. The difference between these two pressures results in a downward force which bends the paper. Use this information to explain what happens in the following experiments.

2. Blow a stream of air between two table tennis balls that are suspended by thread (Fig. 12.21).

3. Hold a table tennis ball in the mouth of a filter funnel, blow hard down the stem of the funnel, and while still blowing let go of the ball (Fig. 12.21). Is it possible to blow the ball out of the funnel?

4. *a*) What have these experimental results in common with the effect of air-flow across a bird's wing?

b) How do these results show that lift increases as the speed of air-flow across a wing increases?

D *An investigation of feather structure*

1. Obtain a primary feather and identify its vane, shaft, barbs, and quill from Figure 12.22.

2. *a*) Pull the feather up and down between the thumb and index finger until its barbs have become separated from each other.

b) Stroke the feather gently upwards (i.e. away from the quill, like the preening action of a bird).

c) Note how this 'locks' the barbs together again so that the vanes once more form an airtight sheet.

d) What advantage is this 'locking' mechanism to a bird?

3. Examine the vane of a feather under a microscope using powerful illumination both on and through the vane. Use Figure 12.22 to help in identifying the parts of the 'locking' mechanism which holds the barbs together.

4. *a*) Examine down feathers (found on the sides of a bird).

b) How do they differ from primary feathers?

c) What is the function of down feathers?

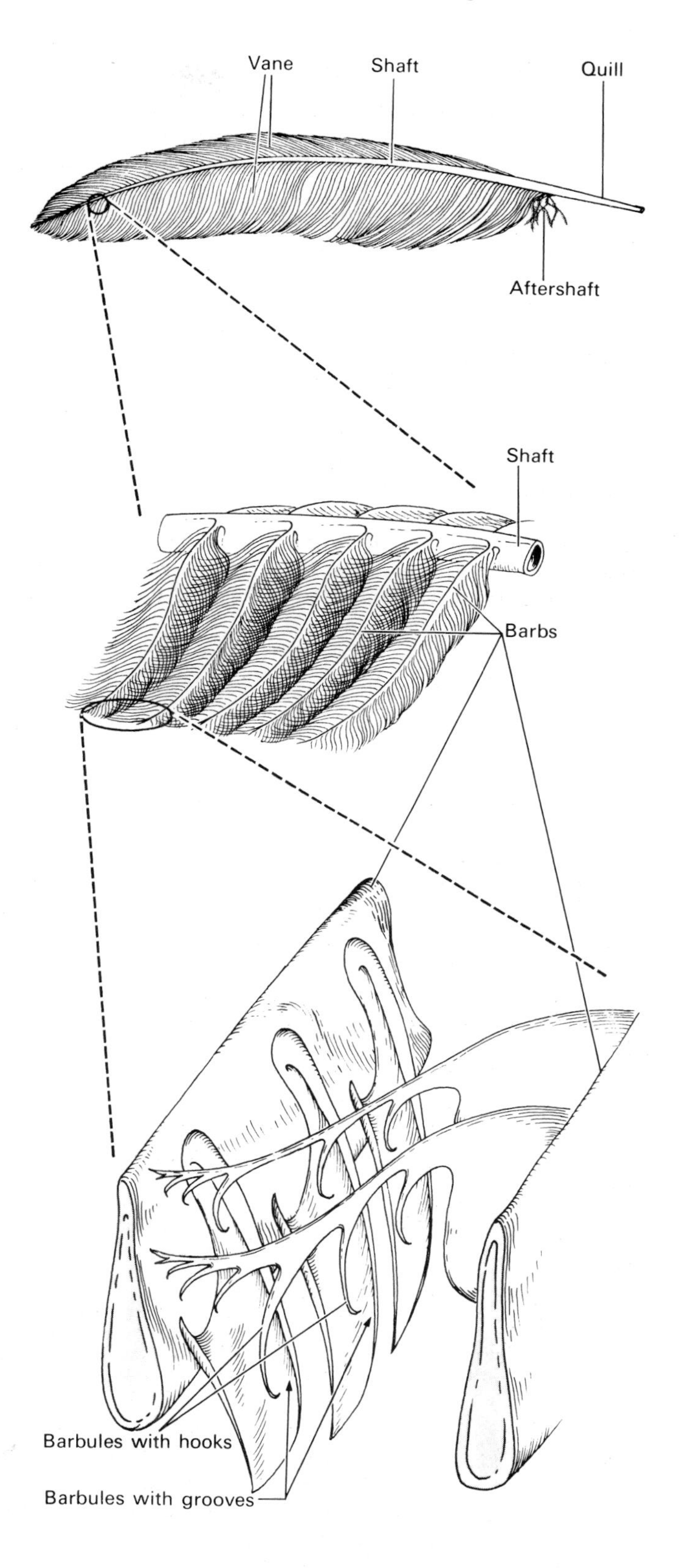

Fig. 12.22 Structure of a feather (exercise D)

Comprehension test

1. Salt water is denser than fresh water. Marine fish generally have smaller swim bladders than fresh-water fish. Use the first of these facts to explain the second.

2. The largest animals in the world (and possibly the largest that have ever existed) live in the sea. Suggest why this is so.

3. What is the significance of each of the following?

a) Birds have a small oil-producing gland at the base of the tail. During preening, a bird dips its beak into this gland and then spreads oil over its feathers.

b) Birds have hollow bones, but they are filled with air and not bone marrow.

c) The sternum (breastbone) of a bird extends outwards from the chest like the keel of a ship.

d) Birds have no external ear.

4. Even if a man were fitted with wings exactly like those of a bird he would still be unable to fly. Suggest how this can be explained by reference to the human skeleton and muscles.

Pigeons in flight. Compare the various wing positions with those illustrated in Figure 12.10, and identify upstrokes and downstrokes

Summary and factual recall test

Soft-bodied animals such as (1–name two) are supported mainly by (2). These animals are said to have a (3) skeleton.

Insects have an (4)-skeleton which forms on the (5) of their bodies. It consists mainly of (6) and is covered by a thin layer of (7). Together these two substances form the (8). During growth this layer is replaced by ecdysis which is (9–describe the process).

In mammals movement occurs at the freely moveable, or (10) joints. Examples of hinge joints are (11–name three). These are called hinge joints because (12). Examples of ball-and-socket joints are (13–name two). These are called ball-and-socket joints because (14). Joints are held together by fibres called the (15), and muscles are attached to bones by fibres called the (16). The origin of a muscle is the end which (17), whereas the insertion is the end which (18). Every joint is moved by two opposing sets of muscles which are called an (19) system. The muscles which bend a joint are called the (20), while the (21) straighten it.

All lever systems consist of three parts: (22). When the head nods these parts are represented by (23–name the parts of the body involved and the parts of a lever system which they represent); when the elbow is moved they are represented by (24), and when standing on tiptoe they are represented by (25).

The shape of a bird's wing seen in cross-section is called an (26). When moving through the air this shape generates lift because (27). On the downstroke the (28) edge of a wing is lower than the (29) edge. This pushes air (30) and drives the bird (31). Next, the wing bends at the (32) and the 'hand' part of it flips upwards because (33).

By means of a (34) bladder a fish can adjust the (35) of its body until it is (36) to water. This makes the fish completely (37). The force which propels a fish comes from (38) blocks on either side of its (39). Depending on body shape these either bend the body into a continuous series of (40) which move down the body thus forming areas of the body which are directed (41) and (42) against the water; or they move the (43) from side to side. The other fins help to reduce to a minimum movements such as (44–name three).

In small insects such as house-flies (45) muscles raise the wings by pulling the roof of the thorax (46). To lower the wings (47) muscles compress the thorax (48), which raises (49) in the thorax and pushes its roof (50). Insect wing tips move through a (51) pattern in the air. On the downstroke the wings are held in a (52) position, and on the upstroke in a (53) position.

Earthworms move by (54) their shape. Circular muscles contract and make their front ends (55) and (56) while their rear ends are (57) to the soil by bristles called (58). Then the longitudinal muscles contract making the body (59) and (60) while the (61) end is anchored to the soil.

13 Co-ordination in animals

The millions of cells and the scores of different tissues and organs in the body of an animal do not work independently of each other; their activities are co-ordinated. This means that they work together, performing their various functions at certain times, and at certain rates according to the needs of the body as a whole.

One of the most familiar examples of co-ordination is the way in which muscles work together during movement. When a boy runs to catch a ball, for example, he uses hundreds of muscles to move joints in his arms, legs, and back. Using information from his sense organs, the boy's nervous system co-ordinates these muscles so that they contract in the correct sequence, with the correct degree of power, and for precisely the correct length of time needed to get him to the spot where he can catch the ball. But this is not all. Muscular activities like running to catch a ball involve many other forms of co-ordination, such as those which increase the rate of breathing and heart-beat; adjust blood pressure; remove extra heat from the body; and maintain sugar and salt levels in the blood. Furthermore, all this co-ordination occurs without a single thought from the boy; it is an unconscious process.

Using the human body as an example, this chapter describes how co-ordination is achieved in mammals through the activities of the **nervous system** and the **endocrine system**. Very simply, the nervous system consists of tissue which conducts 'messages', called **nerve impulses**, at high speed to and from all parts of the body. The endocrine system consists of glands which produce chemicals called **hormones**. These are released into the blood-stream and transported around the body. Unlike nerve impulses, hormones produce effects which are usually slow to appear, and which are often long-lasting.

13.1 The nervous system

The nervous system of mammals, and all other vertebrates, consists of a **brain** and a **spinal cord**, which together form a **central nervous system**. This system is connected to all parts of the body by **nerves**, which are made up of thousands of long thin **nerve fibres**. Figure 13.1 illustrates the arrangement of these structures in the human body. Like all other organ systems the nervous system consists of different types of cells. The cells which conduct the nerve impulses are called **neurones**.

Nerve cells (neurones)

Figure 13.2 illustrates the shapes of the main types of neurones as they appear when separated from various types of nervous tissue. Like all other cells, neurones have a nucleus surrounded by cytoplasm. The region of a neurone where the nucleus is located is called the **cell body**. Here, the cell is about a thousandth of a centimetre in diameter, which is slightly larger than most other animal cells. However, neurones differ from other animal cells in having cytoplasm which extends outwards from the cell body forming long fine threads as thin as 0.005 mm in diameter and, in humans, up to 1 metre in length. These threads are the nerve fibres along which travel the 'messages' made up of nerve impulses to and from all parts of the body.

Nerve impulses pass along nerve fibres in only *one* direction. They pass into the central nervous system along the fibres of **sensory neurones**, and out of the central nervous system along fibres called **motor neurones**.

Sensory neurones Sensory neurones, illustrated in Figure 13.2A, conduct impulses from the sense organs. That is, they conduct impulses from **receptors** in the body such as eyes, ears, nose, taste buds, and touch receptors. Sensory neurones have two long fibres: a **dendron** which conducts impulses from a sense organ to the cell body of the neurone, and an **axon** which conducts impulses from the cell body into the central nervous system.

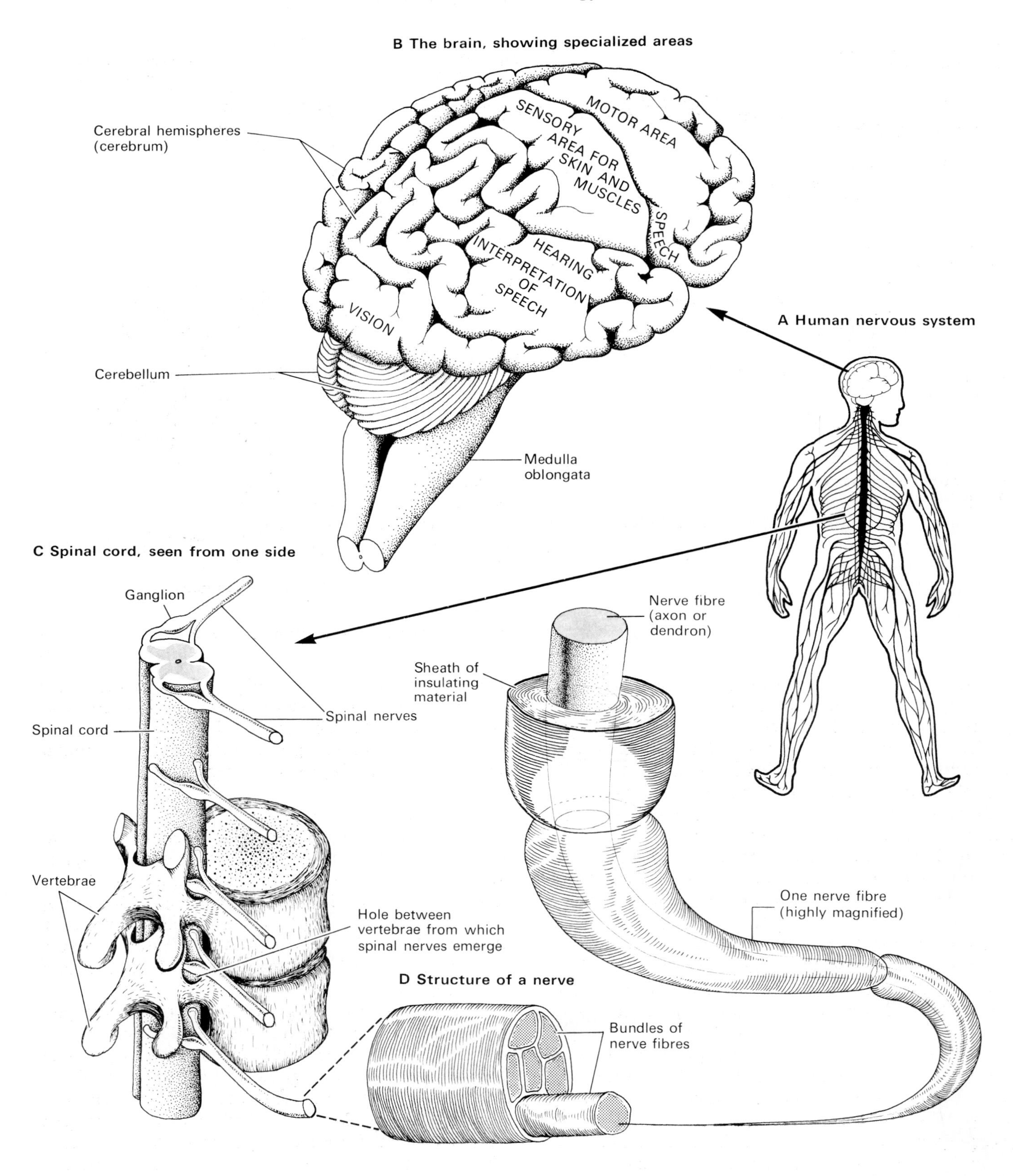

Fig. 13.1 Structure of the human nervous system

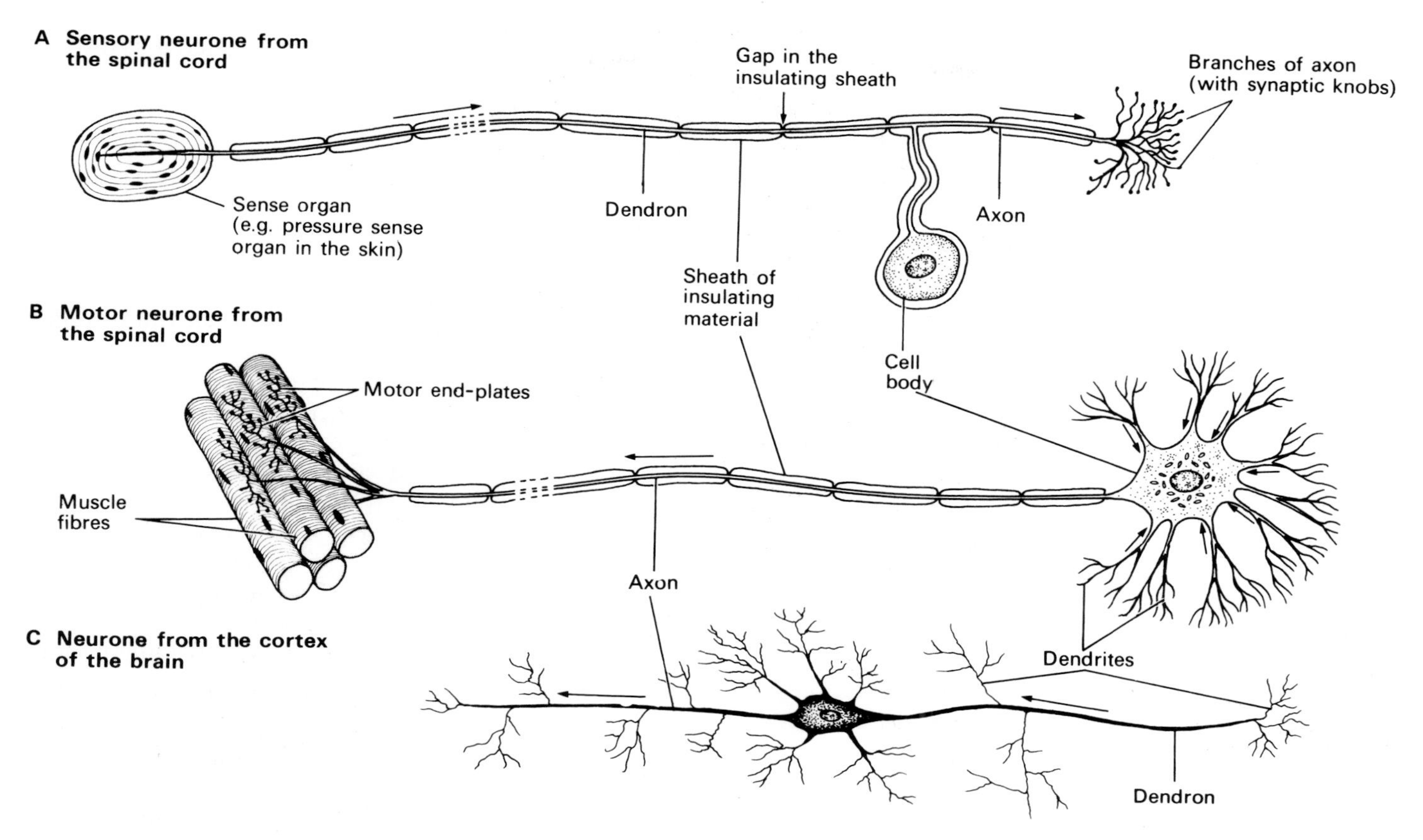

Fig. 13.2 Three types of neurone (nerve cell)

Motor neurones Motor neurones, illustrated in Figure 13.2B, conduct impulses from the central nervous system to the **effector organs**, such as muscles and glands. The cell body of a motor neurone is embedded in the central nervous system. The cell body collects impulses from other neurones through its hundreds of tiny fibres, called **dendrites**. A long, single axon carries these impulses from the cell body to a muscle fibre or a gland. At the point where an axon of a motor neurone enters a muscle fibre there is a structure called a **motor end-plate.** When impulses reach an end-plate they set off chemical reactions which result in muscular contraction.

Most nerve fibres, whether dendrons or axons, that lie outside the central nervous system are encased in a sheath of fatty material which insulates them from one another. The main nerves of the body are, in fact, bundles of these insulated fibres wrapped together in a layer of connective tissue, something like a bundle of insulated wires which make up a large electric cable (Fig. 13.1D).

Nerve fibres which lie inside the central nervous system have no insulating sheath (Fig. 13.2C). The fibres run parallel to each other in thick solid layers, forming the **white matter** that makes up the outer region of the spinal cord (Fig. 13.5) and the inner region of the brain. Neurone cell bodies are also concentrated in certain regions. They form the **grey matter** which makes up the inner core of the spinal cord and the outer layers of the brain.

Nerve impulses

Some nerve impulses originate inside the central nervous system; others come from the sense organs. Indeed, the sole function of a sense organ is to change various forms of stimulus, such as light or sound, into nerve impulses which pass along sensory neurone fibres to the brain. Figure 13.3 illustrates a simple way of explaining what happens during the conduction of an impulse.

Imagine blocks of wood, such as dominoes, arranged in a row. The first domino is knocked over and falls against the next, which falls against the next, and so on to the end of the line. Note two important facts. Nothing, except a certain amount of energy, has moved along the line of dominoes, and this cannot happen again until the dominoes have been stood on end. These events can be compared with a nerve impulse in the following way.

An impulse begins as a change in the arrangement

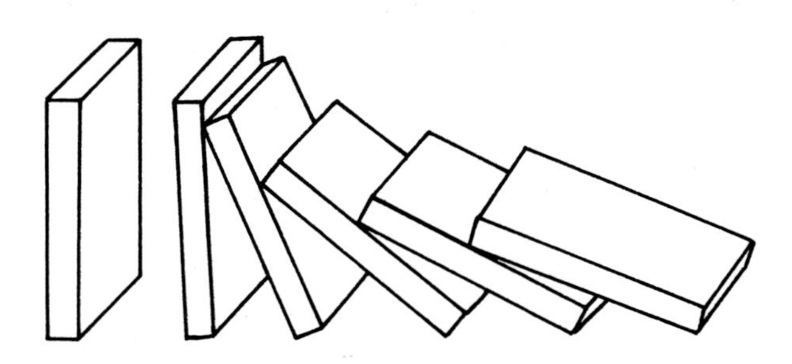

Fig. 13.3 A simple analogy of a nerve impulse

of chemicals in a small area at one end of a nerve fibre. Like one domino falling against the next and knocking it over, this changed area of nerve fibre excites an identical change in the area adjacent to it, which excites the next area, and so on to the end of the fibre. No material object has moved along the fibre, only a wave of chemical rearrangement. Just as the dominoes must be stood on end before they can be knocked down again, so a nerve fibre must recover before it can conduct another impulse. But this recovery period is only a few thousandths of a second.

There is another similarity between falling dominoes and nerve impulses. If the first domino in the line is touched very lightly it may rock back and forth, but will not knock over the next and trigger off the chain of events described above. The first domino must be pushed with a specific amount of force before it falls against the next. Similarly, there are levels of sound and intensities of light which are so weak that they do not stimulate the ears or eyes enough to trigger off impulses to the brain. Stimulation of a sense organ must reach what is called the **threshold level** before the organ sends a nerve impulse to the brain.

Another feature of nerve impulses is that there is no such thing as a weak or a strong impulse. They either occur or they don't, and all are exactly alike no matter where they originate. This is called the 'all-or-none' principle. The only feature of impulses which ever varies is the number of them which pass along a nerve fibre per second. The **frequency** of the impulses depends on the strength of the stimulus. Thus, a strong stimulus, such as a flash of bright light, results in the eyes sending hundreds of impulses per second to the brain, whereas a dim light produces only a few impulses per second, and a very dim one, below the threshold level, produces none at all.

Impulses passing along a nerve fibre eventually reach the end of it, where they encounter an obstacle to their progress: a microscopic gap called a **synapse** between the tip of one fibre and the beginning of the next.

Synapses

Neurones are not continuous with one another. Nerve impulses must cross a synapse where the axon of one neurone meets the dendrites or cell body of another. The dendrites and cell body of a motor neurone, for example, have synapses with hundreds, even thousands, of other neurones (Fig. 13.4).

Synapses, as well as sense organs, each have a certain threshold level. The threshold of a synapse is the number of impulses per second at which the synaptic gap is 'bridged' and impulses begin to flow in the next neurone.

The threshold level of a sense organ, plus the thresholds of every synapse in the chain of neurones connected to it, form a barrier to the movement of impulses between that sense organ and the brain. This barrier is only penetrated by the high frequency impulses which result from strong stimuli. The low frequency impulses resulting from weaker stimuli cannot cross the synapse and so do not reach the brain.

To summarize: the nervous system is a mass of interconnected nerve fibres which conduct impulses from receptors to effectors throughout the whole body.

The nervous system in action

One way of understanding how this system works is to compare it with a telephone exchange. When a caller dials a telephone number the mechanism in the telephone exchange which selects this particular number from all the others, and connects the caller's telephone wire to it, can be compared with events in the nervous system when a person responds to a stimulus. For example, when a boy is trying to catch a ball there is a mechanism in his brain which makes use of sensory information about the ball to select and 'make connections' with those particular motor neurones that extend into his hand and arm muscles. Both telephone exchanges and nervous systems are alike in that they both have mechanisms which select appropriate connections out of an enormous number of alternatives. However, the 'selecting' and 'connecting' mechanisms in nervous systems are infinitely more complex than those which deal with telephone calls.

For one thing, an incoming call to a telephone exchange is connected to only *one* other telephone. But impulses arriving at the central nervous system from a sense organ pass through thousands of interconnections so that they may reach, and bring into action, many different effector organs. These elaborate and complex interconnections in the central nervous

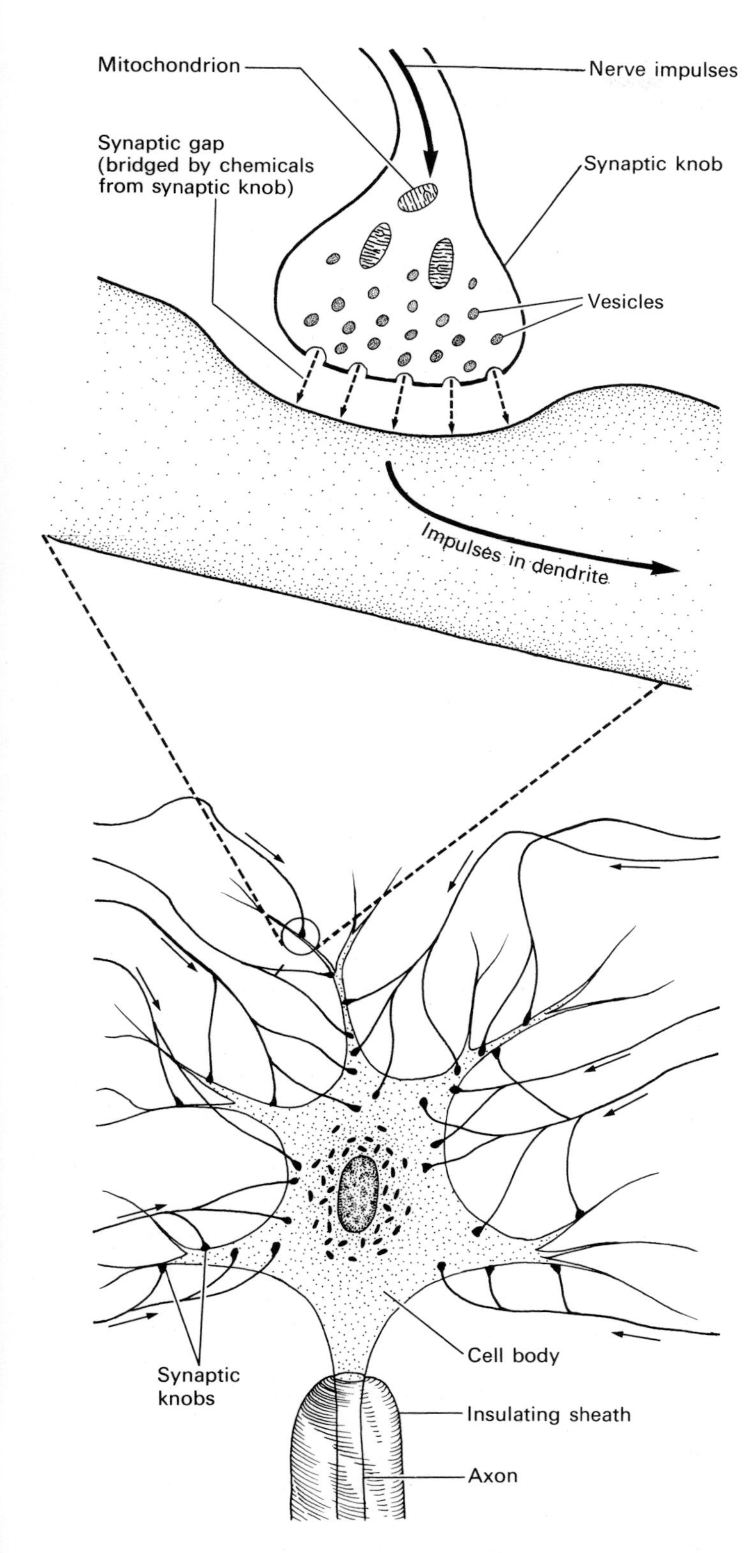

Fig. 13.4 Arrangement and structure of synapses on the cell body and dendrites of a motor neurone. (It is thought that impulses cause the release of a chemical from vesicles in the synaptic knobs which start new impulses in the next neurone.)

system make it possible for a vast number of different responses to be selected, and for several responses to occur together in a co-ordinated pattern of behaviour. The simplest example of these interconnections at work is in a reflex action.

Reflex actions

A reflex action is behaviour in which a stimulus results in a response which does not have to be learned, and which occurs very quickly without conscious thought. For example, a person does not have to learn or even think what to do when his hand accidentally touches a very hot object, he automatically pulls his hand away. Such responses are built in to the nervous system from birth.

Withdrawal from a painful stimulus is an example of a **spinal reflex**. The nerve impulses involved in it pass through the spinal cord along **spinal nerves**. The pathway of these impulses is illustrated very simply in Figure 13.5. From this illustration note the arrangement of sensory and motor neurones, and the synaptic connections between them which form a pathway for impulses from sense organs to a motor end-plate. Such nervous pathways are often called **reflex arcs**, perhaps because of their curved shape.

It is very important to realize that Figure 13.5 is a greatly simplified picture of a reflex action. In fact, there are not three but hundreds of neurones involved in even comparatively simple reflexes like pulling a hand from a hot object. These neurones activate more than a score of muscles to raise the arm and flex the fingers. But even simple arm movements require elaborate muscular co-ordination, the pattern of which differs according to the speed and direction of arm movements. Apparently simple reflex actions are in fact very complex events.

Figure 13.6 indicates nerve fibres which conduct impulses from the sensory side of a reflex arc to the brain, and other fibres which conduct impulses from the brain to the opposite side of the spinal cord. These connections with the brain enable a person to be aware of certain spinal reflexes and, up to a point, exert control over them.

Think what might happen when a housewife picks up a pan of boiling liquid from her cooker to find that the pan handle is very hot. Reflex action may make her drop the pan and spill the liquid. However, if her young child is standing nearby she can deliberately prevent, or **inhibit**, this reflex (using the downward nerve pathways from her brain described in Figure 13.6) and put the pan down safely even though its handle is burning her fingers.

In general, reflex responses to painful stimuli protect the body from injury. Other examples of reflex

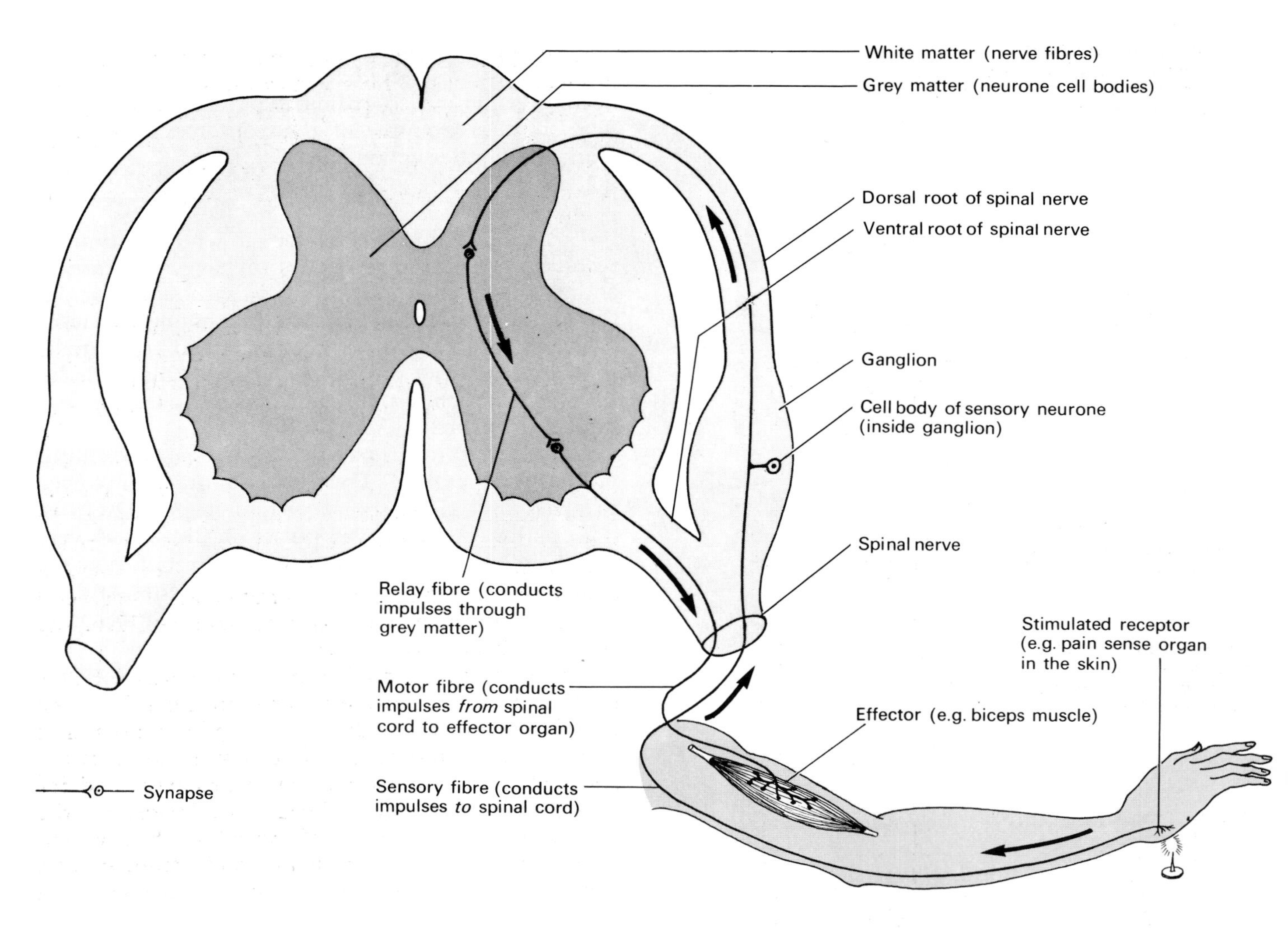

Fig. 13.5 Cross-section of the spinal cord, with a diagram of nerve fibres concerned in a reflex action

responses and their functions are dealt with in comprehension test 1.

The nerve fibres which connect reflex arcs with the brain enable reflex actions to be modified by experience. Conditioned reflexes are examples of such modifications.

Conditioned reflexes

In 1902 the Russian physiologist Ivan Pavlov began to investigate a type of learning behaviour which he later called a conditioned reflex. The most famous of Pavlov's experiments involved the production of saliva by dogs in response to food: a normal reflex action. Pavlov sounded a bell immediately before presenting food to the dogs, and after many repetitions he found that the sound of the bell alone was enough to evoke salivation. The normal reflex behaviour of salivation in response to food had been modified. The dogs salivated at the sound of the bell, even though no food was presented. They were responding to the bell as though it were food. The means by which a reflex is altered so that it occurs in response to a different stimulus is now called **conditioning**. Conditioning is a simple example of learning by association. Pavlov's dogs learned to associate the sound of a bell with the presentation of food, and then the bell alone caused salivation, i.e. they were now *conditioned* to respond to the bell as though it were food.

Humans can be conditioned. The eye-blink reflex normally occurs in response to a puff of air and not to light. But if, over many repetitions, a torch is switched on for a few seconds immediately before air is puffed into the eyes, eventually the light alone will produce an eye-blink reflex.

Some scientists argue that activities such as walking,

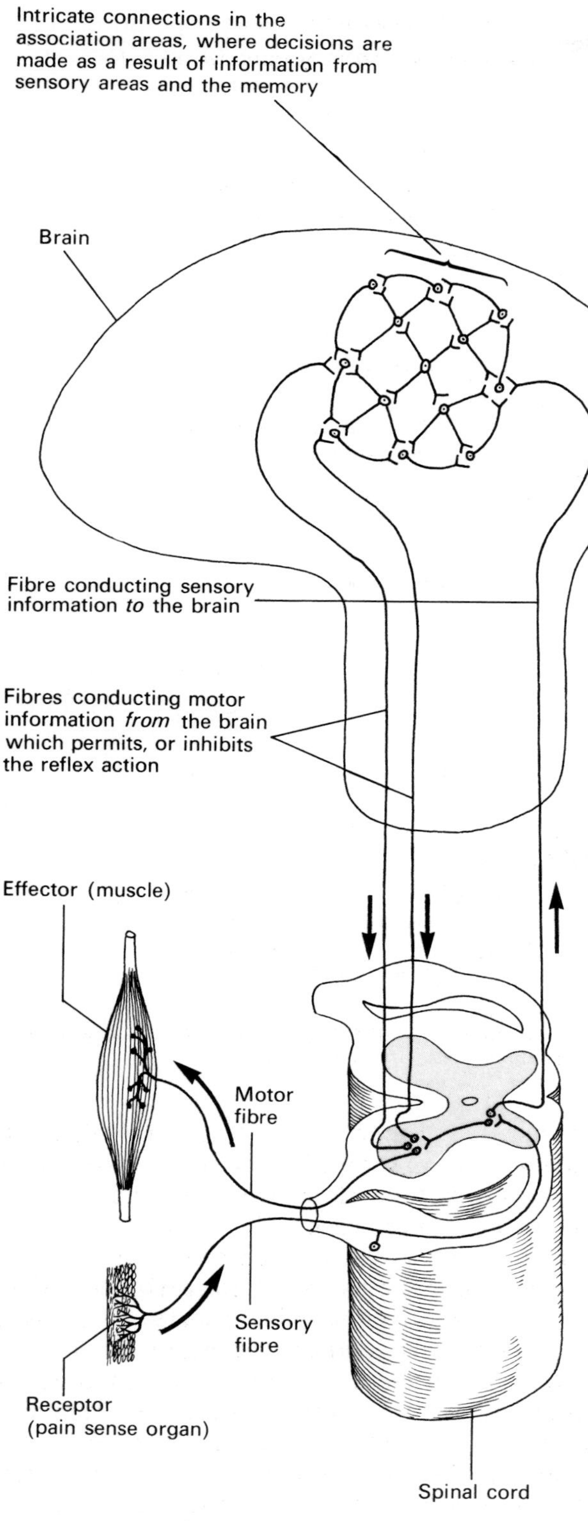

writing, playing the piano, and dancing, are conditioned reflexes. These activities, which are performed consciously at first, become unconscious (reflex) actions after much practice. However, the process of acquiring these skills differs in several ways from the conditioning processes experienced by Pavlov's dogs.

The structure and functions of the brain

During the embryo stage of development a vertebrate's brain appears as a swelling at the front end of the spinal cord. Eventually this swollen region develops thick nerves connecting it to sense organs in the head, i.e. the nose, taste buds, eyes, ears, and organs of balance.

In primitive vertebrates, like fish, amphibia, and reptiles, the brain is little more than a centre for the co-ordination of information from sense organs. In these animals the brain can produce simple reflex behaviour, but it can learn only to a very limited extent. Even so, these 'simple' brains consist of several million neurones and are far more complex than any electronic brain made so far.

The brain in humans also co-ordinates simple reflex behaviour, but is capable of far more than this. The brain's enormous capacity for learning from experience is of foremost importance. It can store information and make use of it later to solve problems and increase the complexity of behaviour. Moreover, the human brain is capable of complex physical co-ordination, such as playing a musical instrument, and complex mental processes like solving mathematical problems.

The human brain weighs about 1.5 kg, and is about the size of two clenched fists held together with the wrists touching. With the fists held in this position look at Figure 13.1A. The wrists are in the same position as the **medulla oblongata**, usually shortened to **medulla**. If the thumbs are bent and tucked into the fingers they are in the same position as the **cerebellum**; and the clenched fingers of each hand represent the two **cerebral hemispheres**, which together form the **cerebrum**.

Estimates of the number of neurones in the human brain vary between ten and fifty thousand million. In the medulla and cerebellum, each neurone has up to a thousand synapses through which it interacts with other neurones. But in the cerebrum each neurone has up to sixty thousand synapses.

It is difficult for the human brain to grasp the extent of its own complexity; but if more statistics help it is estimated that if each brain neurone were equivalent to the electronic units used to make computers, a human brain would be equivalent to more than a

Fig. 13.6 Diagram of the nerve fibres through which the brain becomes aware of, and can control, a reflex action thousand of the largest computers made so far. (Actually, neurones are far more complex than the electronic units used to make computers.)

The 'lower' parts of the brain (i.e. the medulla and cerebellum) control unconscious, involuntary processes, whereas the 'higher' regions (i.e. the cerebral hemispheres, or cerebrum) control conscious, voluntary activity.

Functions of the medulla The medulla forms the base of the brain where it merges with the spinal cord (Fig. 13.1A). This region controls many unconscious processes such as the regulation of temperature, blood pressure, and rates of heart-beat and breathing. Most of the time a person is not aware of these processes. However, excessive stimulation of the medulla, as for example when the weather is very hot, may result in impulses from here passing up to the cerebrum where they activate conscious processes such as opening a window or turning on a fan. On the other hand, during periods of emotional stress, fear, or anger, impulses travel down from the cerebrum to the medulla causing increased rates of sweating, heart-beat, and breathing. This is one reason why prolonged emotional stress can result in disorders such as high blood pressure.

Functions of the cerebellum (Fig. 13.1A). The cerebellum receives impulses from the organs of balance in the inner ear (chapter 14), and from stretch receptors embedded in joints and muscles. (Stretch receptors measure the degree of bending at the joints, and the degree of tension in muscle fibres.) This information is used to achieve balance and muscular coordination in activities like walking, running, or riding a bicycle.

Functions of the cerebral hemispheres (cerebrum) The cerebrum consists of a millimetre-thick outer layer of grey matter (neurone cell bodies), called the **cerebral cortex**, which overlays a thick mass of white matter (nerve fibres). The cerebral cortex, often abbreviated to **cortex**, is folded in a complicated way, which means that it contains a far greater mass of grey matter than it would if it had a smooth surface.

All parts of the cortex consist of exactly the same types of neurone, and yet all parts of the cortex do not have the same functions. The functions of a particular neurone depend more on its location in the cortex than on its structure.

Localization of functions in the cortex One way of discovering the functions of each part of the cortex is to pass tiny electric shocks through the surface of the brain when it is exposed in a surgical operation. This technique is possible because the brain's surface is insensitive to pain and so the patient can be fully awake during the operation to report what happens at each electric shock.

Using this method it is possible to plot a map of the brain's surface to show the functions of each area (Fig. 13.1A). One region of the cortex on each cerebral hemisphere is called the **motor area**, because shocks applied here result in contraction of muscles in various parts of the body. This happens because the motor areas have connections which pass down the spinal cord to motor neurones (Fig. 13.6). It is through these connections that the motor areas direct conscious movements of the body. If certain other areas are stimulated patients experience noises, sight, and sensations of touch. Regions that produce these effects are called **sensory areas**. These areas receive impulses from sense organs all over the body. There are separate sensory areas for vision, hearing, touch, taste, and smell.

Regions other than motor and sensory areas produce no definite response when stimulated with electric shocks. For this reason they were once called 'silent' areas. Now they are known as **association areas**, because it is here that 'association' takes place between information from all the sense organs, together with remembered information from past experience, to produce conscious awareness and understanding of the outside world.

This means that a person does not see with his eyes or hear with his ears. He sees, hears, and experiences all other sense impressions within the sensory and association cortex of his brain. The extraordinary thing about the cortex is that it produces these impressions from nothing more than a pattern of nerve impulses which are transmitted from sense organs like the dots of the Morse code. Nevertheless, these patterns of dots represent information, and what the brain does is to translate the dots into vivid, three-dimensional, full-colour moving pictures of the outside world complete with sounds and smells. Even more extraordinary is that equally vivid impressions can be aroused from the memory (that is, in the 'mind's eye') by thinking back over past experiences. These picture impressions are so realistic that it is sometimes hard to believe they are made of nothing more than fleeting patterns of impulses which flash round and round inside the grey matter of the brain.

Like a child making a complete picture from the scattered pieces of a jigsaw puzzle, the association areas of the brain form an impression of the outside world from the scattered bits of information which pour into the brain from all the sense organs. These impressions, together with memories of past events, are used to solve problems and make decisions.

13.2 The endocrine system

The endocrine system consists of glands, and like all other glands they produce and release (secrete) useful substances. Some non-endocrine glands, like the pancreas and salivary glands, secrete their products through tubes, or ducts, which lead directly to the place where these substances carry out their functions. Endocrine glands, however, have no ducts. They secrete their products–chemicals called **hormones**–directly into the blood-stream which carries them all round the body, where they affect organs sensitive to them.

Both hormones and nerve impulses co-ordinate the body's activities, but they do this in different ways. The difference between the way in which nerve impulses and hormones operate is like the difference between a telephone message and a message broadcast by radio. A telephone message goes along a wire to one person; and nerve impulses go along a nerve fibre to one particular muscle or gland. A radio announcement, however, is broadcast to everybody with a radio set, but only those actually concerned with the message respond to it. Similarly, hormones are 'broadcast' by the blood-stream to every cell in the body, but only certain cells respond to them.

Hormones are produced in minute quantities, but their influence on the body is often profound and long-lasting. For example, hormones affect the size to which the body grows, the development of sexual characteristics, and to a certain extent the development of mental powers and personality. In general, both physical and mental health depend on the endocrine glands producing the right amounts of hormones at the right times.

It is not known exactly how hormones carry out their functions. They may influence the rate at which chemicals become available for metabolic reactions, and they may also speed up or slow down the reactions. Equally important is the ability of some hormones to control the types of chemicals which cells take in or lose through their cell membranes.

The following notes describe only the major endocrine glands, the hormones they produce, and their effects on the body. Other chapters describe the actual mechanisms, such as reproduction, which are controlled by hormones. Figure 13.7 illustrates the position of endocrine organs in the human body.

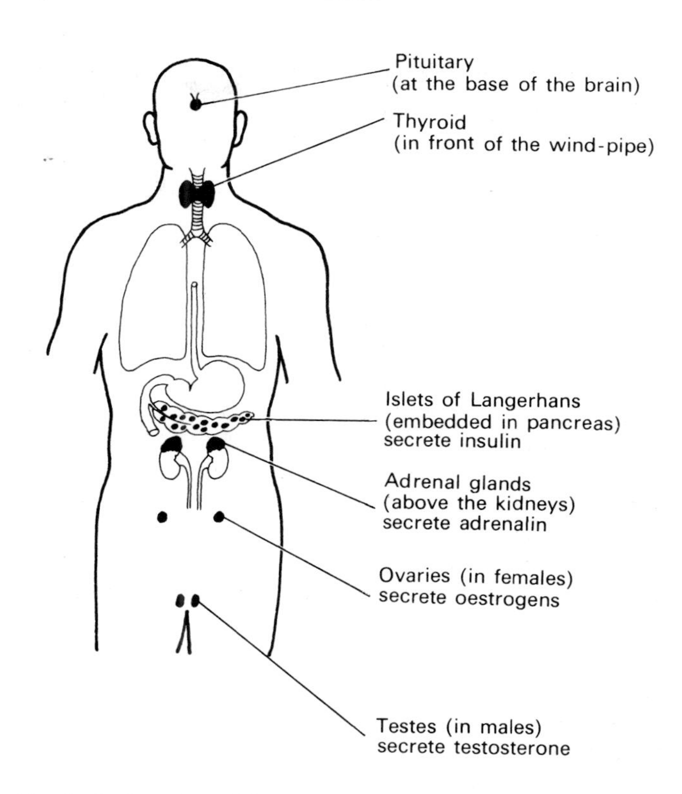

Fig. 13.7 Position of the major endocrine organs in the human body

The pituitary gland

The pituitary gland is situated in a small cavity of the skull 13 mm wide, beneath the brain and above the roof of the mouth. The pituitary secretes several hormones. Some of these affect the body generally, and some have a controlling influence over the other endocrine glands. For this last reason the pituitary has been called the 'master gland', and even 'the conductor of the endocrine orchestra'.

The pituitary strongly influences growth, both directly through its own growth hormone, and indirectly through its influence on other endocrine glands. Pituitary growth hormone controls the size of the bones. If the pituitary produces too much growth hormone the result is abnormal growth, or giantism. Humans, for instance, may grow to two metres or more in height. Too little growth hormone results in delayed or permanently retarded growth.

One pituitary hormone raises blood pressure by causing contraction of muscles in blood vessel walls. Another pituitary hormone causes the strong contractions of the uterus which occur during childbirth, and also the production of milk from mammary glands. A third pituitary hormone controls the amount of water reabsorbed into the blood in the kidneys, thereby influencing the amount of water lost in urine.

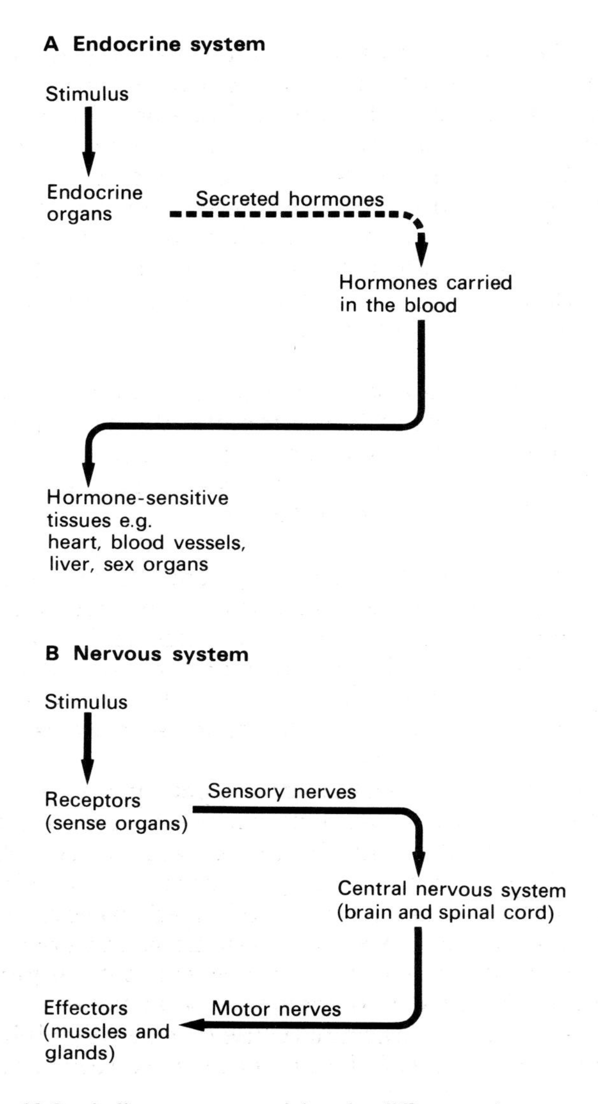

Fig. 13.8 A diagram summarizing the differences between endocrine organs and the nervous system

The pituitary controls other endocrine organs by **feed-back mechanisms**. A typical feed-back mechanism controls production of the hormone **thyroxin** from another endocrine gland called the **thyroid**. Whenever thyroxin production falls below normal the pituitary secretes a thyroid-stimulating hormone (TSH), which stimulates the thyroid to make more thyroxin. The result is an increased level of thyroxin in the blood, and this eventually suppresses TSH production from the pituitary. This in turn causes a slowing down of thyroxin production in the thyroid. The accelerating effect of TSH on the thyroid, and the braking effect of thyroxin on the pituitary, result in balanced hormone production. Similar mechanisms operate between the pituitary and other endocrine organs.

The thyroid gland

The thyroid is a butterfly-shaped gland situated in the neck, in front of the wind-pipe. There are several thyroid hormones, including thyroxin, and they control the rate at which sugar is consumed in cellular respiration. Through these hormones the thyroid controls the rate of metabolism in the whole body.

The thyroid has a major influence upon physical and mental development from birth to old age. In the young, failure of the thyroid to produce its hormones can result in stunted growth and severe mental retardation, a condition known as **cretinism**. Even in adults a slight decrease in thyroxin production can have dramatic results. The victim becomes extremely sensitive to cold, tends to forget his past life, and soon looks much older than he really is. Fortunately, these conditions in both young and old can be treated by ox thyroid in doses of as little as one-hundredth of a gram daily.

In cases of over-active thyroid, the rate of metabolism increases, the body gets thinner, the victim becomes very restless, over-excited, and even mentally unstable. A person's whole physical and mental well-being can depend therefore on the balanced activity of this one tiny gland.

Adrenalin

Adrenalin is a hormone produced by the inner region (medulla) of the adrenal glands. There is one of these glands above each kidney in mammals. The sudden and dramatic effects of adrenalin must be well-known to everyone, for they are experienced whenever a person is faced with a dangerous or very exciting situation.

Imagine a man walking slowly down a street feeling tired after a hard day's work. A car appears from around a corner and suddenly skids out of control towards him. Almost instantaneously the man's tiredness is forgotten and his body becomes super-efficient. He runs faster than he has for years, then jumps effortlessly over a wall which would normally have seemed impossibly high. What has caused this amazing transformation?

The sight of the car hurtling towards him causes a very rapid outpouring of adrenalin from the man's adrenal glands. Within seconds this hormone stimulates his heart to beat much faster; it makes the broncheoles of his lungs increase in diameter, and his breathing rate increases. This charges his blood with extra oxygen and sends it, along with a massive dose of glucose which adrenalin releases from his liver, rushing to his muscles and brain. The flow of blood to these regions is increased in two ways. First, adrenalin constricts blood vessels in his skin and gut, thus diverting blood from these regions. Second, the hormone enlarges blood vessels to his muscles, so that blood flow is greatest where it is most needed.

In short, adrenalin prepares the body for sudden, possibly violent effort. In wild animals, adrenalin is released in situations such as fights between predators and their prey. In humans, situations that require sudden bursts of physical activity are quite rare. Nevertheless, there are many other occasions in which adrenalin is released, such as during emotional excitement, stress, fear, or anger. Familiar symptoms of adrenalin's effects at these times are: a dry mouth, a pounding heart, and an unpleasant 'sinking' sensation in the stomach. (X-rays have shown that the stomach can actually drop several centimetres within the abdomen during emotional stress.)

Insulin

Within the pancreas, a gland which secretes digestive juices (Fig. 4.15), there are patches of tissue with no digestive function. These patches, poetically named the **Islets of Langerhans**, produce a hormone called **insulin**.

Insulin controls the level of glucose sugar in the blood by increasing the rate at which the liver converts glucose into glycogen. Insulin also enables body cells (except neurones and muscles) to absorb glucose, which is their main source of energy. In addition, insulin stimulates the production of fat from glucose, and the synthesis of proteins.

Diabetes is a disease caused by the slowing down of insulin production. This sets off a complex chain of events in the body. The major symptom of diabetes is a massive increase in glucose level in the blood. First, glucose from digested carbohydrate is no longer turned into glycogen or fat. Second, without insulin both fat and protein in the body cells tend to be broken down, which eventually yields even more glucose. Fat breakdown also produces fatty acids, and these are eventually further broken down into poisons such as acetone and acetic acid. Some of the excess glucose in the blood is excreted from the body by the kidneys as part of urine, but this requires large amounts of water and so the diabetic develops an almost insatiable thirst. An added complication is that many body cells can no longer absorb glucose and so their respiration rate slows down. Unless the victim of diabetes is treated with insulin (extracted from cow or pig pancreas) he may become unconscious as a result of dehydration, acetic acid poisoning, and slow cellular respiration.

Sex hormones

The sex hormones are produced by the ovaries of female mammals, and the testes of males (Fig. 13.7).

Female sex hormones The ovaries produce a number of hormones known collectively as **oestrogens**. These have three main functions. First, they control the development of secondary sexual characteristics; that is, the external features of females. In humans these include breasts, wide hips, and a high-pitched voice. Second, these hormones prepare the uterus to receive a ripe, fertilized ovum. Third, they maintain the uterus in a state whereby it can nourish and protect the developing embryo.

Male sex hormones The testes produce the hormone **testosterone**. This promotes the development of secondary sexual characteristics, which in humans include a deeper voice, more body hair, and more powerful muscles than in females.

Hormones and the nervous system affect the whole body and therefore each other in an enormous number of different ways. In fact, the interrelationship between the two is so close that a disturbance in any endocrine organ or any part of the nervous system can have far-reaching effects all over the body, sometimes with disastrous consequences.

Hormones and nerves affect a person's behaviour, but it must also be remembered that a person's behaviour can affect his hormones and nerves. Any behaviour which causes excessive stress, either physical or emotional can, if prolonged, cause damage to the endocrine and nervous systems, and this damage may cause a chain-reaction of events so that symptoms appear which have no apparent connection with the original cause. The original cause may be very difficult to trace. But worst of all the damage may be permanent.

Inquiry exercise

A *An investigation of the effects of thyroxin on the development of tadpoles*

1. Obtain four glass bowls of at least two litres capacity. Label them A, B, C, and D, then fill them as follows:

A 0.000 025 g thyroxin per litre of pond water.
B 0.000 01 g thyroxin per litre of pond water.
C 0.000 001 g thyroxin per litre of pond water.
D untreated pond water.

2. Obtain at least 20 freshly hatched tadpoles by collecting them as they emerge from frog spawn, and place 5 in each bowl. Keep the bowls at room temperature and feed the tadpoles on tiny pieces of boiled lettuce. Remove any uneaten food each day or it will foul the water.

3. Keep separate illustrated records for each bowl, noting any changes, with dates, in the following features:

a) overall body size;

b) disappearance of the tail;
c) appearance of limbs;
d) breathing at the surface.

4. *a*) How does thyroxin affect development?
b) Which concentration of thyroxin affects development the most?
c) In general, what changes occur in methods of breathing, locomotion, and feeding?
d) What would be the likely effect on development if a newly hatched tadpole's thyroid was completely removed?

Comprehension test

1. What is a reflex action, and what reflexes occur:
a) when dust blows into the eyes (2 examples);
b) when a bright light suddenly shines in the eyes (2 examples);
c) when a person runs fast in hot weather (3 examples);
d) when a person moves suddenly from a warm room to a very cold one (2 examples);
e) when a hungry person smells cooking food (1 example);
f) when food accidentally enters the wind-pipe?

2. Explain how each of the reflexes given in answer to 1 protects the body and, in certain cases, conserves energy and body materials.

3. List the receptor and effector organs concerned in each of the reflex actions given in answer to 1.

4. A dog was conditioned to respond to the sound of a bell by salivating. However, after a number of occasions on which the bell was sounded but no food was given, the response of salivation disappeared. What does this result indicate about the difference between conditioned reflexes and normal reflex responses?

5. A piece of thread was tied tightly round an animal's pancreatic duct. The animal subsequently had difficulty in digesting food, but did not get diabetes. Explain.

Summary and factual recall test

The central nervous system consists of (1–name its two parts). It is connected to all parts of the body by nerve (2) which are (3) from each other by a sheath of fatty material. A sensory neurone conducts impulses from receptors such as (4–name five) along a fibre called a (5) to the cell (6) of the neurone and from here along another fibre called an (7) into the (8). The (9) body of a motor neurone is embedded in the (10). Here, it collects impulses through fine fibres called (11) and passes them through a long fibre called an (12) into an effector organ such as (13–name two types).

Sense organs send out impulses only when stimulated above their (14) level. These impulses either occur or they don't, which is called the (15) principle. The only feature of impulses which ever varies is (16). As impulses pass from one neurone to the next they cross a gap called a (17).

A reflex action is (18). An example of a conditioned reflex is (19).

The medulla forms the (20) of the brain where it merges with the (21). This region controls unconscious processes such as (22–name four).

The cerebellum receives impulses from (23–name two places). This information is used to achieve (24) and (25) co-ordination, for example when (26–name two examples).

The cerebrum has a thin outer layer of (27) matter called the (28) which consists of (29). This layer overlays a region of (30) matter which consists of (31). The cerebrum receives impulses from organs such as eyes and ears into its (32) areas, and passes impulses to muscles and (33) from its (34) areas. The association areas are given this name because (35).

The endocrine system consists of glands which have no (36), and which secrete chemicals called (37) into the (38). The pituitary gland is situated beneath the (39). It is sometimes called the 'master gland' because (40). Apart from this, its main functions are (41–list three).

The thyroid gland is situated in the (42). Its main functions are to control the rate of (43) consumption in cellular (44), and influence (45) and (46) development from birth to old age.

Adrenalin prepares the body for (47). In wild animals it is produced in situations such as (48), and in humans during (49–name three situations).

Insulin controls the level of (50) in the blood by (51).

Female sex hormones are known collectively as (52). Their main functions are (53–name three). The (54) produce a male sex hormone called (55) which controls the development of (56).

14 Sensitivity in animals

Sense organs are the **receptors** of the body. They receive information about conditions both inside the body and in the world around it. This information is known as the **stimulus** because it stimulates the sense organs. But no matter how the sense organs are stimulated, whether by the sound of the latest pop tune, the sight of a beautiful face, the smell of bacon and eggs, the smooth touch of a cat's fur, or a pain in the stomach, the stimulus is turned into only one thing: a pattern of nerve impulses which speed down sensory nerves to the brain. Here, the patterns of impulses are interpreted and transformed into **sensations** such as touch, sight, hearing, taste, and smell. Before describing how sense organs work it is necessary to describe a few characteristics of the sensations which they produce.

First, sensations arise in the brain and not in the receptors themselves. Patients with a slipped disc, for example, often complain of pain in their legs when they would have expected pain in their backs. This happens because the protruding disc often rubs against a sensory nerve from the legs at a point between a pain receptor and the brain. The impulses which result from this irritation are interpreted as if they came from the pain receptors in the legs.

Second, the position of sensations in the body can be located immediately. For example, someone who treads on a drawing pin knows without looking which foot has been hurt. This is possible because each sense organ has a separate nerve pathway to the brain, and a separate sensory area of the brain to deal with its particular type of sensation (Fig. 13.1A). Localization of sensation in the brain is so well developed that if it were possible to connect nerves from the ears to the area of the brain which normally deals with impulses from the eyes, a loud noise would be 'seen' as a bright flash of light.

Third when most receptors are first stimulated they send out impulses at a fast rate (high frequency) which soon slows down, or ceases altogether. This happens whenever a stimulus remains unchanged for some time, and is known as **sensory adaptation.** Adaptation is partly responsible for a person becoming unaware for example of the feel of clothing against his skin, or the ticking of a clock. The first time a person wears spectacles he is acutely aware of the unaccustomed pressure on his nose and ears, but owing to sensory adaptation, their presence is almost completely forgotten after a few days.

This chapter describes the structure of sense organs in mammals, particularly humans, and certain equivalent receptors of other more primitive animals. The first section deals with the detection of stimuli from inside the body, particularly from muscles and joints. Later sections describe sense organs which are stimulated by external events. These are receptors in the skin, and the 'special senses' of taste, smell, sight, hearing, and balance.

14.1 **Internal receptors**

The brain receives a constant stream of impulses from receptors buried deep inside the body. Some examples are: osmo-receptors which react to changes in the osmotic pressure of tissue fluid; chemo-receptors such as those which detect changes in the level of certain chemicals in the body like carbon dioxide or sugar; blood pressure receptors which detect stretching and contraction of muscle in blood vessel walls; thermo-receptors which react to changes in body temperature; and a range of proprioceptors which respond to stimuli from muscles and joints.

Proprioceptors

There are proprioceptors, of a type called stretch receptors, embedded in every muscle in the body (Fig. 14.1). Similar receptors are found in muscle tendons, and in the ligaments of joints. These receptors provide the brain with information about the degree of tension in each part of the muscular system, and the angle to which every joint in the body is bent.

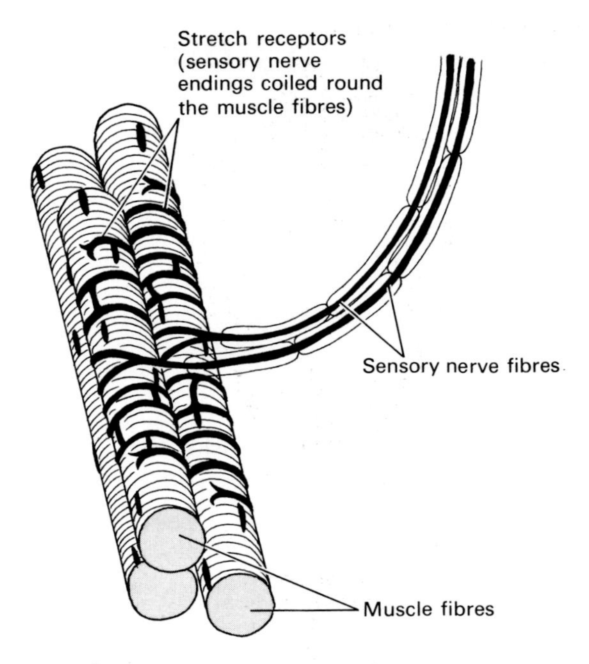

Fig. 14.1 Stretch receptors in muscle fibres

From this information a person knows, either consciously or unconsciously, a great many important facts about his body. Even in complete darkness a person knows the direction and speed of all his movements. He also knows the exact position in space of every limb in relation to the rest of the body, and usually in relation to the ground: this is called 'position sense'.

Try this simple experiment. Close the eyes and try to touch the nose with one finger. Now try a more difficult task. Close the eyes and raise one index finger to any point behind the head, and then try to touch this finger with the index finger of the other hand. Both these apparently simple acts are possible because the brain can calculate the position of each finger almost instantaneously, using a mass of information about constantly changing muscle tensions and joint movements from proprioceptors in the shoulders, arms, and fingers.

Without proprioceptors muscular co-ordination and locomotion would be impossible. Receptors of this type are found in all vertebrate animals, in arthropods, and in all other animals with well developed muscular systems.

14.2 Skin receptors

The skin contains millions of separate tiny sense organs with several different functions. Microscopic examination of the human skin shows that it contains at least five different types of sensory nerve ending. It is not yet certain how these receptors work or exactly what their individual functions are. Nevertheless, they have been named according to their supposed functions, which are: touch, pressure, temperature, pain, and hair movements.

For various reasons the functions of each type of skin receptor are difficult to establish. First, certain parts of the body, like the cornea of the eye (Fig. 14.4), are sensitive to touch, pressure, and pain and yet they have only one type of nerve ending, and these look like the pain receptors illustrated in Figure 10.3. Second, it is possible to find the exact location in the skin of, say, a pain receptor only to discover a few hours later that this exact spot has become sensitive to heat or cold and not pain.

Sense of touch

The sense of touch enables a person to distinguish between a variety of textures, from sandpaper roughness to glassy smoothness, and between hard, soft, and liquid substances. What is more, touch can give a vivid impression of three-dimensional shape, and it is sensitive to fine detail. For example, blind people can read with their finger-tips using a system of raised dots known as Braille. Touch receptors are not evenly distributed over the body surface. They appear to be especially close together in the tongue and at the finger-tips (exercise A).

Sense of pressure

Pressure receptors are also concentrated in the skin of the tongue and finger-tips, where pressure differences as small as $2\,g/mm^2$ can be detected. Humans have the ability to 'project' pressure sense into objects held in the hand. A cook stirring a pudding can feel lumps deep inside it just as if there were live pressure receptors at the end of his spoon. This illusion occurs because pressure changes felt at the handle are projected into the spoon. Projection is important in the skilled use of tools such as files, spanners, and chisels.

Temperature sense

There are separate 'heat' and 'cold' receptors in the skin, but these cannot be used, like a thermometer, to tell the exact temperature of something. Temperature sense is limited to comparing temperature differences. If a person from the Arctic and a person from the tropics both arrive in London by air on a warm spring day, the man from the Arctic is likely to feel hot, while the man from the tropics feels cold. This happens because they are each comparing English spring temperatures with those of different climates. However, after a few days, when their heat and cold receptors have undergone adaptation, they both feel roughly the same temperature sensations.

Temperature receptors in the fingers can distinguish differences as small as 0.5 °C. However, the tongue is surprisingly insensitive to temperature, which explains why drinks that scald the fingers can be sipped without discomfort.

Pain

Pain receptors are more evenly distributed throughout the skin, except that in certain areas they can be partly obscured by thick layers of epidermis, as under callouses which form on the palms and fingers after prolonged manual work.

Pain receptors are not restricted to the skin; they are located inside muscles, tendons, ligaments, in the walls of the digestive system, in fact everywhere except in the brain. Even though pain is unpleasant it is important because it acts as a warning that something is going wrong in the body. Moreover, the brain can usually work out the exact location, and sometimes the nature, of the trouble according to the position of affected pain receptors.

There is a rare abnormality which greatly reduces, or completely removes, a person's sensitivity to pain (in Britain there are fewer than 100 cases). These people are in constant danger because they are unaware of cuts, burns, and pressure which could break their limbs, and unaware of diseases which show no visible symptoms.

Hair movements

Most hair follicles (Fig. 10.3) have a sensory nerve ending attached to their bases. These receptors are stimulated when the hair is moved by objects close to the skin or by air movements. Some animals, such as cats and mice, have very long hairs, or whiskers, which extend outwards from the face to approximately the width of the body. These sensitive whiskers are extremely helpful in avoiding obstacles, especially when the animals move about at night.

14.3 **Taste and smell**

Taste and smell receptors are sensitive to chemicals (i.e. they are chemo-receptors). Smell receptors can detect chemicals in the air given off by objects which can be some distance away. Taste receptors detect chemicals given off from substances touched by the tongue. But both types of receptor can only detect chemicals after they have dissolved in the moisture which always covers these receptors.

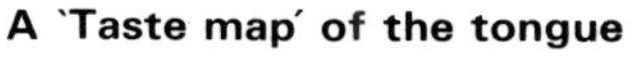

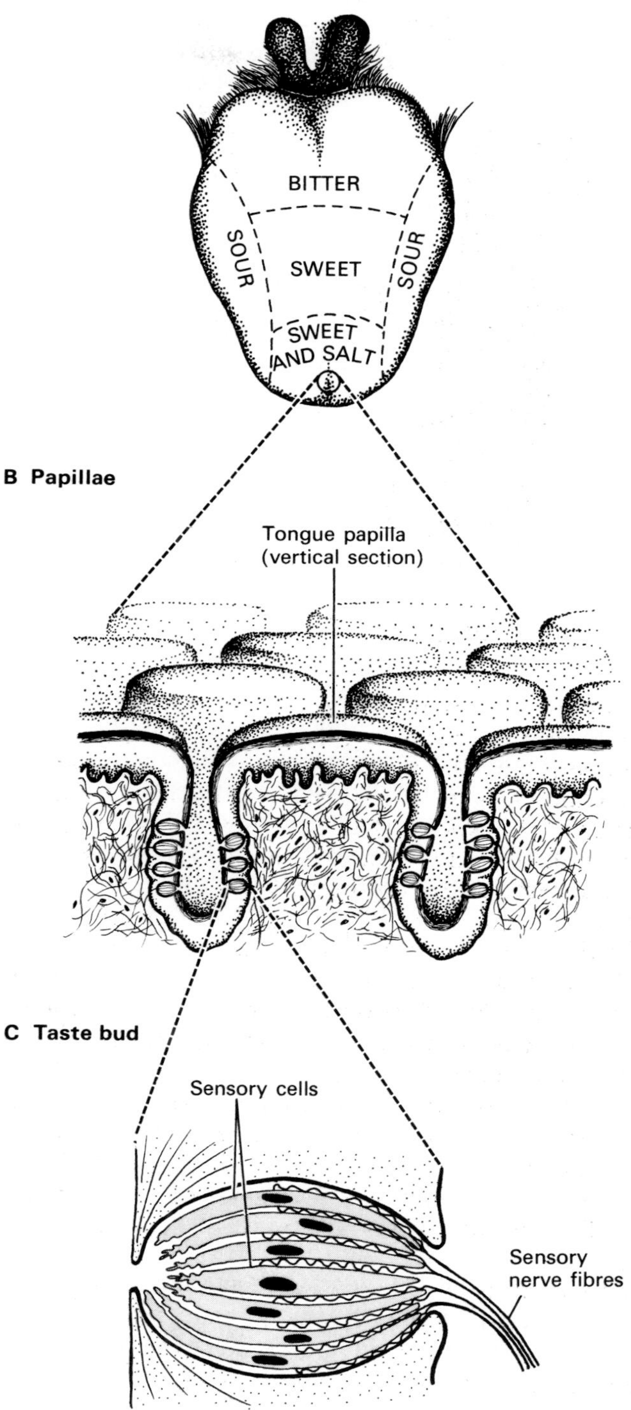

Fig. 14.2 Distribution of taste receptors on the tongue

Taste

Taste receptors, or **taste buds**, are situated around the walls of tiny swellings, called **papillae**, on the upper surface of the tongue (Fig. 14.2). There are four types of taste receptor: separate receptors detect salt, sweet, sour, and bitter-tasting substances. Groups of these specialized receptors are concentrated in certain areas of the tongue.

The countless different flavours encountered in food and drink are identified according to the level of stimulation which they produce in each of the four types of taste receptor. A connoisseur of fine wines does not pour them down his throat in great gulps; he sips the wine and runs it across his tongue so that it stimulates all the different taste areas. However, the function of taste buds is not simply to make eating and drinking pleasurable. Taste sensations stimulate the secretion of gastric juice, thereby preparing the stomach to receive food.

Smell

Smell receptors, or **olfactory organs**, are situated high inside the nasal cavities, well above the main air current through the nose (Fig. 9.1). For this reason, the best way to detect a smell is by quick sniffs, which send small eddies of air directly to the sensitive areas.

A 'cold in the nose' partially reduces the sense of smell. This happens because the nasal membranes react to the cold virus by producing large amounts of mucus, which covers the smell receptors. This also reduces the ability to taste, which shows that many so-called 'tastes' are, in fact, smells.

Certain odours and tastes are extremely unpleasant. On the whole, these unpleasant sensations are produced by poisons, or other harmful substances such as decaying food. Taste and smell receptors therefore protect the body from harm.

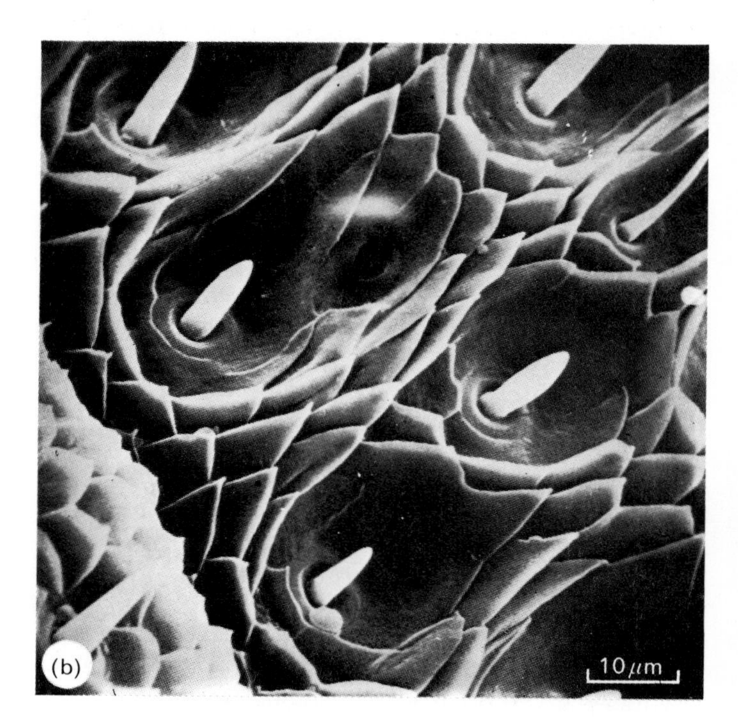

The antenna of a locust. (b) is a highly magnified portion of (a). The white pegs are nerve endings sensitive to chemicals. The pit towards the top of (b) contains a nerve ending sensitive to changes in humidity

14.4 Insect antennae (and other surface receptors)

The thick external cuticle of insects (and arthropods generally) contains an enormous array of sense receptors. The antennae alone have touch, sound, smell, taste, temperature, and humidity receptors. Similar receptors are concentrated on the feet and mouth parts, and many others are scattered over the body surface.

Insects wave their antennae in the air to test wind direction, temperature, and odours. Mosquitoes can detect their warm-blooded prey by slight variations in air temperature. Particularly amazing is the ability of male Luna moths (Fig. 14.3) to detect the scent given

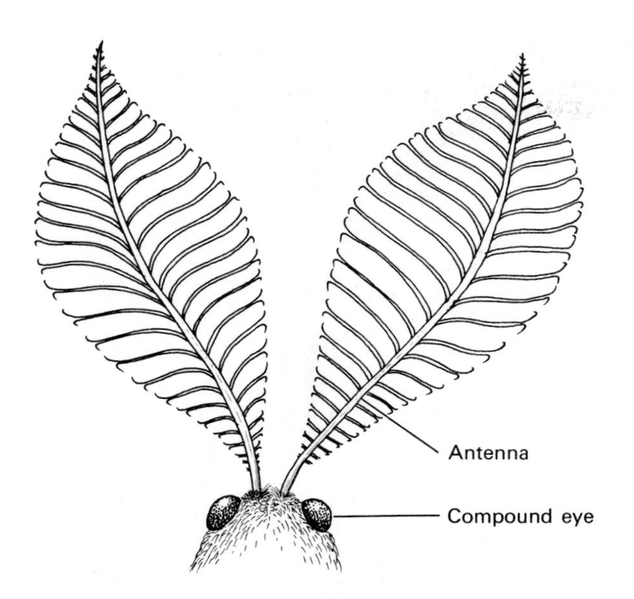

Fig. 14.3 Antennae of a male Luna moth

off by females of their species. The males use their huge feather-like antennae to follow the scent for long distances, in an upwind direction, through the many conflicting odours of town and countryside. They gauge the distance and direction of the female by comparing the strength of scent odour received on each antenna. Incredibly, the male can still detect the female's direction even if only a single molecule of her scent touches one antenna, while none at all touch the other.

Insects can distinguish between a wide variety of substances using taste and smell receptors on their antennae, feet, and mouth parts. House-flies, for instance, are attracted to decaying food by the odour of ammonia which it gives off, and leaf-eating insects appear to distinguish one leaf from another by the presence in them of certain characteristic vegetable oils. Insects locate food not only for themselves, but also as a place to lay eggs so that grubs can begin feeding as soon as they emerge.

14.5 Vision

Most living things, including green plants and even some unicellular organisms, are sensitive to light. But relatively few organisms are capable of vision; that is, few have the ability to form picture images of the outside world. This section describes the image-forming eyes of mammals (e.g. humans), and then compares them with the complicated though less efficient compound eyes of arthropods (e.g. insects).

The human eye

An eye resembles a camera in at least three ways. First, cameras and eyes both have a mechanism which focuses light. In the eye this consists of the transparent **cornea** and **lens**, which act like the glass lens of a camera in forming a clear, upside-down, full-colour image (Fig. 14.4). Second, in the eye this image falls on a layer of receptors called the **retina**, which, like the film in a camera, is sensitive to light. Third, eyes and some cameras have an apparatus called an **iris diaphragm**, which is an opaque disc with a hole at its centre. The size of the hole can be increased or decreased to control the amount of light reaching the light-sensitive surface.

Unlike a camera the eye does not make a permanent record of images which fall on its retina. The retina transforms light into a stream of nerve impulses which pass down the optic nerve to the brain. The frequency and pattern of these impulses vary according to the patches of colour, light, and shade which make up the retinal image. The visual area of the brain (Fig. 13.1A) interprets these impulses to form a three-dimensional, full-colour, moving impression of the outside world.

Protection of the eyes The eyes are protected in many ways. The skull has two deep cavities about 2.5 cm in diameter called the **orbits**, that enclose and protect all but the front of the eyes (Fig. 14.5). The exposed region is covered by a transparent, self-repairing skin called the **conjunctiva**. The conjunctiva is kept moist and clean by a slow continuous stream of liquid from the **tear glands**, and every few seconds it is wiped by the eyelids during their automatic (reflex) blink movements. When dust or chemicals reach the conjunctiva the rate of tear flow and blinking is automatically increased until the eye is clean. The blink reflex also protects the eyes by closing them whenever an object moves quickly towards the face. Finally, the eyelashes form a net in front of the eyes, which traps large airborne particles.

Movements of the eyeballs Each eyeball is held in place, and moved within its orbit, by six **extrinsic muscles** attached to the outer surface of the eyeball (Fig. 14.5). The extrinsic muscles can rotate the eyes to follow moving objects, and direct the gaze to a chosen object. These movements are precisely co-ordinated so that both eyes work together, and at all times are directed at the same spot. This means that the eyes converge inwards to watch objects which move towards the face.

Nourishment and support of eye tissues Eyes are nourished and supplied with oxygen by blood vessels

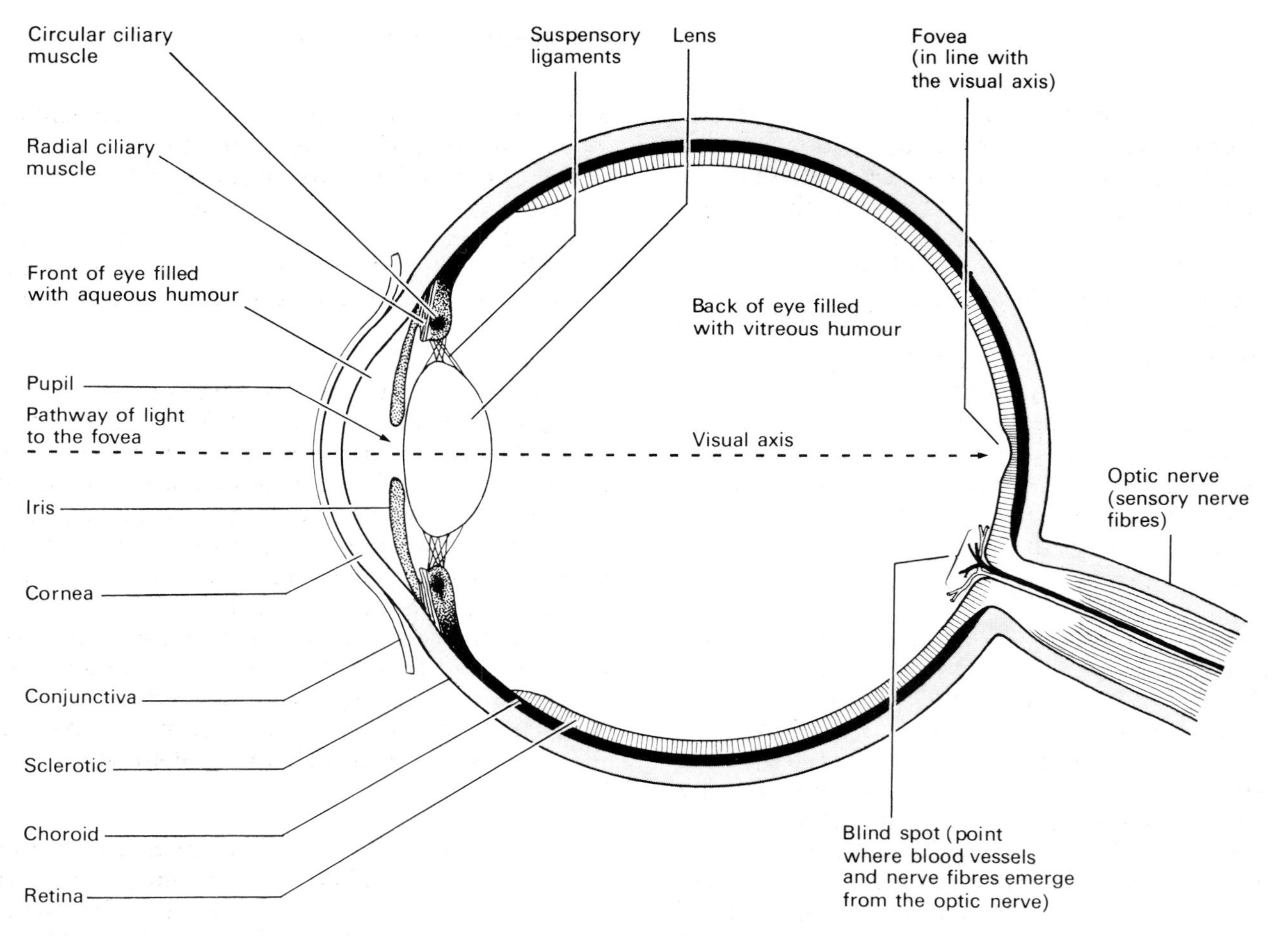

Fig. 14.4 Vertical section through a human eye

Fig. 14.5 An eye and its extrinsic muscles inside the orbit of the skull

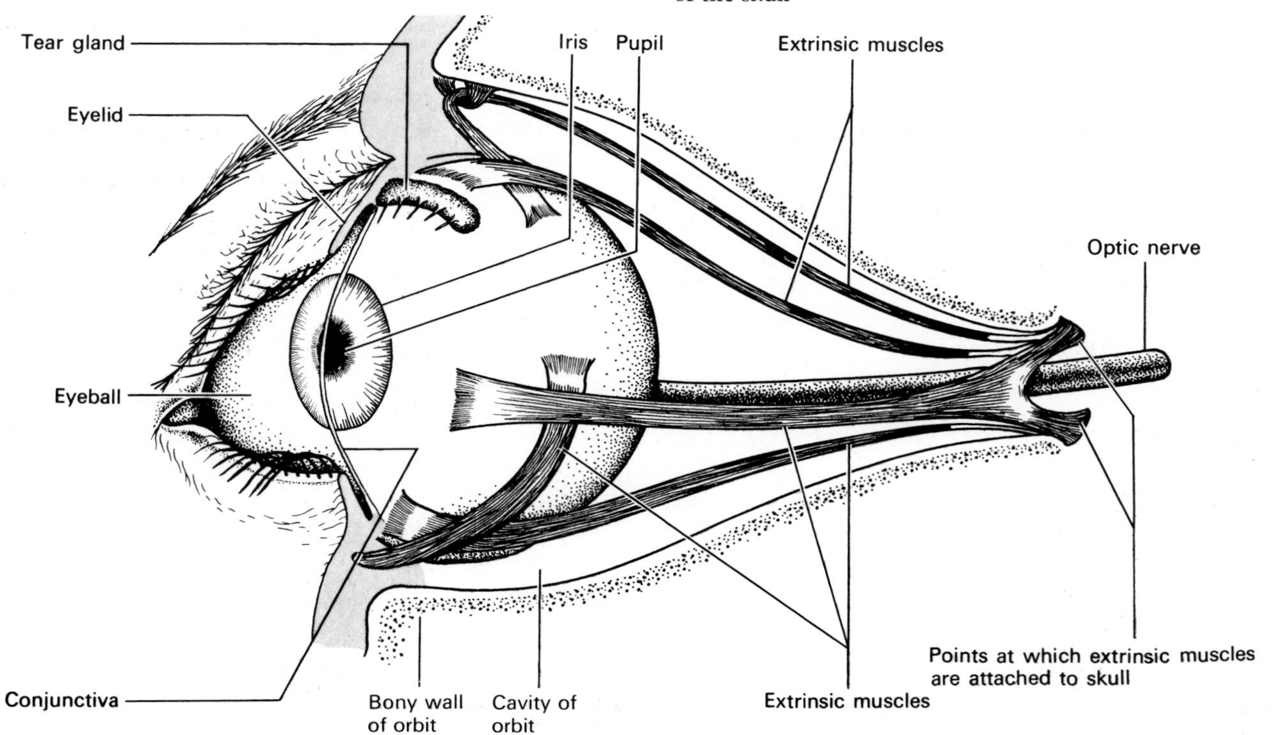

which enter through the optic nerve. These vessels spread out through the **choroid** layer, and over the surface of the retina. It is hardly surprising that the cornea and lens have no direct blood supply, since a network of capillaries would impair their ability to focus light. The cornea and lens obtain oxygen and food by diffusion from blood vessels through a liquid called the **aqueous humour**. This liquid is secreted into, and absorbed from, the front cavity of the eye. It is renewed about every four hours. (Incidentally, tiny particles in the aqueous humour are the cause of spots which sometimes appear to float before the eyes.) The aqueous humour, together with a jelly called the **vitreous humour** in the rear cavity of the eye, exert an outward pressure on the eyeball, and this maintains its rounded shape.

The iris The iris is the coloured part of the eye, and has a round hole at its centre called the **pupil**. The iris consists of radial muscles which contract to enlarge the size of the pupil, and circular muscles which make it smaller in size.

The iris regulates the amount of light which reaches the retina by opening the pupil in dim light and reducing it to pin hole size in bright light (these movements are reflex actions). As all photographers should know, reducing the size of the hole in a camera iris increases the **depth of focus** of the lens, which means that both near and distant objects are in focus at the same time. The same is true of the eye, and this explains a second function of the iris. When the eyes are focused on near objects the depth of focus of the lens is low. Therefore, vision would be poor were it not for an automatic reduction in pupil size (irrespective of light intensity). When the lens is focused on a distant object its depth of focus is far greater, and the iris opens again.

The lens The lens of an eye consists of layers of transparent material arranged like the skins of an onion, which are enclosed in an elastic outer membrane. The whole lens is held in place by **suspensory ligaments** attached to its outer rim, and these ligaments are attached to a ring of muscle fibres, the **ciliary muscles**, which run around the eyeball next to the iris (Figs. 14.4 and 14.6).

The cornea alone is sufficient to form an image on the retina of distant objects. However, the lens makes it possible for this image to be re-focused during shifts of vision from distant to near objects, and back again. The process of changing focus is called **accommodation**.

Accommodation Unlike a camera lens, the lens of the eye is not moved in and out to change focus. In the eye, focusing is accomplished by changing the shape of the lens (Fig. 14.6). This is possible because the lens is made of an elastic substance, and it can be

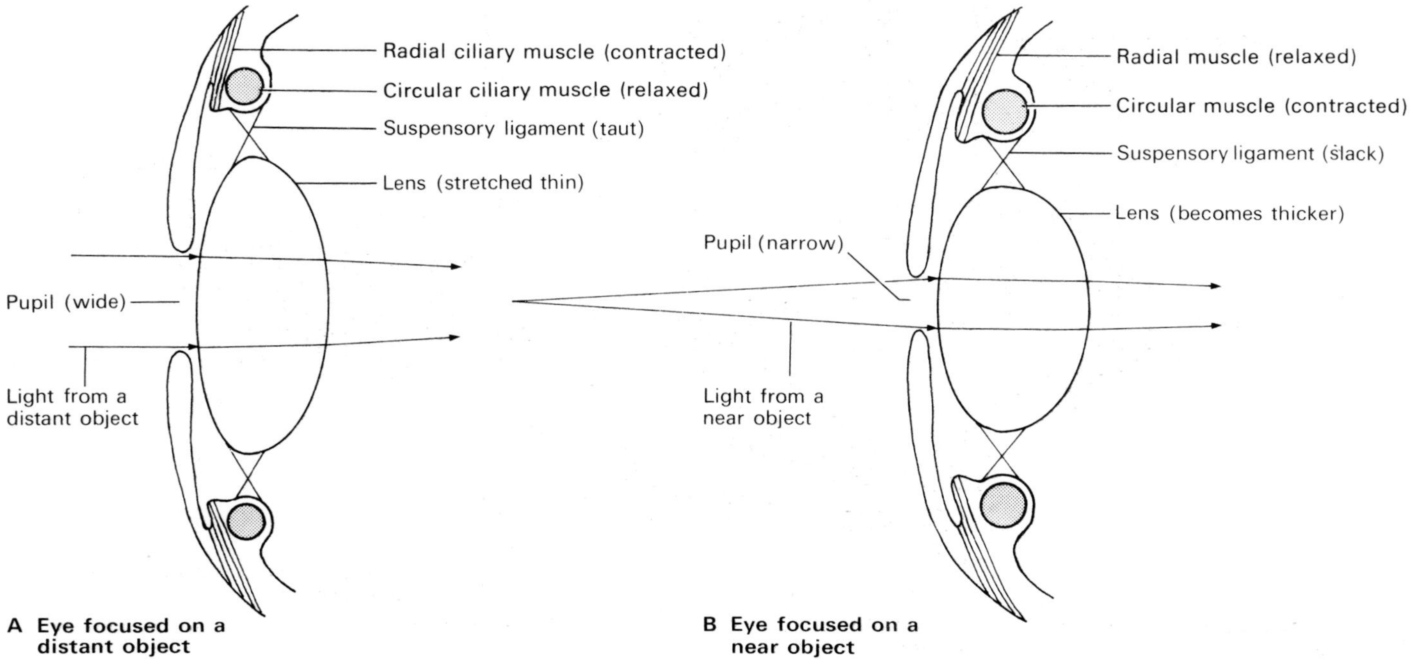

Fig. 14.6 Diagram of the changes which take place during accommodation movements

stretched into a slim (less convex) shape, but when this tension is released it reforms into a fatter (more convex) shape of its own accord. These changes of shape are brought about by the action of the ciliary muscles.

An eye is focused on a distant object by changing the lens to a flattened (less convex) shape. This reduces to a minimum the power of the lens to bend (refract) light. A flattened shape is needed to focus the almost parallel light rays from distant objects (Fig. 14.6A). The lens is flattened by contraction of the radial ciliary muscles. These pull against the suspensory ligaments, which pull against the lens stretching it into a flatter shape.

An eye is focused on a near object by changing the lens to a rounded (more convex) shape. This increases the lens's power to refract light, which is necessary in order to focus onto the retina the diverging light rays from a near object (Fig. 14.6B). The lens is made more rounded by contraction of the circular ciliary muscles and relaxation of the radial ones. Circular ciliary muscles contract to form a circle with a smaller diameter. This reduces tension on the suspensory ligaments and allows the lens to become rounded in shape. Several sight defects may upset this focusing mechanism. Three of them are listed below:

Presbyopia, or old sight The lens continues to grow throughout life, but at a very slow rate after adolescence. By the age of about sixty the centre of the lens is so far removed from supplies of oxygen and food that its cells die. This process reduces the lens's elasticity and it can no longer change in shape. It then becomes more or less fixed into a shape suitable only for distant vision. Therefore, old people usually require 'reading glasses' which have converging lenses to give the eyes extra power for close work.

Hypermetropia, or long sight Long sight occurs when the distance between the lens and the retina is shorter than normal (Fig. 14.7B). An image cannot be focused in so short a distance; in fact the point of clear focus is somewhere behind the retina. Long sight can be corrected by fitting spectacles with converging lenses, which add to the refractive power of the eye.

Myopia, or short sight One cause of short sight is an abnormally elongated eyeball (Fig. 14.7C). That is, one in which the distance between the lens and the back of the eye is so great that the point of clear focus is in front of the retina. This defect can be corrected by fitting spectacles with diverging lenses, which reduce the refractive power of the eye.

The retina The retina consists of 126 million light-sensitive receptors, which are of two types: six

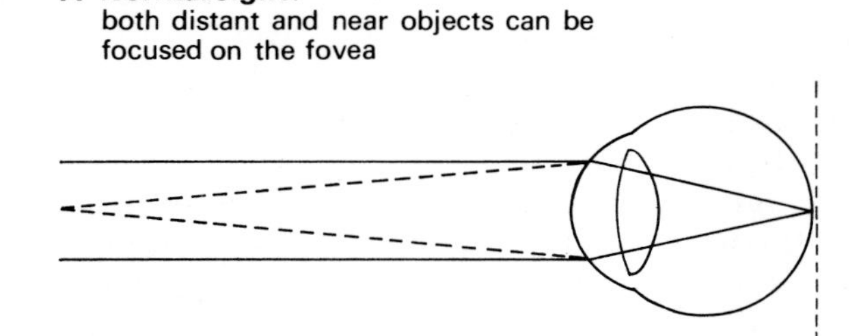

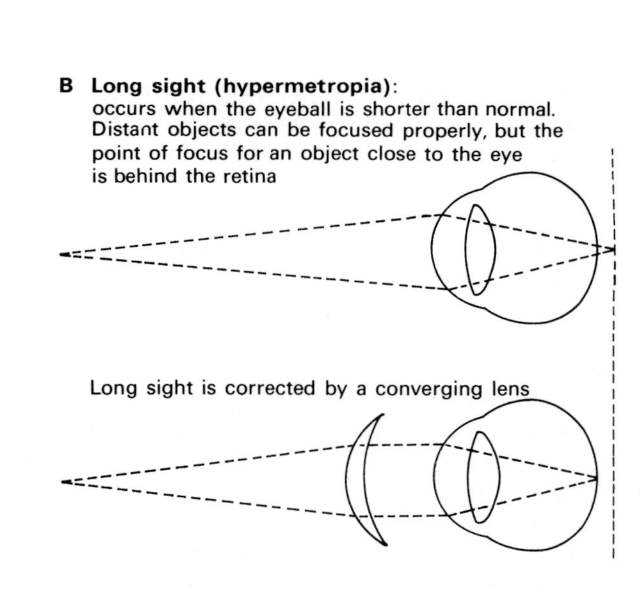

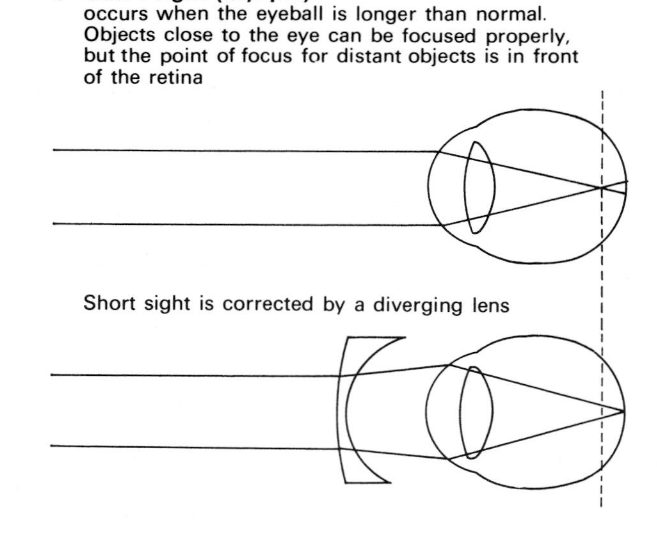

Fig. 14.7 Some eye defects and their correction

million are known as **cones**, and the rest are **rods** (because of their shape).

Surprisingly, rods and cones do not face the light. Like a film which has been put into a camera back to front, the light receptors of the eye are mostly buried under the nerve fibres which conduct impulses from the retina to the brain, and under a layer of capillaries. This results from the way in which the eye develops in the embryo. These obscuring layers are absent from an area directly opposite the lens. This area, the **fovea**, is where the clearest image is formed.

Rods and cones are not evenly distributed over the retina. The fovea consists exclusively of cones, packed tightly together with 125 000/mm^2. In fact, the fovea contains almost all of the 6 million cones in the eye. Elsewhere, the retina consists mostly of rods, packed together with only 6000/mm^2. However, groups of up to 150 rods are connected to the brain by only one nerve fibre, whereas each cone either has its own exclusive nerve fibre connection to the brain or shares it with very few others.

This means that the fovea gives a much clearer visual impression in the brain than the rest of the retina, since an image that falls on the fovea is minutely analysed by the tightly packed cones, which individually or in small groups send separate impulses to the brain. In comparison, an image which falls on rods is less closely analysed, since rods are not so tightly packed as cones, and large patches of them send just one set of impulses to the brain. Accordingly, when a person wishes to examine something carefully he moves his eyes, or the object, until its image falls on the fovea.

Rods at the outermost edge of the retina do not form images at all. They serve only to trigger reflexes which turn the eyes towards objects just beyond the limits of normal vision. This is what happens when something is seen 'out of the corner of the eye'.

There are several other important differences between rods and cones. Rods continue to work in dim light, but they are not sensitive to colour. Cones, on the other hand, work only in bright light, and they are sensitive to colour. These facts explain the differences between day and night vision. Daylight vision is in colour and has precise detail because it results mostly from images which fall on the cones of the fovea. However, the cones stop working as daylight fades, and vision relies more and more on the rods alone. Towards evening, a person gradually loses the ability to see colours, and his vision becomes less distinct.

The **blind spot** in the retina consists of blood vessels and nerve fibres leading to the optic nerve. This part of the eye, as its name implies, is entirely insensitive to light (exercise E).

Stereoscopic vision and distance judgement Stereoscopic or three-dimensional vision is best developed in animals whose eyes face forwards (e.g. humans, apes, cats, owls, and to a certain extent fish such as pike). This type of vision depends on both eyes looking at the same object.

In humans, the eyes are about 6.3 cm apart, and when the eyes are focused on an object they each receive a slightly different view. (This can be demonstrated by closing one eye at a time and comparing the view from each.) The brain puts these two views together and makes from them one three-dimensional impression.

Stereoscopic vision makes it possible to judge distances, but only up to about 50 metres; beyond this point objects produce an almost identical image in both eyes. Distances are also judged by using information from proprioceptors in the eyes. These detect tension changes in the ciliary muscles during accommodation, and in the extrinsic muscles as they swivel the eyes inwards or outwards to look at an object.

Compound eyes of insects

Insects, and arthropods generally, have eyes made up of hundreds or even thousands of separate visual units called **ommatidia**. House-flies have 4000 of these units, and dragonflies have 28 000. For this reason, the visual organ as a whole is called a compound eye (Fig. 14.8).

Each ommatidium has two lenses. The outermost lens is formed by a six-sided, lens-shaped thickening of the transparent cuticle which covers the eye. Below this is another cone-shaped lens. Together, these lenses funnel light rays downwards, concentrating them into a narrow beam which shines on to another transparent structure called the **rhabdom**. The rhabdom is formed by the central region of seven or more light-sensitive cells which, when stimulated, send impulses into the insect's central nervous system.

It was once thought that each ommatidium was a separate image-forming eye, but this is not the case. The ommatidium simply generates impulses at a frequency which depends on the intensity of light coming from an area immediately in front of it (from an area within an angle of 20°). The 'brain' of the insect assembles the signals from each ommatidium to make a visual impression that must look something like a picture made up of mosaic tiles. In a sense, then, each ommatidium of a compound eye is equivalent to a rod or a cone of a mammalian eye. Both register spots of light of varying intensity that are then combined to make a visual impression in the brain. Eyes with many ommatidia give a wider and more complete visual impression than those with only a few. Even so, the best

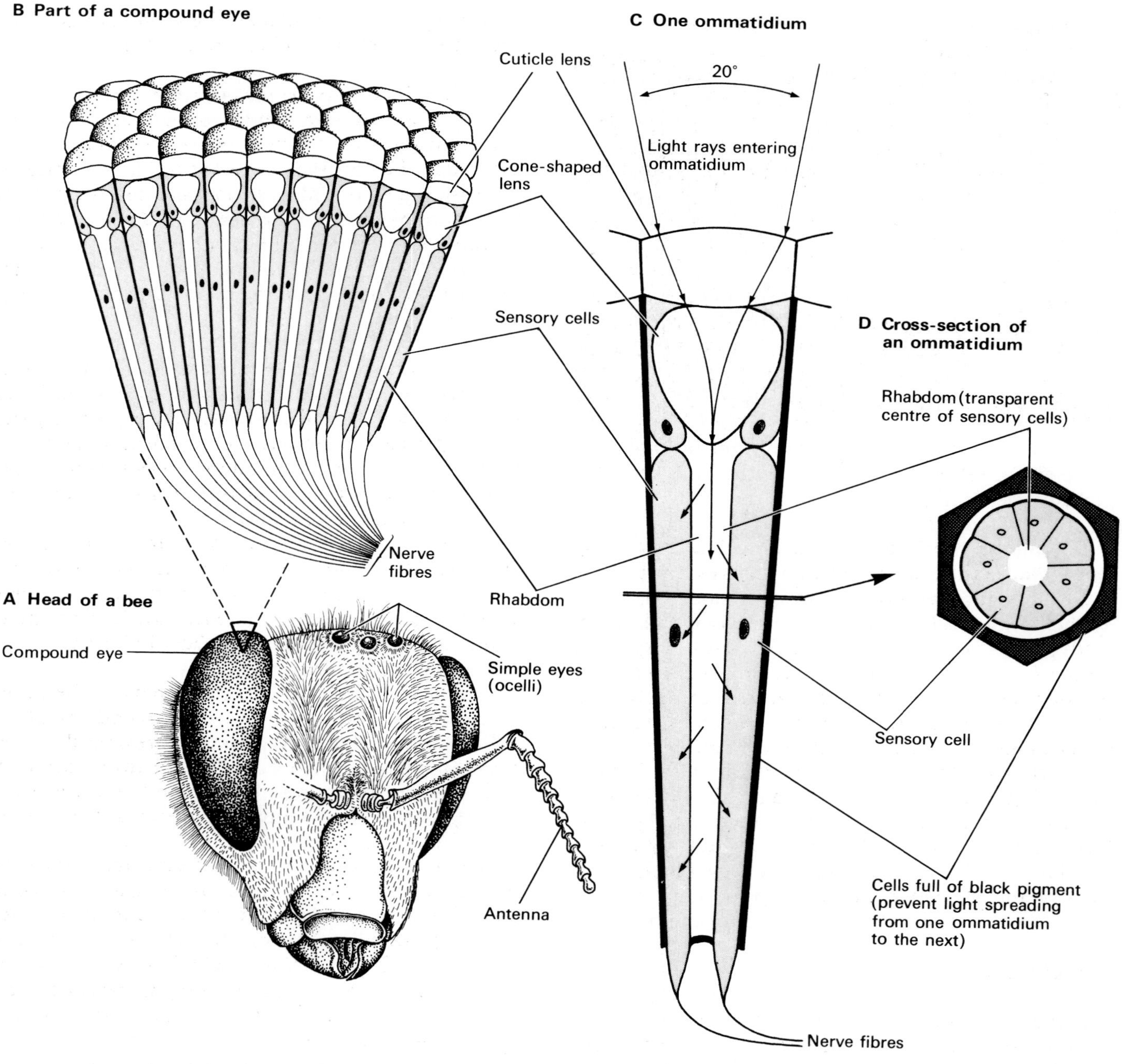

Fig. 14.8 Compound and simple eyes of a bee

compound eye is far less efficient at forming images than a mammalian eye.

Moving objects are detected by one ommatidium after another, thus producing a sequence of sensations in the insect's brain. Consequently, even though objects are not seen in detail, any movement is immediately detected. This is important in predatory insects such as dragonflies and mantids, which have to distinguish between their prey and surrounding non-living objects. It also enables honey-bees to detect swaying, wind-blown flower heads from surrounding stationary objects.

There is evidence that insects can distinguish between certain colours. Bees for instance, appear to be sensitive to blue, yellow, violet, and even to ultra-violet which is invisible to humans. This enables bees

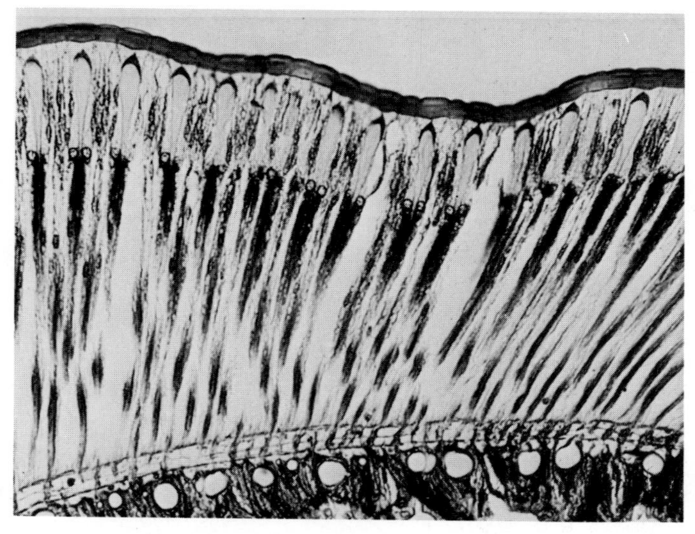

Vertical section through the compound eye of an insect. Identify as many parts as possible using Figure 14.8 as a guide

to detect various types of flower by colour, including many that, although dull to human eyes, must glow with ultra-violet light to the eyes of a bee.

Bees also use ultra-violet light in navigation. Normally, they use the direction of the sun's rays, like humans use a compass needle, to calculate their position and direction of flight relative to their hive. On cloudy days, however, bees locate the sun by its ultra-violet radiations which penetrate the thickest clouds.

Apart from compound eyes, many insects have small, single eyes called **ocelli** (Fig. 14.8A) which are similar in structure to a single ommatidium. The function of ocelli is not clear, but it is known that bees are slower and less certain in their movements when their ocelli are covered with black paint.

14.6 Hearing

Hearing is a sensation in the brain produced by vibrations, or sound waves, that are transformed into nerve impulses by the ear. Sound waves are a form of mechanical energy which can pass rapidly through solids, liquids, and gases. In air they travel at 332 metres per second.

When the frequency of these vibrations (i.e. the number of vibrations per second) is constant, a sound is heard as a musical 'note' with a certain **pitch**. The pitch of a note increases as the frequency increases. In humans, the lowest audible pitch has a frequency of 20–30 vibrations per second; middle C on a piano has a frequency of 256 vibrations per second, top C has a frequency of 4096 vibrations per second, while a frequency of 17 000–20 000 vibrations per second is the highest pitch which humans can hear.

Sound detection in mammals (e.g. humans)

The ear of a mammal consists of two sets of organs, each with distinctly separate functions. One set, called the **semi-circular canals**, are concerned with the sense of balance, and are described in the next section. The other set, described here, transform vibrations in the air into nerve impulses which are interpreted in the brain as sound. This hearing mechanism consists of the following regions (Figs. 14.9 and 14.10).

The outer ear The outer ear consists of those parts which are visible on the outside of the head. The most conspicuous of these is the funnel-shaped **pinna** (plural pinnae), made of flexible cartilage covered with skin. In some animals, e.g. horses, dogs, and cats, the pinnae are moveable, and this helps the animal to detect the source of a sound.

Pinnae collect sound waves and funnel them down a tube, about 2 cm long in humans, the end of which is closed off by a sheet of very thin skin reinforced with fine fibres called the **ear drum**. Behind the ear drum is the air-filled cavity of the middle ear.

The middle ear Across the cavity of the middle ear, opposite the ear drum, is a second membrane called the **oval window**. The ear drum and the oval window are connected by a chain of three tiny bones called the **ear ossicles**. The ossicles transmit sound vibrations from the outer to the inner ear in the following manner.

Sound waves in the air cause the ear drum to vibrate, which causes the ear ossicles to move against each other in such a fashion that they cause the innermost ossicle, the **stapes**, to be levered in and out like a piston (Fig. 14.10). The stapes is attached to the oval window, and so when it moves in and out, it moves the oval window in and out as well, and this sends vibrations pulsing through the fluid in the inner ear.

The ossicles do more than simply transmit sound across to the inner ear, they magnify (amplify) it as well. These tiny bones form a system of levers that magnify the minute vibrations of the ear drum so that they have about twenty times their original force when they reach the oval window.

The ossicles are held in place by muscles. These help protect the ear from damage by preventing excessive vibrations of the ossicles during very loud noises.

The inner ear The inner ear is a complicated series of passageways in the bones of the skull filled

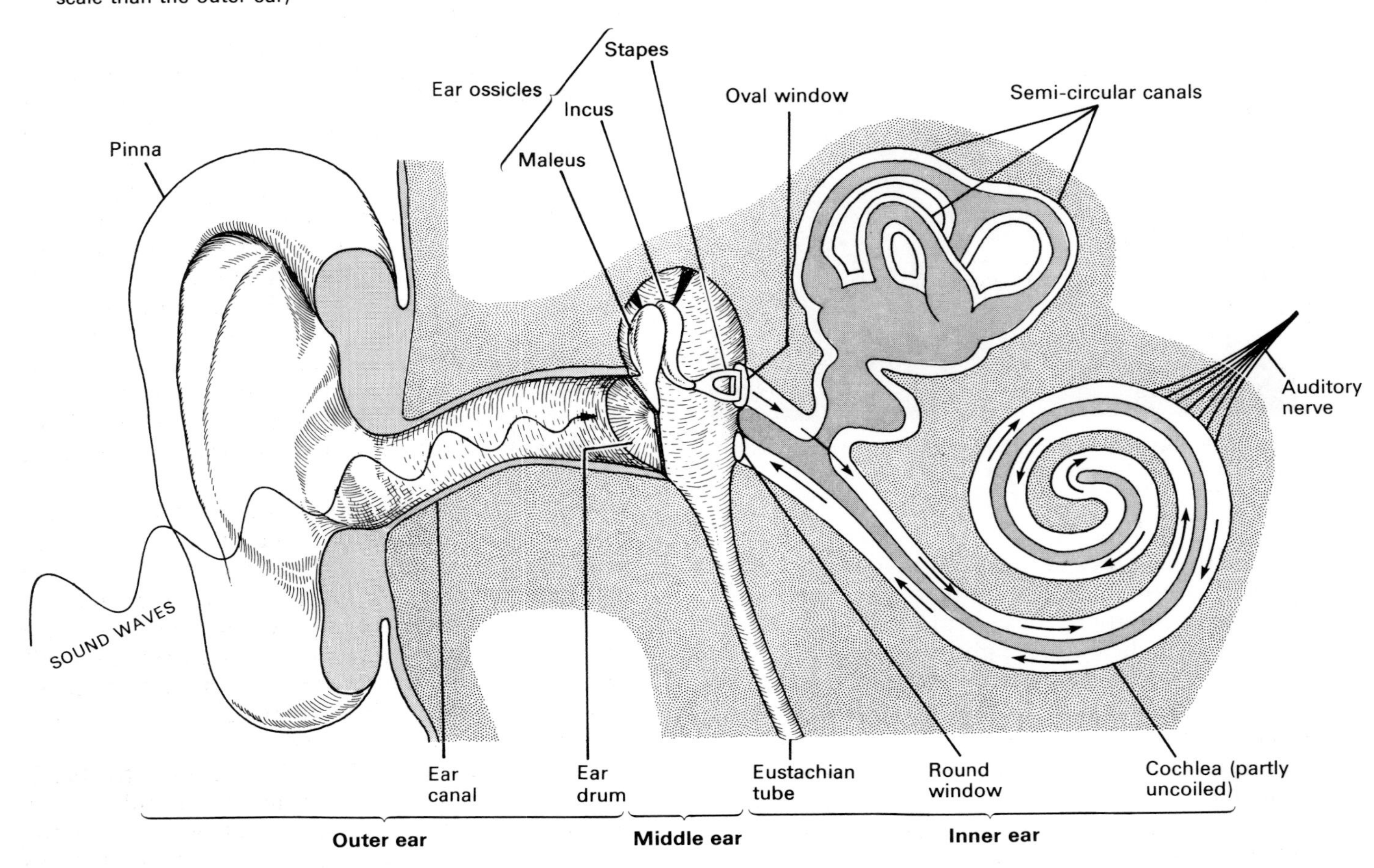

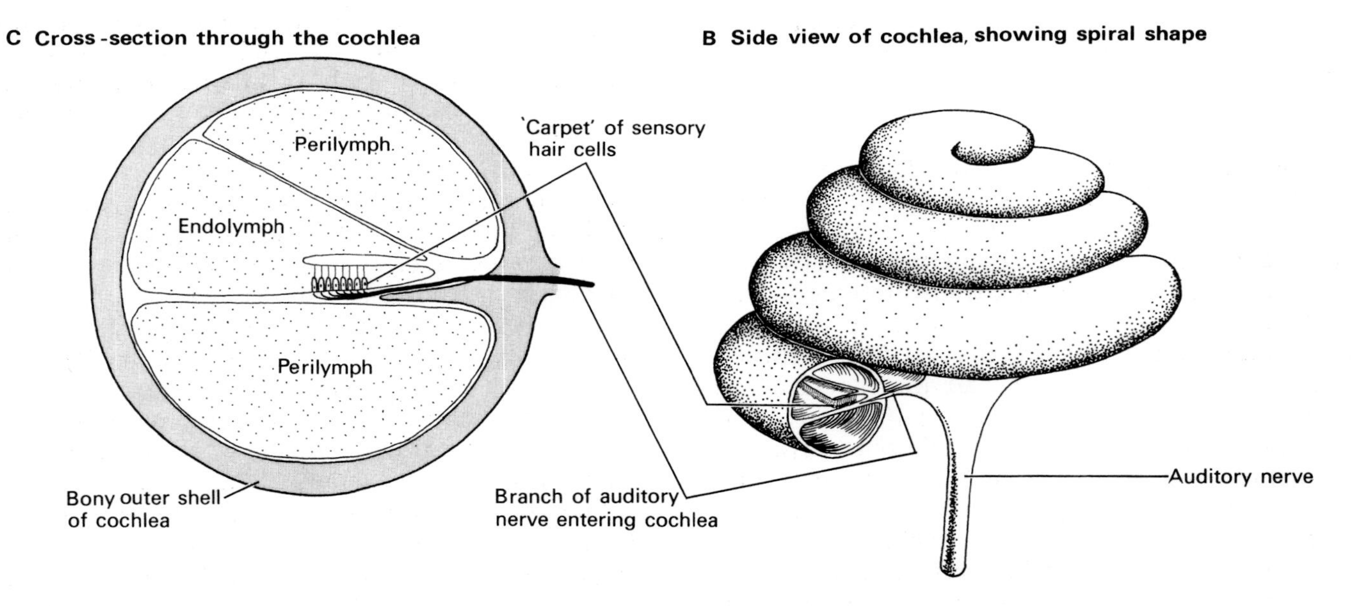

Fig. 14.9 Structure of the middle and inner ears

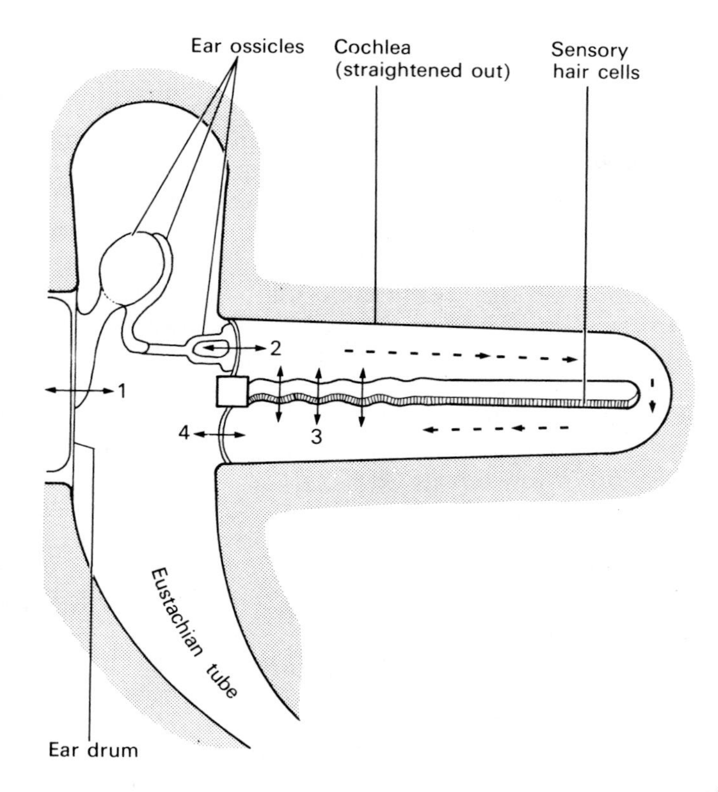

Fig. 14.10 Diagram showing the functions of the middle and inner ears

1 Sound waves vibrate the ear drum

2 This causes the ossicles to vibrate and moves the stapes in and out against the oval window

3 Pressure waves are set up in the cochlea. These vibrate the hair cells, which send impulses to the brain

4 Pressure waves travel around the cochlea and move the round window in and out, after which they pass away

with a liquid called **perilymph**. Inside these passages there are tubes filled with another liquid, the **endolymph**. The part of this system concerned with hearing is the **cochlea**, which is a tube that rises in a spiral through three and a half turns (Fig. 14.9B).

Actually, the cochlea is three tubes in one (Fig. 14.9C). The middle tube contains endolymph and the outer ones contain perilymph. The floor of the middle tube is covered with sensory nerve endings, the **hair cells**, which stand upright like the pile of a carpet. The tip of each hair cell is embedded in a membrane which hangs over it like a shelf jutting out from a wall.

Figure 14.10 illustrates how the cochlea works. The stapes ossicle moves against the oval window and causes vibrations (pressure waves) in the perilymph of the cochlea. These spread into the endolymph and finally cause the 'carpet' of sensory hair cells to vibrate up and down. This stimulates them, and they send impulses to the brain along the auditory nerve, which are interpreted as sound.

There is a theory that high pitched notes cause vibrations only in hair cells nearest the oval window, while lower pitched sounds cause vibrations further up the cochlea spiral. If this is true, the brain must distinguish notes from each other according to the region of hair cells which they stimulate, as well as by the frequency of their vibrations. Whatever the mechanism is, it must be extremely sensitive, for humans not only distinguish between very similar notes at a wide range of different volumes, but between the same note played on many different instruments, even when several of them are heard simultaneously. The response of the ear to high frequency sounds decreases with age, and older people are less able to hear high notes in a musical performance.

Ears as direction-finders Since mammals have two ears, sounds coming from one side will stimulate one ear more than the other. To be more exact, volume will be louder in the ear facing the sound, and since the sound waves will take fractionally longer to reach the other side of the head, impulses from the two ears will be slightly out of step. The brain can analyse these minute differences and from them calculate the direction from which a sound comes.

However, sounds from directly in front of the head, from directly behind it or from above, stimulate both ears equally. In humans, the origin of such sounds can be detected by turning the head from side to side until one ear is stimulated more than the other. Other mammals, such as dogs, horses, and cats, achieve the same effect by rotating or turning the pinnae of the ears, or by cocking the head on one side.

Sound detection in insects

Crickets, grasshoppers, locusts, and certain moths have quite a highly developed sense of hearing. The sound detecting mechanism is usually part of the tracheal system (Fig. 9.9) and in grasshoppers is situated in the front legs. In their front legs grasshoppers have an enlarged tracheal tube. In the walls of this tube is a thin membrane which is supplied with sensory nerve endings. Sound waves in the air make this membrane vibrate, and this stimulates the nerve endings, which send impulses to the insect's 'brain'.

This apparently simple mechanism can not only detect the direction of sounds, but can also detect frequencies several octaves above the limit of human hearing. It seems likely that grasshoppers use this apparatus to detect sounds made by members of their own species, especially the opposite sex. They can

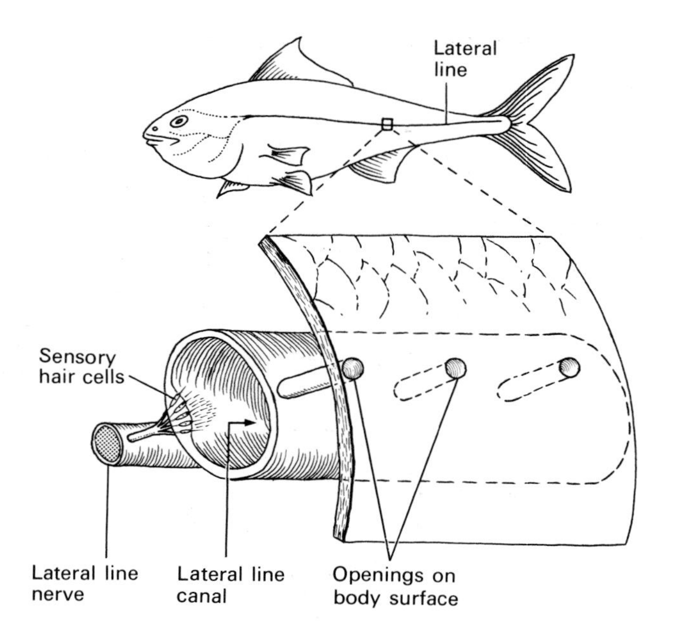

Fig. 14.11 Lateral line system of a fish

distinguish the particular sounds given out by their own species from those made by a host of different species.

Sound detection in fish

Fish detect vibrations in water by means of organs called **lateral lines** (Fig. 14.11). A lateral line is a fluid-filled tube, the **lateral line canal**, which runs along each side of the fish, with branches over the head region. The tube has small openings at intervals along its length, and it has many hair-like sensory nerve endings in its walls which are connected to nerve fibres in the lateral line nerve. These eventually lead to the brain.

Sound waves disturb the liquid in the lateral line canal, which stimulates the hair-like sensory nerve endings, which send impulses to the fish's brain. The region of the lateral line facing in the direction of a sound is stimulated more than the rest. From this information fish can quickly locate the presence and direction of other animals by the sounds which they make.

Lateral line organs are also used like the echo-location equipment found on modern trawlers and submarines, which detect sound waves after they have bounced off solid objects in the water nearby. Fish use these echoes to build up a 'sound picture' of their underwater world, so that they can avoid objects in the dark, and in muddy water.

14.7 Sense of balance

Mammals are sensitive to their position in space (i.e. they know whether their bodies are tilted or upright), and to changes in the speed and direction of movement. This sensitivity results from many different stimuli. The eyes, for instance, supply information about the position and movements of the body, as do proprioceptors in the muscles and joints. Additional information comes from the **semi-circular canals**, the **utricles**, and the **saccules** of the inner ear.

Semi-circular canals

The semi-circular canals detect changes in the direction of movement (Fig. 14.12). There are three semi-circular canals situated at 90° to each other. At one end of each canal is a swelling, the **ampulla,** containing a mound of sensory hair cells attached to nerve endings. The hairs of these cells are embedded in a cone of jelly called a **cupula**.

When a person nods his head, he disturbs fluid in the vertical canal in the same plane as this head movement. In fact, when the head and canal move one way the fluid in it is left behind and so apparently moves in the opposite direction. (The same thing happens to medicine when the bottle is shaken.) The moving fluid pushes against the cupula and bends it over. This stretches the sensory hairs, which send impulses to the brain. The same thing happens in the other canals when movements occur in their particular plane. The brain therefore calculates the direction of a turning movement from the amount of disturbance which it causes in each semi-circular canal.

Utricle and saccule

The utricle and saccule are fluid-filled spaces in the inner ear which detect acceleration, deceleration, and the position of the body relative to the pull of gravity (Fig. 14.12). Both these structures contain patches of sensory cells with protruding hairs embedded in a jelly-like substance which is full of tiny pieces of chalk. These pieces of chalk are called **otoliths**, and they make the blobs of jelly in which they are embedded much heavier than the equivalent mass of jelly which makes up the cupulas of the semi-circular canals.

When the head is upright the weight of the otoliths presses downwards on the sensory hairs. But when the head is tilted sideways the otoliths pull against the sensory hairs, which send impulses to the brain. When the body is accelerated or decelerated in any direction the otoliths are 'left behind' (like the endolymph in the semi-circular canals during turning movements) and so pull against the sensory hairs.

When the body is upright and perfectly still, few impulses reach the brain from cupula or otolith hair cells. But when the body is tilted or swayed, as, for instance, on the deck of a ship in rough seas, the brain receives a continuous barrage of simultaneous messages from every hair cell. This over-stimulation may upset other areas of the brain causing vertigo and vomiting (seasickness). Lying down on the back with the eyes closed minimizes sensory impressions and reduces sickness.

Semi-circular canals, utricles, and saccules are not confined to mammals. They are found in all the vertebrates. Otolith organs similar to those described above are even more widespread. Indeed, they are found in jellyfish, insects, and even certain plants.

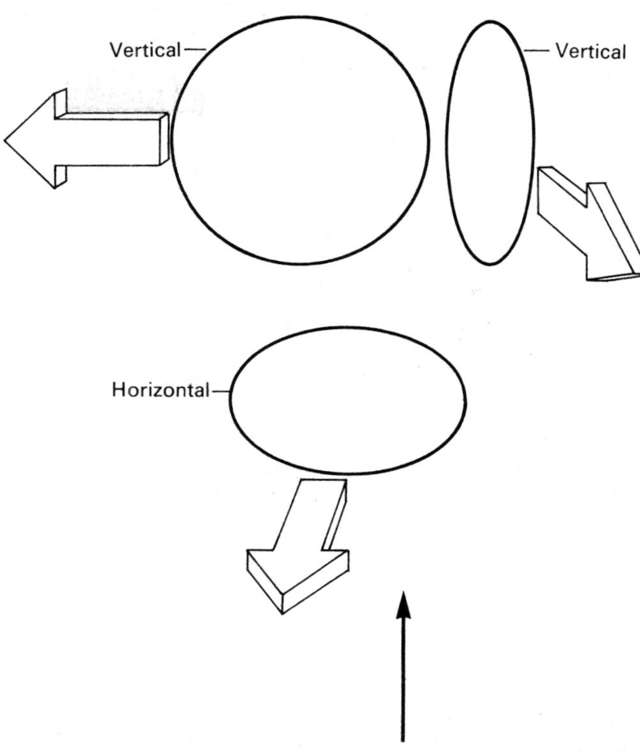

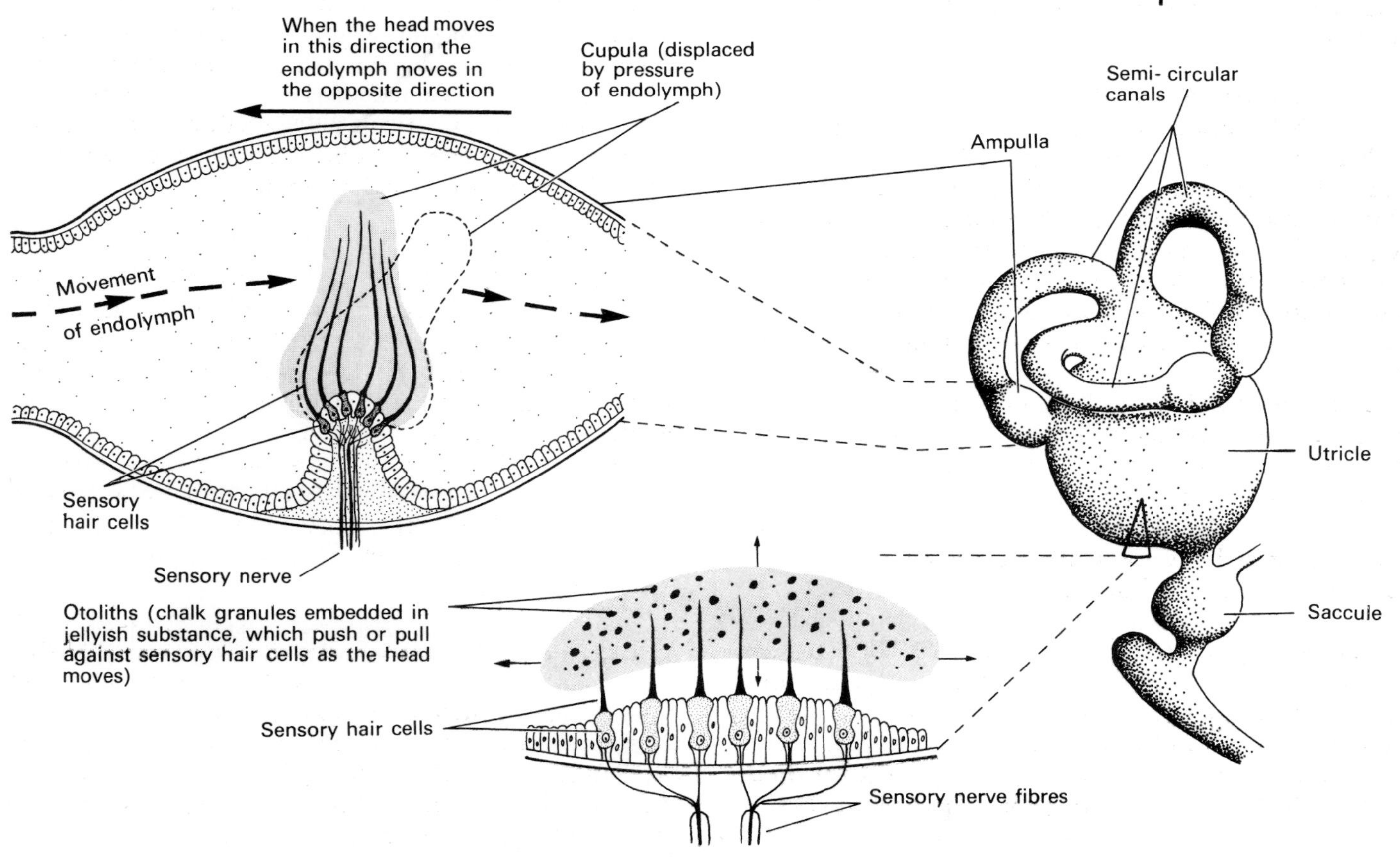

Fig. 14.12 Structure and functions of the semi-circular canals, utricle, and saccule

This chapter has tried to convey some idea of the enormous amount of sensory information which reaches the brain. Literally millions of receptors pour out a continuous torrent of impulses, day and night. The brain is a highly complex and amazingly fast computer which deals with this information consciously or unconsciously by controlling thousands of physical and chemical processes in the body.

But the human brain is far from overloaded by all this work; it can do much more. When someone drives a car, for instance, his brain not only deals with the working of all its controls, it does this while processing a mass of information received at high speed about traffic lights, police signs, road signs, stray cats, children playing in the road, weather conditions, etc. The driver can also talk to his passengers, listen to his car radio, and make complicated business decisions all at the same time. The brain and its sensory equipment is yet another example of the incredible complexity and efficiency of the human body.

Verification and inquiry exercises

A *An investigation of skin receptors*

1. *To discover the distribution of touch receptors in the skin.*

a) Pass two needles through a piece of bottle cork so that their points protrude the same distance and are 1.5 cm apart. Prepare another cork in the same way but with the needle points only 0.5 cm apart.

b) Ask someone to sit with his eyes closed and one arm bare to the elbow.

c) Beginning with the needle points 1.5 cm apart, touch the skin of the person's forearm gently with either one or both points. Ask the subject if he feels one or two points. He must not guess. Repeat, using one or two points in random order until two points have been used five times on the forearm. Record the answers received when two points were used.

d) Repeat this procedure on the palm of one hand and then on the finger-tips. Repeat the whole experiment using needle points 0.5 cm apart.

e) Compare the number of correct answers obtained with each set of needles on the forearms, palms, and finger-tips. What do these results indicate about the density of touch receptors in these three regions?

f) How is it possible for a person to be touched with two needle points but feel only one?

2. *An investigation of heat and cold receptors*

a) Obtain several knitting needles (preferably size 12 or 13; of the type used to knit socks).

b) Place half the needles in hot (not boiling) water, and the rest in ice cold water for at least two minutes.

c) Take a needle from the hot water, dry it, then run its point gently but slowly over the skin on the back of the hand. Place an ink mark wherever the heat sensation is felt most strongly. Continue, changing the needles as they cool, until several 'hot spots' have been found.

d) Use dried cold needles to find out if the 'hot spots' are sensitive to cold, or if there are separate 'cold spots'.

e) Devise an experiment to find out the smallest difference in temperature to which the fingers are sensitive.

3. *An investigation of sensory adaptation*

a) Obtain three 500 cm^3 beakers. Fill one with hot (not boiling) water, a second with ice-cold water, and a third with lukewarm water.

b) Place the fingers of one hand in the hot water, and fingers of the other hand in the cold water. After about 1 minute place the fingers of both hands into the lukewarm water.

c) Does the lukewarm water feel different to each hand? If so, does this difference persist for long? How does this experiment illustrate sensory adaptation?

B *Dissection of an eye*

1. Obtain bulls' or sheeps' eyes from a slaughter-house.

2. Cut away the layers of fat from the back of the eye to expose the extrinsic muscles and optic nerve.

3. Using a new No. 11 scalpel blade in a suitable holder, make two cuts through the cornea at right angles to each other (i.e. like a + sign). Note the flow of aqueous humour from the eye. Fold back the flaps of cornea with forceps and examine the iris.

4. Using scissors, extend the cuts through the iris and into the sclerotic to about half-way around the walls of the eye. Fold back the walls leaving the lens and its black suspensory ligaments floating on the vitreous humour. Examine the ciliary muscles.

5. Remove the vitreous humour and lens and look for the blind spot. Sometimes the fovea can be identified as a tiny yellow spot.

C *Changes of pupil size*

1. Hold a mirror close to the face, close the eyes and cover them with the hands for about 15 seconds. Quickly remove the hands and open the eyes, and observe how the pupils alter in size.

2. Observe changes in pupil size when a person shifts his gaze from a distant object to an object about 1 metre away.

3. What changes in pupil size take place in each situation, and what is the significance of these changes?

D *Observing accommodation movements in the human eye*

1. In a darkened room it is possible to see three reflections of a point of light (such as a candle flame) in a person's eye. These reflections come from the front surface of the cornea, and the front and rear surfaces of the lens.

2. Observe these reflections carefully while the subject shifts his gaze from a near to a distant object and back again.

3. Which of the three reflections change in size? What causes these changes?

E *Detecting the blind spot*

1. Hold this book with Figure 14.13 at arm's length. Close the left eye and stare at the cross with the right

Fig. 14.13 Blind spot experiment (exercise E)

eye. (Note that the black circle is still visible.) Bring the book slowly towards the face. At a certain point the circle will disappear. This happens when its image falls on the blind spot.

2. Why don't the blind areas in each eye interfere with normal vision?

Comprehension test

1. Explain how each of the following facts illustrates a difference between the fovea and the remainder of the retina.

a) When the eyes are fixed on a single letter in the middle of a word in the middle of a printed page, no more than two or three letters on either side of that particular letter are clearly visible.

b) If a very dim star is looked at directly it disappears from view, but it reappears when looked at indirectly 'out of the corner of the eye'.

c) When a person buys a coloured garment in a dimly lit shop he often finds that it appears to be a different colour when seen in daylight.

d) It is difficult to read small print in dim light, and those who persist in doing this often develop defective sight.

2. The Eustachian tube often becomes blocked during a cold in the head. When this happens air in this tube is drawn into blood vessels lining its walls with the result that the ear drum is drawn inwards by reduced air pressure.

a) Why does this cause deafness?

b) Why does the deafness disappear when the cold is better?

c) Why do airline companies sometimes supply their passengers with sweets to eat as the aircraft takes off?

3. Why would it be dangerous for a person with defective semi-circular canals to ride a bicycle?

Summary and factual recall test

Stretch receptors are present in (1–name three structures). They provide information about the degree of (2) in each part of the (3) system, and the (4) to which (5) are bent. From this information a person knows the (6–list three things).

Touch receptors enable a person to distinguish between (7) and (8) substances and (9) and (10) textures. Temperature receptors cannot be used to tell the exact (11) of something; they can only (12). Pain receptors are useful because (13). Taste receptors are called (14). There are separate ones for (15–list the four different tastes). Smell receptors are called (16). Unpleasant tastes and smells are useful because (17).

Insect antennae are sensitive to (18–name six stimuli). There are other receptors on an insect's (19) and (20).

Eyes are protected in the following ways (21–list four). They are held in place by six (22) muscles, which also control eye (23).

The iris has a (24) in its centre called the (25). This controls the amount of (26) entering the eyes, and the (27) of focus of vision in the following way (28).

To focus the eye on a near object the (29) ciliary muscles (30), which (31) the tension on the (32) ligaments, which allows the lens to become (33) in shape. This (34) its power to (35) light. To focus on a distant object the (36) ciliary muscles contract, which makes the lens (37) in shape.

(38), or old sight, is caused by reduced (39) of the lens, so that (40) objects cannot be seen clearly. (41), or long sight, occurs when (42), and it is corrected by (43) lenses. (44), or short sight, occurs when (45), and is corrected by (46) lenses.

An image is formed on the (47) of the eye. This layer consists of light-sensitive cells called (48–two names). These differ from each other in the following ways (49–describe their sensitivity and in what light they work best).

Humans can hear sounds at frequencies between approximately (50) and (51) vibrations per second. When sounds reach the ear they cause the (52) to vibrate. Bones called the (53) transmit these vibrations to the (54) window in such a way that they are amplified about (55) times. The vibrations next pass into a liquid called (56) within a spiral tube called the (57), then they pass into another liquid called (58) and finally cause a carpet of sensory (59) to (60), which results in impulses travelling to the brain along the (61) nerve.

The (62) canals have swellings called (63). Each of these contains a cone of jelly called a (64) attached to (65). This arrangement detects (66) in the following way (67). The utricles and saccules contain blobs of jelly with pieces of (68) embedded in them called (69). This arrangement detects (70–name three stimuli) in the following way (71).

15

The reproductive process and reproduction in protists

Organisms have a limited life span. They all die eventually, either from accidents, diseases, or old age. And yet life continues, because organisms have the capacity to create new organisms by reproducing. Indeed, some organisms from each species must reproduce or that species will die out and disappear from the earth. Reproduction gives rise to new organisms with the same basic characteristics as their parents: sunflower seeds always grow into sunflowers and not cabbages; and cats always produce kittens and not puppies. This and the following two chapters describe many examples of reproduction among protists, animals, and plants. First, however, it is necessary to deal with the general features of the reproductive process.

15.1 The reproductive process

How is a new organism created? To answer this question it is necessary to refer to what was said in chapter 2 about the structure of a cell nucleus, and its division by mitosis.

Every cell in every organism contains a set of 'instructions', in chemical form, for building the whole organism. These instructions exist as a pattern of molecules that make up structures called chromosomes, which are situated in the nucleus of each cell. Since every cell contains these instructions it follows that a new organism can, theoretically, be created from any cell of a living organism. If this were true in practice an organism could produce another one like itself simply by releasing any cell from its body under conditions which would allow that cell to divide by mitosis.

In fact, this is more or less what does happen, except that new organisms do not arise from just any cell from the parent. They usually arise from specialized **reproductive cells**, and these are often produced by specialized **reproductive organs**.

Thus, the answer to the question: 'How are new organisms created?' is that they arise from cells released from the bodies of parent organisms, and develop according to building instructions contained in the chromosomes of these cells. These building instructions are called the **hereditary information** of the species, and, by means of the reproductive process, they are passed from one generation to the next in the chromosomes of the reproductive cells. If a species becomes extinct it can never arise again because its hereditary information is lost forever.

There are two ways in which organisms give rise to new organisms: by **asexual reproduction**, and by **sexual reproduction**.

Asexual reproduction

Asexual reproduction involves only one parent organism. In theory, a new organism produced asexually should be identical to its parent, because the new organism has inherited only *one* set of hereditary building instructions. In practice, however, offspring and parent are not always exactly the same. There are sometimes minor differences, which arise because of a process called mutation (described in chapter 19), and because the offspring may be exposed to environmental conditions which differ from those experienced by its parent. (These are also the reasons why 'identical' twins are never exactly identical.)

In asexual reproduction the parent organism produces by mitosis a number of cells with chromosomes that are precise copies of those in its own body cells (section 2.6). When these cells develop they form new organisms with the same hereditary information as their parent. When the young grow up and reproduce asexually the same hereditary information is copied by mitosis again, and passes unchanged to form the next generation, and so on as long as the organisms continue to reproduce asexually. Consequently, asexual reproduction does not allow a species to change from one generation to the next, except in very minor respects. It simply results in new organisms which are almost identical to their parents. The asexually produced descendents of a single parent are referred to collectively as a **clone**.

Sexual reproduction

Unlike asexual reproduction, sexual reproduction almost always involves two parent organisms. The parents give rise to reproductive cells called **gametes** by a type of cell division called meiosis (section 19.2). A gamete contains a copy of part of the hereditary information of the organism which produced it.

During sexual reproduction a gamete from one parent fuses with a gamete from another parent. This process is called **fertilization** and results in a single cell called a **zygote** which contains two different sets of hereditary information: one set from each parent. When the zygote develops into a new organism it does not use all this information; it selects some from each parent according to 'rules' which are studied in the science of genetics. Consequently, the new organism is not an exact copy of either parent. It shows some features of both.

The fact that sexually produced organisms differ in several ways from their parents is one of the most important factors in the process of evolution because these differences make it possible for a species as a whole to change, in time, and give rise to an entirely new type of organism. Figure 15.1 summarizes these points, and gives the main differences between asexual and sexual reproduction. Several aspects of sexual reproduction should be understood before actual examples are studied. The following are the most important.

Gametes and sex organs In most species there is a clear difference between the male and female organisms, and between the gametes which they produce. Where this is so the female produces gametes which are larger than the male's. The female gamete is larger because it contains stored food in its cytoplasm. This food is used to nourish the embryo which develops from the zygote. Examples of female gametes are the large shelled eggs of birds, the smaller eggs of other animals, and the ovules of plants.

Male gametes move, or are moved, to the larger female gamete. Sperms of mosses, ferns, and animals are examples of male gametes which move. Sperms are equipped with tail-like structures with which they swim to the female gamete. Pollen grains of flowering plants are examples of male gametes that are moved, by wind and insects for instance, to the female.

A species with distinctly different male and female gametes usually has specialized reproductive organs, often called **sex organs**. Examples of sex organs are the testes (singular testis) which produce sperms in male animals, and the ovaries which produce eggs in female animals.

Fig. 15.1 Diagram of the differences between asexual and sexual reproduction

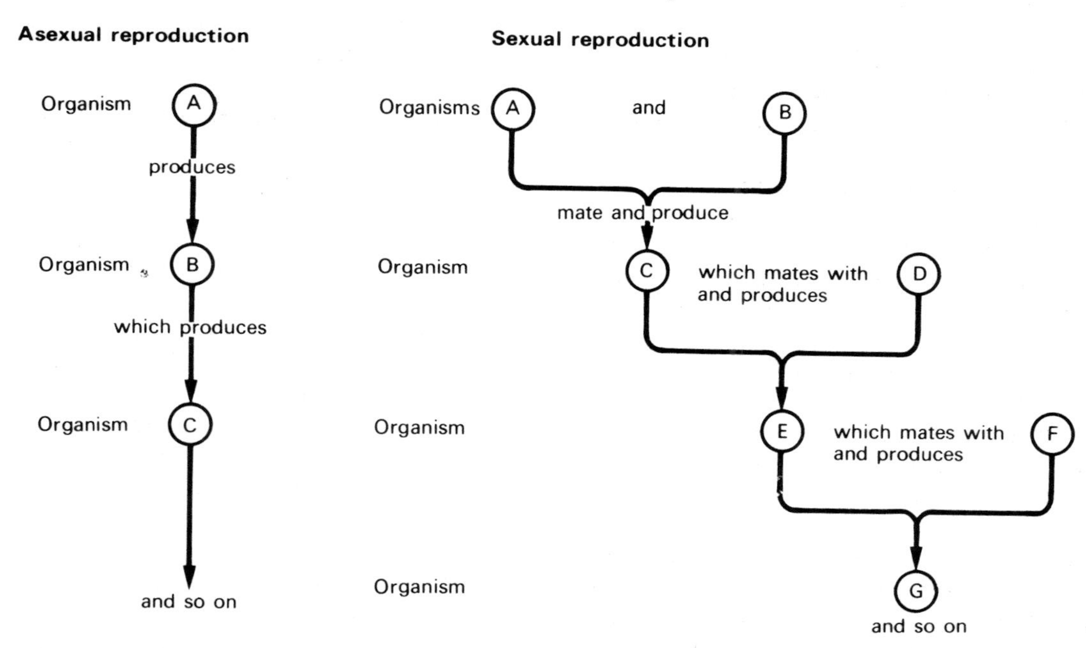

In unicellular and other relatively simple organisms there is often no visible difference between either the gametes or the parents which produce them (although there are differences in the hereditary information contained within the gamete chromosomes). It is not possible in this case to use the terms 'male' and 'female' when describing two individuals which are capable of reproducing sexually. Instead, the two individuals are said to belong to 'plus' and 'minus' strains of that particular species.

Sexuality Sometimes male and female sex organs are found in different parent organisms and sometimes in the same organism. In humans, for example, the sexes are separate: there is a unisexual male and a unisexual female organism. In organisms such as *Hydra*, earthworms, and many types of flowering plants, both male and female reproductive organs occur in the same organism. These are called **bisexual,** or **hermaphrodite** organisms. In most hermaphrodite species there are mechanisms which prevent the fusion of male and female gametes from the same individual (i.e. self-fertilization). This is a safeguard against the production of organisms with the same hereditary characteristics as their parents. If self-fertilization were to occur regularly it would reduce the chances of evolutionary change in the species.

Types of fertilization In some organisms fertilization occurs after gametes have been shed from the parents' bodies. This is called **external fertilization,** and is typical of animals such as fish and amphibia, which breed in water. Animals whose gametes fuse externally generally have patterns of mating behaviour which ensure that sperms are shed either directly on to the eggs, or nearby in the water. Furthermore, sperms do not swim at random through the water until they reach an egg by accident; they swim towards the nearest one attracted by chemicals which it produces.

Internal fertilization takes place inside the female parent's body, usually in her reproductive organs. This is typical of land animals such as reptiles, birds, and mammals, but is also found in certain types of fish and amphibia. Internal fertilization entails a mating behaviour called **copulation,** during which the male passes sperms from his sex organs directly into the female's body. Once inside her body the sperms behave as in external fertilization: they swim to the eggs attracted by chemicals.

Internal fertilization has certain advantages over external fertilization. First, it increases the chance that fertilization will occur, since sperms are shed into a confined space–the female reproductive system–which they share with the eggs. Second, the increased chance of fertilization is probably related to the evolution of females which produce a small number of eggs. A frog, for example, produces an enormous number of eggs, whereas a human female produces only one egg a month, which has the advantage of saving body materials. Third, internal fertilization means that the zygote can be retained in the female's reproductive system where it develops in a protected environment.

Reproduction is part of a sequence of events called a **life cycle** in which new organisms are created, develop to maturity, and repeat the process. In addition to details of asexual and sexual reproduction, this chapter and the following two include information about the life cycles of various organisms. The remainder of this chapter is concerned with the reproduction and life cycles of certain protists.

15.2 Amoeba

Amoeba and many other unicellular organisms reproduce asexually by the simplest of all methods; they divide into two parts. The technical name for this process is **binary fission**. First, the nucleus divides by mitosis. Then the cytoplasm divides into two parts, one part around each nucleus, and thus the organism separates into two daughter cells (Fig. 15.2). When this happens the original parent ceases to exist; and becomes instead two new identical individuals. There is no evidence of sexual reproduction in *Amoeba*.

Fig. 15.2 Binary fission in *Amoeba*

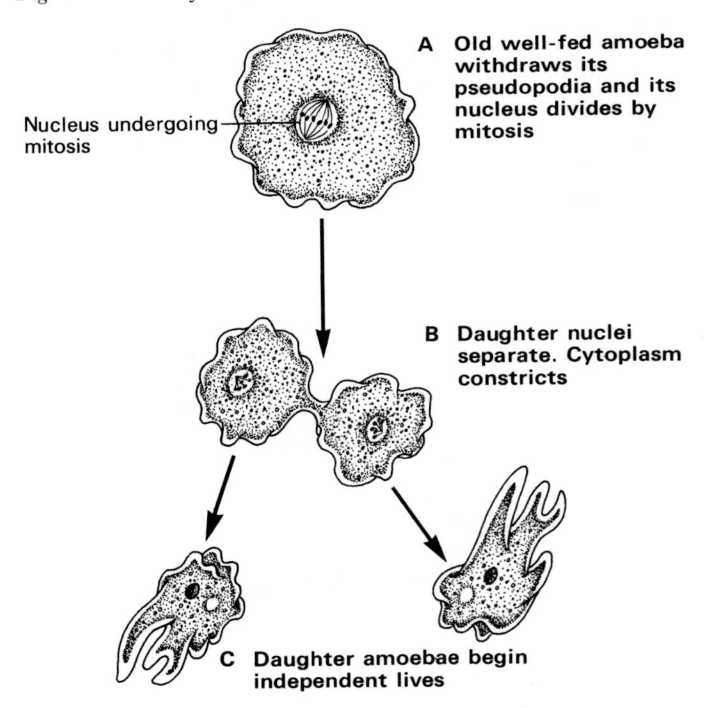

Under unfavourable conditions, such as a shortage of water or food, *Amoeba* becomes spherical in shape and develops a hard outer wall called a **cyst**. In this **encysted** condition it can withstand drought and wide variations in temperature, and may even be blown about in the air as particles of dust. During this time *Amoeba* undergoes another type of asexual reproduction called **multiple fission**. The nucleus of the encysted *Amoeba* divides many times and then each nucleus becomes surrounded by an equal share of the cytoplasm. If such a cyst reaches favourable conditions, its wall disintegrates, releasing many small *Amoebae*. Cyst formation is a means by which *Amoeba*, and several other simple organisms, can both survive adverse conditions and become distributed over a wide area.

15.3 Yeast and Mucor

Yeast

Yeast is a unicellular fungus. Its nutrition and importance to man have been described in chapter 8. Yeast cells reproduce asexually by **budding**. The process is described as budding because new organisms develop as bud-like outgrowths from the parent (Fig. 15.3). A bud forms on the wall of a yeast cell at the same time as the cell nucleus in the parent is dividing into two by mitosis. One nucleus passes into the bud, the other remains in the parent cell. At this point the bud either separates from the parent cell or remains attached and produces another bud, eventually forming a chain of cells. Under favourable conditions budding occurs with great rapidity, and each parent cell can produce thousands of daughter cells in the space of a few days. In budding, the parent cell does not immediately lose its individual identity, as it would if it underwent binary fission.

Yeast can also reproduce sexually, but the details are beyond the scope of this book. Very briefly, two yeast cells behave as gametes, fusing together to form a zygote which divides into eight tiny yeast cells. These grow to adult size, reproduce asexually, and continue the life cycle.

Fig. 15.3 Budding of yeast cells

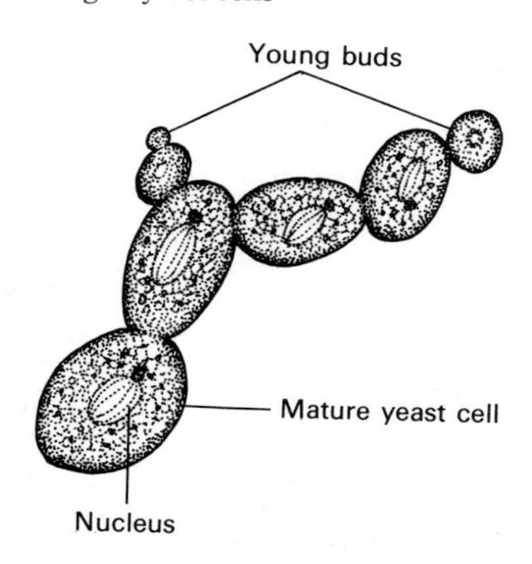

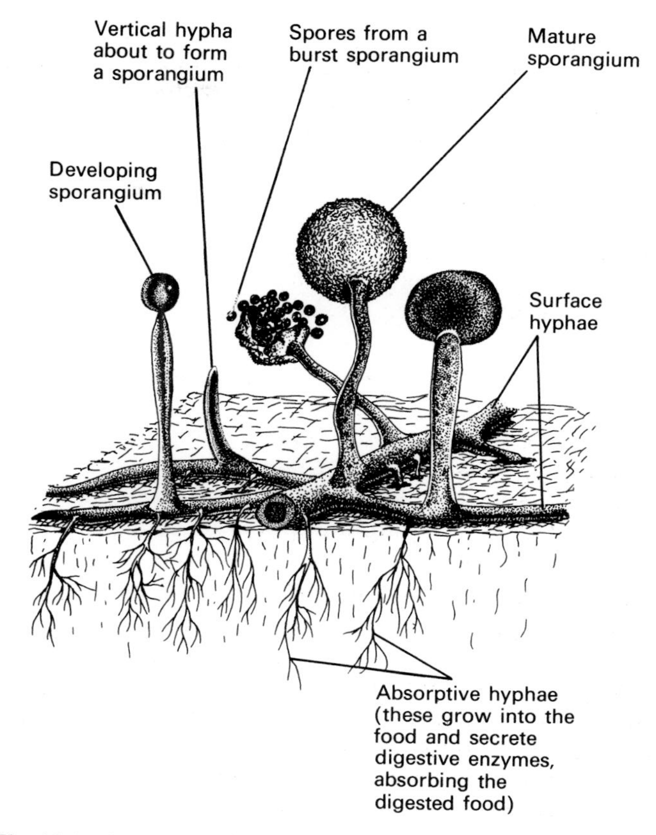

Fig. 15.4 Structure of a *Mucor* colony, showing asexual reproduction by spore formation

Mucor

The mould *Mucor* is commonly known as 'pin mould' because of the shape of its reproductive structures (Fig. 15.4). *Mucor* is very widespread and grows saprotrophically on the decaying remains of dead organisms. Like most fungi, *Mucor* consists of fine threads called **hyphae**, and all the hyphae in a particular mould colony are referred to collectively as the **mycelium**.

Most moulds, *Mucor* included, usually reproduce asexually. They do this by forming enormous numbers of microscopic structures called **spores**. Spores are produced by special hyphae which grow upwards from the substance on which the mould is growing. Cytoplasm and nuclei flow into the tip of each vertical hypha causing it to swell, and eventually form a large spherical spore case, or **sporangium**. Soon, the cytoplasm in the sporangium undergoes a form of multiple fission in which it separates into hundreds of tiny elliptical spores, each containing several nuclei. In some types of *Mucor* spore formation causes an ever-increasing internal pressure in the sporangium so that its wall eventually bursts open, exposing ripe, powdery spores that are quickly carried away by air currents. In other types of *Mucor* the wall of the ripe sporangium dissolves, exposing spores covered by a sticky jelly. It is thought that spores of this type are carried away on the feet and bodies of insects, such as house-flies which feed on the decaying substances on which the mould grows. When a spore reaches a substance on which it can grow, it opens and produces new mycelium.

Spore formation is both an extremely rapid method of reproduction and a very efficient means by which a mould can spread throughout its environment. Countless trillions of mould spores are produced daily, and these are carried by air currents and other means over the entire surface of the earth, and high into the atmosphere. Some moulds produce spores with thick protective walls which can survive extremes of heat, cold, and drought for months or even years. Provided there is moisture and warmth, there is nowhere on earth that organic matter, such as bread, can be left exposed to the open air without it developing a growth of *Mucor* or other types of mould within a few days.

Fig. 15.5 Sexual reproduction in *Mucor*

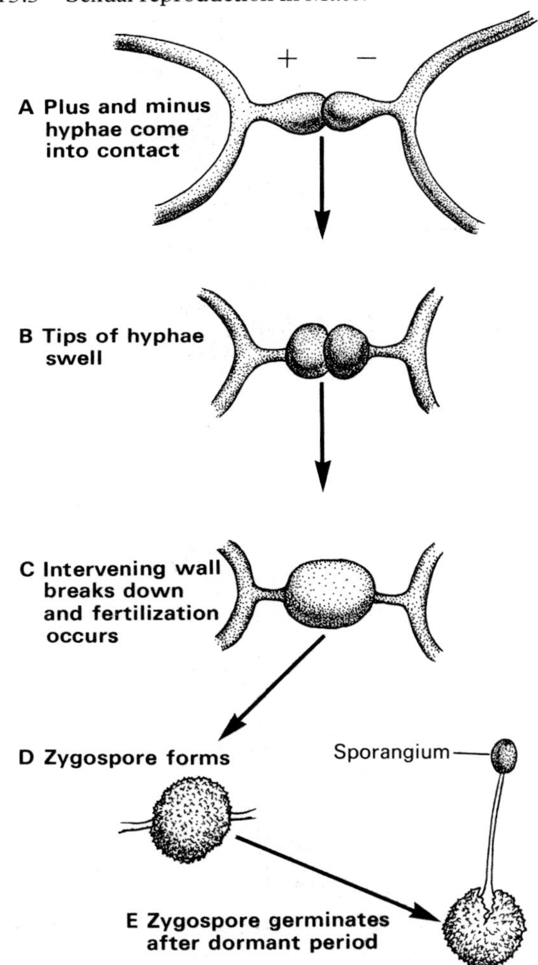

Mucor also reproduces sexually, by **conjugation**. When two hyphae from different mycelia grow alongside each other they often produce short branch hyphae which come into contact (Fig. 15.5). These hyphae are referred to as 'plus' and 'minus' strains, since there is no visible difference between them.

The tips of the two conjugating hyphae swell with cytoplasm and nuclei. Soon, their swollen tips become separated from the rest of the mycelium and then the intervening wall between them breaks down to form a single container, called the zygote. The nuclei in the zygote behave as gametes and fuse together in pairs: each 'plus' nucleus fuses with a 'minus' nucleus. Finally, the zygote develops a thick outer wall. It is now called a **zygospore**, to distinguish it from an asexual spore. It remains dormant for some time, and may be blown about in the wind as dust. Zygospores can probably survive for much longer periods than asexual spores. When conditions are suitable for growth the zygospore breaks open and produces a single vertical hypha, which immediately produces a sporangium. This releases hundreds of asexual spores which continue the life cycle.

15.4 Spirogyra

Spirogyra is a simple plant-like organism consisting of a chain of several hundred cylindrical cells (Fig. 15.6). These cells are joined end to end making a fine, unbranched thread, or **filament**, about 10 cm long and 0.1 mm in diameter. *Spirogyra* shares its filamentous shape with many related organisms known collectively as the **filamentous algae**. These are very common in ponds and streams, and some species are found in the sea.

Around the inside of each *Spirogyra* cell is a thin layer of cytoplasm containing two or more green, spiral, tape-like structures called chloroplasts. These contain chlorophyll and carry out photosynthesis. No other filamentous alga has spiral chloroplasts, which makes *Spirogyra* easy to identify. At the middle of each cell there is a fluid-filled space called the vacuole. The nucleus is suspended at the centre of this vacuole by fine threads running from the layer of cytoplasm around the cell wall.

Spirogyra, and other similar species, increase in length by cell division. Some of the larger threads

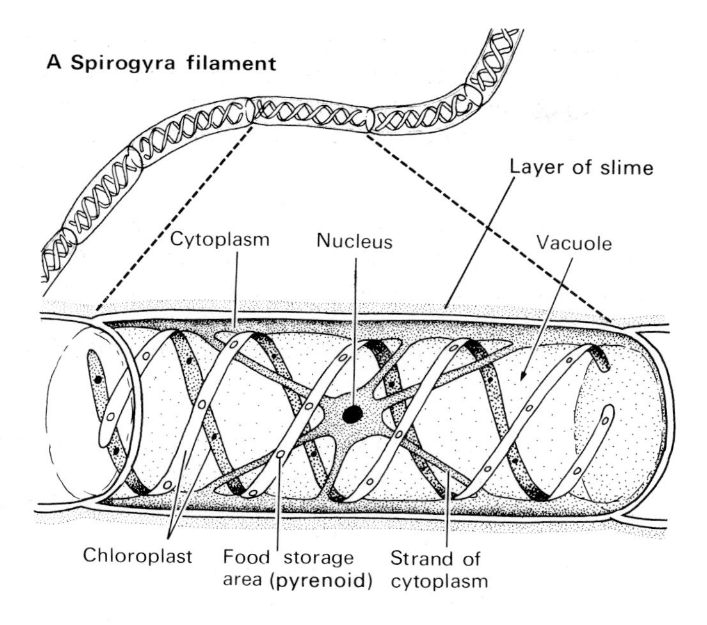

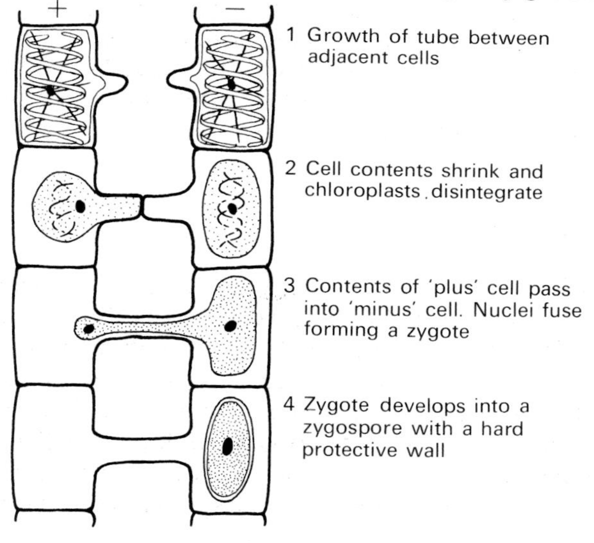

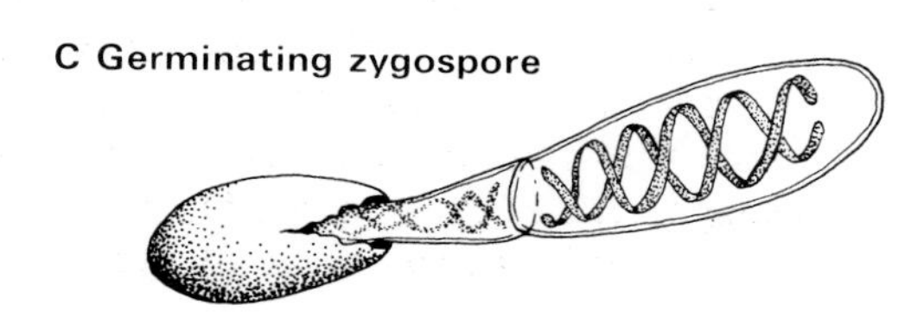

Fig. 15.6 Sexual reproduction in *Spirogyra*

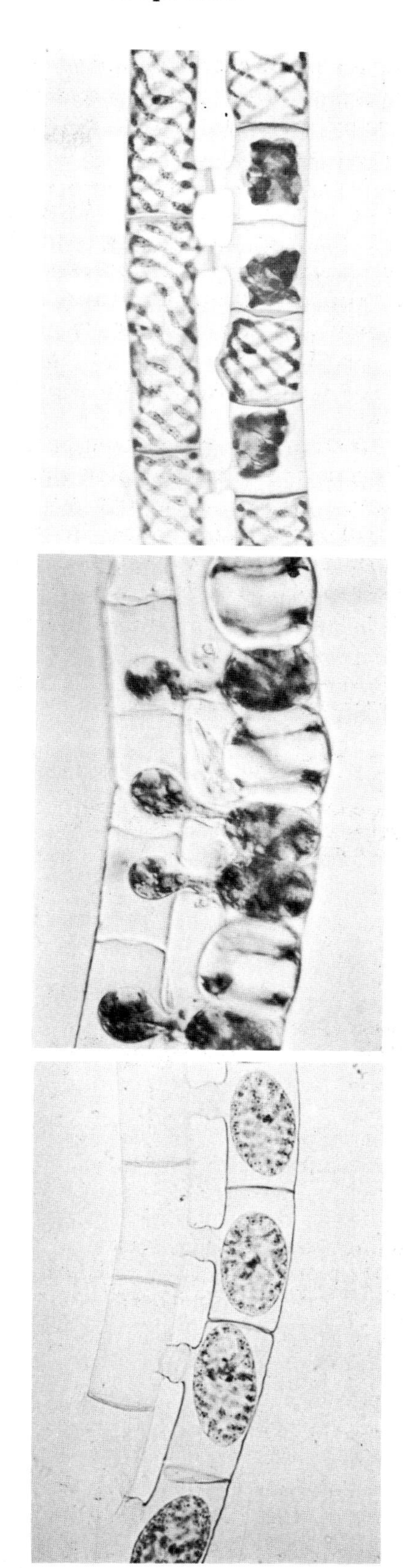

Conjugation in *Spirogyra*. Identify the + and − strains and spiral chloroplasts using Figure 15.6 as a guide produced by growth of this kind may break up into smaller fragments which continue growing in the same way. This fragmentation process is their only means of asexual reproduction. Nevertheless, when conditions are favourable it proceeds at such a rate that these tiny green filaments can practically fill a small pond or choke up a slow moving stream within a month.

Sexual reproduction in *Spirogyra* is by conjugation (Fig. 15.6). Two parallel filaments become attached to each other by tubular outgrowths between adjacent cells, so that they look like a ladder. There is no visible difference between the two conjugating filaments, so they are called 'plus' and 'minus' strains. The chloroplasts in both filaments disappear and then the cytoplasm and nucleus of each 'plus' cell squeeze their way through the connecting tube and fuse with the contents of the opposite 'minus' cell. The result is a series of zygospores in one filament, and empty cells in the opposite filament. Zygospores are set free when the parent filaments die and decay. The zygospores sink to the mud at the bottom of the pond or stream and remain there in a dormant state until conditions are favourable for growth, when they split open and produce a new filament. The process of zygospore formation can occur with such speed that thick masses of filamentous growth can disappear within a few days leaving a pond or stream apparently empty.

The general pattern of the life cycles mentioned so far is common to many other simple organisms.

In spring and summer, or whenever there is plenty of food, the organisms rapidly increase their numbers, usually by asexual reproduction. In addition, some distribution mechanism, such as wind-distributed spores, usually operates to spread the organisms throughout their environment. As summer merges into autumn, or as food supplies dwindle, asexual reproduction ceases, and the organisms may then begin to reproduce sexually, forming some kind of weather-resistant dormant stage, such as a zygospore. Alternatively, weather-resistant asexual spores survive the unfavourable conditions. In favourable conditions these spores germinate and continue the life cycle.

To summarize: the life cycles of many simple organisms often follow a repeated pattern of alternating asexual and sexual reproductive phases. In the next two chapters it will be seen that evolution has led to the gradual disappearance of the asexual stage in the life cycles of higher organisms 'in favour' of more and more complex sexual methods of reproduction.

Summary and factual recall test

In (1) reproduction there is one parent. This parent organism produces cells, by a type of cell division called (2), containing (3) information which is a precise (4) of that contained in the parent's cells. This type of reproduction does not allow a species to (5) from one (6) to the next.

In (7) reproduction there are usually two parents. These produce cells called (8) by a type of division called (9). The fusion of two of these cells is called (10), and results in a single cell called a (11). This cell develops into an offspring which differs from both parents because (12).

Most organisms are unisexual which means (13), whereas some are hermaphrodite which means (14).

Most fish carry out (15) fertilization. Mammals carry out (16) fertilization which takes place within the (17) of the female by means of mating behaviour called (18). The main advantage of this type of fertilization is (19).

Amoeba reproduces asexually by a method called (20), as follows (21). Under unfavourable conditions such as a shortage of (22) an amoeba becomes (23) in shape and develops a (24) outer wall called a (25). In this condition it can withstand (26) and can reproduce by another method called (27), as follows (28).

Yeast is a unicellular (29) whose cells reproduce asexually by a method called (30), as follows (31).

Mucor consists of fine threads called (32), known collectively as the (33). During asexual reproduction one of these threads grows in an (34) direction and then (35) and (36) gather at its tip causing it to (37). It then forms a spherical (38) in which (39) takes place producing hundreds of (40) shaped (41). Each of these contains several (42). They are dispersed in two ways (43). *Mucor* reproduces sexually by a process called (44). Two hyphae from (45) and (46) strains come together and form a (47) in which (48) fuse together in pairs. Under favourable conditions this breaks open releasing asexual (49) from a single (50).

Spirogyra reproduces asexually by (51). It reproduces sexually by a process called (52) in which two parallel filaments form (53) outgrowths between (54) cells. The (55) and (56) from a (57) strain cell moves across into the (58) strain cell where (59) takes place, resulting in the formation of a (60). Under favourable conditions this grows in the following way (61).

16
Reproduction in plants

16.1 Mosses

Mosses belong to a group of plants called the Bryophytes, whose general features are summarized in section 1.3.

In Britain there are about six hundred species of mosses. The majority consist of a stem rarely more than 5 cm long which is anchored to the ground by root-like **rhizoids**. The stem is surrounded by tiny spirally-arranged leaves which are extremely thin, and whose only distinctive feature is a single mid-rib (i.e. there is no leaf stalk or pattern of veins as in the leaves of flowering plants). These leaves usually dry and curl up very quickly when mosses are taken from the relatively wet and humid conditions in which most species thrive. Mosses are particularly common on the banks of streams, forest floors, dead tree stumps, and old buildings.

Mosses, and Bryophytes in general, can be thought of as the 'amphibia' of the plant world because all of them, even the few which can live in dry climates, are dependent on water for sexual reproduction. This is because their fertilization involves swimming sperms. Bryophytes also have an asexual stage in their life cycles, and this is better suited to dry conditions since it involves the large scale production of dry, wind-distributed spores.

All Bryophytes have a life cycle made up of alternating sexual and asexual reproductive stages. The sequence of events in this cycle is summarized below, and in Figure 16.1, using a moss as an example.

Sexual reproduction

During the spring, and often in the autumn as well, mosses produce male and female sex organs. These grow on the same or different plants depending on the species, and in many cases they occur in the centre of a rosette of leaves at the tip of the stem.

The male sex organs, the **antheridia**, are tiny capsules which produce microscopic sperms. When released, the sperms swim through rainwater or dew to the flask-shaped female organs, the **archegonia**. A sperm swims down the neck of each archegonium and fertilizes a single ovum at its base. One fertilized ovum (zygote) in each moss plant develops into a structure which is one of the most characteristic features of mosses: a **spore case** on a long thin stalk. This structure lives as a semi-parasite, absorbing food from the leafy moss plant on which it grows. Its function is to produce spores asexually.

Asexual reproduction

When a spore case is ripe the lid at its top drops off, but the spores inside are not always released immediately. There is a mechanism at the opening of the spore case consisting of tiny flaps, or teeth. These teeth curve inwards in humid conditions closing the opening, and curve outwards in dry conditions exposing the opening and releasing the spores.

A single spore case may contain more than 700 000 tiny spores which are shaken free by the slightest breeze. The advantage of releasing them in dry conditions is that at these times there is often greater air turbulence close to the ground and this helps to disperse the spores over a wide area.

Under warm damp conditions a spore germinates by splitting open and producing a long green thread which branches many times. Eventually the green thread produces tiny buds which grow into a tightly packed mass of new leafy moss plants.

Alternation of generations

A moss plant with its semi-parasitic spore case looks like a single plant, but it is actually two different plants. The spore case and stalk have been produced sexually and so, for reasons explained in chapter 15, have different hereditary information from the leafy plant on which they grow. Consequently, the life cycle of a moss involves two different plants: a leafy gamete-producing plant called the **gametophyte generation** and a spore-producing plant called the **sporophyte generation**. This type of life cycle, in which a

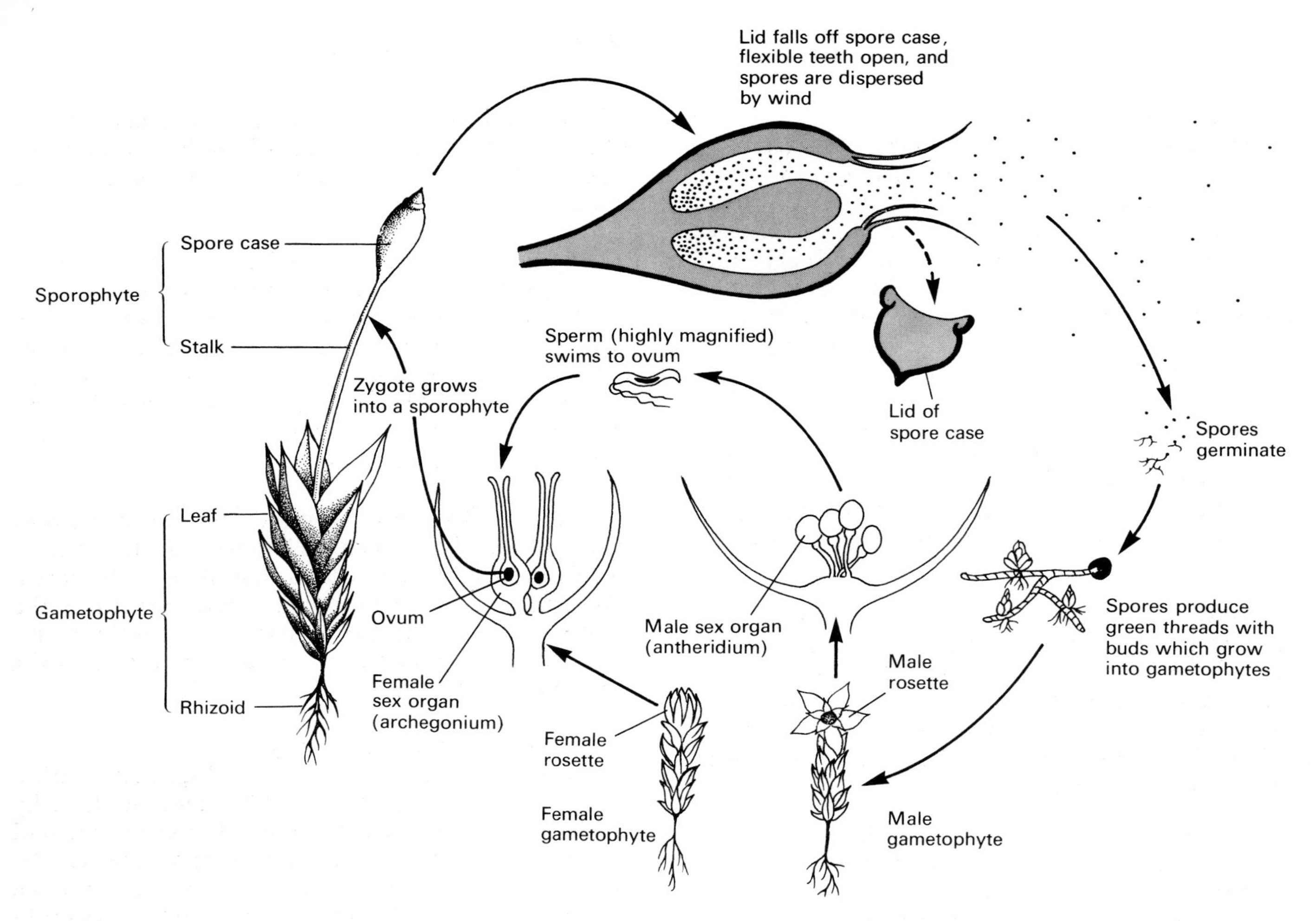

Fig. 16.1 Life cycle of a moss plant
The general pattern of this life cycle is typical of all Bryophytes

generation of gametophytes alternates with a generation of sporophytes, is called an **alternation of generations**. It is typical of all Bryophytes, and also of the Pteridophytes (ferns).

16.2 Flowering plants

At first sight Angiosperms, or flowering plants, appear to have a simpler life cycle than the mosses and ferns, since flowering plants lack the ability to produce asexual spores and so there is no obvious alternation of generations. However, there is evidence that a flower is in fact an extremely elaborate version of a sporophyte generation possessed by the ancestors of modern plants which lived millions of years ago. The pollen and ova produced by a flower are thought to contain the remains of the old gametophyte generation.

Polytrichum communae, the common hair moss. A moss found in very damp places where it can grow up to 20 cm high. Identify the gametophyte and sporophyte regions of each plant, and the spore cases

Flower structure

Flowers are the most advanced and complex reproductive structures in the plant kingdom. Their function is sexual reproduction, as a result of which they form fruits and seeds that give rise to the next generation. There is an immense variety of shape and structure among flowers, but all of them have certain features in common. These features, with their scientific names, are described below, and they are illustrated in Figure 16.2.

Pedicel Flowers grow on a specialized reproductive shoot called the flower stalk or pedicel. In some plants, such as the tulip, there is only one pedicel bearing a single flower. In others, such as the lilac, the pedicel is branched in various ways and bears many flowers.

Receptacle The receptacle is the region of a flower stalk to which the parts of the flower are attached. Receptacles can be flat, dome-shaped, or concave depending on the species. The parts of a flower are attached to the receptacle in rings, or **whorls**. Starting with the innermost whorl the floral parts are arranged as follows:

Carpels Carpels are situated at the centre of the receptacle, and are known collectively as the **gynoecium** or **pistil**. Carpels are the female reproductive organs. Each carpel consists of an expanded hollow base called the **ovary**, above which is a narrow region called the **style** which ends in a pointed, flattened, or sculptured region called the **stigma**. The stigma receives pollen grains from the same or another flower during pollination.

Within the ovary are varying numbers of **ovules**. At the centre of each ovule is a large cell called the **embryo sac**, and this contains several nuclei, one of which is the female gamete, or **egg nucleus**. It is this egg nucleus which is fertilized by a 'male' nucleus from a pollen grain, after which the whole ovule develops into a seed, and the ovary wall becomes the **pericarp**.

Some plants have pistils made up of separate carpels (e.g. buttercup), but in most the pistil is a number of carpels fused together (e.g. poppy, Fig. 16.9C).

Stamens Surrounding the gynoecium is a whorl of stamens. Stamens are known collectively as the **androecium**. These are the male reproductive organs. Each consists of a stalk, or **filament**, bearing an **anther** which is made up of four **pollen sacs** in which **pollen grains** are formed. Pollen grains contain the male gametes.

The structures described so far are characteristic of all flowers, but this is not true of the following parts. One or all of them may be absent, according to the species.

Petals In the majority of flowers the reproductive organs are surrounded by a whorl of petals. Petals are known collectively as the **corolla** of the flower. Some flowers have coloured and scented petals with a **nectary** at the base which produces sugary nectar. Petals of this type attract insects which come to collect the nectar and by doing so transfer pollen from one flower to another.

Sepals In many flowers there is an outermost whorl of sepals, known collectively as the **calyx**. Sepals are usually green and look like small leaves. They enclose and protect the central region of the flower when it is in the bud stage of development. The calyx and corolla together are referred to as the **perianth** of a flower.

Flower shape and symmetry

Compare a buttercup flower (Fig. 16.2) with a sweet pea (Fig. 16.3) and a white dead nettle (Fig. 16.4). Buttercup petals are about the same size and shape as each other, and the flower as a whole can be cut in two along many vertical planes to produce identical halves. Such flowers are said to be **regularly** shaped, and **radially symmetrical**. Other regular flowers are the rose, tulip, poppy, and daffodil.

In contrast the petals of sweet pea and white dead nettle flowers are of several different shapes and sizes, and some are fused together. Such flowers are said to be **irregularly** shaped, or **zygomorphic**. They are also **bilaterally symmetrical**, which means they can be cut into identical halves along only *one* vertical plane. Other irregular flowers are the foxglove, and antirrhinum (snapdragon).

Another very distinctive flower shape is found in the largest group of flowering plants: the Compositae, or **composite** flowers, e.g. daisy and dandelion (Fig. 16.5). They are called composite flowers because they consist of many tiny flowers, the **florets**, packed together on a large flattened receptacle. Dandelion florets, for instance, have five petals but these are fused together making a corolla which looks like one long petal. This arrangement is called a **ray floret**. In other composite flowers, e.g. thistle, the florets are simple tubes called **disc florets**. Daisies and sunflowers have ray florets around the edge, and disc florets in the centre.

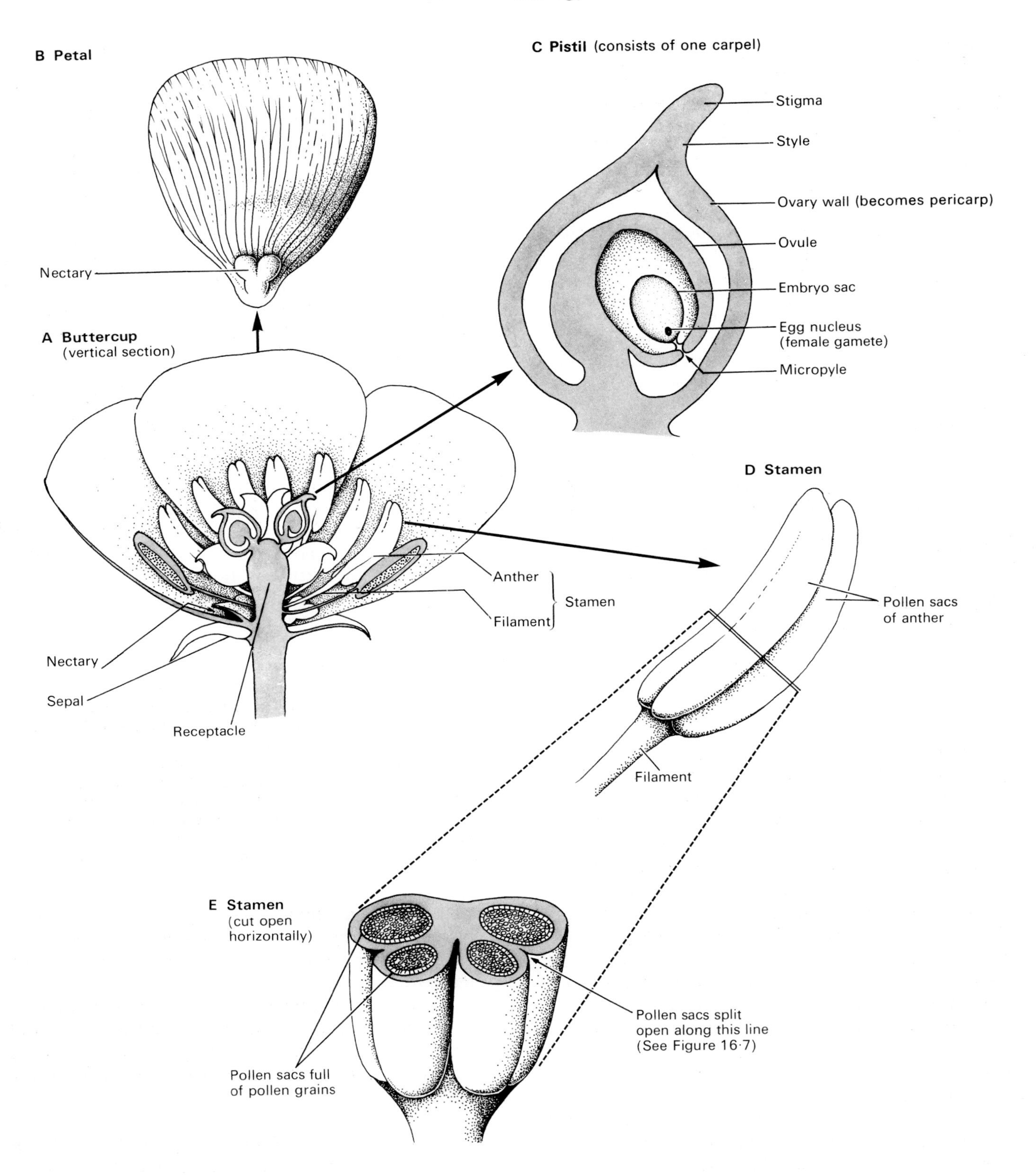

Fig. 16.2 Structure of a buttercup flower

Pollination is by a variety of insects, e.g. bees, wasps, and butterflies. These collect and transfer pollen as they crawl over the flower in search of nectar

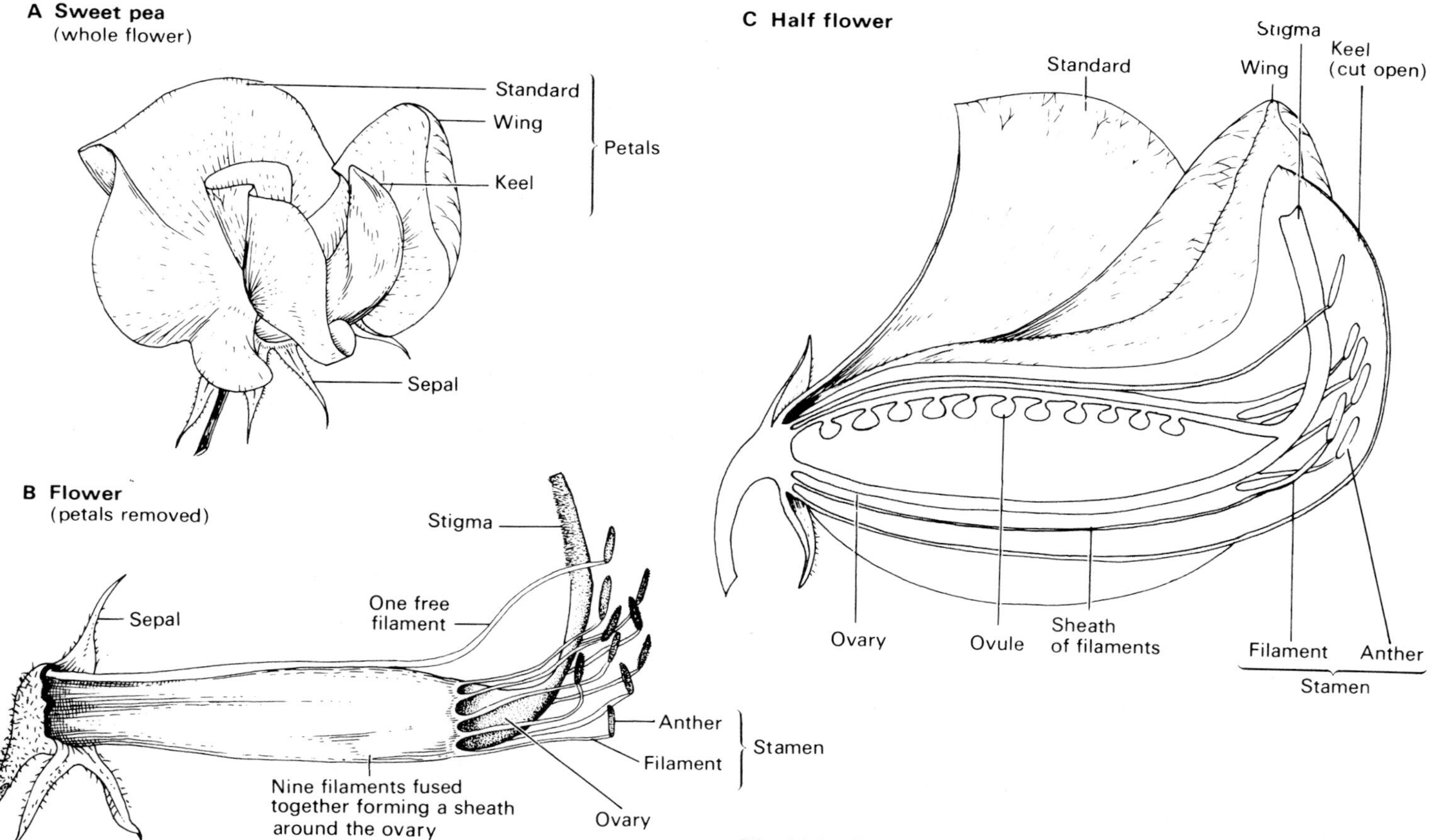

Fig. 16.3 Structure of a sweet pea flower

Similar in structure to lupin, gorse, and flowers of other leguminous plants. Pollination is mostly by bees, which land on the 'wings' and 'keel'. A bee inserts its tongue into the keel in search of nectar which is produced at the base of the nine fused filaments, and gathers in the trough formed by this structure. The insect's weight depresses the petals so that the anthers and stigma protrude from a hole at the end of the keel, and touch the insect's underside. The flowers are protandrous, and so fertilization occurs only when bees visit young flowers and then older flowers

Pollination

Pollination is the transfer of pollen grains from anthers to stigmas, and eventually leads to fertilization. **Self-pollination** is the transfer of pollen from anthers to stigmas in the same flower, or between flowers on the same plant. **Cross-pollination** is the transfer of pollen from one plant to the stigmas on another plant of the same species. Fertilization does not result when pollination occurs between different species.

In the course of evolution, plants have developed characteristics which help to prevent self-pollination and favour cross-pollination. This ensures that, more often than not, fertilization involves the inter-mixing of hereditary information from two plants, thereby increasing the chance of further evolutionary change. Self-pollination is obviously impossible in unisexual species: that is, species in which some plants have flowers containing stamens, while other plants have flowers containing carpels. Bisexual plants (that is, plants with flowers containing both stamens and carpels) have more elaborate methods of preventing self-pollination. In the dandelion, for example, stamens ripen and release their pollen before the carpels in the same flower are fully developed. This is called **protandry**. In the plantain, however, the carpels mature before the stamens. This is called **protogyny**.

There are two main ways in which pollen is transferred from anthers to stigmas: on wind currents, and on the bodies of insects. Most flowers are so adapted that they can be pollinated by one or the other of these methods.

Wind pollination Grass (Fig. 16.6), stinging nettle, hazel, and willow are examples of plants pollinated by wind. Wind-pollinated flowers usually have no petals, but some have small inconspicuous petals which are often white or green. Nectaries are absent altogether. Most wind-pollinated flowers have all or

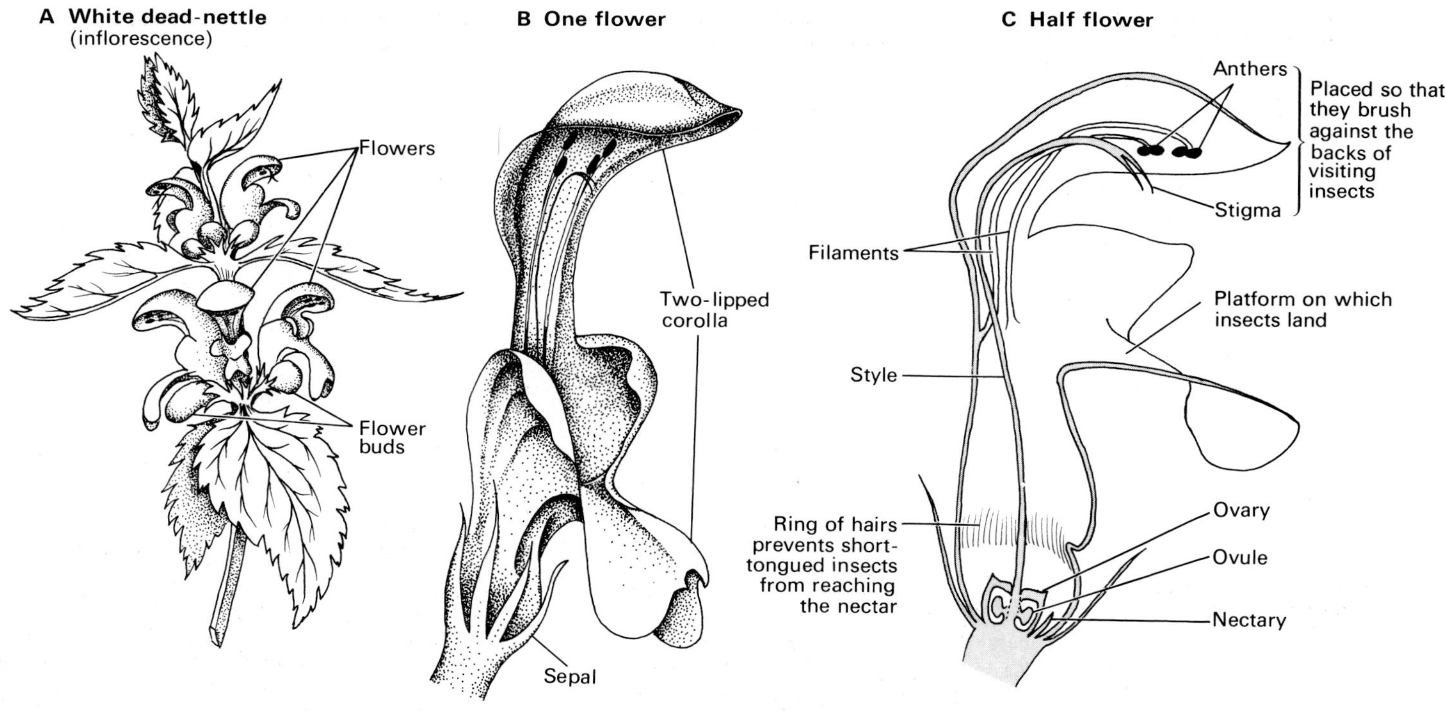

Fig. 16.4 Structure of white dead nettle flower

Pollinated mostly by bees, which land on the lower lip and search for nectar at the bottom of the tubular corolla. In this position the anthers and stigma touch the insect's back. The flowers are protandrous, and the forked stigma remains closed until the pollen has been shed

some of the following characteristics which increase the chances that pollen will reach their stigmas:

a) Abundant pollen production. This compensates for the fact that each grain has an extremely small chance of reaching a stigma;

b) Small, light pollen grains, which float on the lightest breeze;

c) Large anthers which often have long filaments so that they hang well outside the flower, and are attached so that they sway and shake out pollen in the lightest breeze;

d) Spreading, 'feathery' stigmas which act like a net, catching pollen as it floats through the air;

e) Flowers which are either on long stalks well above the leaves, or which develop from flower buds that open before the leaf buds. Both these features increase the flowers' exposure to air currents.

Insect-pollinated flowers Buttercup, sweet pea, dead nettle, and dandelion are some examples of flowers pollinated by insects. These flowers have some or all of the following characteristics which attract insects, and ensure pollination:

a) Large, often brightly coloured and scented petals with nectaries. Some flowers have petals with grooves or dark lines leading from the petal border to the nectaries. These are **honey guides**, and are thought to 'guide' insects to the source of nectar, e.g. pansy and foxglove;

b) The production of a few large pollen grains. In some plants the grains are smooth and sticky, in others they are covered with spiky hairs. Both of these features help to make the grains cling to the insects' bodies;

c) Anthers tend to be small, and situated inside the flower where insects are likely to brush against them;

d) Stigmas are also situated so that visiting insects brush against them. Stigmatic surfaces often produce a sticky, sugary fluid to which pollen grains become attached.

Fertilization

Of the thousands of pollen grains released from the anthers of a flower only a tiny fraction reach the stigmas of another flower of the same species. But even when a grain has made this long journey and become firmly attached to a stigma, the male gamete within it is still separated from the egg nucleus (female gamete) by seemingly impassable barriers.

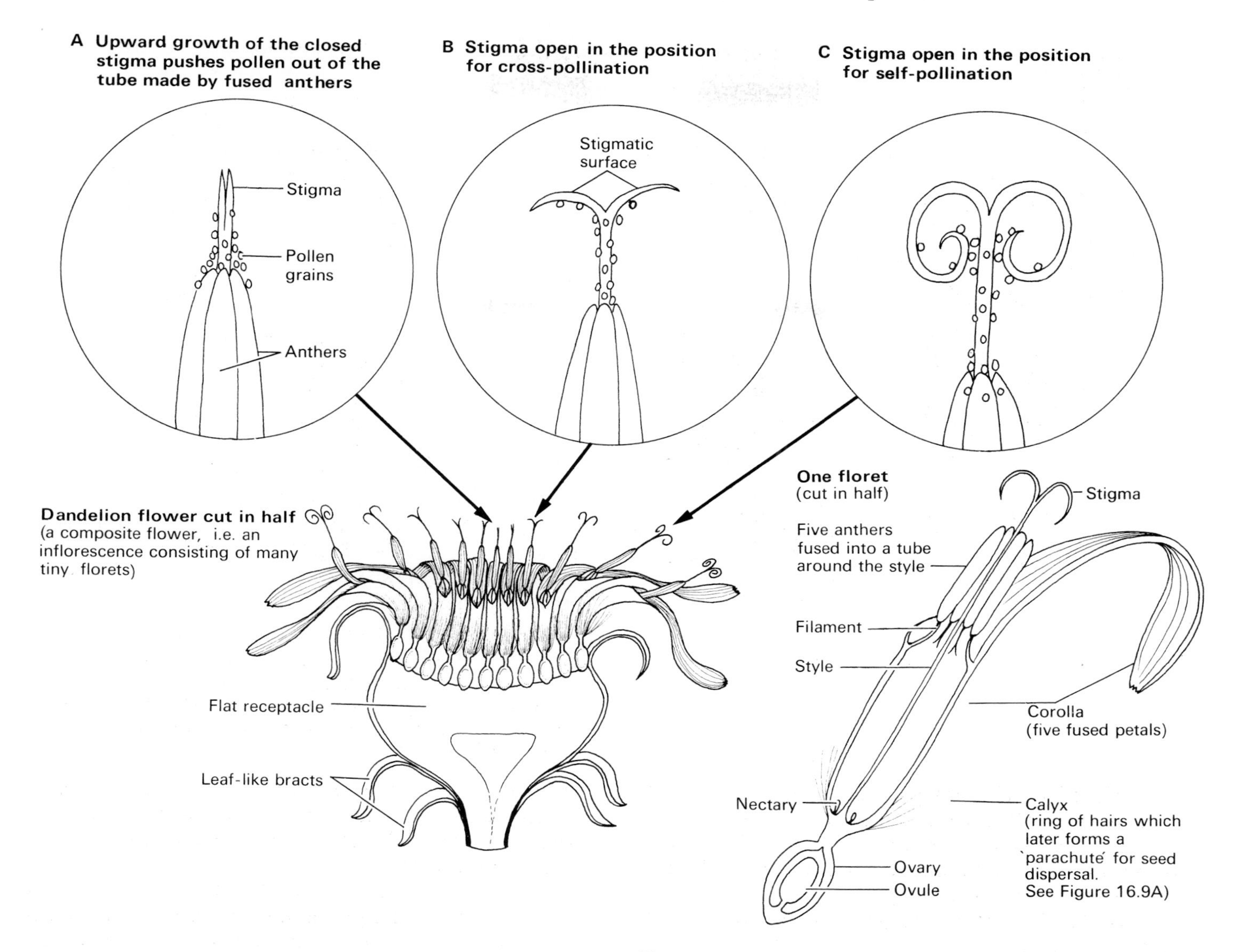

Fig. 16.5 Structure of a dandelion flower head

This is a composite flower. Pollination is by a variety of insects. which crawl over the florets in search of nectar. There is also provision for self-pollination: the stigmas curve over until they touch pollen attached to their own styles

First, there is the hard wall of the pollen grain, then there is the tissue of the stigma and style, and finally the egg nucleus itself is embedded in the embryo sac of the ovule, both of which are suspended in the empty space of the ovary. Even if the male gamete were able to swim like a sperm it could not overcome these obstacles. Its pathway to the egg is cleared in a most remarkable fashion (Fig. 16.7).

The stigma produces a sticky fluid which nourishes the pollen grains and stimulates each one to burst open and develop a long, hollow, tubular outgrowth called a **pollen tube**. This tube pushes its way between cells of the style from which it absorbs more nourishment. It grows towards the ovule, probably guided by chemicals, then enters it by a tiny hole, the **micropyle**, through which it reaches the embryo sac. Here, the tip of the pollen tube bursts open forming a clear pathway through which the male gamete reaches the female. In some species the whole process takes only a few minutes.

The male gamete is a nucleus from one of the cells inside the pollen grain. This nucleus passes down the pollen tube into the embryo sac and brings about fertilization by fusing with the egg nucleus. After

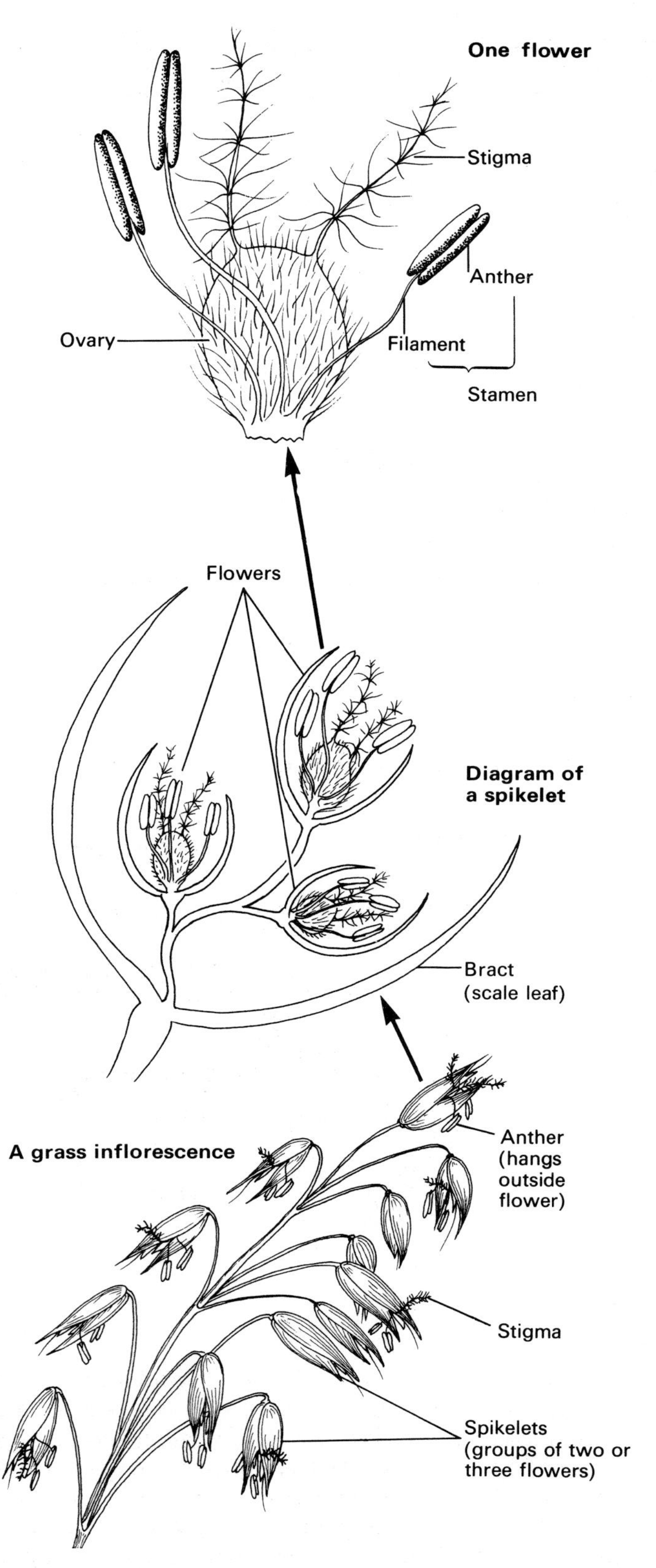

Fig. 16.6 Structure of a typical grass flower
Note the features of wind pollination

fertilization the stamens, petals and eventually the sepals shrivel up and drop off. At the same time, the ovule develops into a seed, enclosed within the ovary wall which becomes the fruit.

The seed The fertilized ovule divides by mitosis, forming a tiny embryo plant. This embryo consists of a young root, called the **radicle**; a shoot, or **plumule**; and one or two 'seed-leaves', or **cotyledons**. The embryo also develops a supply of stored food. Depending on the species, this food is deposited either inside the cotyledons, or in a mass of cells called the **endosperm** which surrounds the embryo. These two methods of storing food are illustrated in Figure 16.8.

While this is happening the embryo sac and the membranes around it, called **integuments**, expand as the embryo grows inside them. The integuments become much thicker and form a tough protective 'seed coat', or **testa**, around the embryo. A seed, then, is made up of an embryo plant and stored food, enclosed within a protective testa.

Lastly, most of the water is withdrawn from the seed making it hard and extremely resistant to cold and other adverse conditions. In this state a seed is said to be **dormant**. Dormant seeds can survive for months or years. There are unconfirmed reports that corn seeds germinated after being stored for thousands of years in Egyptian tombs.

The fruit The fruit develops from the ovary after fertilization; therefore a fruit consists of the ovary wall and the seeds it contains.

In **true fruits** the ovary wall changes in various ways to form a protective layer called the **pericarp** which surrounds the seeds. For example, a bean pod; the skin, flesh, and stone of a plum; and the hard shell of a nut are all types of pericarp.

In **false fruits** the ovary wall may grow a little but otherwise remains unchanged. For example, in strawberries the fruits are the tiny seed-like objects attached to a large fleshy structure which develops from the receptacle of the flower. In apples and pears the receptacle becomes fleshy and completely encloses the whole ovary, which remains as the core (Fig. 16.9E and F).

In both false and true fruits, the development of the ovary wall or receptacle is usually associated with a mechanism which disperses the seeds.

Dispersal

It is important that seeds are carried away from the parent plant. This avoids overcrowding. It can also lead to the spread of plants into new and different environments. There are three main types of dispersal mechanism (Fig. 16.9).

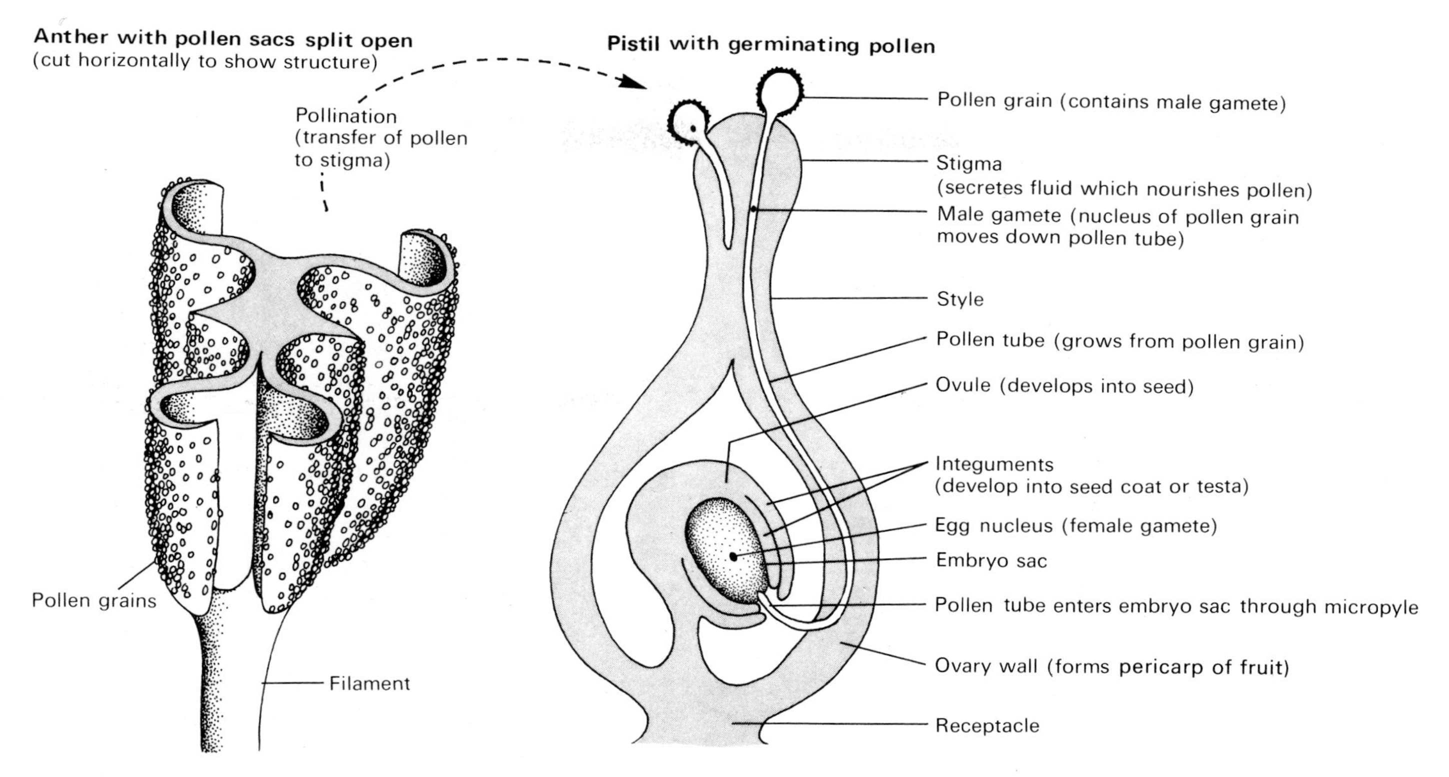

Fig. 16.7 Pollination and fertilization

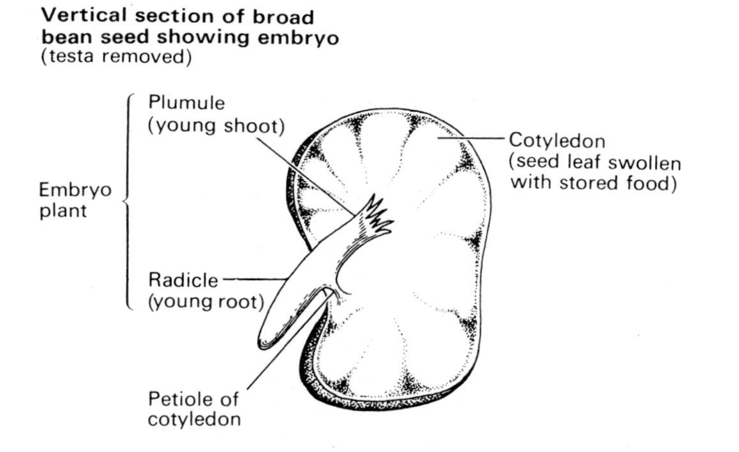

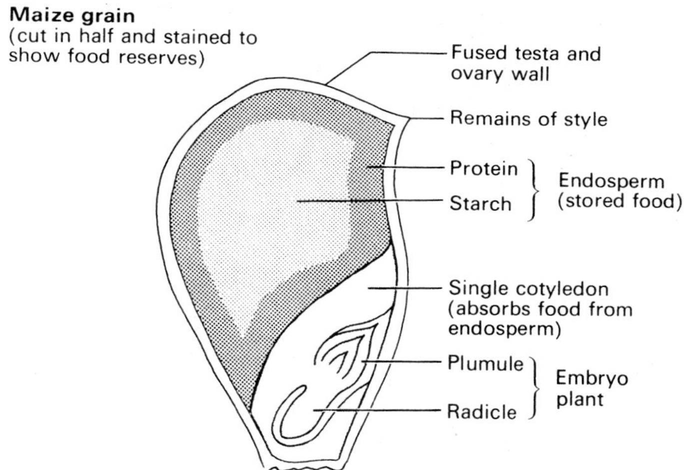

Fig. 16.8 Seed structure

Wind dispersal (Fig. 16.9). In some plants the surface area of the fruit or seed is increased in various ways, as in the 'parachutes' of a dandelion, or the wings of the sycamore and ash. In a poppy the seeds are shaken through pores in the ovary wall as the plant sways back and forth in the wind.

Animal dispersal (Fig. 16.9). Hooked fruits and seeds, e.g. burdock and goosegrass, can be carried long distances attached to animals' fur. Succulent fruits and nuts attract animals as a source of food. Small hard-coated seeds, e.g. blackberry, strawberry, and rose hip, can pass through an animal's digestive system unharmed and can therefore be carried some distance before being deposited on the ground–along with a convenient supply of fertilizer. In other plants, such as apple, plum, and cherry, the seeds in their hard coats are discarded after the soft fruit has been eaten by animals.

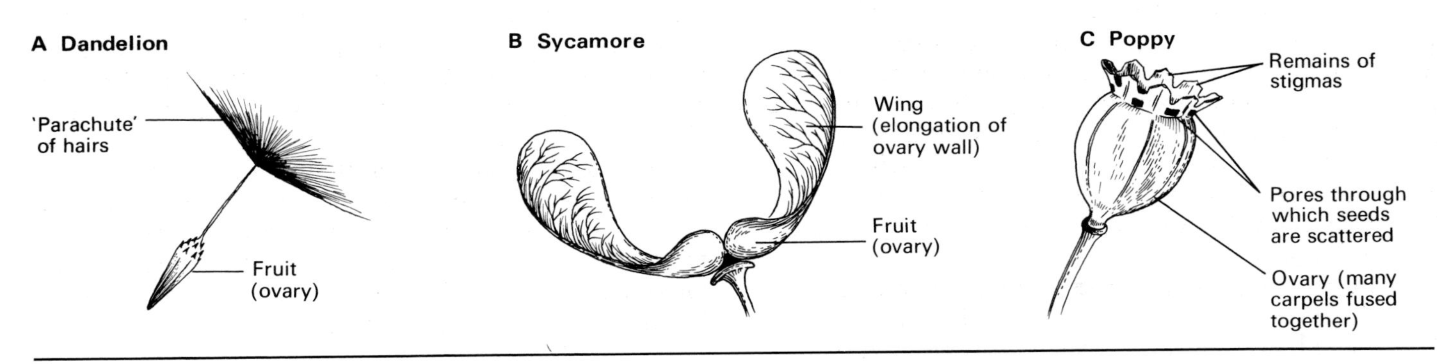

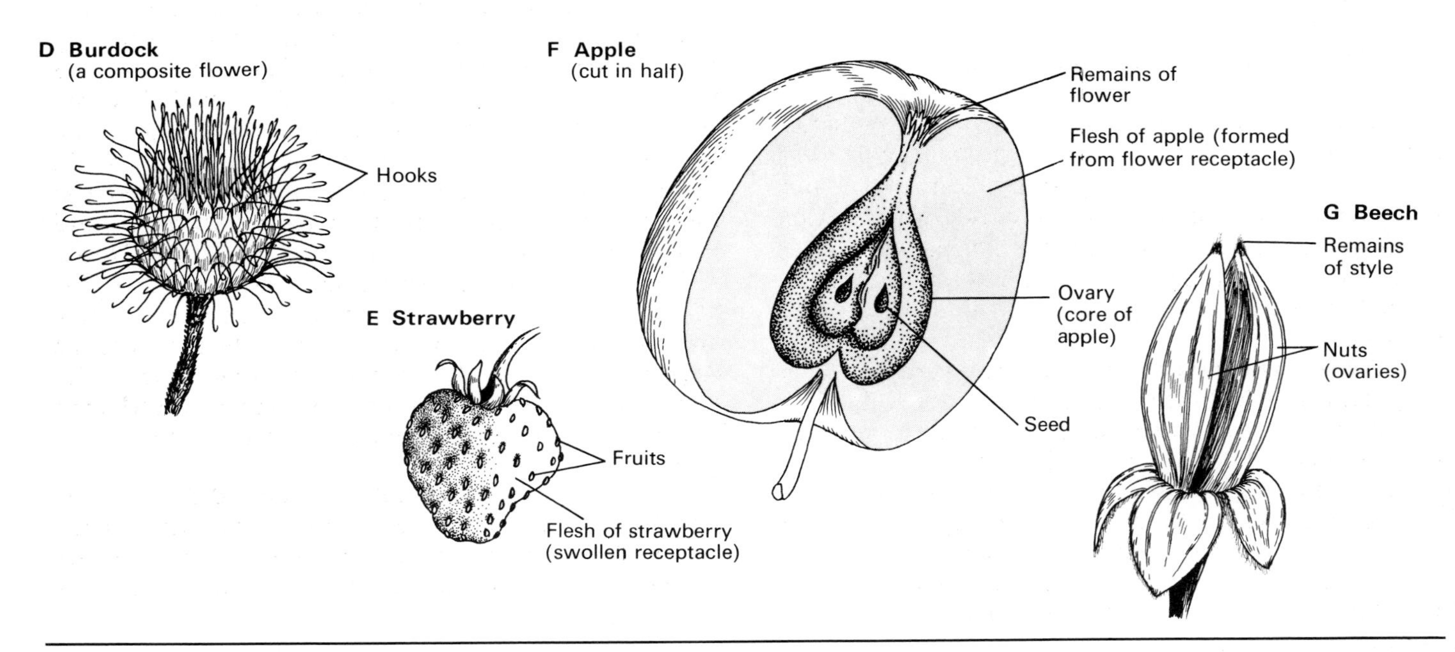

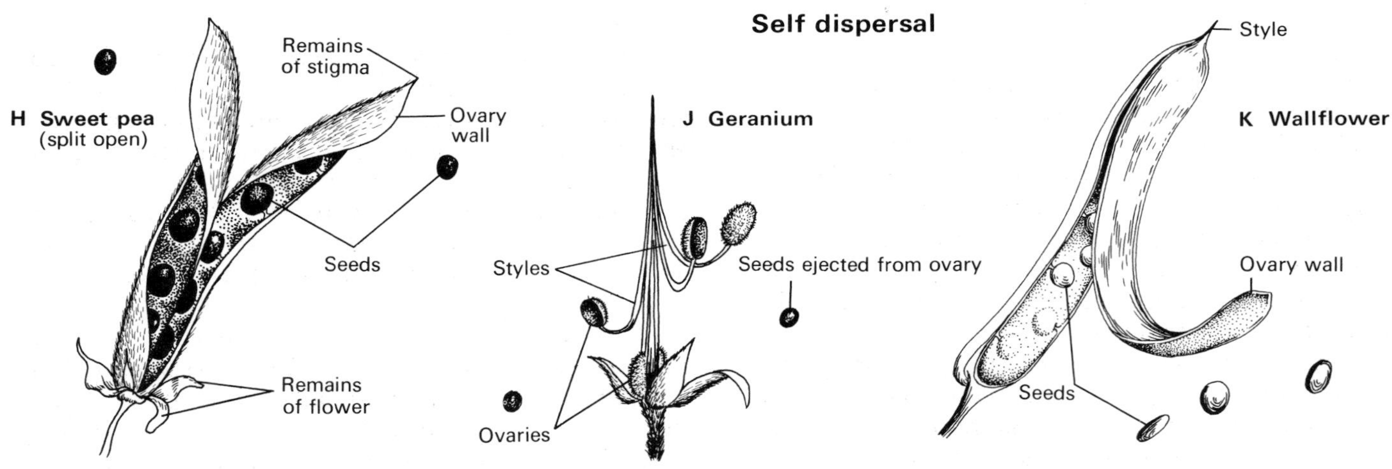

Fig. 16.9 Dispersal mechanisms of fruits and seeds

Self-dispersal (Fig. 16.9). Several plants have mechanisms which throw seeds some distance from the parent. Most of these depend on tension caused by the drying of the fruit wall. The ripe pods of sweet pea, gorse, broom, and lupin, suddenly split open and the two halves curl outwards, scattering the seeds. In the geranium, the styles curl up and out, throwing seeds from the cup-shaped ovaries. A similar mechanism occurs in the wallflower fruit, which splits open from the base upwards.

Under the right conditions seeds germinate and form a new generation of plants. The conditions necessary for germination to take place are investigated in exercise B.

16.3 **Life cycles of flowering plants**

The basic life cycle common to all flowering plants consists of two phases. As a plant grows from a seed it passes through a period of vigorous growth, called the **vegetative phase**, in which its body rapidly increases in size and weight, and accumulates stored food. This is followed by the **reproductive phase**, in which the stored food is used in the production of flowers, fruits, and seeds.

This basic pattern varies in several ways from species to species. For instance, the length of time taken by vegetative and reproductive phases varies considerably. In many species parts of the root or shoot have evolved into organs which store the surplus food produced in the vegetative phase. Frequently, these food storage organs become dormant and survive beneath the ground during adverse climatic conditions. The following are the most common examples of variations in the basic life cycle.

Ephemeral and annual plants

There are many species in which the parent plant is completely exhausted by the reproductive phase, and it quickly dies and disappears altogether after its fruits and seeds are ripe. Chickweed and groundsel are examples of this type of plant. They are called **ephemerals** because they grow, reproduce, and die at such speed that the sequence is repeated several times in one season. Ephemerals include some of the most prolific garden weeds.

Annuals are plants which grow and reproduce only once in a growing season. Some examples are shepherd's purse, marigold, and nasturtium. Both ephemeral and annual plants survive the winter as seeds.

The effects of the reproductive phase on annual and ephemeral plants is easily demonstrated by removing their flower buds as soon as they appear. This prolongs their vegetative phase beyond its normal span by preventing the enormous expenditure of energy needed to produce flowers and seeds. Provided the plants are kept warm in winter, disbudding keeps them alive for many years.

In temperate climates annual plants which cannot produce seeds by the end of the growing season die off and disappear from the vegetation of the region. However, this does not happen in plants whose vegetative phase can lie dormant during the cold season. Biennials fall into this category.

Biennial plants

Biennials have a two-year life cycle. Their vegetative phase takes up the whole of the first growing season and results in a build-up of food. This is stored in various parts of the plant body, which become swollen in the process. For example, the root is the food storage organ in carrots and parsnips; and in radishes, turnips, and sugar beet the root and the lowest part of the stem store food. Cabbages and brussels sprouts consist of enlarged buds full of food.

Cultivated biennials are gathered after the first year of development, and their food stores are eaten. If they are left in the ground, all parts except the food storage organs die and the plant remains dormant during the winter. In the second year, the stored food is used up in producing flowers and seeds, after which the parent plant dies.

Many cultivated biennials are derived from plants which are naturally annuals. Their life cycles have been adjusted through selective breeding so that the vegetative phase is prolonged until the autumn, when they are harvested.

Perennial plants

Perennials are plants which can survive the great expenditure of energy and food reserves needed to produce flowers and seeds. They build up enough food reserves in their vegetative phase to undergo reproduction and still have some to spare. The parent plant lives on almost indefinitely to reproduce again and again at regular intervals. Broadly speaking there are two types of perennial: woody and herbaceous.

Woody perennials As the name implies, these plants have hard, woody stems which persist year after year. Trees are woody perennials with a typically erect stem, or trunk, forming the main axis of the plant, from which grow many spreading branches. Every year the trunk and branches grow longer and higher, and new branches are formed. The additional weight so produced is supported by yearly increases in the circumference of the trunk, a process known as **secondary thickening**. Trees continue their potentially unlimited life span by growing bigger and bigger, until high winds or disease eventually destroy them.

Unlike trees, shrubs and bushes have many woody stems. These cease to grow in length and circumference after one or two seasons. At the beginning of the next season new stems are produced from buds at the bases of the old stems.

Evergreen trees and shrubs shed leaves and grow new ones continuously and are therefore never bare of foliage. **Deciduous** plants shed their leaves all at once. In tropical climates the leaves fall off just before the dry season, which helps the plant to avoid an excessive loss of water. In temperate climates leaves fall off in the autumn, and the plant is dormant throughout the winter. The mechanism of leaf-fall is explained in Figure 16.10.

In temperate climates leaf-fall is accompanied by the development of winter buds. These are extremely short stems with small overlapping leaves which are protected by an outer layer of scale leaves. A typical example of a stem with winter buds is illustrated in Figure 16.11.

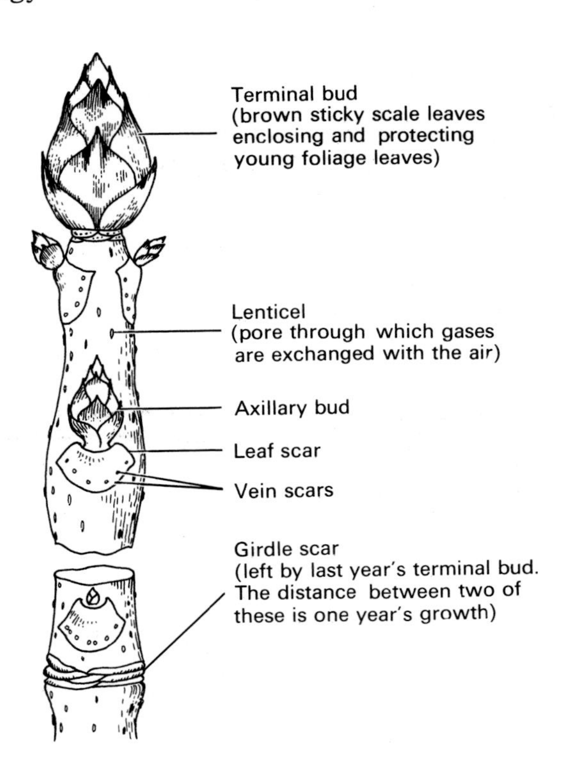

Fig. 16.11 Winter twig of horse chestnut

Herbaceous perennials Perennials of the herbaceous type die away completely above ground level in the autumn, leaving dormant, food-filled organs protected under the soil. These grow again in the spring. Lupin and delphinium are common examples. The food-filled underground structures are called **perennating organs**, and consist of parts of root or shoot which are filled with surplus food at the end of the growing season. Rhizomes, tubers, corms, and bulbs are perennating organs, and are described in the next section.

In addition to sexual reproduction by means of flowers, herbaceous and woody perennials also undergo a type of asexual reproduction. This involves either the parent plant separating into parts which continue to grow independently, or the production of outgrowths from the main plant body which eventually separate and grow into new daughter plants. Since these processes involve the vegetative phase of the life cycle they are called **vegetative reproduction**, or **propagation**. Examples are: stolons, runners, and suckers, together with the perennating organs (rhizomes, tubers, corms, and bulbs) already mentioned.

Fig. 16.10 Diagram of leaf-fall

Leaves do not simply die and fall off. Their shedding results from changes in the petiole base, as follows: 1. Loosely attached cells, called the abscission layer, form where the petiole joins the stem. 2. Useful substances are absorbed from the leaf, which then changes colour (develops autumn tints). 3. Cork cambium produces a layer of cork beneath the abscission layer. The cork seals off the leaf vascular system. 4. The leaf is dislodged by wind

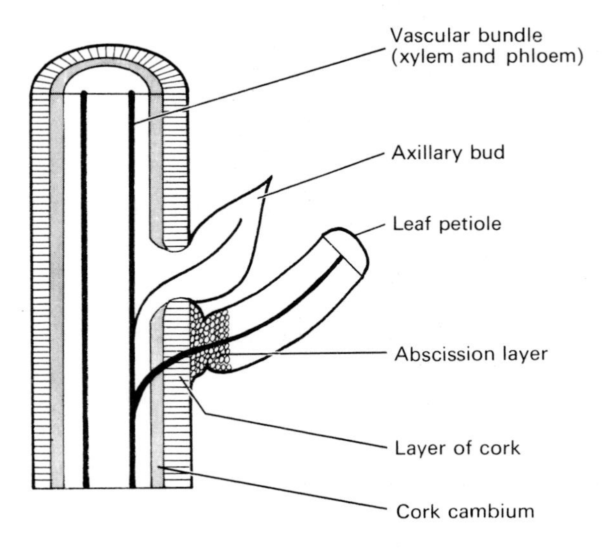

16.4 Vegetative reproduction

Before reading the following explanatory notes it is best to revise the technical names for the parts of a plant from Figure 5.3.

Fig. 16.12 Stem tubers of potato

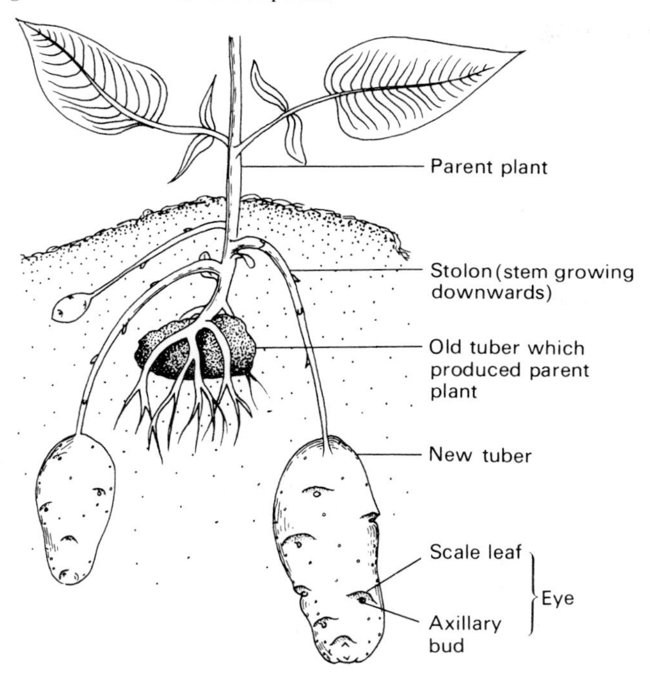

Stem tubers (e.g. potato, Fig. 16.12)

Stem tubers are formed by outgrowths from the lowest axillary buds which turn downwards into the soil. Eventually the tip of the underground stem fills with food (mainly starch) and swells rapidly to form a tuber. Stem tubers are distinguished from root tubers by their origin, and the presence on their surfaces of scale leaves, and axillary buds, which form the 'eyes'.

Tubers lie dormant until spring, when shoots arise from one or more of the eyes, using the stored food to establish a new plant. Stem tubers are formed on potatoes and artichokes.

Runners (e.g. creeping buttercup, Fig. 16.13)

Runners grow out horizontally from axillary buds, and 'run' over the surface of the soil forming several new plants. The terminal bud of a runner turns upwards and roots form behind it to produce a daughter plant some distance from the parent. The daughter plant, in turn, produces runners and more daughters, until a long chain of plants is formed. When the daughter plants are well established their connections with the parent plant die away.

Runners are formed on creeping buttercups, strawberries, and houseleek (which forms a very short runner sometimes called an offset).

Rhizomes (e.g. iris, Fig. 16.14)

In some plants, the entire stem is underground and grows horizontally through the soil. Stems of this type are called rhizomes. A rhizome is usually more persistent than a stem tuber or runner; its older parts do not shrivel and die away for several years. Consequently, rhizomes become large branched structures, and it is a considerable time before they separate into a number of plants.

In spring, the terminal bud and some axillary buds of a rhizome grow upwards and form shoots above the soil. These shoots produce flowers, and then pass surplus food down into the rhizome which continues a

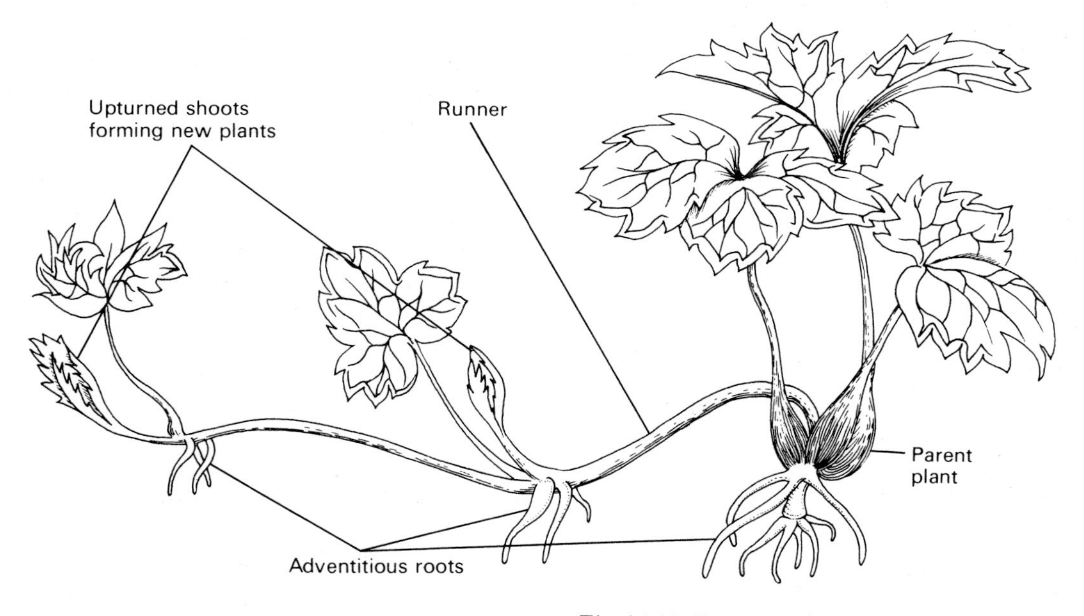

Fig. 16.13 Runners of creeping buttercup

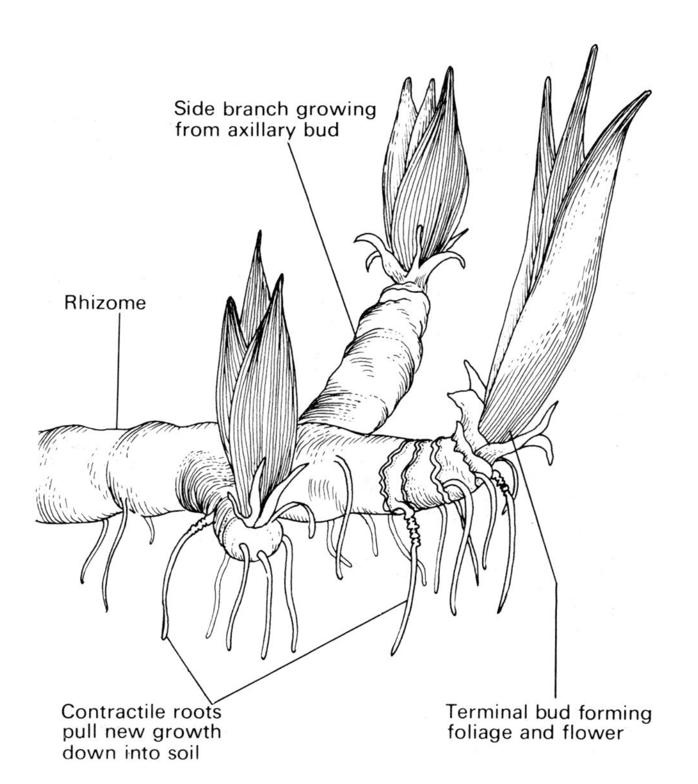

Fig. 16.14 Rhizome of iris

branching growth through the soil from underground axillary buds. A tendency for newer regions of the rhizome to rise up out of the soil is prevented by the contraction of certain adventitious roots. Rhizomatous stems are found in iris, Solomon's seal, coltsfoot, and many types of grass.

Corms (e.g. gladiolus, Fig. 16.15)

Corms are also underground stems but, unlike rhizomes, they do not grow horizontally. Corms are very short vertical stems which are swollen with stored food, and enclosed within overlapping fibrous leaf bases.

During spring and summer the terminal bud of a corm grows upwards, forming leaves and eventually flowers. This uses up all the food stored in the corm. After flowering, the leaves remain and manufacture more food which passes down to form a new corm in the region of stem immediately above the shrivelled remains of the old corm. In this way a new corm forms on top of the old one each year. This process would lead to the structure growing out of the soil were it not for special adventitious roots which contract and hold it below ground level. Vegetative reproduction in corms involves the formation of daughter corms from axillary buds. These develop on the sides of the parent and eventually separate, forming independent plants.

Bulbs (e.g. tulip and daffodil, Fig. 16.16)

Bulbs are large underground buds, whose leaves are thick and fleshy with stored food and water. The whole plant appears as if it were compressed, so that its stem is only about a centimetre high with leaves at each node, and with one or two axillary buds. In bulbs, the internodes do not elongate and so the leaves and axillary buds are packed tightly against each other. Tulip, daffodil, and onion are plants which produce bulbs.

Fig. 16.15 Structure and appearance of gladiolus corm

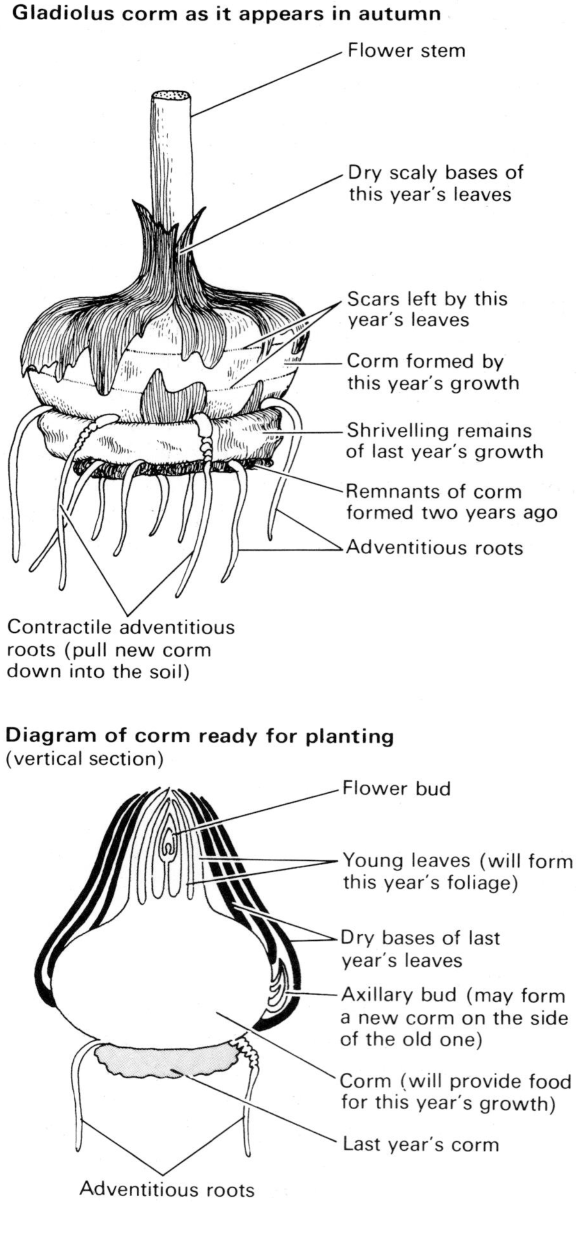

Tulip (Fig. 16.16) In spring, the bulb bursts open and the flower stem emerges complete with large green leaves attached along its length. Growth of this stem uses up all the food in the bulb's underground food-storage leaves so that they dry up into paper-thin brown scales.

After flowering, the green leaves above the ground persist for some months producing food which passes down the stem and accumulates in one or more axillary buds, which swell and form new bulbs.

Advantages of vegetative reproduction and food storage

Vegetative reproduction produces a tightly packed group of plants which spread outwards from a parent. This leaves little room for other plants which would compete for light, water, and minerals.

Many additional advantages are gained by plants with access, throughout the year, to stores of food and water. They can survive drought more easily than plants which grow from seed each year and, since their food and water are stored underground, they can survive destruction of their leaves by fire, insects, or humans. Furthermore, perennial food supplies enable plants to produce new growth earlier in the spring than plants growing from seed. This allows the flowers and seeds of these plants to be produced before competition with other plants for light, water, and minerals reaches its height.

Couch grass and other plants with an extensive system of rhizomes can resist being cut into small pieces by ploughing and digging because, provided they bear a bud, even small fragments can grow again.

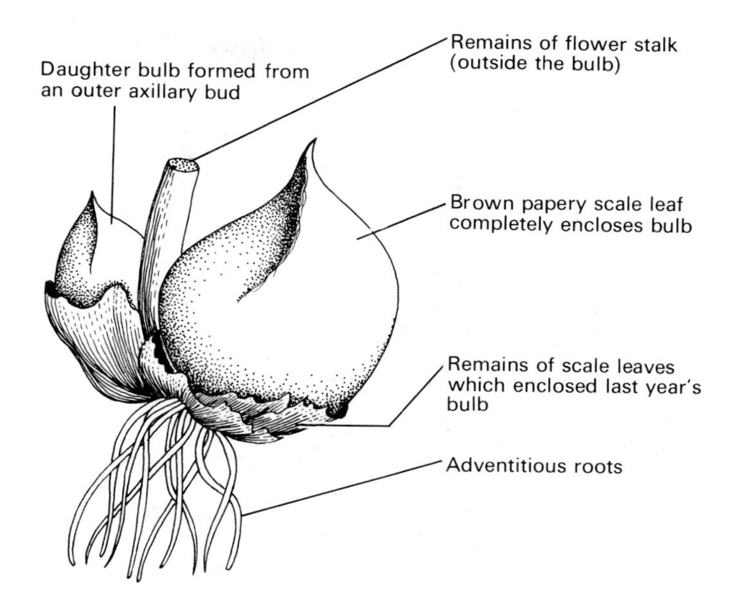

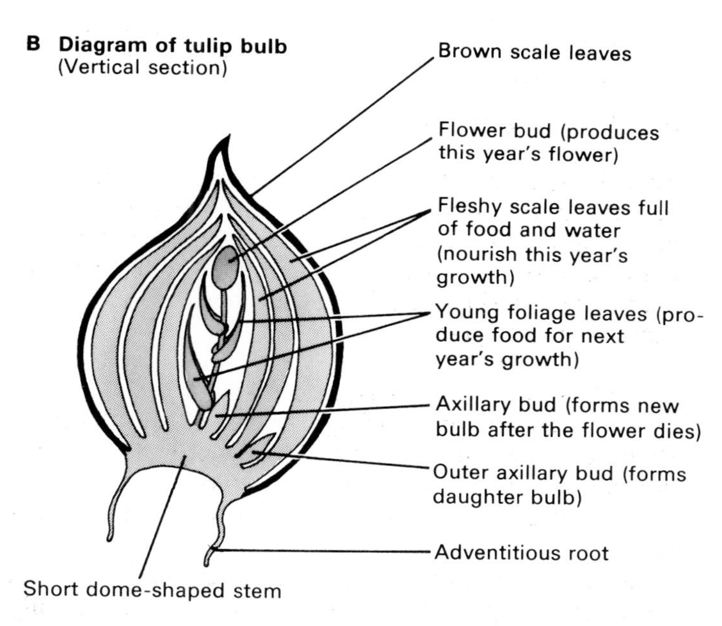

Fig. 16.16 Structure and appearance of a tulip bulb

16.5 Artificial propagation

There are several ways in which new plants can be produced, or propagated, artificially from parent plants.

Cuttings

A cutting is any portion of a root or shoot which, after being severed from a parent, can be induced to grow into a new individual. Propagation by cuttings is an easy and inexpensive method of obtaining many plants with the same characteristics as a parent.

Chrysanthemums, coleus, and geraniums can be propagated by **stem cuttings.** These are taken from non-flowering side shoots. A sharp knife is used to cut off a shoot just below a node (Fig. 5.3). Leaves are trimmed off the part of the stem to be inserted into moist rooting compost. Root growth can be encouraged if, before planting, the end of the cutting is dipped into powder containing rooting hormone.

Phlox and primulas can be propagated from **root cuttings**, and begonia rex plants can be produced from a single leaf.

Budding and grafting

It is possible to cut part of the stem from one plant and join it to a cut surface on another plant of the same genus in such a way that the two parts unite and become one plant.

The portion with roots is called the **stock**, and the portion transferred to the stock is either a single bud with a small piece of bark attached (Fig. 16.17), or a **scion**, which is a whole shoot (Fig. 16.18).

Care is taken to have the cambium (Fig. 7.2) of the bud or scion in contact with the cambium of the stock.

The type of bud or scion used decides what the flowers and fruit will be like. The type of stock used decides the ultimate size of the plant and the time it takes to mature.

Budding is a popular way of producing large numbers of rose bushes from a parent with desired characteristics. Grafting is used to propagate apple, pear, and plum, and ornamental trees like laburnum and rhododendron, some of which cannot be grown from seed.

Advantages of artificial propagation

Seeds from a red rose will not necessarily grow into plants with red roses; and seeds from a Cox's Orange Pippin apple are likely to grow into trees with entirely different apples. This happens because seeds nearly always result from cross-pollination between plants with different characteristics. Consequently, seeds grow into plants with characteristics from two parents. By using artificial propagation, however, gardeners can produce any number of plants with exactly the same characteristics.

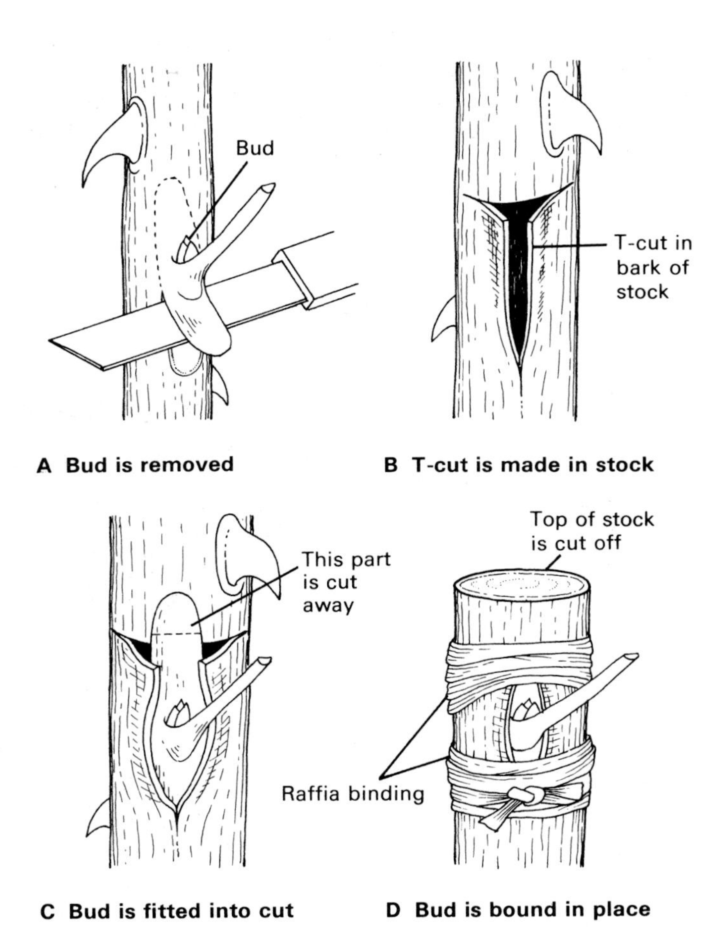

Fig. 16.17 Budding of roses. **A** A well-developed bud is cut from one parent. **B** A T-cut is made in the stock (root and stump of other parent). **C** and **D** The bud is fitted into the T-cut and bound in place with moist raffia. Further growth from the stock is cut away. Only growth from the grafted bud is allowed to remain.

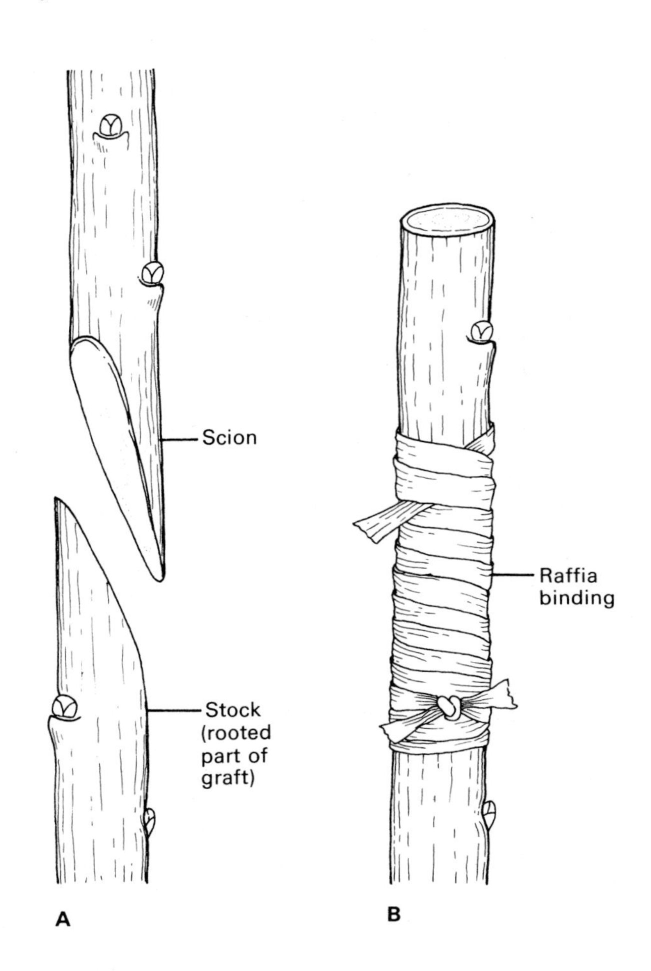

Fig. 16.18 Splice grafting (roses, clematis, and broom). Stock and scion should be equal in diameter. **A** Slanting cuts are made in scion and stock. **B** The two cut surfaces are fitted together and bound with raffia.

Verification and inquiry exercises

A *An investigation of moss life cycles*

1. Collect specimens of moss in the early spring and late summer. Try to find specimens of the very common moss *Mnium hornum*. This moss is found in damp regions of shady woodland.

2. *Mnium's* large male rosettes are conspicuous in early spring. Female organs are on separate plants. Use fine mounted needles to dissect out archegonia and antheridia. Mount on a slide in water and observe under a microscope.

3. Obtain ripe sporophytes in late summer. Spread the spores on damp blotting paper, then cover them with a glass dish to prevent them drying up. Observe spore germination.

B *An investigation of the conditions necessary for the germination of seeds*

1. Obtain a quantity of mustard seeds, and five test-tubes numbered 1–5.

a) Tube 1. Put a few seeds on dry cotton wool in the bottom of the tube. Leave the tube in the light, in a warm place.

b) Tube 2. Put a few seeds on wet cotton wool in the tube. Leave the tube in the dark in a warm place.

c) Tube 3. Put a few seeds on wet cotton wool in the tube. Leave the tube in a refrigerator.

d) Tube 4. Put a few seeds in the tube and cover with boiled and cooled water, then cover with a layer of olive oil. Leave it in the light in a warm place.

e) Tube 5. Put a few seeds on wet cotton wool in the tube. Leave them in the light, in a warm place.

2. This experiment is designed to investigate the influence of five different factors on germination: light, dark, warmth, cold, and oxygen.

a) Which tube is designed to investigate which factor?

b) Note the changes in each tube after 2 or 3 days. Which of the five factors must be present for germination to occur?

C *An investigation of germination*

1. Obtain seeds of broad bean and french bean, and some maize grains. Obtain three jam-jars. Roll short lengths of blotting paper into cylinders and drop one into each jam-jar. Fill each jar with dry sawdust or clean sand by pouring it into the blotting paper cylinder. Place three seeds between the glass and blotting paper of each jar. Use a separate jar for each type of seed. Lastly, pour enough water into each jar to dampen the paper thoroughly.

2. Make drawings each day to record the stages of germination up to the establishment of root and shoot.

3. *a*) Germination in which the cotyledons remain below ground is called **hypogeal**, and where they rise above the ground it is called **epigeal**. Which of the seeds has hypogeal, and which has epigeal germination?

b) Of what advantage to the plants are the hook-shaped plumules which emerge from broad beans and french beans, and the coleoptile sheath in maize?

D *An investigation of seed formation*

1. Plant a few runner beans in well-manured soil in the second half of May. Support them with canes as they grow.

2. Dissect a mature flower, and compare it with Figure 16.3.

3. Pick a flower every day or two from the moment they begin to shrivel (i.e. after fertilization). Cut out the ovaries and open them to observe the seeds.

4. Draw the changes which take place between fertilization and the formation of a large bean pod.

E *An investigation of winter twigs*

1. Obtain winter twigs of horse chestnut, ash, oak and sycamore. Compare their appearance with Figure 16.11.

2. Use a sharp knife or scalpel to dissect away the scale leaves on a horse chestnut bud. Take off one leaf at a time and lay them in a row. Note how their shape changes towards the middle of the bud. How many young foliage leaves are there at the centre of the bud?

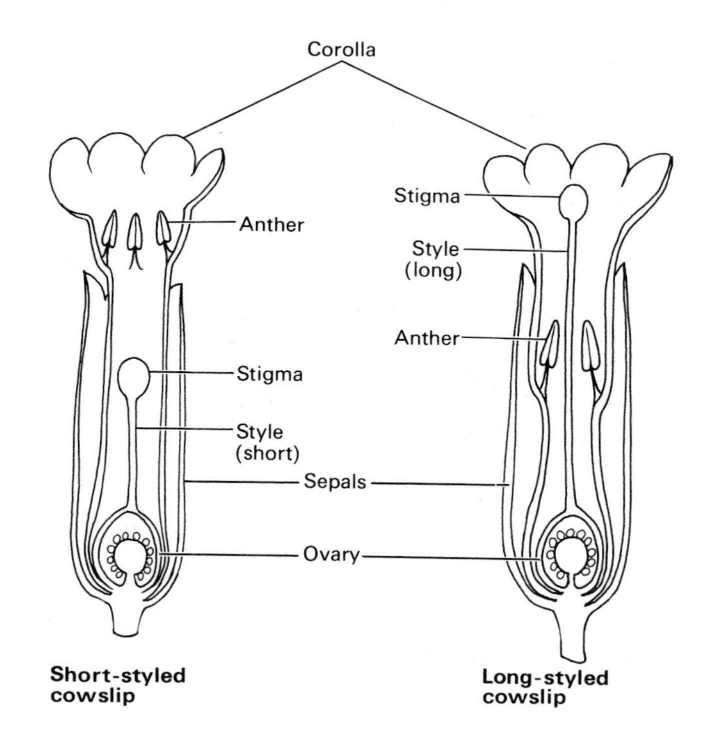

Fig. 16.19 Structure of cowslip flowers

See comprehension test 3

Comprehension test

1. Why do botanists class the following as fruit: vegetable marrows, maize grains, cucumbers, apple 'seeds'?

2. Which of the illustrations in Figure 16.9 show examples of fruit dispersal, and which show seed dispersal?

3. Cowslips have two types of flower, and these occur on separate plants (Fig. 16.19).

a) When a bee visits a short-styled flower it collects pollen mostly on its abdomen. Why is this bee more likely to cause pollination if its next visit is to a long-styled flower, and less likely if it visits another short-styled flower?

b) Why will a bee collect pollen on a different part of its body when it visits a long-styled flower? On what part of its body will this pollen collect, and why will this increase the chances of pollinating flowers with short styles rather than those with long styles?

c) How do these arrangements increase the chances of cross-pollination in cowslips?

Summary and factual recall test

The male sex organs of a moss plant are called (1). These produce (2) which swim to the flask-shaped (3) organs, called (4), and (5) the ovum which each of these contains. The ovum then develops into a (6) which produces (7). These are distributed by (8), and under (9) conditions germinate producing new moss plants as follows (10). This sequence is called an alternation of generations because (11).

The stalk or (12) of a flower ends in a region called the (13) to which the flower parts are attached. These parts are arranged in rings called (14). Carpels are the (15) reproductive organs, and are known collectively as the (16). Each carpel has a hollow base called the (17). This contains one or more (18) each of which has a cell called the (19) at its centre with several nuclei, one of which is the (20) nucleus, or female (21). After fertilization, the (22) becomes a seed while the (23) becomes the fruit surrounding the seeds. Stamens are the (24) reproductive organs, and are known collectively as the (25). Each consists of a stalk called the (26) bearing an (27) with four (28). These produce (29) grains which contain the male (30). Petals make up the (31) of a flower, and can attract insects in three ways (32). As a result the insects (33) the flower. Sepals make up the (34) of a flower. Their function is (35).

A buttercup flower is said to be regularly shaped because (36), and (37) symmetrical because (38). The white dead nettle is said to be irregularly shaped because (39), and (40) symmetrical because (41).

Self-pollination is the (42–define the term); whereas cross-pollination is the (43). Many flowers have characteristics which favour (44)-pollination, which ensures that (45), and thereby increases the chance of (46) change. Three examples of these characteristics are (47).

Wind-pollinated flowers usually have no (48), and never have (49). Most of them have some or all of the following features (50–list five). These features (51) the chances of (52) reaching their (53). Three examples of wind-pollinated flowers are (54). Insect-pollinated flowers usually have the following characteristics (55–list at least three). These features (56) insects and ensure that (57). Three examples of insect-pollinated flowers are (58).

After pollination pollen grains produce a (59) outgrowth called a (60). This grows towards the (61) which it enters through a hole called the (62). When it reaches the (63) sac its tip (64), releasing the (65) gamete which fuses with the (66) gamete.

A seed consists of a small root or (67), a shoot or (68), and one or two (69)-leaves called (70). There is also a store of food either in the (71) or in a mass of cells called the (72). The whole seed has a tough outer coat called a (73).

Seed dispersal is important for at least two reasons (74). The three main ways in which seeds are dispersed are (75).

The vegetative phase of a plant life cycle is the one in which (76). This is followed by the (77) phase in which (78).

Ephemerals are plants which (79). Two examples are (80). Annuals are plants which (81) in one growing season. Two examples are (82). Biennials have a (83) year life cycle. Two examples are (84). Perennials can live and reproduce year after year because (85).

Vegetative reproduction is a form of (86) reproduction. Examples of vegetative reproduction are (87–name seven). Some of these are called perennating organs because (88). Examples are (89–name four).

17

Reproduction in animals

17.1 Hydra

Hydra is a member of a group of animals called the Coelenterates. In common with many other members of this group, *Hydra* reproduces asexually by **budding** but it can also reproduce sexually. Two stages of bud development are illustrated in Figure 17.1, which also shows what the organs concerned with sexual reproduction look like.

Budding

A small area of the body wall grows rapidly by cell division to form an outgrowth, the bud, on the body surface. This bud eventually develops tentacles, and then a circular constriction at its base detaches it from the parent.

After leaving the parent *Hydra*, buds usually float to the surface of the water where they remain for a day or two carried along by currents. In this way the animals are distributed throughout their environment and over-crowding is avoided.

In some other Coelenterates, such as certain corals, buds do not completely separate from the parent. They live alongside the parent, attached to it by a strand of living cells, and may eventually form their own buds. In this way enormous colonies of interconnected animals can be built up, often containing hundreds of thousands of individuals, all from one parent. A colony of this sort is an example of a clone (section 15.1). Colonies of this size can only form where there is a plentiful supply of food.

Sexual reproduction

At the onset of winter *Hydra* usually begins to reproduce sexually. Most species are hermaphrodite, and produce male sex organs (testes) towards the mouth end of the body and female sex organs (ovaries) towards the base (Fig. 17.1). A very simple mechanism makes it impossible for a sperm to fertilize an ovum on the same animal: the testes develop and release their sperms long before the ovaries of the same *Hydra* are ripe. The sequence of events is as follows (Fig. 17.2).

The testes swell and then burst, releasing an enormous number of sperms into the water where they live for only a few hours. One or two days later ovaries on the same animal each produce a single, large ovum filled with stored food. These ova remain attached to the parent where they are fertilized by sperms released from another *Hydra*. Each fertilized ovum (zygote) divides many times until it forms a ball of cells. This ball stops growing, forms a hard protective wall around itself, and drops from the parent to the bottom of the pond or stream. Eventually the parents are killed off by the cold weather and lack of food, but the balls of cells survive within their protective covering until they are stimulated into cell division and growth by the warmth of spring. The protective wall then bursts open and a new *Hydra* emerges.

Fig. 17.1 *Hydra* with buds and sex organs. (Note that buds and sex organs are not normally present at the same time.)

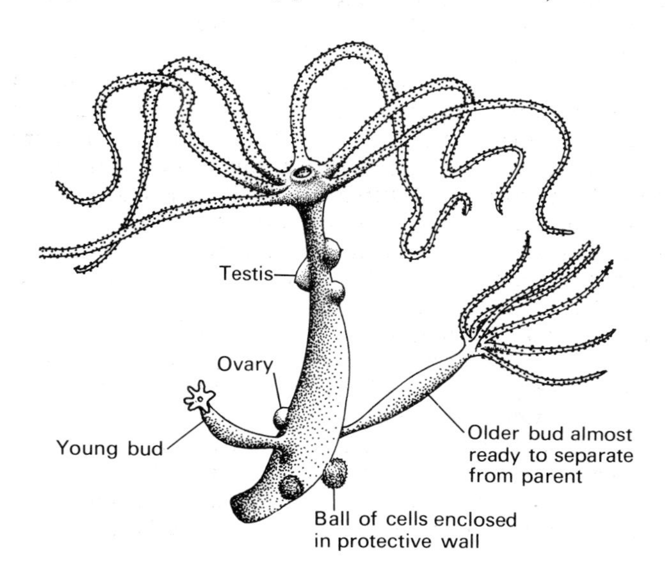

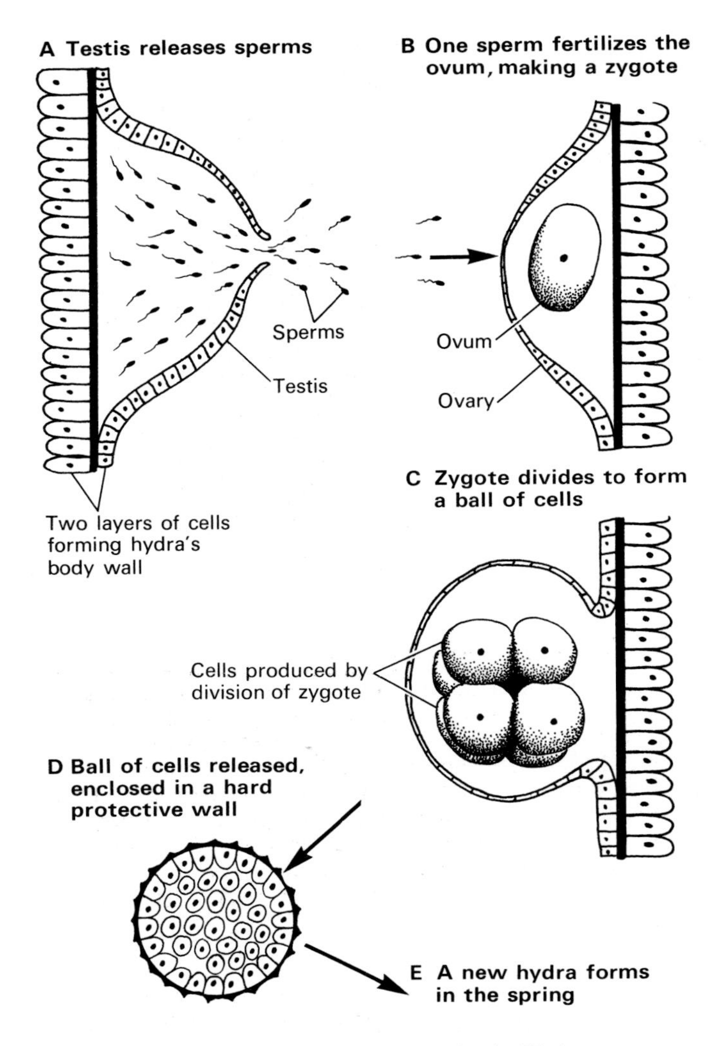

Fig. 17.2 Diagram of sexual reproduction in *Hydra*

17.2 Insects

Insects reproduce sexually, mainly by internal fertilization followed by the production of yolk-filled eggs. The reproductive organs are not described in this book. The following description is confined mainly to the life cycles of various insects. The stages of growth by which a fertilized egg becomes an adult insect, or **imago**, follow one of three different patterns.

First, in the most primitive group of insects such as the tiny wingless silverfish and bristle-tails, the young which emerge from the eggs are miniature replicas of their parents. They grow to adult size, periodically shedding the tough cuticle which forms their external skeletons as these become too small for them (section 12.4).

Second, in the groups of winged insects which include dragon-flies, locusts, and cockroaches, the newly hatched young are called **nymphs**. A nymph resembles its parents in most ways, except that it is wingless, sexually immature and, of course, smaller. Nymphs eat voraciously and shed their cuticles several times during their development into mature winged adults. This pattern of development, in which there are relatively few changes between young and adult stages, is called **incomplete metamorphosis**. Figure 17.3 illustrates this sequence of events in the cockroach.

Third, in the most highly evolved insects such as butterflies, moths, flies, and bees, the young which emerge from the egg differ from the parents so much they do not even look like insects. The newly hatched creatures are called **larvae**, and their development into an adult is called **complete metamorphosis**, because they change completely in shape, form, and sometimes in their way of life. For example, a legless and sightless bluebottle maggot eating its way through rotten meat later changes into a flying insect with legs and eyes; and a fat, elongated caterpillar chewing leaves with massive jaws later changes into a butterfly which sips nectar through a slender tongue-like proboscis.

Larvae, like nymphs, spend almost their entire existence eating, and quickly outgrow several cuticles, which they shed by ecdysis. They continue to grow until they are several thousand times larger than when they emerged from their eggs. Yet this growth does not involve any cell division; it is achieved entirely by enlargement of the larva's body cells. These creatures literally inflate themselves with digested food. Eventually this constant feeding stops and the larva becomes a **pupa**, such as the cocoon of a moth, or the chrysalis of a butterfly.

A larva does not exactly 'turn into' a pupa; it is replaced by a pupa. The bloated larval cells begin to die, and slowly give up their stored food to a number of cells which have remained unchanged in the larva's body since it emerged from the egg. These cells now begin to divide rapidly and start to form the adult insect's body organs. Pupae seem to be dead, or resting, when in fact their cells are engaged in the complex activity of constructing the adult insect. This is the time of metamorphosis.

Eventually the pupa splits open and the soft, helpless adult struggles out. At first its wings are crumpled and folded, but blood is quickly pumped into them making them expand to full size. This takes about thirty minutes, and it takes at least another hour before the wings are hard and dry enough to use. Figure 17.4 illustrates this process in the butterfly.

The most complex insect life cycles are found in groups such as bees, termites, and ants. These are

Egg case (ootheca)

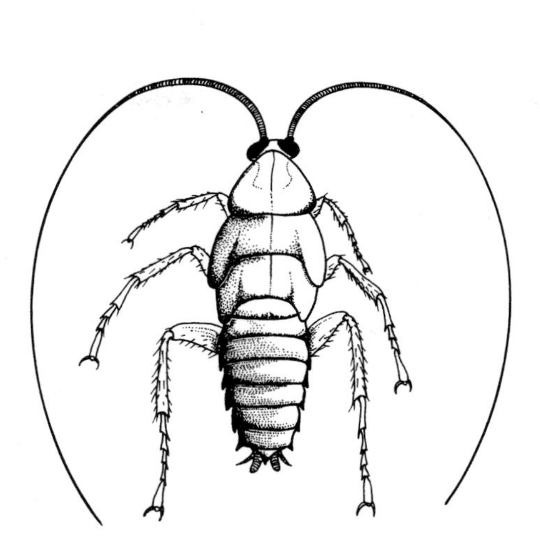

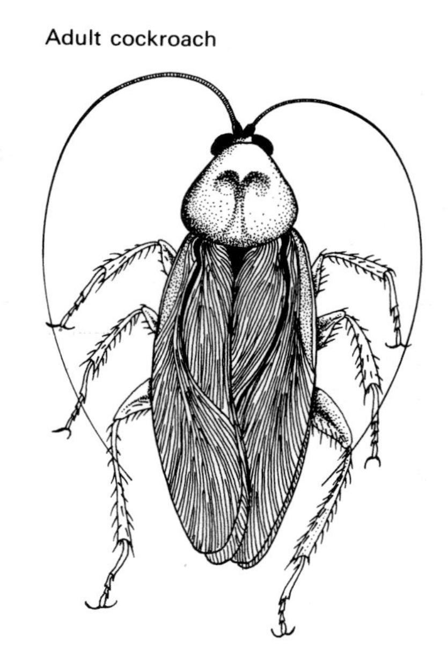

Fig. 17.3 Life cycle of the cockroach.

This is an example of incomplete metamorphosis. After mating females produce egg cases about 10 mm long, containing sixteen eggs. Each case is carried by the female for a few days in a pouch at the tip of her abdomen and then concealed in a warm dry place near food. Eggs hatch about a month later. The nymphs, like adults, eat almost anything organic (e.g. paper, leather, seeds, scraps of discarded human food). Nymphs take about three years to grow into adults with wings and reproductive organs. During this time they undergo ecdysis six times.

called **social insects** because they live in highly organized societies in which each member serves the group as a whole, rather than its own individual needs.

Honeybees

Honeybees are chosen here as an example of social insects, partly because it is comparatively easy to study their life cycle, and partly because of their usefulness to man in pollinating the flowers of many crop plants and producing honey.

Where food is plentiful and the weather favourable, a hive may contain a population of 50 000 bees. In every colony there is one **queen bee**, which is the only egg-laying female. In addition there are two or three hundred male bees, or **drones**. The remainder are sterile female bees, or **workers**.

Workers The body of a worker bee is equipped to perform many tasks. The worker's mouthparts include a long tongue, or proboscis, which is used to suck up liquids such as nectar from flowers; and mandibles which are used to manipulate wax during comb construction (Fig. 17.5A). The front pair of legs carries a row of hairs which clean the eyes, a notch lined with hairs which clean antennae, and another row of hairs, the **pollen comb**, which cleans pollen off the head and mouthparts and passes it to the hind legs (Fig. 17.5B). The hind legs are equipped for collecting and carrying the pollen which adheres to a worker's body as it emerges from a flower. The inner surface of the metatarsus of each leg is covered with rows of hairs which form another pollen comb (Fig. 17.5C). The outer surface of the tibia has long stiff hairs which curve over a concave area forming a **pollen basket** (Fig. 17.5D). The joint between the metatarsus and the tibia opens like pincers forming a pollen press which acts in the following way. The pollen combs on the hind legs gather pollen from the body and middle legs. The pollen so collected is scraped off by rakes made of stiff hairs situated on the tibia edge of each pollen press (Fig. 17.5C and D). The press is then closed, forcing pollen into the pollen basket along with a drop of nectar to hold it in place.

Inside the worker's body there is a region of intestine called the **crop**, or **honey stomach**. Nectar is stored here during flights between flowers and the hive. Workers have a sting which consists of two lancets armed with barbs (like fish hooks), and a sac of venom located at the tip of the abdomen.

During their forty days of life worker bees are constantly active and they perform different tasks as they get older. The first five days of a worker's life are spent cleaning out empty cells in the comb, making them ready for the queen to lay eggs, and for older workers to deposit stores of honey and pollen. The

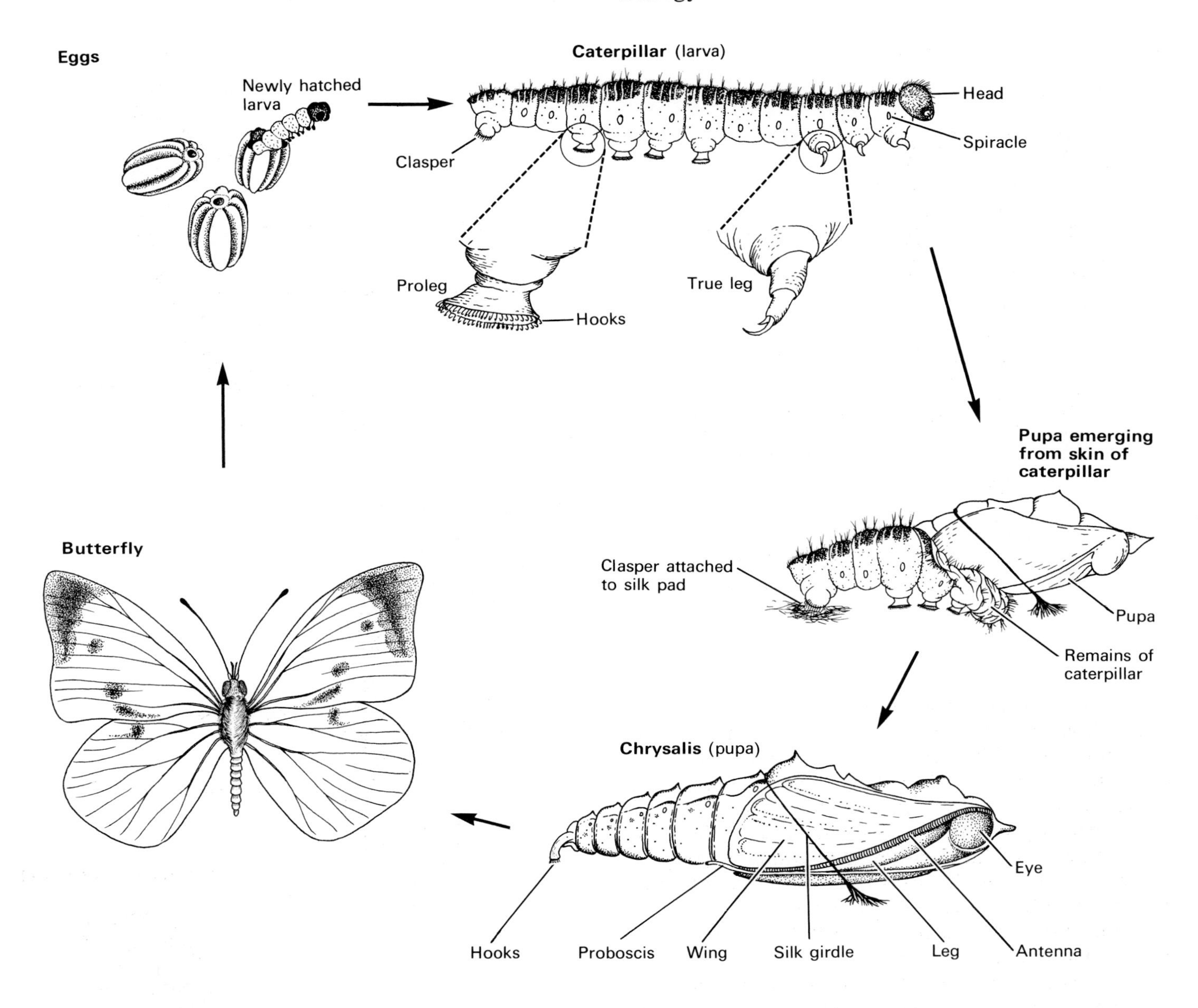

Fig. 17.4 Life cycle of the cabbage white butterfly.

This is an example of complete metamorphosis. Females lay eggs 2 mm long in batches of about 100 on the underside of leaves such as cabbages and nasturtium. Larvae eat the egg shell after hatching, and then eat the leaves on which they were laid. They feed for about a month, shedding their cuticles six times, and then crawl off the food plant into some dry sheltered place such as a tree trunk. Here, the larva spins a pad of silk to which it attaches its claspers, and then spins a girdle of silk around the middle of its body and attaches it to the tree on either side. The caterpillar cuticle splits and is pushed off backwards through the silk girdle. The pupa's last segment has hooks which grip the pad of silk formerly held by the claspers. About 14 days later the pupa's cuticle splits and the adult emerges.

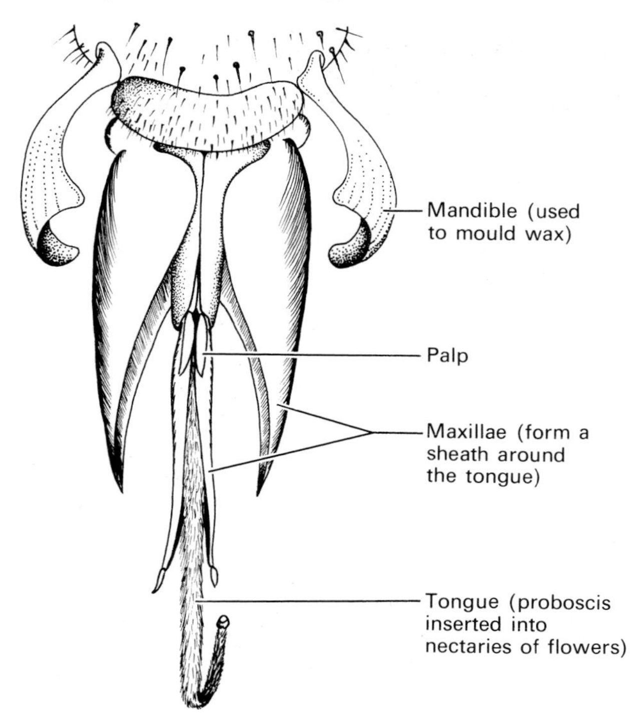

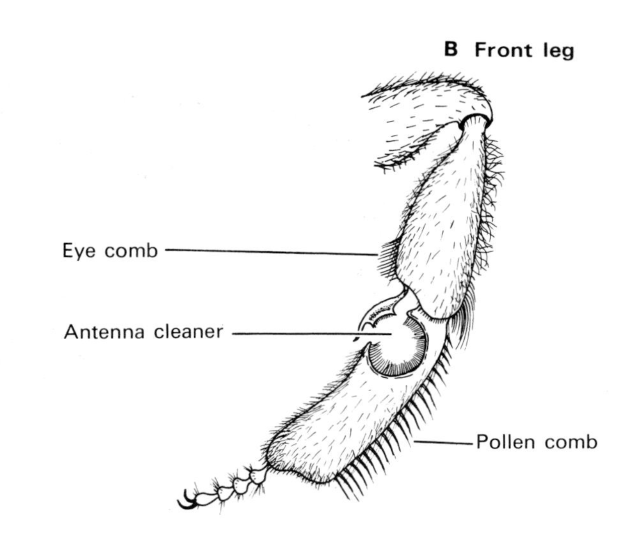

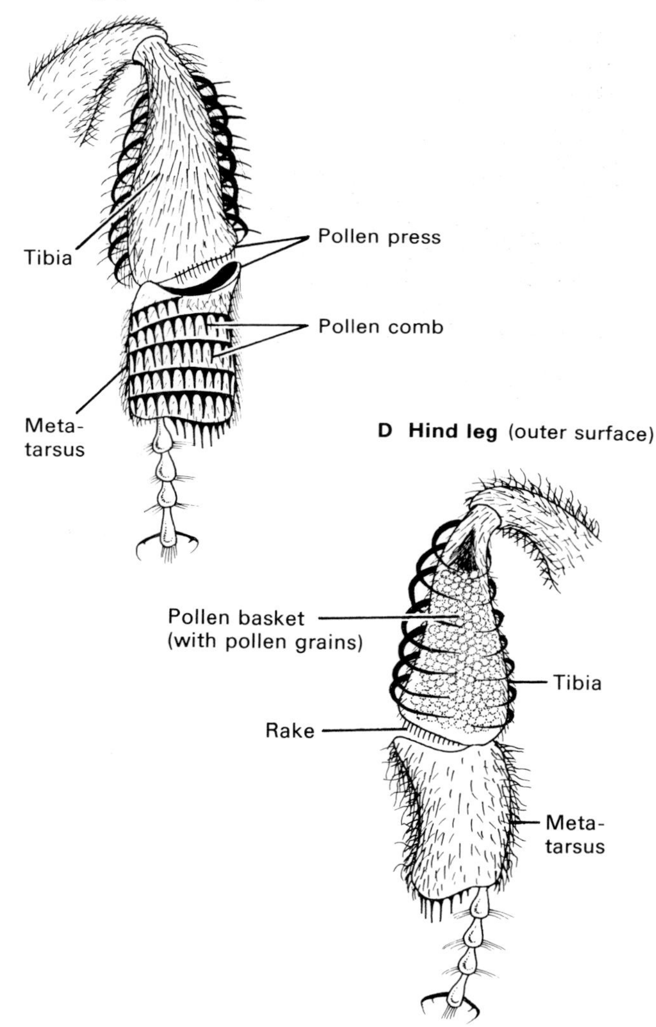

Fig. 17.5 Mouth parts and legs of a worker bee

young workers also drag the corpses of dead bees out of the hive before the bodies can decompose and cause infection. The decomposing remains of larger animals, such as mice which have been stung to death inside the hive while trying to steal honey, pose a more serious problem because they are too large to carry away. The older workers embalm these corpses by covering them with a substance which their bodies manufacture from tree resin. After this, the corpses soon become mummified in the hot dry air of the hive, and are then quite harmless.

When workers are about six days old they develop a **nurse gland** in their mouths which secretes a rich food called **royal jelly**. For about the next seven days these workers wander over the combs using this gland to feed larval bees. The length of time a larva is fed on this substance determines whether it will become a worker or a queen. Larvae which become drones and workers are fed royal jelly for $2\frac{1}{2}$ to 3 days. Larvae which become queens receive it for $4\frac{1}{2}$ to 5 days.

By the time a worker is thirteen days old its nurse glands have disappeared and it starts producing wax from glands on its abdomen. The worker uses this wax to help in constructing combs. Combs consist of hollow chambers called cells, arranged back to back and built so that they slope downwards from their entrances (Fig. 17.6). This helps prevent honey and

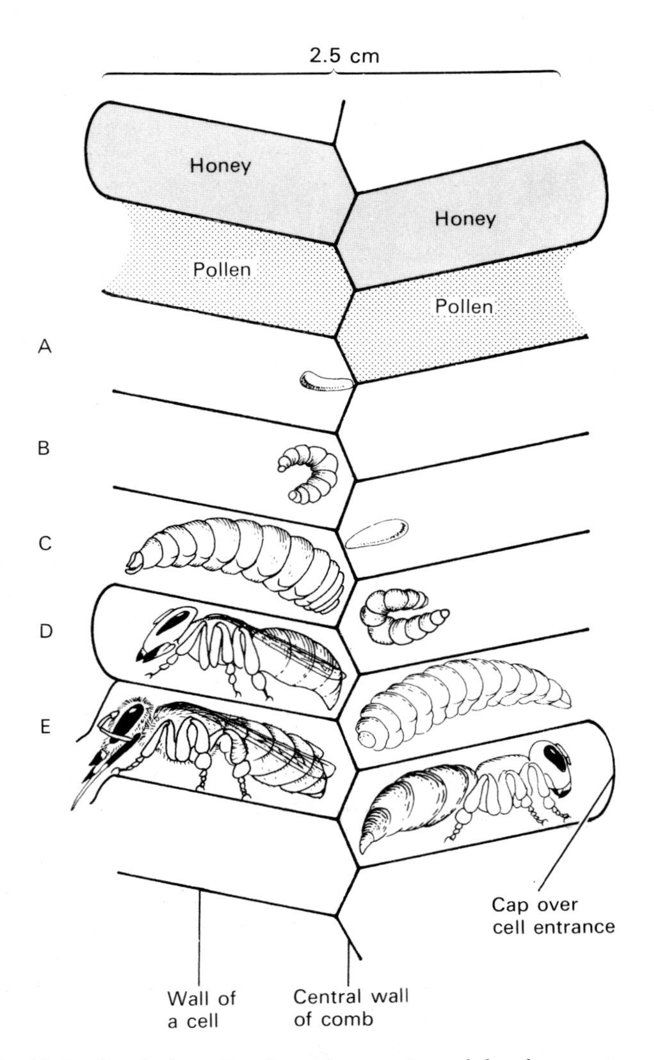

Fig. 17.6 Vertical section through a comb, and development of a worker bee.

A The queen lays a fertilized egg at the bottom of a cell. **A** legless larva **B** hatches after three days. It is fed on royal jelly for three days and a mixture of honey and pollen for two days, growing rapidly all the time **C**. A pupa **D** forms eight days after hatching. The adult **E** bites its way from the cell one week later.

larvae spilling out. In a man-made hive there are wooden frames in which the bees construct about 6000 cells. The workers take tiny flakes of wax from their wax glands, chew them with their mandibles until they are soft, and then use the wax to build up the hexagonal walls of a cell.

Between the sixteenth and twentieth days of their lives, workers guard the entrance to the hive. Guards either fly back and forth in front of the landing board or stand side-by-side like rows of soldiers. They inspect every arriving bee to discover if it carries the colony's particular scent. If not, the intruder is either driven away or stung to death. Guards are essential because bees from other colonies often come in large numbers to steal honey. Wasps and hornets are also honey thieves.

The last twenty days of a worker's life are spent visiting flowers for nectar and pollen.

Inside the hive, pollen is placed in unsealed cells and covered with a layer of honey. Nectar is treated differently. It is first pumped back and forth between the tip of the proboscis and the nectar sac until it has lost some water by evaporation. Next, it is stored in a cell with an enzyme from a gland in the worker's head, which changes nectar into honey. Pollen and honey are stored in this way as a source of food.

The queen bee A queen differs from workers and drones in practically every way. She is reared in a larger cell than other bees; she receives royal jelly throughout her larval life; her adult body is longer than that of other bees; and she lives for up to five years.

A honeycomb with worker bees. The open cells contain larvae; the closed cells contain honey or pupae; and the enlarged cells which protrude from the surface of the comb contain developing queens

Under circumstances explained later, a young queen bee takes off on a mating flight during which she is pursued by drones from her colony. Several drones mate with her (usually six), depositing sperms in tiny sacs in her abdomen called **spermathecae**. She is now ready to begin laying eggs.

Every day during the summer months a queen lays up to two thousand eggs, and in her lifetime she lays at least half a million. Most of these eggs are fertilized by sperms stored in her spermathecae, but some are laid unfertilized. The fertilized eggs develop into workers or queens depending upon how they are fed as larvae. Unfertilized eggs develop into drones.

The queen is always surrounded by workers who lick and groom her. These workers absorb a chemical known as **queen substance** from glands on her head, and this substance is passed on to all other members of the hive, probably as they lick and groom each other. In some mysterious way this chemical holds the bees together as a colony. If the queen is killed the queen substance disappears from the hive and this is the signal for workers to raise new queens from the larvae.

The queen substance also runs out when the colony is so large that there is not enough of the chemical to spread from bee to bee throughout the entire hive. In these circumstances an event called **swarming** occurs. This is the departure of a queen and a number of workers from a hive in order to establish a new colony.

A swarm of bees hanging from the branch of a tree. The majority of the swarm remain together like this while a few workers search the surrounding countryside for a place to build new combs. When a suitable place is found the workers return and guide the swarm to it. When domesticated bees swarm, the bee keeper waits until they settle like this before collecting them in a basket and taking them to an empty hive

Establishing a new colony A few weeks prior to swarming, some workers construct extra large cells and feed their occupants continuously on royal jelly. When these larvae change into pupae their cells are sealed off (capped). As soon as this happens the old queen leaves the hive taking most of the bees with her. She and her swarm gather on a nearby tree, where a bee keeper can catch them in a basket and take them to a new hive. Here, workers build new combs in the frames provided. When this is done the queen lays eggs and the workers search for nectar and pollen.

Back in the old hive the virgin queens begin breaking out of their cells. The first queen to emerge does one of two things. She may fly off with another swarm and establish a new colony, or she may kill the other queen larvae in their cells, take off on her mating flight, and return to take over the hive and begin her egg-laying career.

17.3 **Fish**

The pattern of reproductive behaviour in fish varies enormously from species to species, and there is room to describe only one fish here. The stickleback has been chosen because it can easily be caught and studied in the classroom.

Life cycle of the stickleback

Three-spined and ten-spined sticklebacks are common in small streams and rivers all over Britain, and are of particular interest because their life cycle involves nest building, courtship, and parental care of the young (Fig. 17.7).

In the early spring male sticklebacks take over and defend a small area of stream bed in shallow water which will become their nest site. Other fish, especially

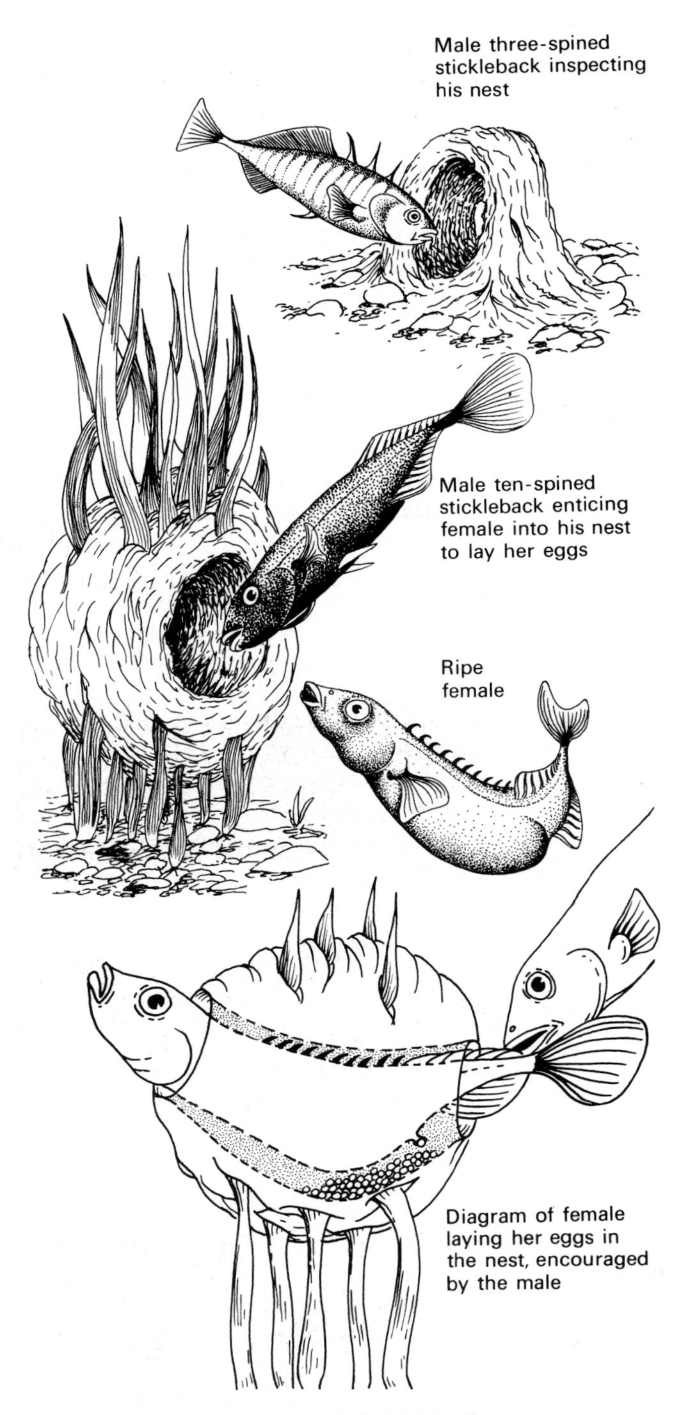

Fig. 17.7 Breeding behaviour of sticklebacks

males of their own species, are driven away from this territory. Nest building now begins.

A three-spined male stickleback clears an area of stream bed and digs a shallow hole in it by vigorously fanning it with his tail fin, and scooping up mud in his mouth and spitting it out some distance away. He then builds a roof over this hollow area with strips of vegetation stuck together with a glue secreted from part of his kidneys. The result is a small tunnel open at both ends. A ten-spined male builds a spherical nest above the stream bottom by glueing together vegetation, and then forcing his way through it to make a tunnel.

Female sticklebacks whose bodies are swollen with ripe eggs are approached and courted whenever they enter a male's territory. Other females are chased away. The male performs a kind of dance around the ripe female by bobbing up and down, and by swimming around her with his head held downwards. Finally he rushes off in the direction of his nest. Once at the nest the male remains still with his nose pointed at the entrance. If the female is ready to lay her eggs she pushes her way into the tunnel until her nose protrudes from one end and her tail from the other. The male now pushes his nose against her tail, which appears to be the signal for her to lay the eggs. This is over within a few seconds and the female leaves the nest to be replaced immediately by the male who sheds sperms over the eggs and so fertilizes them.

The female plays no further part in rearing the young, and is driven from the nest territory. The male remains on guard over the eggs, and later over the newly hatched young, which are called **fry**, fiercely defending them against all intruders. He also spends long periods 'fanning' the nest with his tail fin. This creates a current of water which aerates the eggs or fry.

The eggs hatch about ten days after fertilization, and the fry are kept in or near the nest by the ever-vigilant male parent for at least another three weeks. After this time the male relaxes his guard and the fry disperse.

17.4 **Amphibia**

Amphibia are considered 'higher', or more advanced than fish because their bodies have several features which enable them to live out of water for long periods in conditions that fish cannot tolerate. Nevertheless, amphibia are not fully adapted land animals. Most amphibia fertilize externally and this must take place in water, or in very moist conditions, otherwise the sperms and eggs dry up and die. Amphibia also go through a larval stage when they are called tadpoles. These live in water and are similar in certain ways to fish. Eventually the fish-like tadpole undergoes metamorphosis into an adult amphibian. The frog is chosen as an example of an amphibian because it is easily collected and studied (Fig. 17.8).

One day after hatching

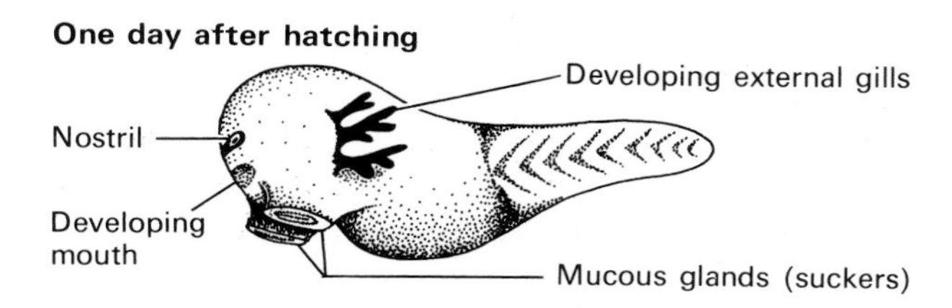

Three weeks old

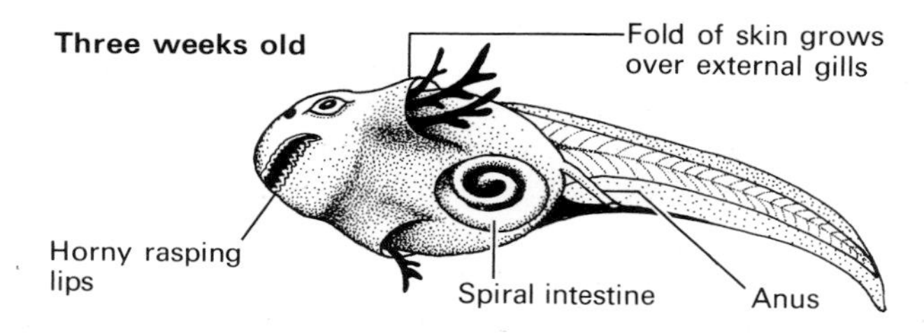

One month old

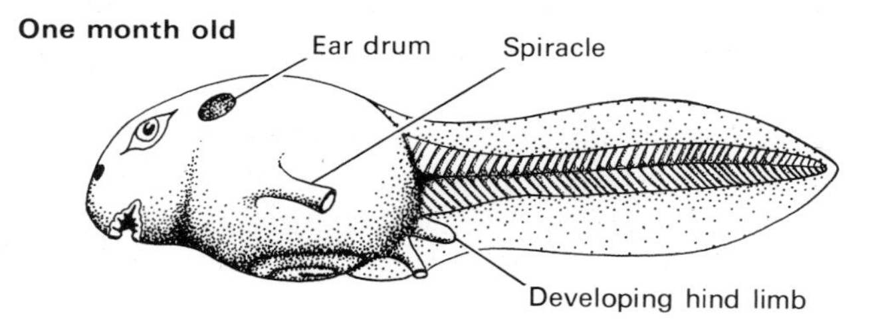

Two months old

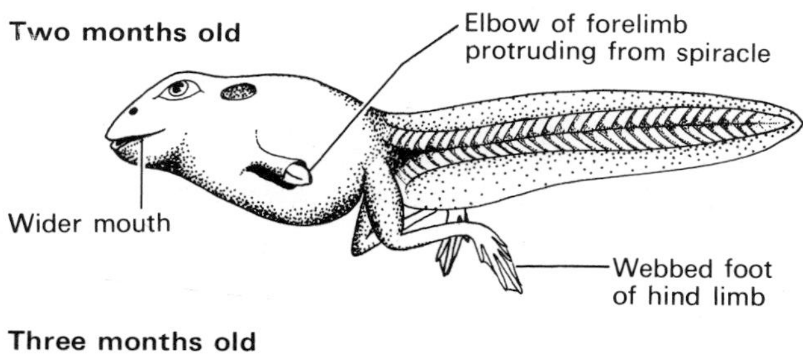

Three months old

Fig. 17.8 Development and metamorphosis of a frog

Life cycle of the frog

Frogs spend the cold months hibernating. They spend this period buried in mud or dense vegetation close to water. They do not move or feed but lie dormant with their mouths, eyes, and nostrils shut, absorbing a little oxygen through their moist skins. At the beginning of spring frogs wake and gather in the shallow water of ponds and slow moving streams. At this time the males can be heard croaking loudly.

Mating By this time the females are swollen with eggs. The female is mounted by a male who climbs on her back and holds on with the thick, horny pads on his thumbs, which grip the female behind her forelimbs. Males remain in this position for two or three days prior to mating.

The female lays about a thousand eggs through an

Frogs mating

Frog spawn

opening between her back legs called the **cloaca**. While this is happening the mounted male sheds sperms from his cloaca directly on to the eggs, fertilizing them as they emerge from the female. Fertilization must take place immediately because the eggs are covered in a layer of jelly, the **albumen**, which swells when it comes in contact with water and forms a thick protective layer.

Development and metamorphosis After a sperm nucleus has fused with an egg nucleus, the resulting zygote begins rapid division by mitosis and develops into an embryo. The embryo grows using food called **yolk** which is contained in the original egg cell.

About ten days after fertilization the embryo has developed into a tadpole which struggles out of the albumen but remains attached to its outer surface by sticky mucous glands below its mouth. At this stage its mouth has not opened, and it is still living on the remaining yolk now located in the cells of its intestine. Until its gills develop it breathes through its skin.

Two or three days later the tadpole swims to nearby water weeds, attaches itself to them by its mucous glands, and feeds by rasping and scraping microscopic algae from the leaves with the horny lips of its newly opened mouth. It now has three pairs of branched external gills through which blood circulates and absorbs oxygen from the water.

Over the next few weeks the mucous glands disappear, and the tadpole becomes increasingly active among the water plants in search of algae. It swims with its tail, like a fish.

About three weeks after hatching, several important changes occur in the tadpole's breathing mechanism. The external gills start to shrivel and are absorbed into the body. Simultaneously, internal gills develop which are more like those of a fish, that is, they are situated in gill slits which penetrate the sides of the body, opening into the mouth cavity. While all this is happening a fold of skin called an **operculum** grows backwards from the front of the head, covering the new internal gills, and fusing with skin near the base of the tail. This operculum has a hole called a **spiracle** on the left side of the tadpole. When the tadpole breathes it draws water into its mouth, passes it through the gill slits and over the gills, then squirts it out of the spiracle. By this time the animal has developed a long coiled intestine in which it digests its vegetable diet.

About two months after hatching a pair of hind legs develop but are not yet used for locomotion. The front legs also develop, but are inside the operculum and can be seen as bulges at the sides of the body. A pair of lungs develop, and the tadpole frequently comes to the surface to inflate them with air. In addition, the coiled 'vegetarian' intestine begins to change into a shorter one, more suited to a carnivorous diet. The tadpole loses its rasping mouth and develops jaws with which it eats small water animals.

Metamorphosis into an adult amphibian begins about three months after hatching, and takes about four weeks. The following changes progress more or less simultaneously. The tail shortens; in fact it is absorbed into the body as the sole source of nourishment during metamorphosis. The forelimbs burst out of the operculum, the left one appearing first by pushing out through the spiracle. All four legs are now used in locomotion. The hind legs are used for swimming in water, and for jumping when on land. The front legs are used to support the weight of the body, especially when landing from a jump. The animal is now a miniature frog, about 2 cm long. It feeds by using its long sticky tongue to catch gnats, midges, and grubs, until the autumn when it goes into hibernation. Four years later, if the frog manages to avoid snakes, birds, and otters, it is old enough to begin breeding.

17.5 Birds

Unlike frogs, birds do not have to return to water to breed. Birds reproduce by means of internal fertilization, and eggs with a hard protective shell. But this does not mean that bird reproduction is independent of water. Sperms from a male bird still swim through 'water' to reach the eggs, but it is not pond water, it is the liquid inside the female bird's reproductive system. Similarly, bird embryos still float in an environment of albumen and water, but this liquid is contained, like a small private pond, within the protective walls of a shell, and so is safe on dry land. In short, whereas frogs go to water to breed, birds breed in the water contained in their own bodies.

Other characteristic features of bird reproduction include their varied and often colourful plumage, their elaborate courtship displays which lead to copulation, and the extended care often given by both parents to the fledgelings until they can fend for themselves.

Courtship and nest building

Even in the tropics the breeding habits of birds are usually seasonal. When they are not breeding, bird reproductive systems are inactive and only partly developed. In some male birds, for instance, the testes are 1000 times heavier in the breeding season than during the rest of the year. In Britain the breeding season begins in spring, and as the period of daylight increases the sex organs mature and bird behaviour, song, and sometimes plumage begin to change. This is

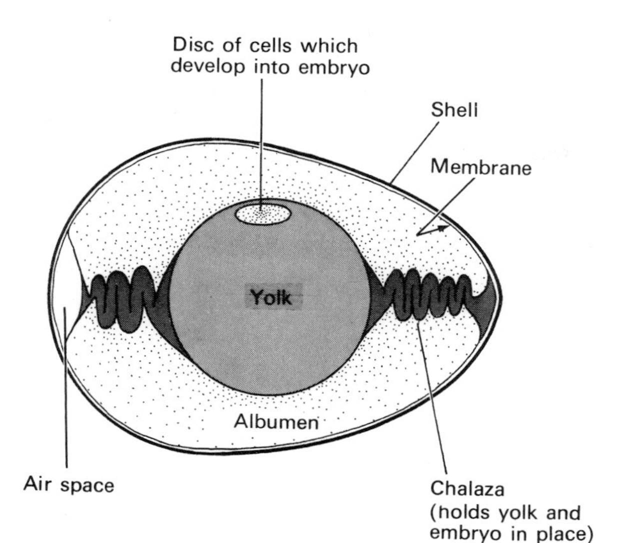

Fig. 17.9 Vertical section through a bird's egg

followed by a pattern of courtship behaviour which includes breeding songs, body movements, and displays of feathers, all of which have four main functions.

First, each species has its own characteristic courtship display, which is usually performed by the male. This behaviour enables birds to recognize members of the opposite sex of the same species, with which they can pair off either briefly while copulation occurs, or for the whole season. Second, male displays usually include threatening behaviour, which drives away other males of the same species. Third, courtship initiates nest building, which may amount to anything from scratching a hollow in the ground, to weaving a complex structure out of vegetation and feathers. One or both parents may carry out this work. Fourth, since the male bird has no front legs and fingers with which to grasp the female, he must have her full co-operation in the act of copulation. Courtship stimulates the female sexually to the point at which she is a willing partner.

Copulation

The reproductive systems in both male and female birds lead to an opening between the legs called a cloaca. During copulation the male presses his cloaca directly against the female's, and sperms are ejaculated from his body directly into a long tube called the oviduct inside the female's body. Sperms swim up the oviduct until they come to the region where newly formed eggs, without shells, are situated. Here, fertilization occurs. Fertilized eggs then pass down the oviduct where they receive albumen and a shell. Finally, they are laid, pointed end first, into the nest.

Fig. 17.10 Chicken egg with ten-day-old embryo

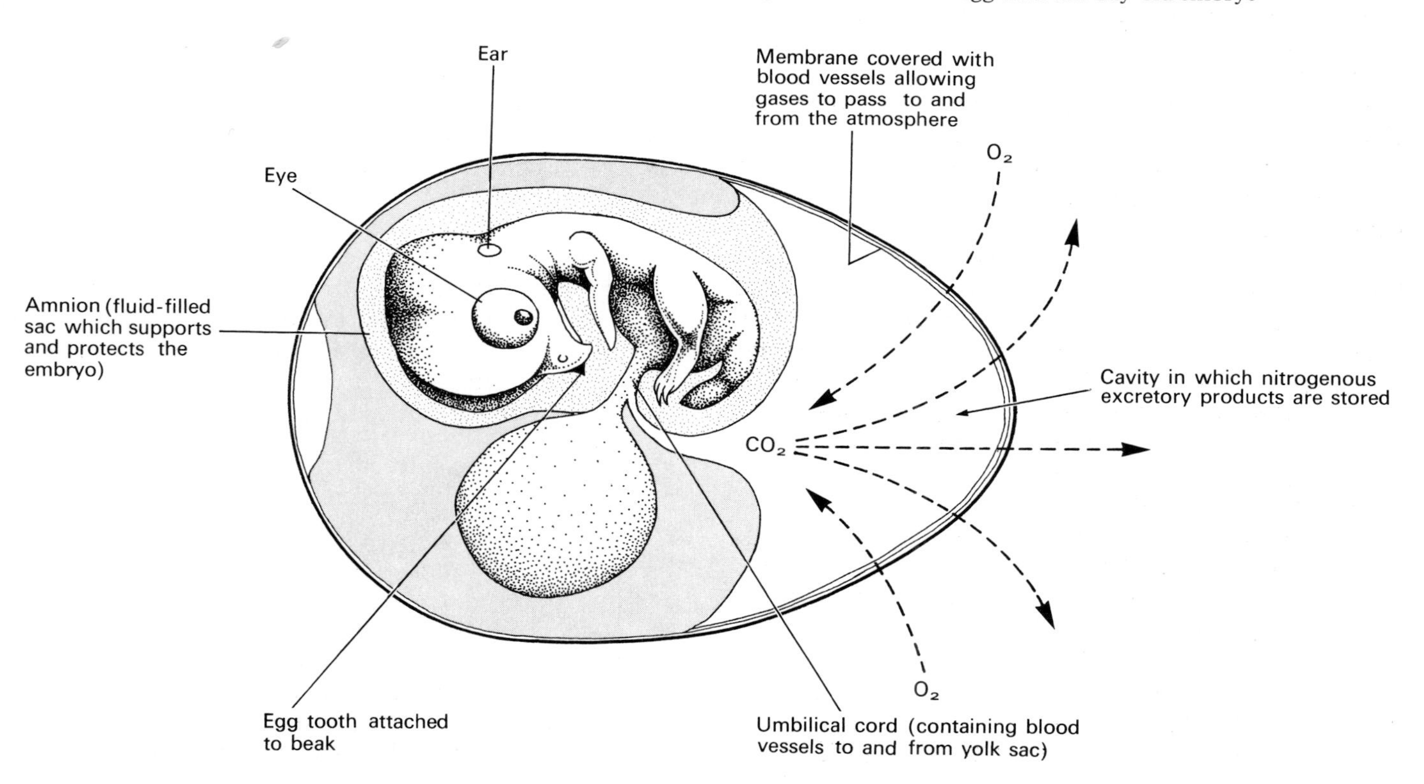

Robin fledgelings 'begging' for food from the female

Incubation and parental care

Courtship behaviour now gives way to an even more complex behaviour pattern concerned with incubating the eggs (keeping them warm) and looking after the young. This behaviour is essential: first, because in the beginning the embryo in the egg is cold-blooded and dies if it is left cool for long periods, and second, because most newly hatched fledgelings are blind and helpless and take some time to grow feathers, learn to fly, and feed themselves.

Usually the female incubates the eggs. She may lose some of her breast feathers which exposes an area of skin, the **brood patch**, which she presses against the eggs, giving them extra warmth. In some species the male takes over incubation for a while.

Depending on the species, incubation may continue for two or three weeks. At hatching time the chick breaks out of the egg shell using an egg tooth–a temporary projection on top of its beak (Fig. 17.10).

In most species the female continues to sit on the nest for long periods after hatching, keeping the nestlings warm at least until their feathers have grown. The young are fed by one or both parents on caterpillars, grubs, worms, and insects. Nestlings of most species react strongly to the approach of a parent, stretching their necks upwards, opening their beaks and exposing the orange colour inside their mouths. This stimulates the parent to give them food.

Parent birds display many types of protective behaviour whenever danger threatens their young. Certain plovers, for instance, walk away from the nest dragging one wing as if they were injured. This distracts predators from the nest area. Frequently, birds with nestlings will attack intruders.

When their feathers have developed, young birds take off on short practice flights around the nest. Until their strength and flying techniques improve they are easy prey for cats, kestrels, and hawks. Only the strongest survive to begin breeding in the following years.

17.6 **Mammals**

Mammals, like birds, are independent of water for reproduction in the sense that they breed in the water contained in their reproductive organs. But in many other ways mammalian reproduction is both different, and more advanced than that of birds.

In addition to internal fertilization mammals also have internal development. This means that the embryo develops inside the female's body, where it lives as if it were a parasite absorbing food and oxygen from its mother's blood. Unlike birds' eggs, which are exposed to cooling and other dangers whenever the female leaves the nest, mammalian eggs have the advantage of developing in the continuous warmth of their mother's body, where they are fed and protected from injury and predators. Mammals have the added advantage that since the embryo develops inside the female she does not have to remain on a nest incubating the eggs. She is free to lead a reasonably normal life until a few hours before the birth of her young.

For some time after birth mammals remain entirely dependent on milk produced by their mother's mammary glands. Milk is a complete and balanced diet, and is used to nourish the young until they are old enough to take solid foods. In the following description the human body is used to illustrate the structure and functions of mammalian reproductive systems.

Female reproductive system

The reproductive system in female mammals consists of ovaries, oviducts, a uterus, and a vagina (Fig. 17.11).

Ovaries The ovaries are oval-shaped structures, about 3 cm long in humans, which are attached to the back wall of the abdomen below the kidneys. The ova, or female gametes, begin to develop inside the ovaries while the female is still a developing embryo, and a newly born baby girl has several hundred thousand partly developed ova in her ovaries. After birth no new ova are produced; in fact the majority disintegrate. The remaining ova gradually complete their development, and between the ages of about eleven and fifteen years, when a girl reaches sexual maturity or

puberty, the ova are released from the ovaries one at a time by a process called **ovulation**. Between puberty and about fifty years of age when ovulation ceases, a woman releases about five hundred ova from her ovaries. At ovulation an ovum is released from an ovary into a tube called an oviduct.

Oviducts The uppermost part of each oviduct is funnel-shaped and is lined with cilia which create a current that draws the released ovum inside. A narrow tube leads from the funnel to a wider, thick-walled tube about 7.5 cm long called the **uterus** or **womb**. It is here in the uterus that a fertilized ovum undergoes its embryonic development. A ring of muscle, the **cervix**, closes the lower end of the uterus where it joins another tube, the **vagina**. The vagina extends for about 10 cm before reaching the exterior at an opening called the **vulva**.

Between puberty and about fifty years of age the female reproductive system passes through a regular monthly sequence of events called the **menstrual cycle**. These events are controlled by hormones produced by the ovaries and the pituitary gland.

The menstrual cycle During one menstrual cycle an ovum is released from an ovary, and the uterus is prepared to receive this ovum should it be fertilized and begin to develop into a baby.

The ovaries of a sexually mature female contain ova at various stages of development (Fig. 17.12). The final stage is a structure called a **Graafian follicle**, which consists of an ovum, and a mass of follicle cells which enclose a large bubble of liquid. A fully developed Graafian follicle can reach 1 cm in diameter and often bulges from the surface of the ovary.

During the last days of its formation the cells of a Graafian follicle produce a hormone (oestrogen) which causes a layer of cells lining the uterus to grow rapidly and develop a dense network of blood vessels. This is the first stage of preparations for the reception of a fertilized ovum.

Ovulation now takes place. Fluid pressure in the Graafian follicle increases to a level which bursts the follicle, shooting the now ripened ovum onto the surface of the ovary. From here it is drawn into an oviduct, and begins its journey to the uterus. At this stage, the remains of the follicle in the ovary collapse

Fig. 17.11 Reproductive system of a human female

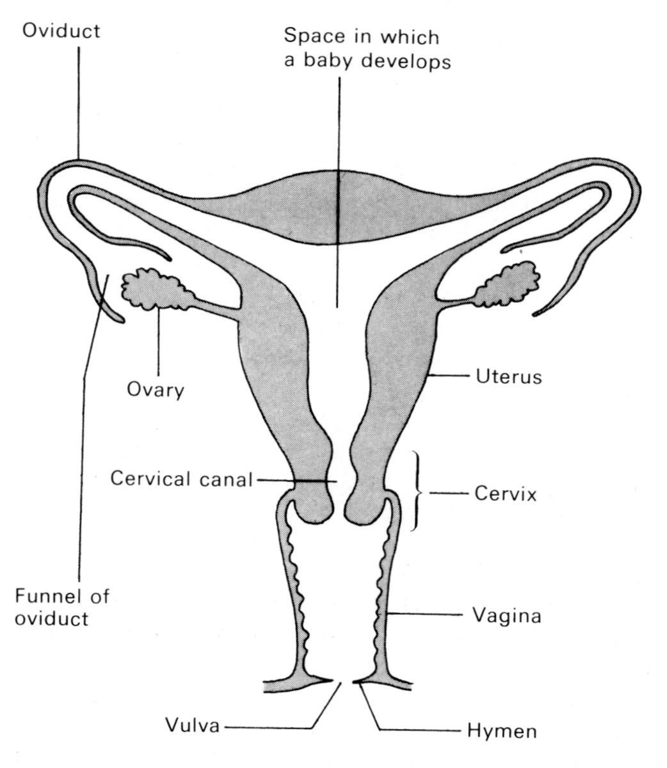

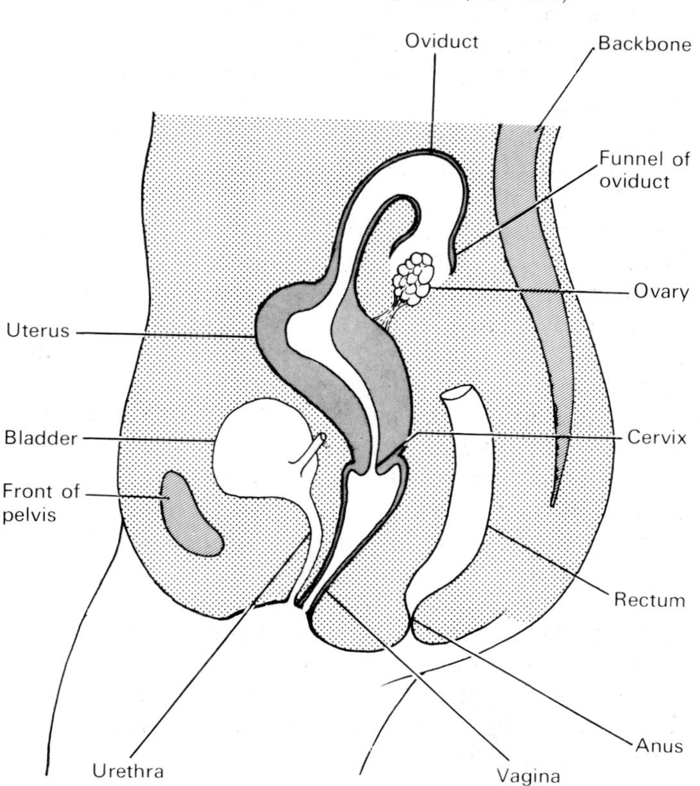

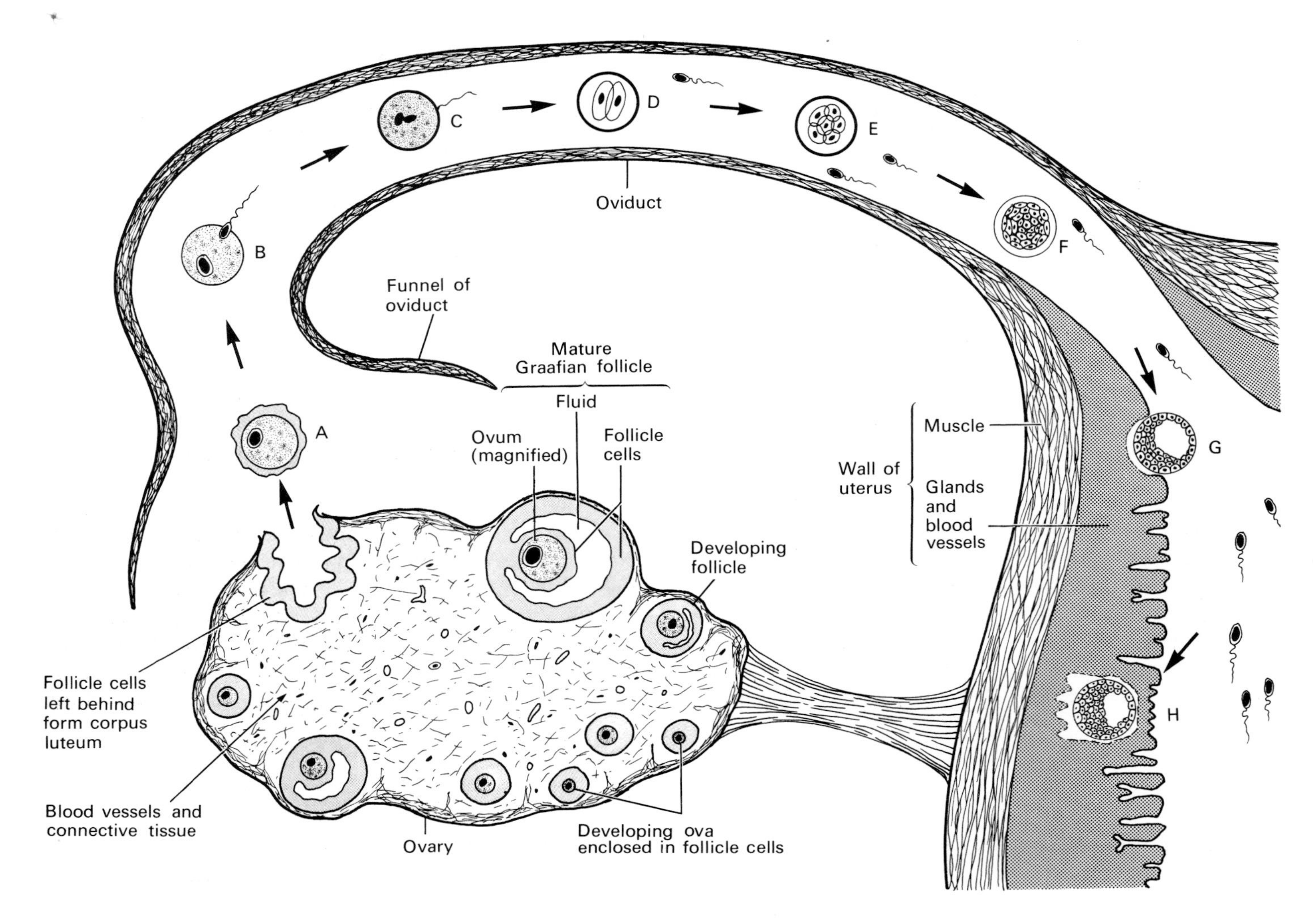

Fig. 17.12 Ovulation, fertilization, and first stages of development.

A Ovulation. **B** Sperm penetrates ovum. **C** Sperm nucleus fuses with ovum nucleus (fertilization). **D**, **E**, and **F** Cell division of zygote produces a ball of cells (embryo). **G** and **H** The embryo digests its way into the uterus wall and becomes completely embedded

and its cells become yellow in colour forming a solid structure called a **corpus luteum**. The corpus luteum produces hormones which stimulate cells lining the uterus to complete the preparations described above ready for a fertilized ovum.

If the ovum is not fertilized within thirty-six hours after ovulation it dies. This is followed by a slow disintegration of the thickened lining of the uterus, and about twelve to fourteen days after ovulation the dead ovum together with the uterus lining and a quantity of blood are passed out of the body through the vagina. This process is called **menstruation**.

Counting the onset of menstruation as day one, ovulation usually takes place on about day fourteen, but it may occur on the thirteenth or the fifteenth day. If the ovum is not fertilized menstruation begins again on about the twenty-eighth day. The menstrual cycle is summarized in Figure 17.13.

Male reproductive system

The reproductive system of male mammals consists of testes and sperm ducts (Figs. 17.14 and 17.15).

Testes The testes are oval in shape, about 5 cm long, and are located in a sac called the **scrotum**. Here, they are at a slightly lower temperature than the

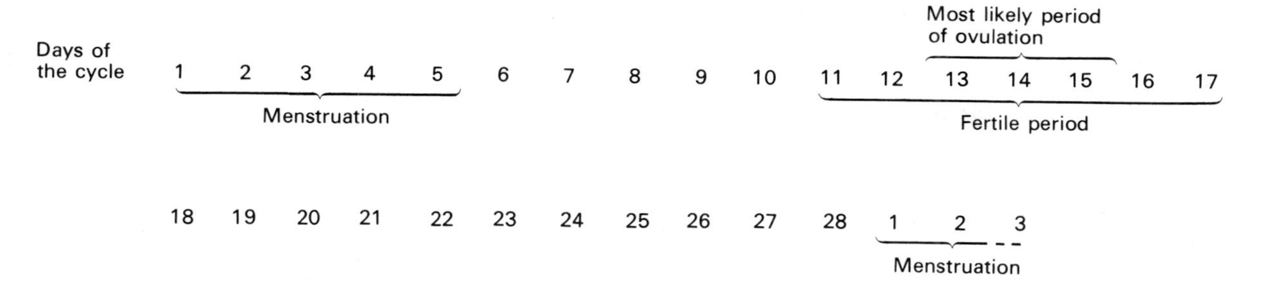

Fig. 17.13 Menstrual cycle in the human female

rest of the body, which appears to be necessary for proper formation of sperms. Unlike ovulation, sperm production is not a monthly event; it proceeds continuously from about twelve to seventy years of age.

The inside of a testis is divided into about three hundred compartments each containing three coiled and twisted tubules about 50 cm long. These coiled tubules are lined with rapidly dividing cells which produce sperms (Fig. 17.15). The spaces between the sperm-producing tubules are packed with cells which produce male sex hormone (testosterone). The hundreds of sperm-producing tubules join together forming a smaller number of collecting ducts which convey sperms out of the testis into a single coiled tube,

Fig. 17.14 Reproductive system of a human male

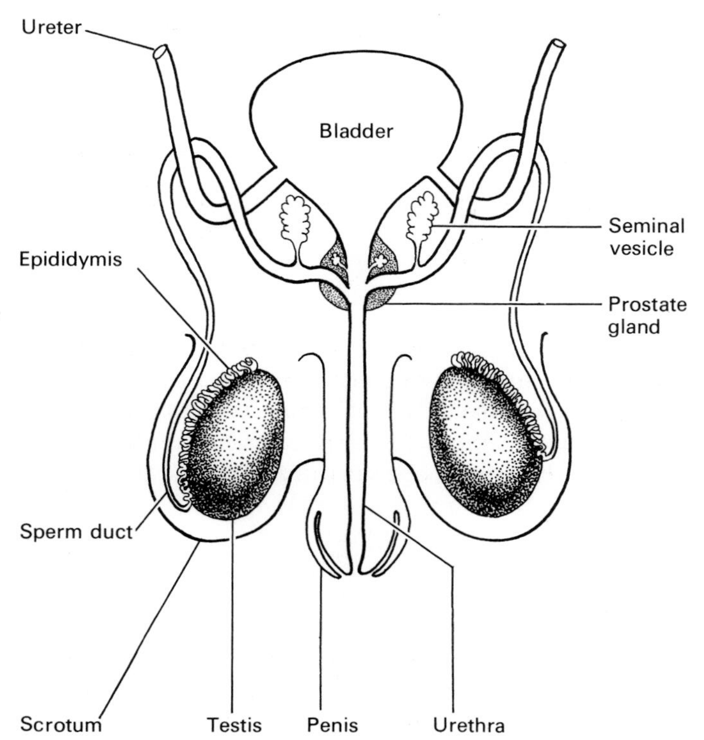

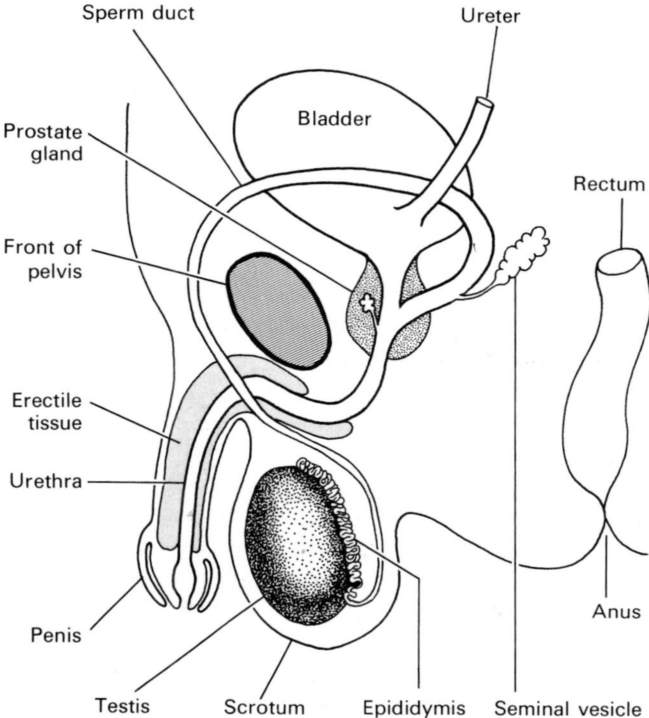

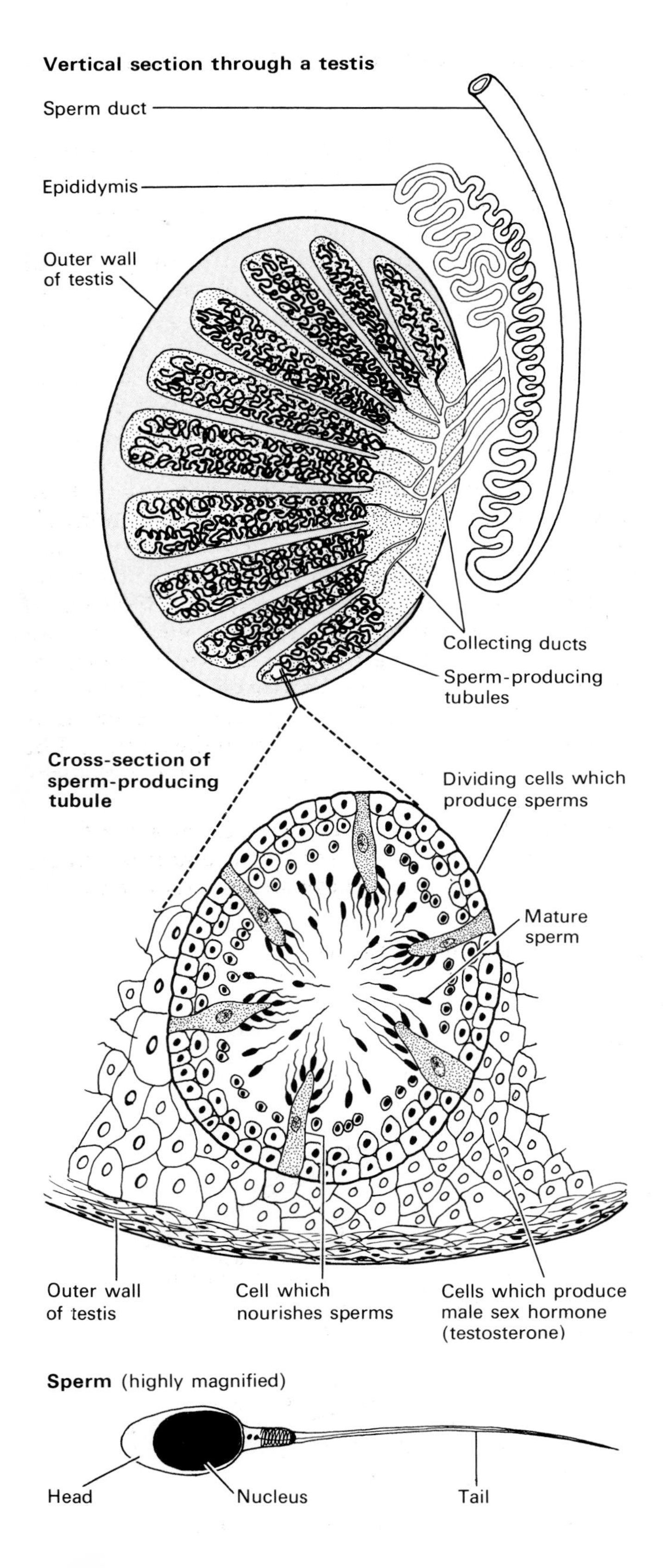

Fig. 17.15 Structure of a testis

called the **epididymis**. This tube is 6 m long and forms a temporary storage area for sperms, which at this stage are completely dormant and immobile.

Sperm ducts Each epididymis leads into a sperm duct, which has thick muscular walls. The sperm ducts, one from each testis, rise up the body and are joined by a duct from a **seminal vesicle**, and finally join a single tube called the **urethra** near the base of the bladder. At this point the urethra and sperm ducts are surrounded by the tissues of the **prostate gland**. The urethra leads to the outside of the body through an organ called the **penis**. During copulation the urethra carries sperms, and when the bladder is emptying during urination it carries urine.

Copulation and fertilization

The thick walls of the penis contain sponge-like spaces which fill with blood whenever a male is sexually stimulated. This makes the penis erect and firm. During copulation the erect penis is inserted into the vagina of the female and moved back and forth. These movements stimulate sense organs in the penis and eventually cause an ejaculation in which about 5 cm^3 of a liquid called **semen** is passed from the epididymis and sperm ducts into the female reproductive system. Semen is forced out of the penis during an ejaculation by rhythmic contractions of the sperm ducts and other muscles. The reflex action of ejaculation and the physical excitement associated with it are known as an **orgasm**.

The semen ejaculated into the female contains up to 100 million sperms from the epididymis, together with liquid produced by the seminal vesicles and prostate gland. This liquid contains chemicals which stimulate swimming movements of the sperm tails, and other chemicals necessary for the nourishment and survival of sperms inside the female's body.

Copulation usually causes the female to experience an orgasm, during which various muscular contractions draw a little of the semen into the uterus. From here a few thousand sperms may manage to swim up into the oviducts, and if a ripe ovum is present at the same time fertilization may occur (Fig. 17.12).

Out of the millions of sperms which enter the female only one brings about fertilization. As this sperm penetrates the ovum it instantly triggers off the formation of an extra membrane around this cell so that no other sperms can enter. The sperm tail is left behind and the head, which contains the nucleus, moves towards and then fuses with the nucleus of the ovum.

Sperms can live for two or three days after entering the female, and therefore copulation two days before ovulation can still result in fertilization.

Pregnancy

The period of development between fertilization and birth is called pregnancy. In humans pregnancy lasts 280 days, plus or minus about 7 days (i.e. approximately 9 months). One of the shortest periods of pregnancy is found in hamsters, which give birth only two weeks after fertilization.

The fertilized ovum begins rapid cell division immediately after fertilization, and during its seven-day journey down an oviduct to the uterus it develops into a hollow ball containing hundreds of cells. It is now called an embryo. Meanwhile, the uterus wall has developed its thick inner lining of cells and blood vessels, and in the next four days the embryo produces enzymes and digests its way into this layer, using broken-down cells and substances produced by the uterus wall as food. In this way the embryo becomes firmly embedded in the uterus wall.

This process is called **implantation**. When it is finished, hormones appear in the female's blood which prevent the menstrual cycle from taking place until well after the baby has been born. In a sense, the mother's body is 'aware' of the presence of the embryo, and is now preparing to care for it.

Clearly, the first requirement of the embryo is oxygen and food since, unlike the eggs of birds, mammalian eggs are not exposed to the air and contain very little yolk. From the beginning the embryo absorbs food and oxygen by diffusion from nearby capillaries in the uterus wall, and the efficiency of this system is greatly increased by the development of a **placenta**.

The placenta The placenta develops partly from the embryo's tissues and partly from the uterus wall. The whole structure takes the form of a large disc-shaped mass of tissue which spreads over and deep into the lining of the uterus as the embryo grows (Fig. 17.16). The embryo is attached to this disc by a tube called the **umbilical cord**, which carries an artery and a vein from the embryo's developing circulatory system. These blood vessels lead to an immense network of capillaries which extend throughout the disc of the placenta, and into millions of finger-like villi which grow into the uterus wall. The capillaries in these villi carry blood from the embryo to within a fraction of a millimetre of the mother's blood supply; in fact the two blood systems are separated by only the capillary walls and the membrane covering each villus. Food and oxygen diffuse across these membranes from the mother's blood into the embryo's blood, and carbon dioxide and nitrogenous wastes diffuse out of the embryo's blood into the mother's blood supply. It is important to realize that the mother's blood does not flow into the embryo, but it does flow close enough to it for these vital exchanges to take place (Fig. 17.17).

To a limited extent, then, the placenta has functions which will be taken over by the embryo's lungs, digestive system, and kidneys after it has been born.

Fig. 17.16 An embryo and its placenta

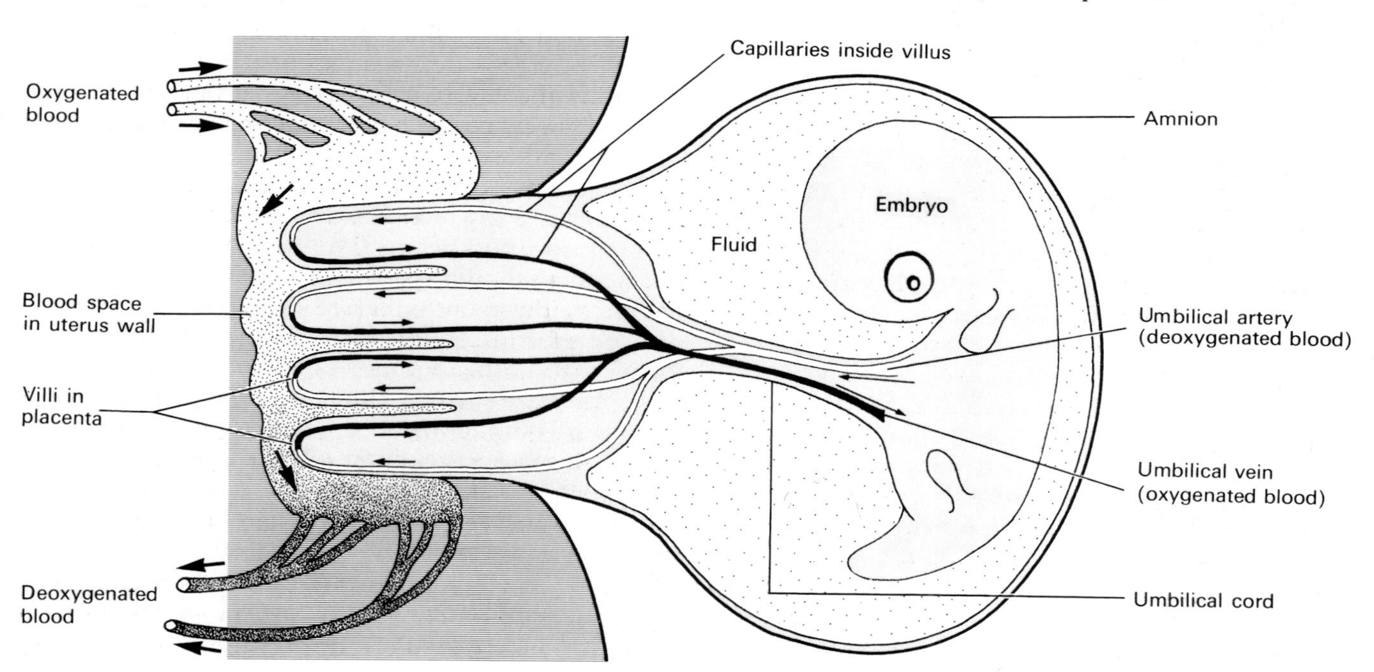

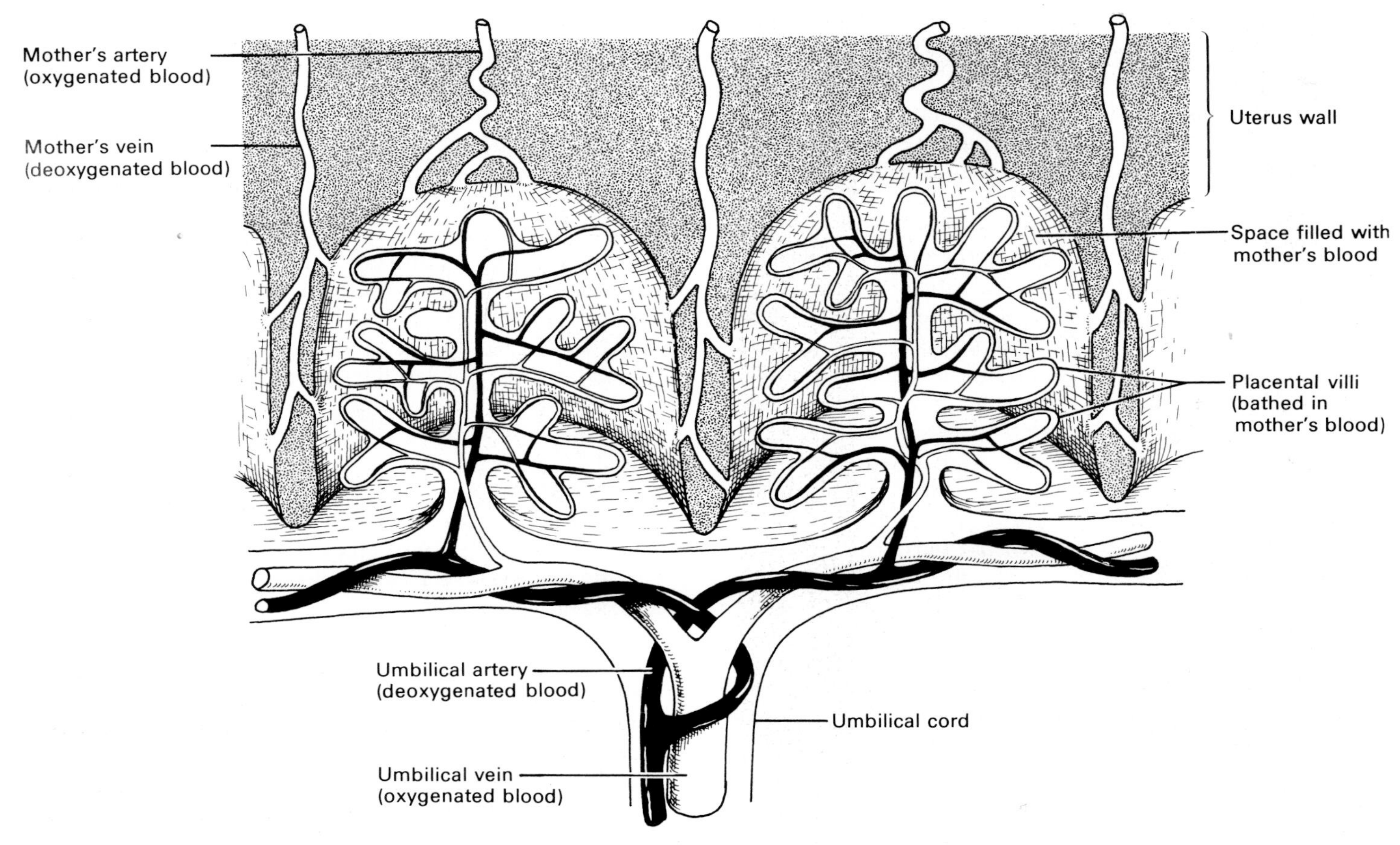

Fig. 17.17 Part of a placenta

At the same time the placenta is an endocrine gland; that is, it produces hormones. These hormones stimulate further growth of the uterus, and stimulate milk-producing glands in the breasts (mammary glands), ready for the infant's birth.

Development At fertilization the ovum is the size of a full stop on this page (and the sperm is 20 000 times smaller). After five weeks the embryo is almost 10 mm long, its brain has begun to develop, and its heart is pumping blood through the placenta and around its body. After two months the embryo is recognizably human. It is nearly 6 cm long, has fore-limbs with fingers and toes, and a well formed face. After six months the embryo is almost 30 cm long, has hair, finger- and toe-nails, and milk teeth are developing in its jaws. At birth a baby weighs, on average, between 3 and 3.5 kg, and is about 50 cm long.

From the first few weeks of development the embryo is enclosed in a water-filled sac, called the **amnion**. The fluid in the amnion acts as a shock-absorber and helps protect the embryo from damage should anything hit or press against the mother's abdomen.

Birth

Birth is a dangerous time for any baby mammal. It is the moment when it leaves the warm protection of its mother's womb, emerges into the cold air, and is suddenly cut off from its supply of food and oxygen through the placenta. Within seconds its lungs must inflate with air for the first time, and soon afterwards its digestive system must absorb the first meal of milk. Thus, birth is the moment when a mammal ceases to be a parasite, and begins the process of becoming an independent organism.

During the months before birth the uterus walls develop the muscle fibres which will be used to expel the baby from the mother's body. A few weeks before birth the baby turns within the uterus until its head lies towards the cervix (Fig. 17.18).

Exactly what causes birth to begin is not fully understood, but it is almost certainly controlled by changes in the mother's hormone output. The uterus walls begin rhythmic muscular contractions, which are intermittent at first, but become increasingly more powerful and frequent. The cervix opens and the baby's head passes into the vagina. This bursts the

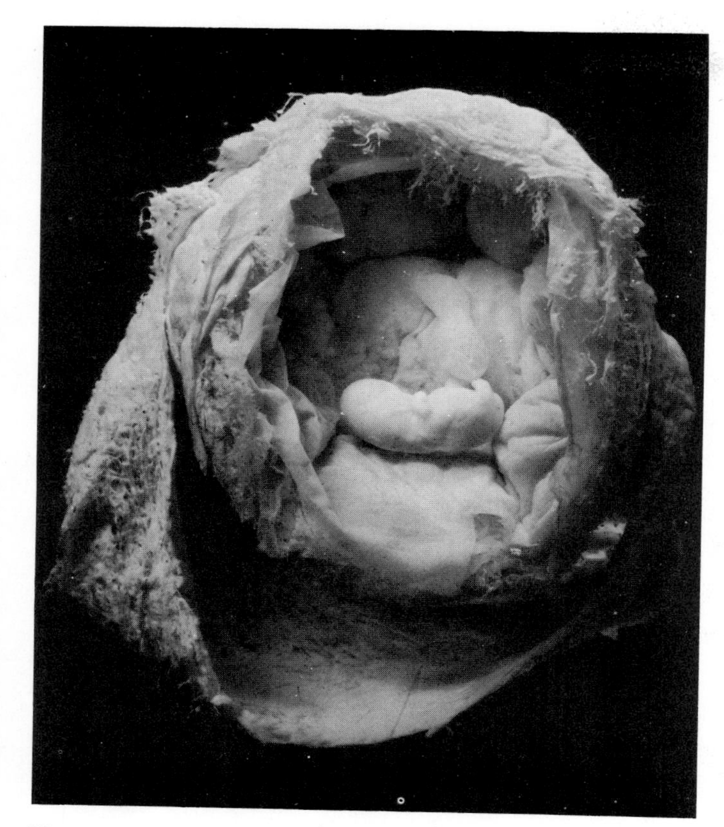

Human embryo approximately six weeks old attached to the placenta

amnion and its fluid escapes. Soon, contractions of the uterus, aided by voluntary contractions of the abdominal muscles, propel the baby out of the mother's body.

In humans, the umbilical cord is cut and tied to prevent excessive bleeding and infection, but in wild mammals the mother simply bites through the cord. In any case, bleeding stops within a few seconds. Shortly after the baby's appearance further contractions of the uterus expel the placenta from the mother's body. This is called the **after-birth**.

On leaving the mother the baby experiences a sudden drop in temperature, and this stimulates the reflex action of its first breath. Human babies are ready to take their first food after about 24 hours. Normally, they are breast-fed on the mother's milk which contains an ideal, balanced diet. Gradually, the baby is weaned on to more solid foods.

17.7 Birth control

Birth control methods allow people to limit the size of their families. They are also playing an increasingly important part in helping to solve the world's population problems.

In the world as a whole about 187 babies are born every minute but, in the same period of time, only 100

Fig. 17.18 Birth

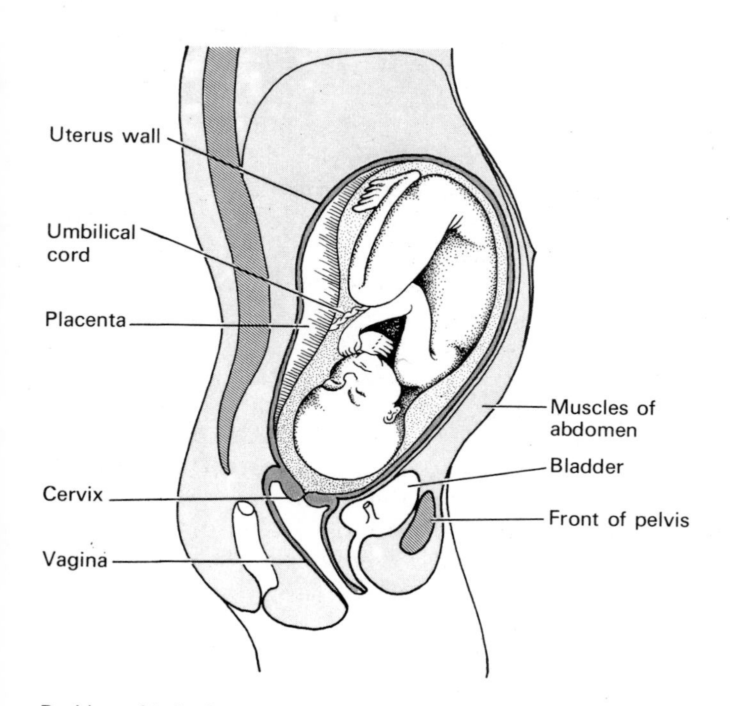

Position of baby immediately before birth

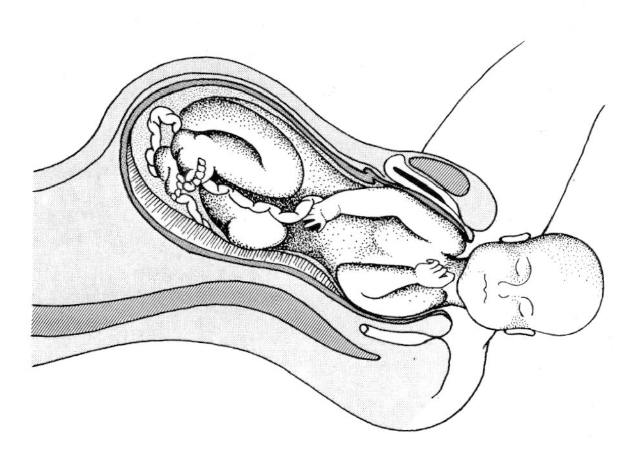

Baby emerges head first

people die. Consequently, the world's population is increasing by approximately 125 000 every day. Obviously, many problems arise from this rapid increase. Birth control techniques are a 'biological' method of population control which many people find more acceptable than the more horrific alternatives of death by starvation and war which may result if population increase remains unchecked. The commonest methods of birth control involve preventing sperms from fertilizing ova.

The condom

A condom is a sheath of thin plastic or rubber which is pulled on to the erect penis just before it is inserted into the vagina. Ejaculated semen is retained in the condom, and does not enter the female reproductive system.

The diaphragm or cervical cap

A diaphragm is a thin disc of plastic which is inserted into the vagina so that it covers the cervix, and prevents sperms from entering the uterus and oviducts.

Both condoms and diaphragms are quite effective when properly fitted, but both are easily damaged, and may even be defective at the time of purchase.

The contraceptive pill

These pills are a mixture of female hormones, and have several effects on the female body.

1. They stop the mechanism of ovulation so that ova are not released into the oviducts.

2. They stop the mechanism which transports ova along the oviducts so that even if an ovum were to be released it would never reach the uterus.

3. They interfere with the mechanism which causes the inner lining of the uterus to develop, and so preparations for the reception of a fertilized ovum are incomplete.

4. They cause mucus in the cervix and vagina to become thick and sticky, so that sperms are unlikely to penetrate into the uterus.

Contraceptive pills are usually taken every day from the 5th day after menstruation up to the 25th day of the cycle. After this, the effects of the hormones quickly wear off, and the lining of the uterus is shed as in normal menstruation.

These pills are almost 100 per cent effective when taken properly. However, they must not be taken by women with liver diseases or diabetes, and there is evidence that in very rare cases they may cause blood clots to develop in arteries and veins.

Verification and inquiry exercises

A *Reproduction in insects*

1. *Flies and bluebottles* Expose moist raw meat to the air until flies have laid their eggs on it, then cover the meat with a glass dish. Through a hand lens observe the feeding habits of maggots. When they stop eating transfer a few to a clean saucer under another glass dish and observe the stages by which they become adult flies. Alternatively, obtain maggots from a shop which supplies angler's bait.

2. *Butterflies* Examine cabbage and nasturtium plants for eggs and caterpillars. Use a hand lens to observe the caterpillars hatching and feeding. Keep some under a large glass dish and supply them with fresh leaves daily. Observe pupation, and the emergence of adults.

3. How do maggots differ from caterpillars in both structure and feeding methods?

4. How do fly pupae differ from butterfly pupae?

B *Reproductive behaviour of sticklebacks*

The mating behaviour of sticklebacks can be studied even in a small aquarium provided the fish are placed in it in the early spring.

1. Prepare the aquarium with plants from the fish's home environment. Introduce one male fish on his own and watch for nest-building behaviour. If this occurs wait until the nest is complete before introducing one female swollen with eggs. If mating occurs remove the female immediately afterwards.

2. The male's territory defence behaviour can be observed if a large aquarium is available. Place one male in it and wait until he has claimed his territory. Then introduce another male, or even a carefully shaped and coloured wooden model of a male attached to a length of wire.

C *Reproduction in amphibia*

1. Obtain frog, toad, or newt spawn from a pond in the early spring.

2. After they have hatched tadpoles can be fed by placing them in an aquarium of pond water which has been planted with weeds from their natural environment. Later they can be fed on scraps of raw meat suspended on thread so that it can be removed before it decomposes.

3. At metamorphosis, provide rocks which rise above the surface for the young frogs to climb on. Finally, release them in the area in which the spawn was found.

D *Development of bird embryos*

1. Obtain fertile eggs from a poultry farm which have been incubated for 1, 10, and 15 days.

2. Use fine pointed scissors (or a scalpel with a No. 11 blade) to cut a circular hole about 2 cm in diameter in the upper surface of the shell.

3. Observe the developing embryos through a lens, or stereoscopic binocular microscope, keeping the eggs at about 20°C if possible.

Comprehension test

1. What are the main differences between the eggs of amphibia, birds, and mammals, and what is the significance of the differing amounts of yolk which they possess?

2. 'A placenta performs functions which are taken over by the lungs, digestive system, and kidneys after the baby is born.' How far is this true?

3. Identical twins are produced when a zygote separates into two cells which then develop independently, whereas fraternal (non-identical) twins are produced when two ova are released from the ovaries at the same time. Explain why fraternal twins have fewer features in common than identical twins.

Summary and factual recall test

Hydra reproduces asexually by (1). These form as outgrowths from the (2) of the parent. *Hydra* is hermaphrodite which means (3). Self-fertilization does not occur because (4). A fertilized ovum divides to form a (5) protected by a (6) and then it drops from the parent. In (7) conditions it forms a new *Hydra*.

Insects such as (8–name three types) have young called nymphs. These resemble an adult insect in most ways except (9–describe three differences). Its development into an adult is called (10) metamorphosis because (11). Insects such as (12–name three) have young called larvae. Their development into adults is called (13) metamorphosis because (14). Larvae spend all their time (15) and shed their skins several times by (16). Next, they change into (17) in the following way (18). A newly emerged adult must pump (19) into its wings because they are (20).

A male stickleback builds his nest out of (21) stuck together with (22). Only a female whose body is (23) will be coaxed into the nest. Here she (24), and is immediately replaced by the male who (25). It is the (26) fish who rears the young, as follows (27).

Frogs spend the winter months (28) in places such as (29). Prior to (30) a male mounts the female and holds on with the (31). He sheds (32) directly on the (33) as they emerge from the female's (34). Tadpoles hatch after about (35) days. At first they do not feed because (36–give two reasons). After about (37) days they begin feeding in the following way (38). After about three weeks they develop (39) gills covered by a fold of skin called an (40) with an opening called a (41). The legs develop after about (42) months, together with (43–describe two other important developments). The process of (44) into an adult begins about (45) months after hatching. The following changes occur (46–summarize these events).

Birds do not have to return to water to reproduce because (47). Bird courtship behaviour has four functions which are as follows (48). During copulation sperms are (49) from the male into a tube called the (50) in the female's body. Fertilization takes place and then (51) and (52) are deposited around the eggs. Eggs must be incubated because (53) and is usually done by the (54) bird who develops a (55) patch. Parental care of the young is essential because (56).

In mammals internal development of the young occurs, which means (57). Compared with bird reproduction this has at least four advantages (58). For a time a newborn mammal is fed on (59) from the (60) glands of the female. At sexual maturity or (61) a girl's ovaries release ova by a process called (62) about once every (63). An ovum is released from a structure called a (64) follicle which afterwards turns yellow and is called a (65). This structure produces hormones which (66). The released ovum is drawn into a tube called the (67) and from there passes to the (68) or womb. If fertilization does not occur within (69) hours after the ovum is released it (70) and a process called (71) occurs about (72) days later.

Sperms are produced in organs called (73), and then pass into a temporary storage tube called the (74). During an ejaculation sperms, together with chemicals from the (75) and the (76) gland are propelled by contractions of the (77) through an organ called the (78) into the female. The sperms and chemicals together form a liquid called (79). The chemicals are necessary to (80–describe two functions).

The period from fertilization to birth is called (81). In humans it lasts about (82) months. The fertilized ovum divides to form an (83) which becomes embedded in the (84) wall in the following way (85). A large disc-shaped organ called the (86) now develops through which the embryo absorbs (87) and (88) from the mother's (89), while (90) and (91) pass in the opposite direction. Birth is a dangerous moment for a baby because (92). Contractions of the (93) and (94) muscles push the baby (95)-first from the mother's body. The (96) cord is then cut and tied to prevent (97).

18 Evolution and natural selection

One of the most obvious characteristics of living things is their enormous variety. About two million different species are known and more are being discovered all the time. In addition, several million more species existed in the past but are now extinct.

But this variety is not a haphazard jumble of totally different organisms. Living things can be sorted into groups according to shared features. Members of the largest groups (kingdoms) have few features in common, but these groups can be divided into smaller groups (phyla, classes, etc.) whose members have more common features until, at species level, there are numerous similarities.

The fact that organisms can be sorted into groups according to shared features suggests that the members of each group are related in some way, and that they are more closely related to each other than to members of other groups. Mammals and insects illustrate this point. Despite differences of size and shape elephants and humans must be related because they both have hairy skins, and young which are born alive and suckled on milk. Butterflies and ants must also be related because they have three pairs of legs, antennae, a body divided into three parts, etc. However, the fundamental differences between insects and mammals suggest only a remote relationship between these two groups.

Several questions arise from these facts. Have living things always existed in their present variety or have they become varied with time? Furthermore, is it possible to explain the fact that this variety is not haphazard but displays many inter-relationships? This chapter describes two scientific theories which suggest answers to these questions. These are the theories of evolution and natural selection. Opposed to these theories is the theory of special creation, which in its earliest form proposed that all organisms on earth were created simultaneously at some time in the distant past and have remained the same ever since.

18.1 The theory of evolution

The theory of evolution proposes that species change with time. To be more precise, the theory states that the first living things were quite simple in structure and much less varied than at present. It is argued that these simple creatures gave rise to successive generations, some of which were slightly different and sometimes slightly more complex than their ancestors. Over hundreds of millions of years this process is thought to have produced a gradual sequence of changes leading from a simple state to more and more variety and complexity. Put simply, the idea of evolution implies a slow development over long periods, rather than a once-and-for-all creation of every living species.

The theory of evolution also explains the presence of groups with shared features. If evolution occurred it must have involved a number of stages. Thus, organism A produced organism B and this produced C and so on. It is unlikely, however, that when organism B appeared all the type-A organisms died off. It is more likely that they continued reproducing, and perhaps evolving, up to the present day, giving rise to a group with the basic features of their ancestor (Fig. 18.1A).

If this is so, it should be possible to arrange modern organisms into a sequence from relatively simple types (modern representatives of the earliest living things) to highly complex types (modern representatives of more recently evolved organisms). Moreover, it should be possible to construct an evolutionary tree showing how present-day groups can be traced back to ancestors whose features they share. This has been done in Figure 18.1B.

The theory of evolution also offers an explanation for the existence of **fossils** which represent the remains of extinct organisms. Indeed, fossils form one of the most important pieces of evidence in support of the theory.

18.2 Evidence in support of evolution

Fossil evidence

Normally, when an organism dies its remains decay and quickly disappear. However, under certain circumstances its remains are preserved as fossils, usually in the following way.

If the body of an animal or plant is washed into a river or comes to rest at the bottom of a shallow sea, it may become covered with sediment such as sand and other minerals which settle on top of it. The soft parts of the organism will probably decay, but the hard parts, such as animal bones or the cellulose and lignin of plant tissues, may survive long enough to absorb minerals from the water. These minerals gradually replace some or all of the organic materials in the body's remains, literally turning them into stone. When the body has been completely changed into stone, a fossil has been formed.

In time the fossils become covered by additional layers of sediment containing more trapped fossilizing remains. This causes a build-up of weight that presses on the deeper layers, hardening them into **sedimentary rock**. Millions of years later these rocks and the fossils in them may be pushed upwards during movements of the earth's crust. Later still, the fossils may be exposed by cracking or faulting of the rocks, and by the action of water which carves out gorges and valleys through them.

Since layers of sedimentary rock are laid down by the slow accumulation of material, the lowest layers are usually the oldest and contain the oldest fossils, while the topmost layers are the youngest and contain the most recently formed fossils. If life has evolved gradually through time it is reasonable to assume that fossils of the simplest organisms will be found only in the lowest and oldest layers, and that progressively higher layers will contain fossils of more complex organisms. This shows both the order in which organisms developed and the structural changes which they underwent.

This assumption has been confirmed by the discovery that only the most primitive organisms are found in the lower, older layers of rock, and a sequence of progressively more varied and advanced animals can be found in successively higher, younger layers of rock.

Recently developed techniques make it possible to determine the age of rocks and fossils with great accuracy, so that they can be placed in a precise order according to age. These techniques have been used to establish with even greater certainty that the most primitive fossilized organisms are indeed the oldest.

A The theory of evolution Life began with simple creatures (**A**), which produced more complex ones (**B**), which produced (**C**) etc thus giving rise to an evolutionary sequence from simple to complex types. In addition, each organism in the sequence continued reproducing and evolving. This increased the variety of life by producing many groups of organisms each of which share the features of their ancestor.

B An evolutionary tree Present day organisms can be placed in a sequence from simple types (e.g. bacteria) to more complex types (e.g. mammals). This suggests that they are the products of an evolutionary sequence. Moreover, organisms can be arranged in groups which share certain features, and this suggests that each group is evolved from a common ancestor.

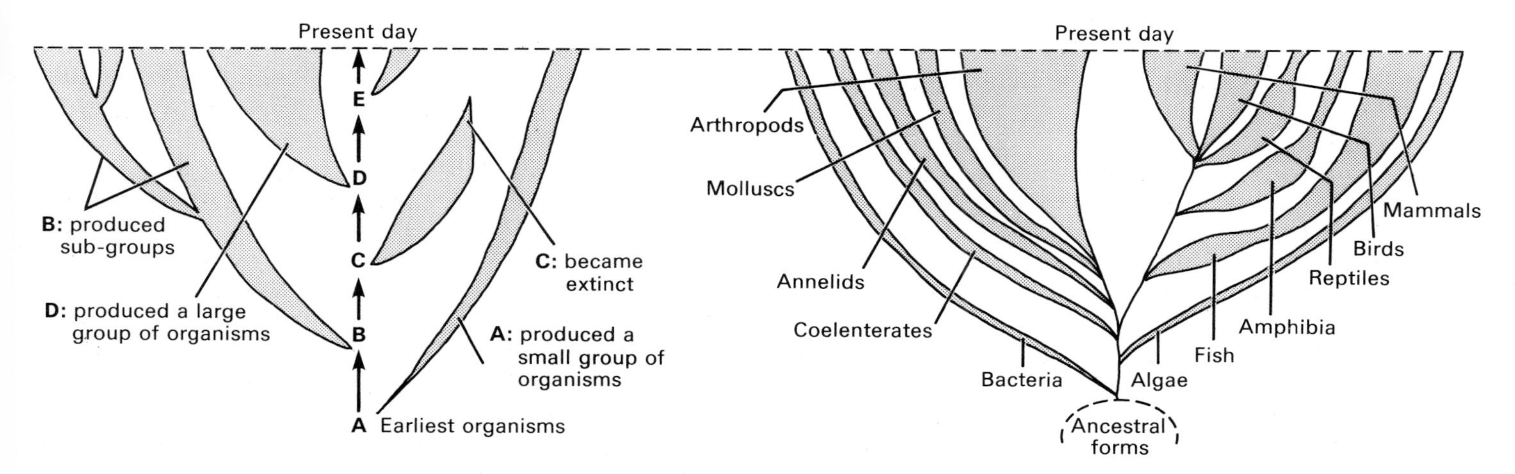

Fig. 18.1 Diagrams of evolutionary change

Sometimes these 'fossil records' as they are called, yield a sequence of fossils showing a possible line of descent to a present-day organism from an ancestor which lived millions of years ago. Such fossil records have been found for horses (Fig. 18.2A), elephants, giraffes, and camels. Each fossil in the sequence appears to represent a stage in the series of changes right up to the modern animal.

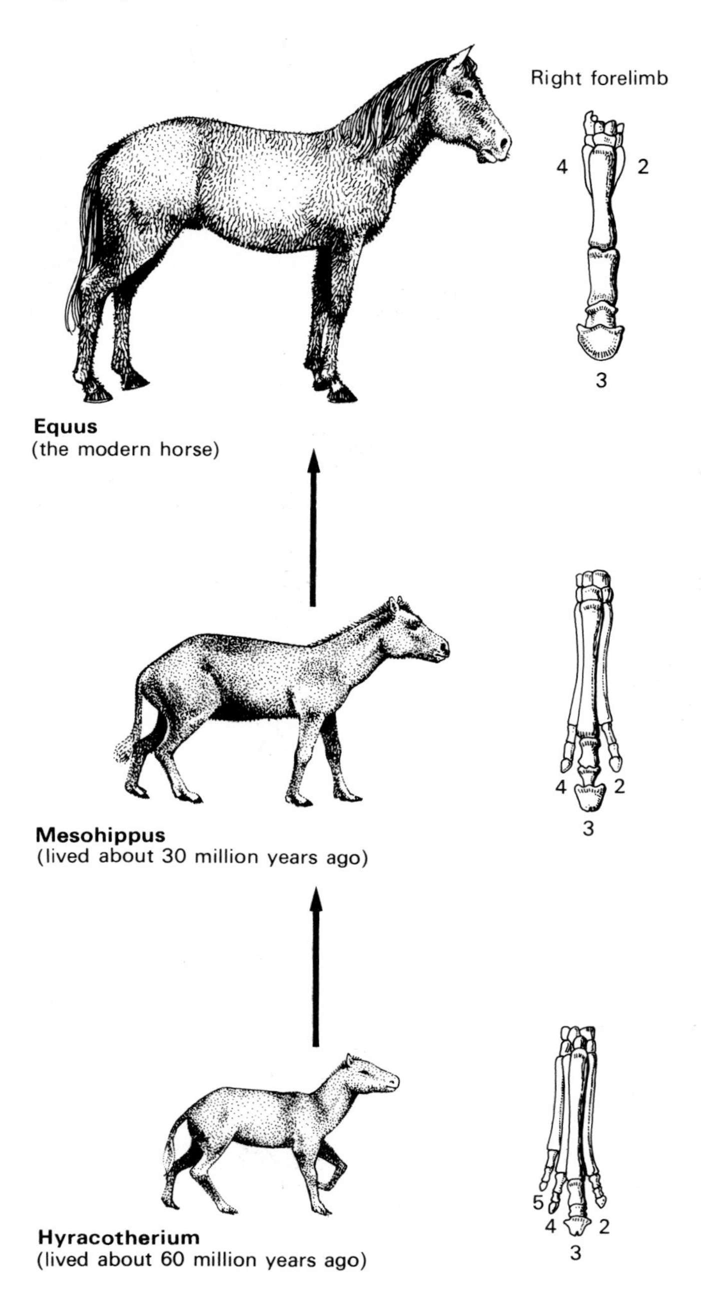

Fig. 18.2A Two of the many fossils which are thought to represent stages in the evolution of modern horses. Note the gradual loss of toes, and the development of toe 3, which forms the hoof of modern horses

The study of fossils shows that certain ancestral organisms from the past could have given rise to not one, but many different present-day forms. This is known as **divergent evolution**. If this is true, then it would be reasonable to assume that a group of organisms with a common ancestor will share certain features, all of which they inherited from this single ancestor, although these shared features will have been extensively modified in the course of evolution. This assumption is supported by comparing the anatomy of various organisms in search of similarities.

Evidence from comparative anatomy

Figure 18.2B is a diagram of the limb bones of several vertebrates, showing that they are all constructed on the same basic pattern with similar bones arranged in about the same order and position. This suggests that they are descended from a common ancestor which had relatively unspecialized limbs, and that these have been modified in form and function as the divergent groups became more and more specialized for a particular form of locomotion.

Snakes are also vertebrates but they have no limbs. Some types, such as boas, have bones which are obviously the remains of a limb skeleton. These bones are entirely useless for locomotion. It makes sense to assume that snakes are a specialized offshoot of some other vertebrate ancestors which lost their limbs during an evolutionary change.

Figure 18.3 is a greatly simplified version of the 'family tree' which can be constructed from fossil evidence to show how modern vertebrates may have evolved from common ancestors. When looking at this chart several important factors must be kept in mind.

First, modern vertebrate groups (e.g. amphibia and reptiles) can be traced back to their ancestors with some detail, but there are several 'missing links' where one group diverges from another. For example, it is almost certain that birds and mammals had a reptile-like ancestor, but the fossils of certain early links in the sequence are missing. This does not mean they do not exist; they have simply never been discovered, though they may be in the future.

Second, it is incorrect and misleading to say: 'Birds and mammals evolved from reptiles', because this suggests that their common ancestors were like modern reptiles. In fact their reptile ancestors were far less specialized than the modern types. Eventually these ancestors gave rise to 'reptiles' with the beginnings of bird characteristics, another group with the beginning of mammal characteristics, and another group on the way to becoming modern reptiles. It is therefore more accurate to say that birds and mammals share a common ancestor with the modern reptiles.

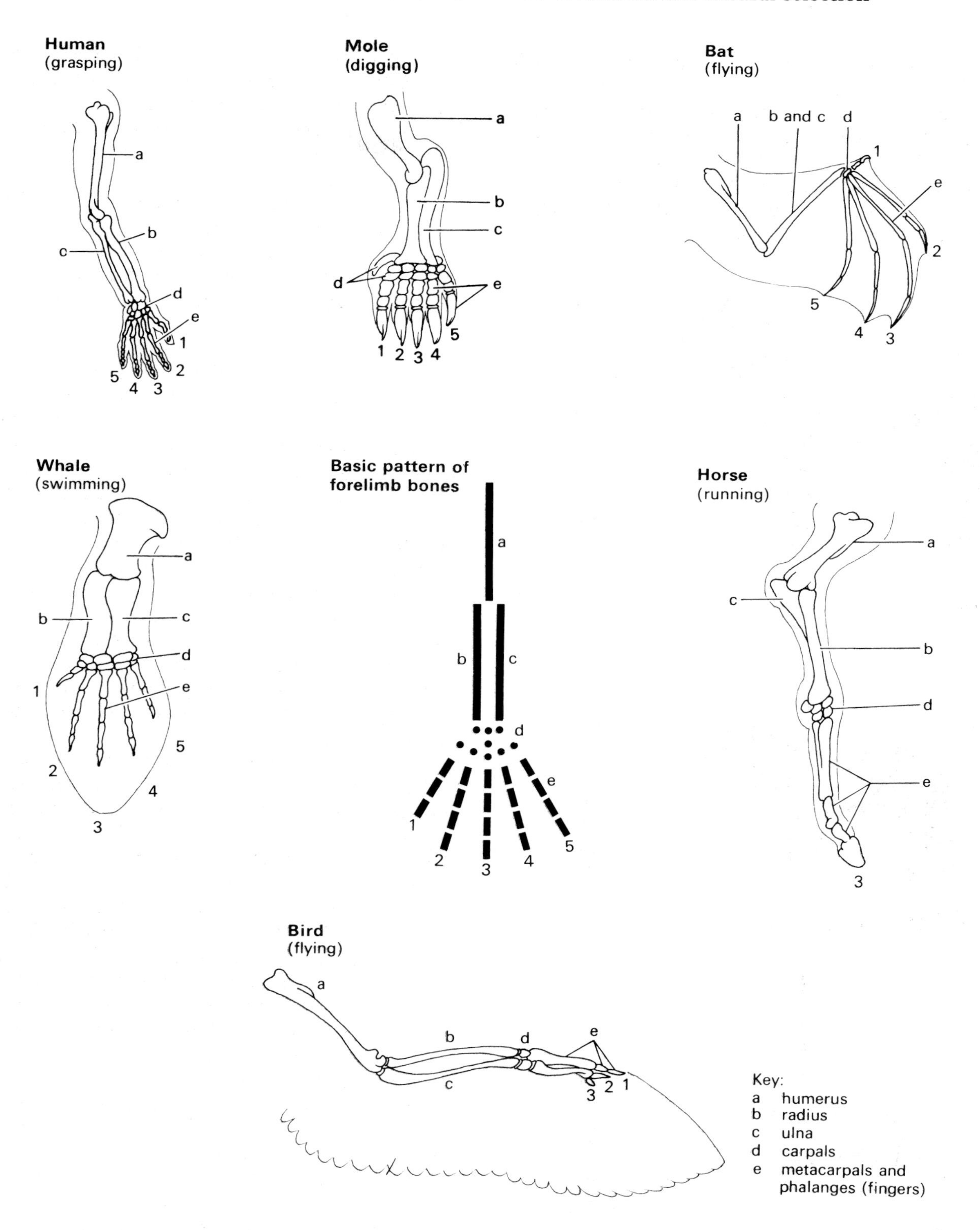

Fig. 18.2B Diagram illustrating the similarity in structure of various vertebrate limbs. Either these limbs were created independently, or they have evolved from a less specialized vertebrate ancestor

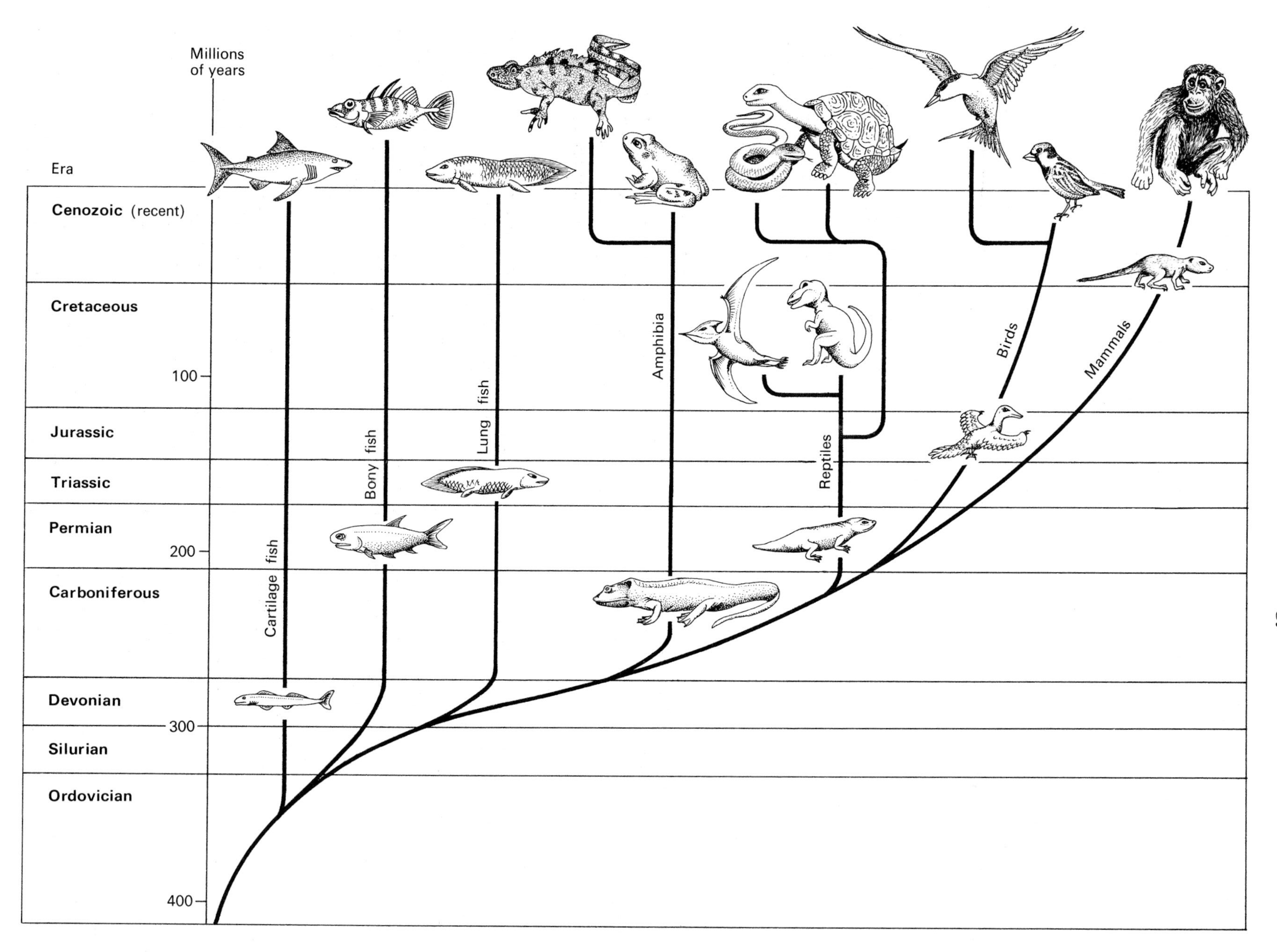

Fig. 18.3 An evolutionary tree of the main vertebrate groups, illustrating how vertebrates are thought to have evolved from common ancestors. All except animals along the top line are extinct

Third, the chart shows that evolution sometimes appears to stand still. Lungfish are an example; they are descended from organisms which have changed relatively little over more than a hundred and fifty million years.

Fourth, many organisms have become **extinct**. The large reptile dinosaurs are an example of this. They developed features which permitted a very specialized way of life, and when conditions in their environment suddenly changed they were unable to survive.

The theory of evolution explains the variety of life without actually suggesting how one species could have given rise to another slightly different species, and thereby produce an evolutionary sequence. In other words, it does not suggest a mechanism of evolution.

18.3 The search for a mechanism of evolution

Many people have been involved in the search for an explanation of how species change, but among the most famous are Lamarck, Wallace, and Darwin.

The French biologist Jean Baptiste Lamarck suggested that evolution may have come about by the inheritance of **acquired characteristics**. By this he meant that the young of a species may inherit certain physical characteristics which their parents acquire in the course of their daily lives. In time this would produce organisms different in structure from their ancestors.

According to Lamarck, giraffes evolved their long necks in this way from short-necked ancestors. He believed that, in constantly striving to reach the leaves of trees, the ancestors must have stretched their necks. This acquired characteristic was then inherited by their young who through the same activity stretched their own necks still more and had young with even longer necks. Lamarck believed that this sequence, repeated from generation to generation, would have resulted in the very long necks of modern giraffes.

Lamarck's theory assumes that the body of an organism is 'plastic' in the sense that it will change in shape and form if some part, such as the neck, is put under strain or is required for more constant use than other parts. It also assumes that these changes will be inherited by the young. The first assumption is true to a limited extent. Men who have developed their muscles with prolonged, strenuous exercise have massive bodies. But the second assumption is not true. 'Muscle men' do not necessarily produce stronger or bigger children.

To be inherited an acquired characteristic would have to become incorporated into the hereditary information in an organism's chromosomes. At present there is no clear evidence how this could occur (this subject is discussed in Section 18.6). Consequently Lamarck's theory of evolution cannot be accepted without modification.

In 1858 the British biologists Alfred Russel Wallace and Charles Darwin published an essay in which they stated an hypothesis which attempted to explain the mechanism of evolution. Unfortunately, their work aroused very little interest. In 1859 Darwin repeated this hypothesis in his book *The Origin of Species*, and almost immediately raised a sensational response which included both the highest praise and the severest criticism.

But Darwin's many critics served a useful function. By pointing out the weaknesses and loopholes in his arguments they stimulated many people to test his work thoroughly and examine all its implications. This research continues to the present day. The original hypothesis, now called **the theory of natural selection**, has been modified and extended into a form which is accepted by the majority of biologists as the most likely explanation of how species change with time and give rise to new species.

18.4 The theory of natural selection

Darwin based his theory of natural selection on a number of observations made in many parts of the world. These observations and the reasoned arguments which he derived from them are summarized below.

First observation

Each generation of a species has more offspring than parents. In fact most species reproduce at a rate which could at least double their numbers at each generation. An amoeba, for instance, divides into two, these two divide to produce four, and then eight, sixteen, thirty-two, sixty-four, and so on. The same is true of reproduction in higher organisms. Each set of parents has only to produce four offspring for the total population of the species to double at each generation.

Second observation

Despite their high rate of reproduction the total population in most species remains about the same once they are established in a particular environment. Of course there are seasonal variations–good years and bad years–but there are no instances of wild species which regularly double their population year after year.

First argument

Darwin argued that since populations do not increase as would be expected from their rate of reproduction, something must be controlling their numbers. He assumed that there must be a 'struggle for survival', especially among the young, so that many die before reaching reproductive age. This may happen through competition for food, water, light, warmth, and other factors affecting growth; through failure to escape from predators; death from disease or accidental injury; and all manner of hazards which wild organisms face.

Third and fourth observations

Darwin observed the variation which exists within a species, especially among those which reproduce sexually. Individual organisms in a species differ from each other in many small ways (this is easily verified by looking at any group of people); offspring differ slightly from their parents; and successive offspring of the same parents are different from each other. In fact, no two members of the same species are exactly alike.

At the same time, Darwin noted that although offspring differ from their parents in certain ways, they still inherit many parental characteristics. Darwin did not understand the mechanism of heredity. Indeed, like Lamarck, he believed in the inheritance of acquired characteristics, but this does not invalidate his arguments. Briefly, the third and fourth observations are: variation, and the inheritance of variations.

Second argument

Darwin reasoned that certain variations help an organism to survive in the struggle for existence while other variations do not. For example, in each batch of offspring those with favourable variations, such as strength and stamina, are more likely to win the competition for food, escape from predators, and withstand disease, than their weaker fellows. Consequently, organisms with favourable variations are likely to survive longer and reproduce more often than those with unfavourable variations.

Darwin called this 'the survival of the fittest', meaning that in the struggle for existence the fittest, i.e., those with favourable variations, will survive while the others die or are limited in numbers.

Darwin also argued that the survival of the fittest is a selection process. That is, 'nature' (i.e. the hazards of life in the wild) 'selects' those organisms best fitted for survival. Hence the phrase 'natural selection'. Furthermore, after natural selection has taken place all the favourable variations are not lost when their owners eventually die; some at least are inherited by the next generation. In other words, nature selects the favourable variations and inheritance preserves them by transmitting them to the young.

But inherited variations are not the only ones which play a part in evolution. Darwin noted that organisms sometimes display unique variations possessed by neither parent. These new variations are now called **mutations**, and their evolutionary significance is discussed more fully in section 19.3. Here it is enough to say that they add more characteristics to those upon which the forces of selection can operate.

Darwin now used his arguments to show how a new species can originate. He pointed out that variation, natural selection, and inheritance, operating on generation after generation of a species for millions of years, could limit unfavourable variations and lead to the accumulation of more and more favourable variations within a species. Moreover, the accumulation of variations with survival value could lead to a process of change, and probably of improvement. Ultimately, change and improvement could give rise to a new species which is, in a sense, an advancement on its ancestors, because it possesses characteristics with survival value that its ancestors lacked. This represents one step in the long sequence of evolutionary changes from simple to complex organisms.

The main points of Darwin's arguments can be summarized as follows:

1. Organisms reproduce at a rate which could potentially more than double their numbers at each generation.
2. Despite this, populations of wild organisms normally remain fairly constant in the numbers of their individuals.
3. Therefore something must be controlling population numbers. Darwin suggested that life involves a struggle for survival in which many of the young die before reaching reproductive age.
4. There is variation within every species and new variations are appearing all the time. No two members are exactly alike.
5. Offspring tend to inherit some of their parents' characteristics.
6. Some variations help an organism to survive in the struggle for existence; that is, these variations have survival value. Organisms with these variations will tend to grow and reproduce while their less favoured fellows will die off or be limited in numbers. This process is called natural selection.
7. Inheritance ensures that features with survival value are passed on to the next generation which will also have its own unique set of variations.
8. Over millions of years variation, natural selection, and inheritance may lead to the accumulation within a species of many features with survival value. Thus, a species may slowly change for the better, and may eventually produce an entirely new species.

It is easy enough to argue that one species may have evolved from another by a number of small changes, but it is another matter altogether trying to *prove* that this is so. In fact, concrete proof is not, and may never be, available. This is why it is necessary to speak of the *theories* of evolution and natural selection. Nevertheless, there is an immense amount of evidence available: Darwin collected it for twenty years before writing *The Origin of Species*, and since his death a vast amount of additional evidence has been discovered. Here is a small portion of it.

18.5 **Evidence in support of natural selection**

One of Darwin's most difficult tasks was to convince people that something as simple as selection can produce change within a species. He supported his argument by pointing out that man himself employs **artificial selection** to change animals and plants for his own use.

Evidence from artificial selection

Man produces completely new varieties of animals and plants by deliberately selecting and breeding those which possess the characteristics he desires.

A Bulldog and a Toy Terrier. Note the many differences between the two breeds remembering that man has produced them from a common ancestor by artificial selection. This is clear evidence that selection can change a species

Starting with the wolves and hyenas which visited his camp sites at the dawn of civilization, man has produced hundreds of varieties of dog. Similarly, by breeding the largest and strongest wild horses, man has produced carthorses which can pull heavy loads; and by breeding only the fastest and sleekest animals he has produced racing and hunting horses. In the same way, by breeding sheep with the longest coats he has increased their yields of wool, and by breeding cows with the highest milk yield he has greatly improved the average yield per cow. Again, by selecting and breeding plants with the largest and tastiest fruit man has produced plums, apples, grapes, etc. of far better quality than those found on wild plants.

Man changes animals and plants by selecting certain specific qualities which fit his requirements, thereby giving them an artificial survival value. In this way man accelerates evolution. Nevertheless, it must be remembered that artificial selection has not yet produced an entirely new species, and the selective efforts of man do not prove that selection occurs normally in nature.

Evidence of selection in nature

If selection operates in nature it should be possible to find organisms with variations which are favoured by selection in their environment, and so have become commoner than their less favoured fellows. A now famous example of such an organism was investigated by Dr. H. B. D. Kettlewell of Oxford University in 1960.

At that time it was well known that a certain insect, the Peppered Moth, had a pale variety (white wings with black spots) and a dark variety (pure black wings). It was also known that over the past hundred years there had been a marked increase in the number of dark moths in the industrial areas of Britain and Europe. Dr. Kettlewell tested the hypothesis that the black variety was favoured by natural selection in smoke-blackened industrial areas because its colouration made it practically invisible to insect-eating birds. He collected large numbers of pale and dark-coloured Peppered Moths and put some of each on trees in a clean country area, and others on blackened trees near factories. In clean woodland the dark moths were quickly taken by birds and the pale ones were overlooked, while in woods near the factories the pale moths were quickly picked off and the dark ones were protected by their camouflage.

It is now known that over eighty species of moths have produced dark varieties which are common in the industrial areas of Britain. In all these insects the principle of survival of the fittest seems to involve the natural selection of those varieties with the best camouflage. These favoured varieties become common

Dark and light varieties of Peppered Moth on a lichen-covered tree in unpolluted woodland. Note that the light variety merges with its background while the black is conspicuous and easily seen by insect-eating birds

Dark and light varieties of Peppered Moth on the smoke-stained bark of a tree near an industrial area. Note that the black variety is now camouflaged and the white one is now easily visible to bird-predators

simply because they are overlooked by birds, and so produce more offspring.

In this example a change in the moth population resulted from a change in their environment. When any environment alters there is a corresponding change in the processes of natural selection within that environment. Under changing circumstances a different set of variations within local organisms may suddenly prove to be an advantage. As a result the characteristics of the local organisms will probably alter. Environmental changes may therefore accelerate evolution.

There are many examples of environmental changes in the history of the earth which have probably influenced evolution: ice ages, mountain building, continental drift, and the spread of deserts are a few of them. But there are other ways in which organisms may be confronted with a new environment. The constant spread of plants and animals by migration and chance dispersal into new areas such as deserts, mountain-sides, and arctic regions may have resulted in the evolution of the specially adapted species found there today. Probably the most difficult migration of all took place when organisms spread from water on to land, and met a totally new environment with completely different characteristics, such as lack of water, lack of support against the force of gravity, and large variations in temperature. Over millions of years evolution gradually produced new organisms able to tolerate these new conditions, and then life on land began.

In the case of Peppered Moths a relatively simple environmental change favoured one variation and changed the population. During the millions of years since life began, large-scale geological and climatic changes, together with migration and dispersal, must have influenced countless variations in countless organisms, resulting in widespread evolutionary changes and the production of new species.

Darwin discovered an example of migration which appears to have produced new species. On a voyage around the world he visited the isolated Galapagos Islands located off the coast of South America. Here, he found several species of finch-like birds, obviously closely related but with differently shaped beaks and different feeding habits (Fig. 18.4). These particular

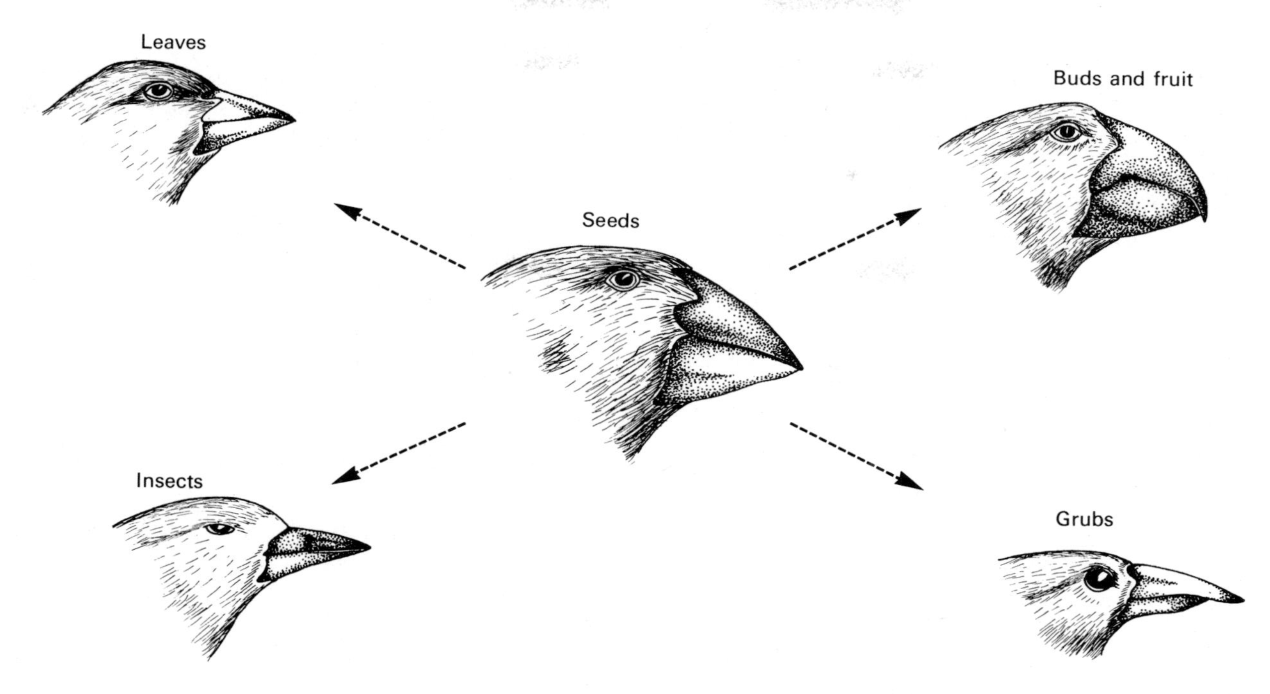

Fig. 18.4 Five of the sixteen types of Galapagos finch. The seed-eating species (centre) is thought to be closest to the seed-eating ancestors from which other types may have been derived by divergent evolution

birds appeared to be unique to the Galapagos Islands. Darwin suggested that long ago finches from the mainland, where only seed-eating species are to be found, accidentally reached these islands and found a number of different foods. Then, owing to the selective influence of competition for seeds, variations evolved with modified beak shapes suitable for eating fruit, leaves, grubs, and insects. These varieties eventually became separate species.

The Galapagos finches are an example of what may happen when organisms invade an isolated area. The invaders are physically separated from their main group by barriers such as water or mountains, and cannot cross-breed with its members. In their new environment the invaders are influenced by many selective forces so that not one but many variations are favoured. The organisms evolve, and become adapted in different ways which permit them to take advantage of all the new opportunities their isolated environment affords. In this way a single invading species may produce many types of organism all adapted in different ways so that eventually they cannot cross-breed with each other. They have now become separate species. This is how divergent evolution probably occurs.

To summarize: the inheritance of variations provides the raw material of evolution, because it produces new candidates to face the hazards of natural selection. It is the death or survival of these candidates which determines the future course of evolution. The weakest point in Darwin's account of this process was that he could not show how variations originate. This was unfortunate because variation is the source of all evolutionary change. Darwin was also unable to understand the way in which variations are inherited, and was unaware that only six years after the publication of *The Origin of Species* an unknown Moravian monk called Gregor Mendel had published an account of the first principles of inheritance. Mendel's laws of inheritance are introduced in the next chapter.

18.6 **Scientific critics of Darwinism**

Natural selection is not a fact. It is a working hypothesis, which means that while it solves many problems there are others which remain unsolved.

One unsolved problem is the presence of large gaps in the fossil record of some major events in evolution: like the emergence of birds and mammals from reptile ancestors. Some gaps will occur in any fossil sequence but if, as Darwin proposed, evolution involves many small steps, it seems odd that we lack so many intermediate stages.

A modern viewpoint, called the **jump theory of evolution**, states that these intermediate stages have not been found because they do not exist. The jump theory proposes that, rather than evolution by many small steps, species change very little for millions of years and then abruptly give rise to something quite different and yet clearly related, in one large jump (i.e. with no intermediate steps).

The essence of Darwin's theory is that evolution occurs by natural selection of *chance* variations, or **mutations** as they are now called. Some people think that chance alone is not enough to produce the enormous variety and complexity of life. Is it possible, for instance, that a structure as complex as the human eye could have evolved by chance? The eye consists of many inter-related parts, most of which are useless without the others. Consequently, for any one of these parts to have survival value they must have appeared at the same time as other related parts. In other words, several useful mutations must have occurred *simultaneously*. In fact useful mutations are so rare that such an event is almost inconceivable. Perhaps other factors are involved in evolutionary change.

Recent investigations suggest that there may be some truth in Lamarck's belief that acquired characteristics can be inherited. One experiment has shown that changes in the height and weight of flax plants produced by different growing conditions can be passed on to subsequent generations. In more recent experiments mice of one strain were given treatment which stopped them rejecting skin grafts from another strain of mice. The treated males were then mated with untreated females of the same strain and about half their young were able to accept the skin grafts without treatment. Confirmation of these and similar results may make it necessary to rethink the role of chance in evolution, and allow for changes brought about by inheritance of acquired characteristics.

Comprehension test

1. Lamarck believed that ducks developed their webbed feet through using them as paddles over a long period. Describe the reasoning on which this belief is based. How would Darwin have explained the evolution of webbed feet?

2. A dairy farmer found that his stables were infested with flies so he sprayed the area with D.D.T. (an insecticide). Nearly all the flies were killed. A few weeks later the number of flies was again large so the stables were sprayed again with the same chemical. Most of the flies were killed. Again the fly population increased and again the same spray was used, the sequence being repeated over several months. Eventually it was clear that D.D.T. was becoming less and less effective in killing the flies.

a) Construct as many different theories as possible to account for these facts.

b) Chose one theory which you consider to be correct and devise a controlled experiment which could be used to test it.

3. In man the muscles which in other mammals move the ears are present but poorly developed and useless. Man has more than 180 such structures, which are useless to him but are functional in other vertebrates. Explain how these structures suggest that man and other vertebrates share a common ancestor.

Summary and factual recall test

The theory of evolution states that species (1) with time. Fossil records contain evidence of divergent evolution (2—define this term). The bones of the human arm and the (3) bones of a horse and a bird have the same basic (4). This suggests that (5).

Lamarck believed in the inheritance of (6) characteristics. By this he meant that (7). Darwin proposed the theory of (8). This is based on the following observations and arguments. Organisms reproduce at a rate which could (9) their numbers at each (10). Nevertheless, populations normally remain fairly (11) in numbers of individuals. According to Darwin this is because of a (12) for survival in which the young die before reaching (13) age. There is variation in every species, and some variations are said to have (14) value, because (15). (16) ensures that some of these variations are passed on to the next generation. Over millions of years (17), (18), and (19) may lead to the accumulation of (20) within a species, and as a result a new (21) may be produced.

Man has produced new breeds of dog by (22) selection. Other examples of this work are (23—name three).

Environmental changes can accelerate evolution because (24). Examples of such changes in the history of the world are (25—name four).

19

Variation, heredity, and genetics

People, and all other living things, vary in many ways. In fact no two living things are exactly alike: even 'identical' twins differ in certain ways. Humans, for example, all have the same general shape and body organs. However, characteristics such as height, weight, shape of the face, knowledge, skills, body scars, etc., differ from one person to the next. These characteristics are examples of **variation**.

During sexual reproduction parents pass certain characteristics to their children. But not all characteristics can be inherited. A child inherits the shape of its face, nose, and ears from one or other of its parents, but there is no chance whatsoever of a child inheriting a knowledge of mathematics from its mother for example, or a scar which its father received in an accident.

Characteristics which can be inherited from parents are called **hereditary characteristics**. They include colouring of the hair, eyes, and skin; shape of the face, ears, nose, and mouth; and all the other characteristics which develop as a child grows from a fertilized egg. Characteristics such as knowledge, skills, scars, etc., are called **acquired characteristics**, because people acquire them during their lives.

Some hereditary characteristics show what is known as **continuous variation**. This means that there are many intermediate forms of the characteristic so that they can be graded from one extreme to the other. People, for example, occur in so many different sizes that it is possible to arrange even a small group, such as a class of students, into a continuous line from the smallest to the tallest. A block graph of a characteristic showing continuous variation has the shape shown on the left below.

Other characteristics have few or no intermediate forms, and so cannot be arranged in a continuous sequence. These characteristics are described as showing **discontinuous variation**. For example, humans are either male or female. Intermediate forms are very rare. Another example is the ability some people have to roll their tongues into a U-shape. Either people can do it or they cannot; there are no halfway stages.

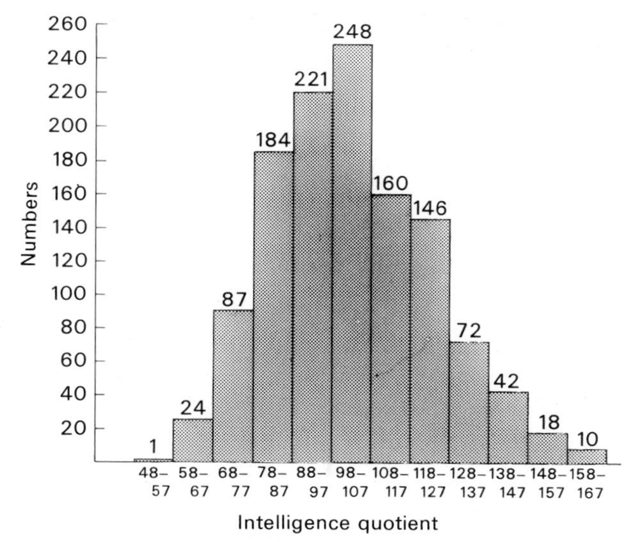

Intelligence shows continuous variation. Note that the majority of children given this intelligence test scored around the average mark of 100, while only a few scored very low or very high marks. A distribution of this type is typical of characteristics showing continuous variation.

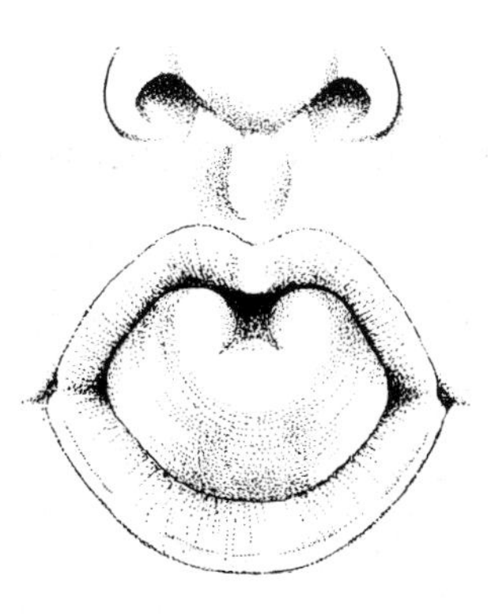

Tongue-rolling is an example of discontinuous variation. Either you can do it or you can't

One of the most important facts about hereditary characteristics is that they are inherited according to certain rules or natural 'laws'. If an African man married a European woman with blond hair and blue eyes, their children would almost certainly have black hair and brown eyes (their father's features).

Why are brown eyes more likely to be inherited than blue eyes? Has the chance of inheriting blue eyes disappeared altogether or could it reappear in future generations? But most important of all, how is it possible to discover the laws which govern this, and the other ways in which offspring inherit parental features?

19.1 Mendel's experiments

Mendel was looking for the laws which govern the way in which hereditary characteristics pass from parents to their offspring. The quest for natural laws is the primary aim of all science, because a knowledge of them makes it possible to predict future events. In this case, a knowledge of the laws of inheritance would make it possible to predict the outcome of a particular mating, thereby helping man to breed particular types of animals and plants. Mendel was successful in this quest where others had failed because of the experimental methods he used.

First, whereas others studied the inheritance of several different characteristics simultaneously, Mendel studied only one characteristic, or character, at a time. For example, using pea plants he chose characteristics such as the length of the stem, i.e. tallness; the shape or colour of the pods; or the shape or colour of the peas (Fig. 19.1).

Second, and most important of all, the characteristics which he chose showed what is now called discontinuous variation. This means that they possessed variations which were distinctly different, and with no confusing intermediate forms. In the case of stem length for example, he used tall varieties and dwarf varieties which, when bred together, produced either tall or dwarf offspring with no intermediate forms that were difficult to classify. Similarly, he used plants with green pods and others with yellow pods which when bred together produced offspring with either green or yellow pods and no confusing intermediate colours. In other words, he used characteristics with *contrasting* variations. This was extremely important because the characteristics were easy to recognize during the course of the experiments. Had he used characteristics with many intermediate variations, i.e. with continuous variation, his results would have been extremely difficult to interpret.

Third, Mendel's use of pea plants in his experiments was a good choice for several reasons. Peas grow quickly and mature in one season; their pollination mechanism is easily controlled by hand so that they can be cross-pollinated–**crossed**–or self-pollinated–**selfed**–easily; and they are inter-fertile, which means different varieties produce fertile seeds after crossing.

Fourth, Mendel always began his experiments with **pure lines**. These are plants which, when self-

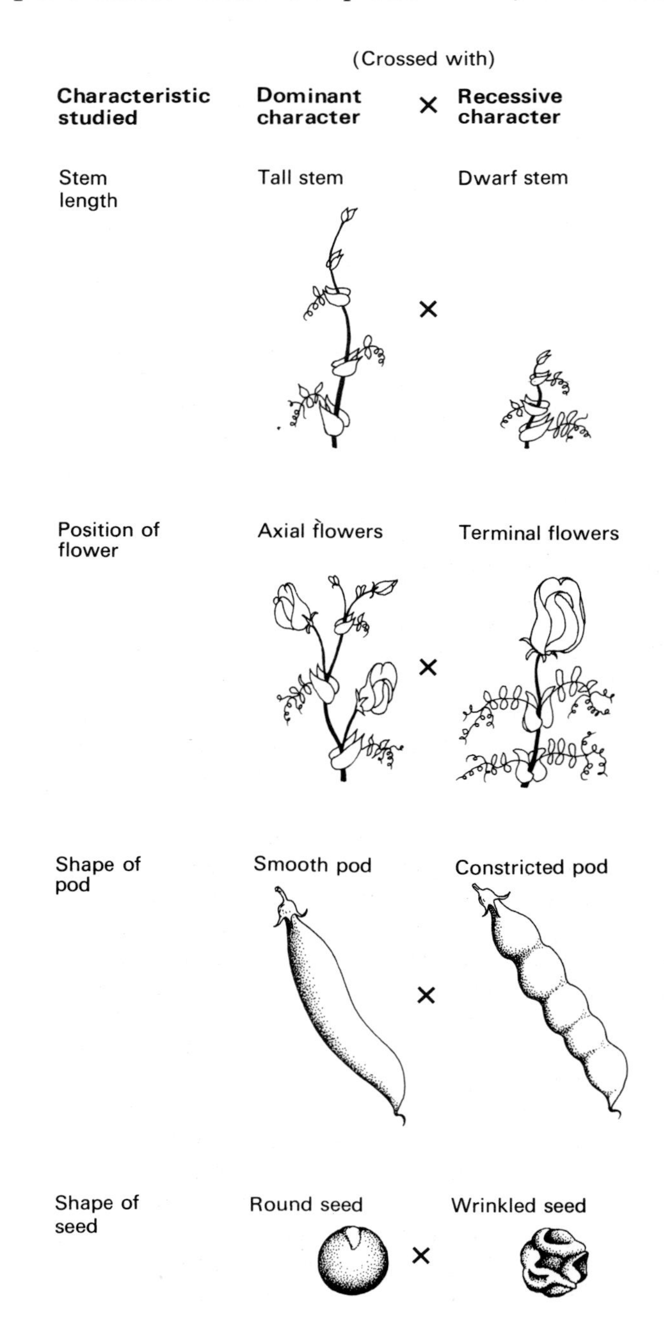

Fig. 19.1 Some of the hereditary characteristics investigated by Mendel in his early experiments.

pollinated generation after generation, always produce plants with the same characteristic, such as a tall stem. In other words, they breed true for a certain characteristic.

Mendel's basic method involved **hybridization**, which involves crossing two organisms which differ in some way and results in offspring called **hybrids**. In his earliest experiments Mendel produced hybrids by crossing plants which differed in only *one* way, such as stem length, or colour of pod, or shape of seed. These are called **monohybrid crosses**. The experiment described below was designed to investigate the inheritance of stem length, and involved a monohybrid cross between tall and dwarf plants.

Results of a monohybrid cross

In order to investigate the inheritance of stem length Mendel obtained pure lines of tall and dwarf plants and crossed them in the following manner, and with the following results.

1. In this, and all other experiments involving cross-pollination, Mendel had first to prevent self-pollination, and then ensure that cross-pollination took place between only the chosen opposite varieties. He prevented self-pollination by removing the anthers from a number of tall and dwarf plants before they had produced mature pollen grains. Later, when their carpels were mature, he transferred pollen to the stigmas of the same flowers from the anthers of the *opposite* variety, i.e. the stigmas of flowers on tall plants were dusted with pollen from dwarf plants, and the stigmas of flowers on dwarf plants were dusted with pollen from tall plants. These pollinated plants, which form the starting point of the experiment, are called the **first parental generation**, or P_1.

2. Mendel collected all the seeds from the P_1 plants, set them in soil and awaited the results. The new plants which grew from these seeds were called the **first filial generation**, or F_1. Without exception all the F_1 plants were tall, no matter whether the seeds which produced them had come from tall or dwarf parents (Figs. 19.2 and 19.3). There were no plants of intermediate size, and no dwarf plants. Clearly, the ability to produce the dwarf characteristic had either disappeared, or it had been suppressed in some way by the tall characteristic. Mendel said that since tallness had 'dominated' dwarfness in this way tallness should be called the **dominant characteristic**. Other examples of dominant characteristics (or dominant characters) are given in Figure 19.1.

3. Next, Mendel self-pollinated all flowers of the F_1 plants, and then covered the flowers to prevent any possibility of cross-pollination. Subsequently, he collected seeds from the F_1 plants, set them in soil and waited for the F_2 generation to grow from them. In the F_2 generation, out of a total of 1064 plants, 787 were tall and 277 were dwarf. That is, roughly three-quarters were tall and one quarter were dwarf, giving a ratio of approximately 3:1 (Fig. 19.2). The ability to produce the dwarf characteristic had reappeared again,

Fig. 19.2 Diagram of a monohybrid cross between pure line tall, and pure line dwarf pea plants

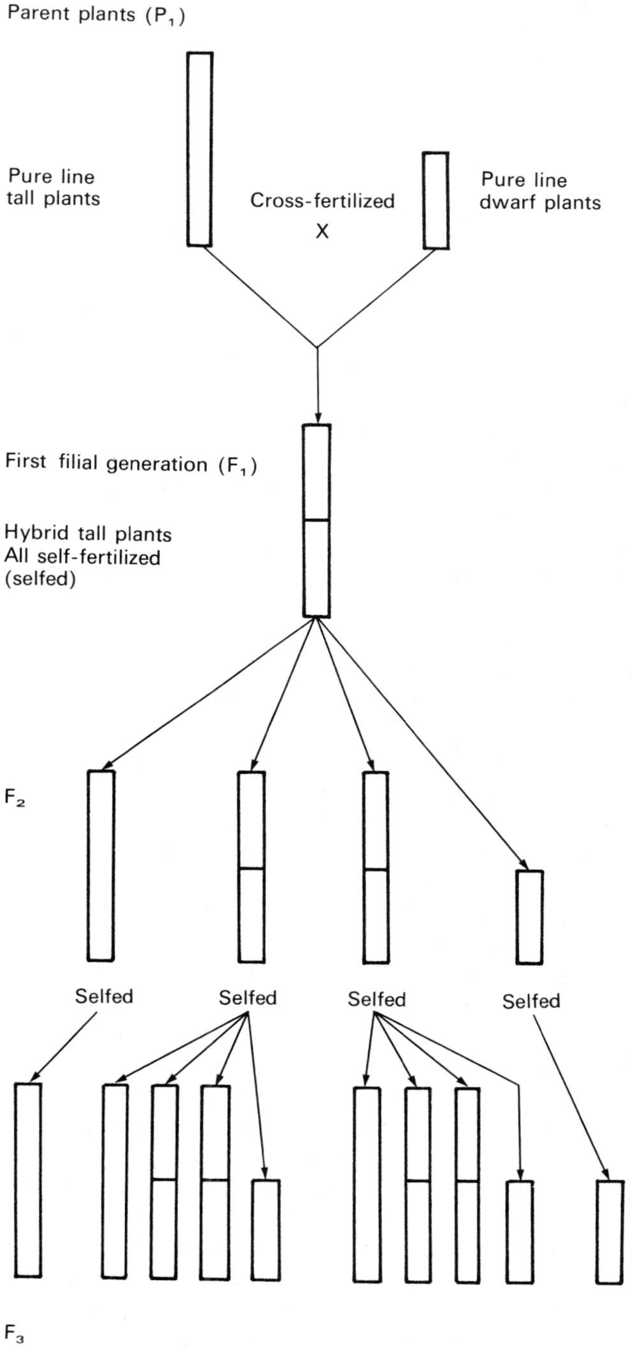

Parent plants (P_1)

Pure line tall plants

Cross-fertilized

X

Pure line dwarf plants

First filial generation (F_1)

Hybrid tall plants
All self-fertilized
(selfed)

F_2

Selfed

Selfed

Selfed

Selfed

F_3

Table 1 Results of Mendel's monohybrid crosses

P_1 crosses			*F_1*	*F_2*	*Ratios*
tall	X	dwarf stems	all tall	787 tall 277 dwarf 1 064 total	2.84:1
round	X	wrinkled seeds	all round	5 474 round 1 850 wrinkled 7 324 total	2.96:1
yellow	X	green cotyledons	all yellow	6 022 yellow 2 001 green 8 023 total	3.01:1
coloured	X	white seed coats	all coloured	705 coloured 224 white 929 total	3.15:1
smooth	X	constricted pods	all smooth	882 smooth 299 constricted 1 181 total	2.95:1
green	X	yellow pods	all green	428 green 152 yellow 580 total	2.82:1
axial	X	terminal flowers	all axial	651 axial 207 terminal 858 total	3.14:1

to a limited extent, and so Mendel called dwarfness a **recessive characteristic** because it had 'receded' in the F_1 generation only to reappear in the F_2. Other recessive characters are given in Figure 19.1.

4. Mendel now self-pollinated the F_2 plants and produced F_3 plants from their seeds. All the dwarf F_2 plants produced only dwarf F_3 plants. This meant that they were pure lines, producing nothing but dwarfs henceforth. However, the F_2 tall plants were found to be of two types: one third of them were pure lines giving rise to only tall F_3 plants, while two thirds proved to be hybrids producing tall and dwarf F_3 plants in the ratio of 3:1.

Mendel used this procedure with seven pairs of contrasting characteristics, and in every case one characteristic was found to be dominant, completely excluding the recessive one from the F_1 generation, while the F_2 plants displayed both dominant and recessive characters in a ratio of approximately 3:1. These results are summarized in Table 1.

Mendel was now faced with the task of interpreting his results. It must be remembered that Mendel worked entirely alone and at a time when nothing was known about chromosomes or cell division, and yet he explained his results in a way which ties in almost exactly with modern knowledge of how hereditary mechanisms work.

Interpretation of a monohybrid cross

Remember, Mendel was looking for a law of inheritance which would enable him to predict the outcome of a particular mating, or cross-pollination, and he had greatly simplified this task by studying the inheritance of only one pair of contrasting characteristics. In fact, by this technique he had achieved his aim, for it had revealed a predictable pattern of events from among the seemingly chaotic process of inheritance. This pattern occurred whenever he crossed plants bearing a dominant character with plants bearing a recessive character: the F_1 plants displayed only the dominant character, while the F_2 plants displayed dominant and recessive characters in the ratio of 3:1.

This means that when two contrasting characters are brought together, as in a monohybrid cross, they do not fuse or blend together and so produce offspring with many intermediate forms, such as medium-sized stems or children with one brown eye and one blue. The contrasting characters retain their individual identity and separate, unchanged, in the F_2 generation in a certain predictable order. Mendel summarized all this in a simple statement now known as Mendel's First Law. He said: 'In a cross between plants bearing contrasting characters, the characters segregate (separate) in the second filial generation'. But how is it possible for a character to survive unchanged from a cross with its opposite character? To answer this question Mendel proposed his theory of **hereditary factors**.

Mendel suggested that the body of an organism contains a number of microscopic particles which he called factors, and that these factors control the appearance of the organism's hereditary characteristics. According to Mendel, each contrasting characteristic has its own separate factor. For example, the ability to produce a tall stem is controlled by one factor, and the ability to produce a dwarf stem is controlled by another, separate factor. Furthermore, in order to account for the appearance of F_1 generations and the observed ratios of F_2 generations Mendel proposed that factors must operate in pairs. The reasoning behind this assumption can be explained by employing Mendel's system of using different letters of the alphabet to represent the hereditary factors. In the following explanation 'T' represents the dominant factor controlling tallness, and 't' represents the recessive factor controlling dwarfness. Figure 19.3 summarizes this explanation diagrammatically.

Starting with the parent (P_1) plants in the monohybrid cross described above, Mendel argued that the pure line tall plants must contain pairs of factors for tallness – written in symbols as TT – whereas the pure line dwarf plants must have pairs of factors for dwarfness – written tt. Mendel assumed that when P_1 plants produce gametes (pollen and ovules) the factors in each pair must separate so that each gamete receives *one* factor. Thus, when tall plants produce gametes, members of their TT pairs will separate and each of their pollen grains and ovules will receive one T factor. Similarly, all the pollen grains and ovules of the dwarf plants will receive one t factor.

Mendel argued that when a pollen grain from a tall plant fertilizes an ovule from a dwarf plant, and vice versa, a T and a t factor will come together making a Tt zygote. When this zygote grows into an F_1 plant the T will dominate the t and the plant will grow tall. Since all the zygotes from the P_1 cross are Tt this explains why all the F_1 plants are tall. During gamete production in the F_1 plants members of the Tt pairs will separate and each pollen grain and ovule will receive either a T or a t factor. Since there are equal numbers of T and t factors an equal number of T and t pollen grains and ovules will be produced.

The next step in the experiment is to self-pollinate the F_1 plants. To understand what happens it must be remembered that pollination and fertilization are

Fig. 19.3 Segregation of hereditary factors in a cross between pure line tall and dwarf plants

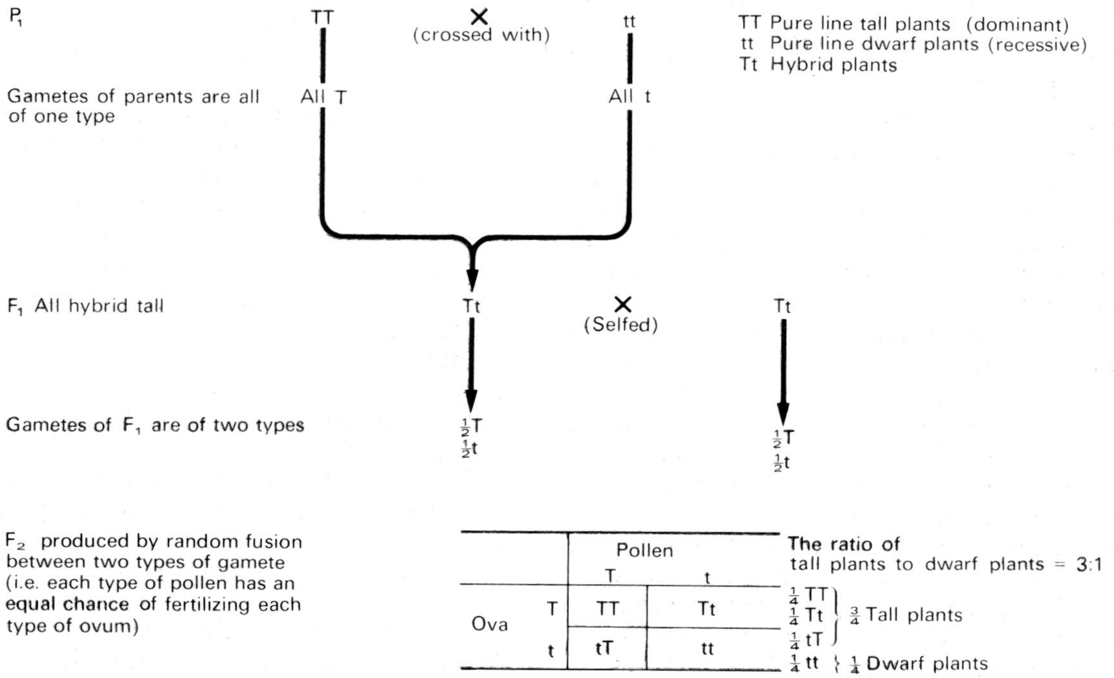

P_1 TT × (crossed with) tt

TT Pure line tall plants (dominant)
tt Pure line dwarf plants (recessive)
Tt Hybrid plants

Gametes of parents are all of one type — All T, All t

F_1 All hybrid tall — Tt × (Selfed) Tt

Gametes of F_1 are of two types — $\frac{1}{2}$T $\frac{1}{2}$t; $\frac{1}{2}$T $\frac{1}{2}$t

F_2 produced by random fusion between two types of gamete (i.e. each type of pollen has an equal chance of fertilizing each type of ovum)

		Pollen T	Pollen t
Ova	T	TT	Tt
	t	tT	tt

The ratio of tall plants to dwarf plants = 3:1

$\frac{1}{4}$TT, $\frac{1}{4}$Tt, $\frac{1}{4}$tT } $\frac{3}{4}$ Tall plants
$\frac{1}{4}$tt } $\frac{1}{4}$ Dwarf plants

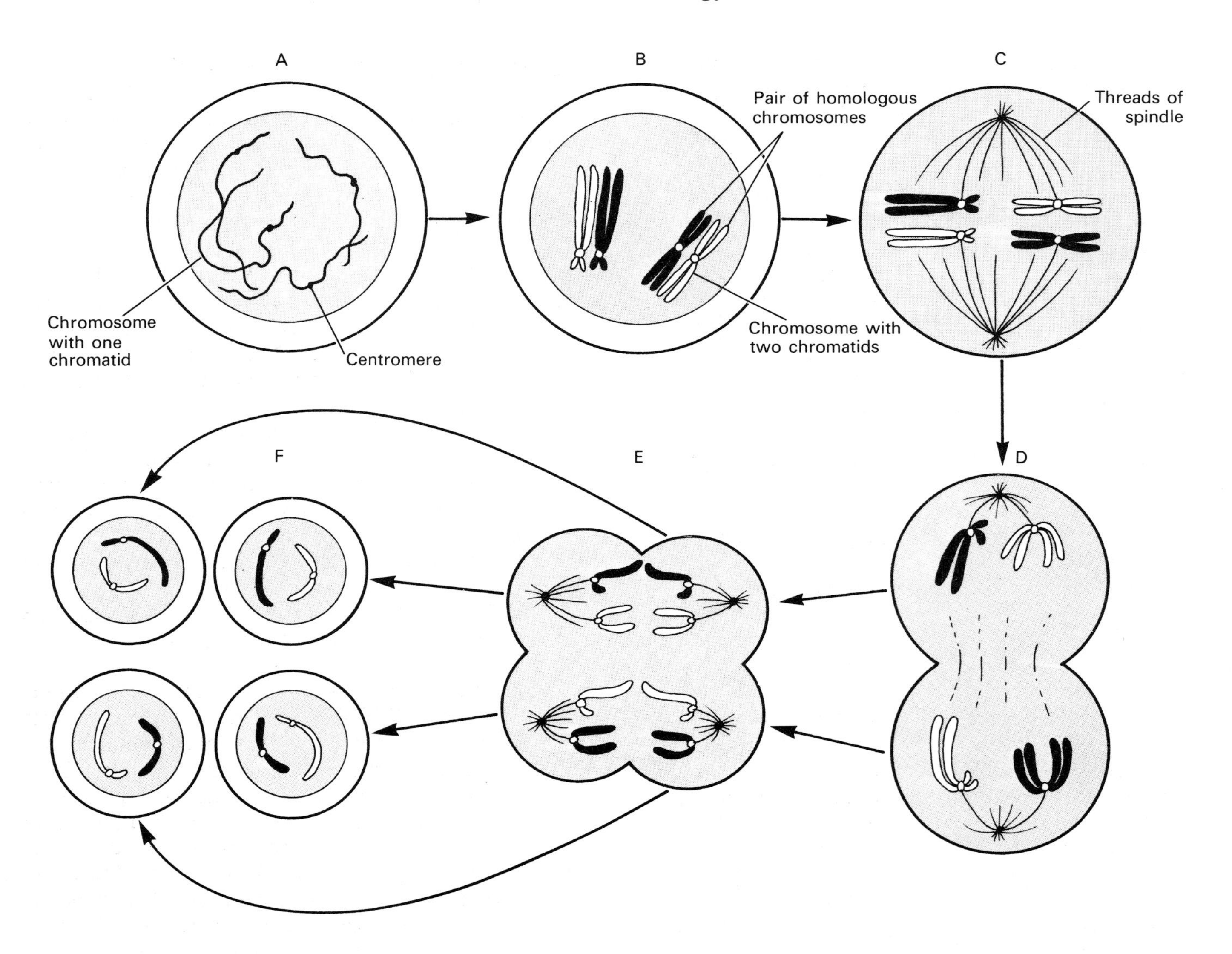

purely random events. That is, either of the two types of pollen grain have an equal chance of fertilizing either of the two types of ovule. This fact is extremely important because it accounts for the 3:1 ratio in the F_2 generation. This can be verified in two ways.

First, look at the matrix at the bottom of Figure 19.3. This shows what happens if random pollination actually results in each type of pollen grain fertilizing each type of ovule. In this experiment, three-quarters of the F_2 plants develop from zygotes with a Tt factor (or a tT factor) or from zygotes with a TT factor. These are tall plants. One quarter of the F_2 plants develop from zygotes with a tt factor. These are dwarf plants. This gives a ratio of 3:1.

Second, the probability that random fertilization will produce such results can be demonstrated by means of the coin-tossing test described in exercise A.

Fig. 19.4 Meiosis (reduction division).

A Chromosomes become shorter and thicker until they are visible as single strands consisting of one chromatid. **B** Chromosomes pair off (i.e. form homologous pairs). The members of each pair are identical in shape and size (one of each pair is drawn in black simply to distinguish it from its partner). Each chromosome manufactures new DNA forming a second chromatid. **C** Homologous pairs become arranged along the equator of the cell; the nuclear membrane disappears; and the spindle threads appear. **D** Members of each homologous pair separate and move in opposite directions. The cell begins to divide into two. **E** The chromatids of each chromosome separate and move in opposite directions. **F** The cell divides into four parts, each containing two chromosomes. The four parts become gametes.

To summarize: Mendel argued that his results can be explained if it is assumed that hereditary characteristics are controlled by pairs of hereditary factors. During reproduction the members of each pair separate and move into different gametes, i.e. a gamete receives one factor from each pair. At fertilization the factors come together again and the pairs are restored. The separation of factors into different gametes and their restoration into pairs as a result of random fertilization is the mechanism behind Mendel's First Law, since it eventually leads to the segregation of contrasting characters in the F_2 generation.

Segregation has now been demonstrated in all the major groups of organisms which reproduce sexually and has been found to operate more or less as it does in pea plants. However, many new discoveries have been made since Mendel's time, and several have revealed exceptions to his laws of inheritance (comprehension test 3). Nevertheless, his principles generally hold good even today, and the next section will show how closely they relate to modern discoveries about the cell.

19.2 Chromosomes and heredity

Since Mendel's death the invention of improved microscopes and techniques for staining cells have made it possible to examine the cell nucleus and revealed the presence of objects called **chromosomes**. The study of how chromosomes behave during cell division and reproduction has led to the realization that they must carry the hereditary factors described by Mendel. To understand how this conclusion was reached it is necessary to study a type of cell division called meiosis.

Meiosis

Unlike mitosis (Fig. 2.7) meiosis occurs only within the reproductive organs (except in protists where reproductive organs are absent). Meiosis is concerned with the production of gametes. Another difference is that whereas mitosis produces two daughter cells which have exactly the *same* number of chromosomes as were present in the original parent cell, meiosis produces four daughter cells with *half* the normal number of chromosomes. For this reason meiosis is sometimes called **reduction division**. In humans, for instance, there are 46 chromosomes in every body cell, but meiosis produces sperms and ova which contain only 23 chromosomes each. The normal number of 46 is restored at fertilization when a sperm fuses with an ovum. The stages of meiosis, in greatly simplified form, are illustrated in Figure 19.4 and this should be studied carefully before reading further.

Mendelian factors and chromosomes

Hereditary factors described by Mendel (i.e. Mendelian factors) are now believed to be part of the structure of chromosomes. The evidence for this comes from observing many parallels between the behaviour of chromosomes during meiosis and fertilization, and Mendel's theory of how factors behave during reproduction.

1. Mendelian factors operate in pairs and, in the first stage of meiosis, chromosomes are seen to form homologous pairs which are composed of chromosomes identical in length and shape. (The 'sex' chromosomes of some animals are an exception, and will be described later.)
2. Mendelian factors separate from each other at gamete formation so that only one from each pair enters each gamete. A gamete therefore contains half the total number of factors. Homologous chromosomes also separate at meiosis, and one from each pair enters each gamete so that a gamete possesses half the normal number of chromosomes.
3. The normal number of both Mendelian factors and chromosomes is restored at fertilization.

The three points show that chromosomes 'behave' in the way that Mendel supposed his factors must behave, and this suggests that chromosomes are factors. However, more recent investigations have disclosed that there are far more factors than there are chromosomes, and therefore it seemed likely that chromosomes contained a number of different factors. This has now been confirmed, and the Mendelian factors have been renamed **genes** (which is why the scientific study of heredity is called genetics). Chromosomes are in fact made up of many different genes.

A gene is a unit of heredity. Each gene controls the development of a set of hereditary characteristics in an organism. In man it is estimated that there are about 10 000 genes contained in the 46 chromosomes of each cell. Among this number there are genes which control the development of all the body organs, and others which control visible characteristics such as the shape of the face, and the colour of the eyes and hair. Characteristics such as height and intelligence are said to be **polygenic**, meaning that they result from the activity of many genes.

Genes affect each other, and in turn are usually affected by the environment of the organism. It has been said of most genes that they provide a 'promise' of a certain result. For example, it is certain that a man who inherits the genes which produce brown eyes will in fact have eyes of that colour, but a man who inherits genes which could produce high intelligence may never develop this characteristic to its full extent

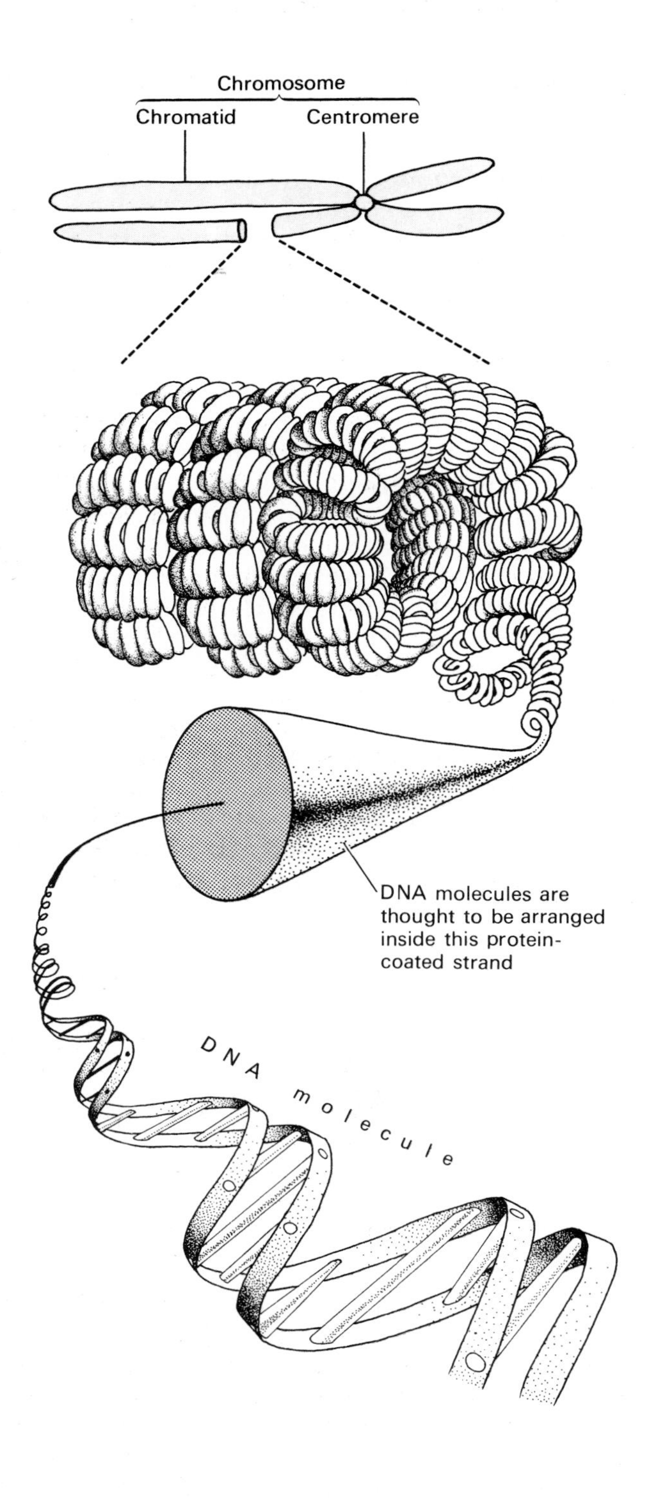

Fig. 19.5 Structure of a chromosome during the early stage of meiosis. A chromosome consists of a protein coated strand which coils in three ways during the time when a cell prepares to divide by meiosis (or mitosis). This strand contains DNA molecules, arranged along its length. The DNA molecules are coiled into a shape known as a double helix

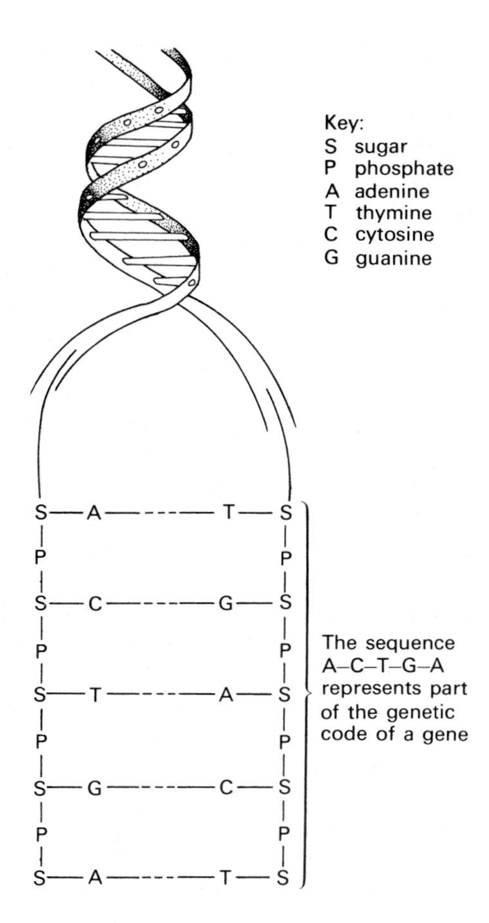

Fig. 19.6 The chemical structure of DNA

unless he is brought up in an intellectually stimulating environment.

Structure of a gene

A gene is a short length of chromosome. In fact, a sequence of genes, all with different functions, are arranged along the length of each chromosome, like beads on a necklace.

In 1944 the American biologist Oswald Avery transferred a substance called **deoxyribonucleic acid**, or DNA, from one type of bacterium to another and, in so doing, discovered that he had transferred the organism's hereditary information as well. It was already known that the cells of all organisms contain DNA, and that in animals and plants DNA is contained in the chromosomes. But as a result of Avery's experiments and numerous others, it became clear that hereditary information is 'stored' in DNA as part of its chemical structure.

DNA can be compared with an architect's plans for a house. Both contain a set of instructions for building something, which in the case of DNA is the body of a living organism. The popular way of describing the building instructions contained in DNA is to call them the **genetic code**, because they are 'translated' into another form when put to use inside a cell. But what is the 'language' in which these coded instructions are written, and how is it used to 'write out' the instructions contained in each of the genes?

To answer these questions it is necessary to study the structure of the DNA molecule. Figures 19.5 and 19.6 illustrate this structure, greatly simplified, and also show how the long coiled DNA molecules are thought to be organized in the chromosomes. The twisted arrangement of a DNA molecule is called a **double helix**. The ribbon-like strands of the helix are made of alternate sugar and phosphate molecules joined into a chain (Fig. 19.6), and the cross-bridges (like the rungs of a ladder) are made of substances called **bases**, of which there are four types: adenine, cytosine, guanine, and thymine.

Incredible as it may seem it is almost certain that the language of the genetic code is no more than the sequence of these bases along DNA molecules. If each base is represented by its initial letter, it can be seen that the 'words' of the language in which the genetic code is written are derived from a four letter alphabet: A, C, G, T.

Although the nature of the genetic code appears simple, its translation into living matter is very complicated. The accepted theory is that the code is arranged into 'words' of three letters, that is, into combinations of three adjacent bases. These groups of three bases are arranged in a specific sequence along the DNA molecule, and this sequence appears to control the linking up of amino acids inside a cell into an equally specific sequence which results in a particular type of protein molecule. For instance, the base sequence of AAT (adenine, adenine, thymine) in the code is translated into an amino acid called leucine; while CAA (cytosine, adenine, adenine) is the code for the amino acid valine.

A gene probably consists of 1000 or more groups of three bases which together represent the code for one complete protein molecule. The function of genes is to control the types of protein made in a cell. In this way genes control both the structure and the functions of the whole organism because the proteins made according to their instructions form both the building materials of life, and the enzymes which control the chemistry of life.

19.3 **Modern genetics**

Having outlined a little of the mechanism of heredity by studying Mendel's First Law, it is now possible to introduce some technical terms and methods used in modern genetics.

Technical terms

The position which a gene occupies on a chromosome is called the **locus** of that gene. On the whole genes remain in the same locus, although there are circumstances in which a gene moves from one locus to another.

Chromosomes occur in pairs (homologous chromosomes) and so genes occur in pairs. In fact the opposite partner of any particular gene occupies the same relative locus on the corresponding homologous chromosome. The genes of such a pair are said to be **alleles** of each other, or an **allelomorphic pair**. The term allele is more often used to refer to all the genes which could occupy a particular locus. The genes which control human blood groups are an example.

There are four main blood groups A, B, AB, and O. These are controlled by three alleles: the genes A, B, and O. Since genes operate in pairs only two of these alleles are present at the same time in a person's cells. Gene O is recessive to both A and B, and A and B are said to be **co-dominant**, because neither can dominate the other. Table 2 shows how different combinations of these alleles produce various blood groups.

Table 2 Inheritance of human blood groups

Gene pair	Produces	Blood group
OO	⟶	O
AA	⟶	A
AO	⟶	A
BB	⟶	B
BO	⟶	B
AB	⟶	AB

The genes of an allelomorphic pair may be identical, in which case they are said to be **homozygous**. For instance, the genes TT in a pure line tall pea plant, and tt in a pure line dwarf plant, are both examples of homozygous genes. Pure lines of any organism are said to be homozygous for the characteristic in question, which is a short way of saying that the allelomorphic pair which controls the characteristic is homozygous.

When genes in an allelomorphic pair are opposite in nature, i.e., when one is dominant and the other recessive, the pair is said to be **heterozygous**. An example of heterozygous genes is the pair Tt found in hybrid pea plants.

An albino and a normal dark-skinned native of the Trobriand Islands (near New Guinea). Albinism results from a mutation in the genes which control skin colouration, so that the colour fails to develop. The skin of an albino is almost transparent and is quickly damaged by exposure to sunlight. The mutation is recessive

The nature and arrangement of genes in an organism is called the **genotype** of that organism. As far as stem length is concerned in pea plants there are three genotypes: TT, tt, and Tt. As far as genotypes are concerned tT is the same as Tt.

It is not always possible to tell the genotype of an organism by looking at its external features. For instance, plants with genotypes TT and Tt look alike, (i.e. tall) but are genetically different. On the other hand, if two plants with the genotype TT are grown on different soils the one on the better soil may grow taller than the other on poorer soil. The word **phenotype** is therefore used in reference to the *visible* characteristics of an organism, as contrasted with genotype (i.e. the genes which it possesses). In a broader sense the term phenotype refers to all the characteristics of an organism which result from the action of the genes, the environment, and the interaction between the two.

A **mutation** is a sudden change in a gene, or a chromosome, which alters the way in which it controls development. The majority of mutations are changes in the DNA of a single gene: the sequence of its bases may be altered, or some may be omitted altogether. The gene will then produce a different type of protein with corresponding effects in the organism. Some mutations affect all or part of a chromosome: parts may break off; parts may break and rejoin in a different way, or become attached to another chromosome; and sometimes a whole chromosome is either lost or gained.

The majority of mutations are harmful. This is because they alter and upset the delicately balanced mechanisms of an organism. Very rarely mutations are lethal, and equally rarely they produce changes which are advantageous.

If mutations occur in body cells they may spread by mitosis but are usually restricted to one organism. On the other hand, if mutations occur in the reproductive organs within cells which produce gametes the mutations may be passed on to the next, and subsequent generations.

It is not certain what causes mutations in nature, but several ways have been discovered of artificially increasing the rate at which mutations occur. In some organisms the mutation rate increases as temperature rises: the rate appears to double with each rise of 10 °C. Several chemicals cause a rise in mutation rate when applied to living organisms, e.g. mustard gas and formaldehyde. By far the most powerful means of increasing the mutation rate is by exposing an organism to high energy radiation such as X-rays, beta, and gamma rays.

This effect of artificial radiation is a clue to the possible source of natural mutations because radiation is present in all natural environments. The earth is constantly bombarded with high energy radiation in the form of cosmic rays from outer space, and the earth's crust contains radio-active elements such as uranium. Furthermore, in recent years the advance of technology into the use of atomic, or nuclear, energy has increased the amount of artificial radiation. Whether or not this will greatly increase mutation rates remains to be seen. The danger is there, however, and precautions must be taken to guard against an accidental outflow of radiation into the environment.

Mutations are comparatively rare events because DNA is a stable molecule, and chromosomes are stable structures. For several reasons, most mutations do not reach the next generation, or have much effect even if they are inherited. First, only mutations carried by gametes are inherited, and only a tiny proportion of gametes take part in fertilization to produce offspring. Consequently, inheritance of mutations depends primarily on the proportion of gametes which carry mutated genes. Second, most mutations are recessive and confined to only one gene of an allelomorphic pair. Therefore, the majority of inherited mutations do not appear in the phenotype because they are suppressed by their dominant allele. Third, since mutations are largely harmful, and sometimes lethal, natural selection operates against them when they do appear in the phenotypes, and this limits their spread through the population.

At the same time, it must be remembered that the majority of species consist of millions of organisms producing millions of offspring yearly. Consequently, the rare event of an advantageous mutation appearing in one phenotype can occur a significant number of times in a species as a whole. In other words, a rare event multiplied millions of times can become a common event.

Mutation and evolution

There is a widely held theory that evolution progresses by natural selection of the mutations which occur at random in every species. The theory argues that unfavourable mutations disappear while the rare favourable mutations survive and multiply. This could happen by means of a type of mutation in a species occurring time and time again. These periodic mutations may be unfavourable for generations until suddenly the environment changes, or the species migrates to a different environment where the mutation is favoured by natural selection.

Consider the Peppered Moth (chapter 18). The mutation which produces the dark variety of this moth probably occurred in most generations prior to the industrial revolution but was not favoured by natural selection. Then industrial pollution blackened parts of the moth's environment favouring the dark colouration because of the camouflage which it provided, thereby allowing this variety to multiply.

It seems possible, then, that mutations provide the raw material of evolution in the form of randomly occurring variations, and the mechanism of heredity distributes these variations throughout the species if they are favoured by natural selection.

Use of Drosophila in genetics

Modern geneticists work with many organisms, but the most famous, and one of the most valuable, is a small fly, about 2 mm long, called *Drosophila melanogaster* (Fig. 19.7). *Drosophila* owes its fame to the American geneticist T. H. Morgan and his pupil H. J. Muller, who chose to use this fly in their experiments because it is easily and quickly bred (it has a 10-day life cycle) and gives results in a few weeks which could have taken Mendel at least two or three years using pea plants. Another advantage of *Drosophila* is that when exposed to radiation the wild types yield many distinctly different mutations with, for example, red eyes, white eyes, crumpled wings and vestigial

Fig. 19.7 Comparison between male and female *Drosophila melanogaster*. Females are longer than males, and have a more pointed abdomen with widely spaced black bands. Males have a rounded abdomen with closely spaced bands (giving the abdomen a black tip), and they have a sex comb on each foreleg

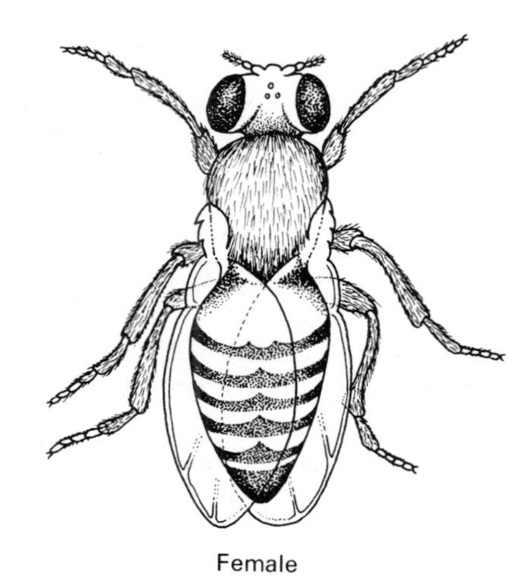

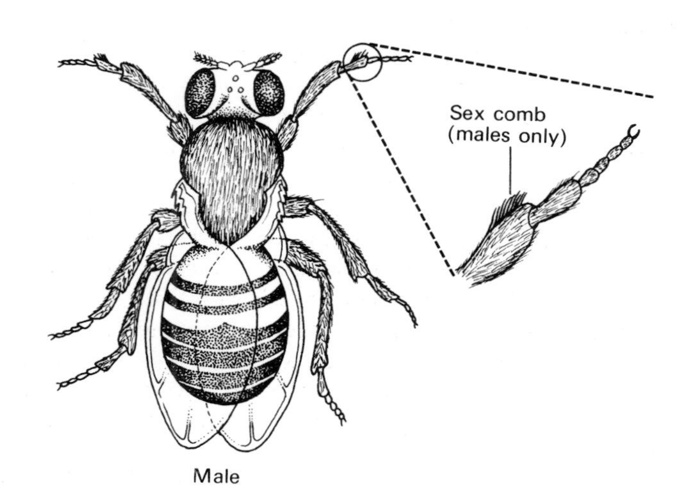

A Changes of phenotypes

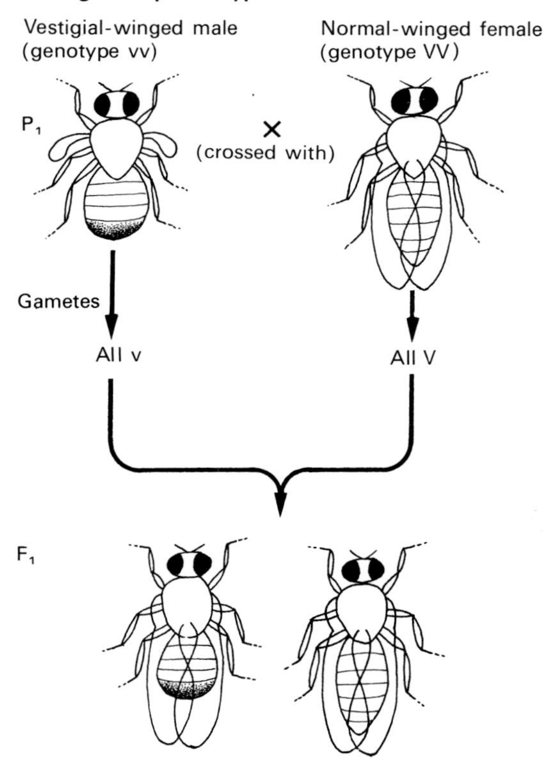

Roughly equal numbers of males and females, all with normal wings (genotype Vv)

Cross between F_1 flies

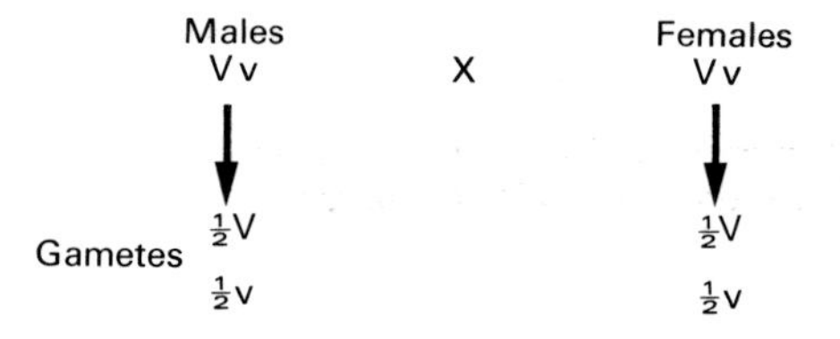

F_2 produced by random fertilization

		Sperms V	Sperms v
Ova	V	VV	Vv
	v	vV	vv

Phenotype ratio of normal-winged to vestigial-winged flies is 3:1

$\frac{1}{4}$ VV homozygous normal-winged
$\frac{1}{4}$ Vv heterozygous normal-winged } $\frac{3}{4}$
$\frac{1}{4}$ vV heterozygous normal-winged
$\frac{1}{4}$ vv homozygous vestigial-winged } $\frac{1}{4}$

B Movements of chromosomes and genes

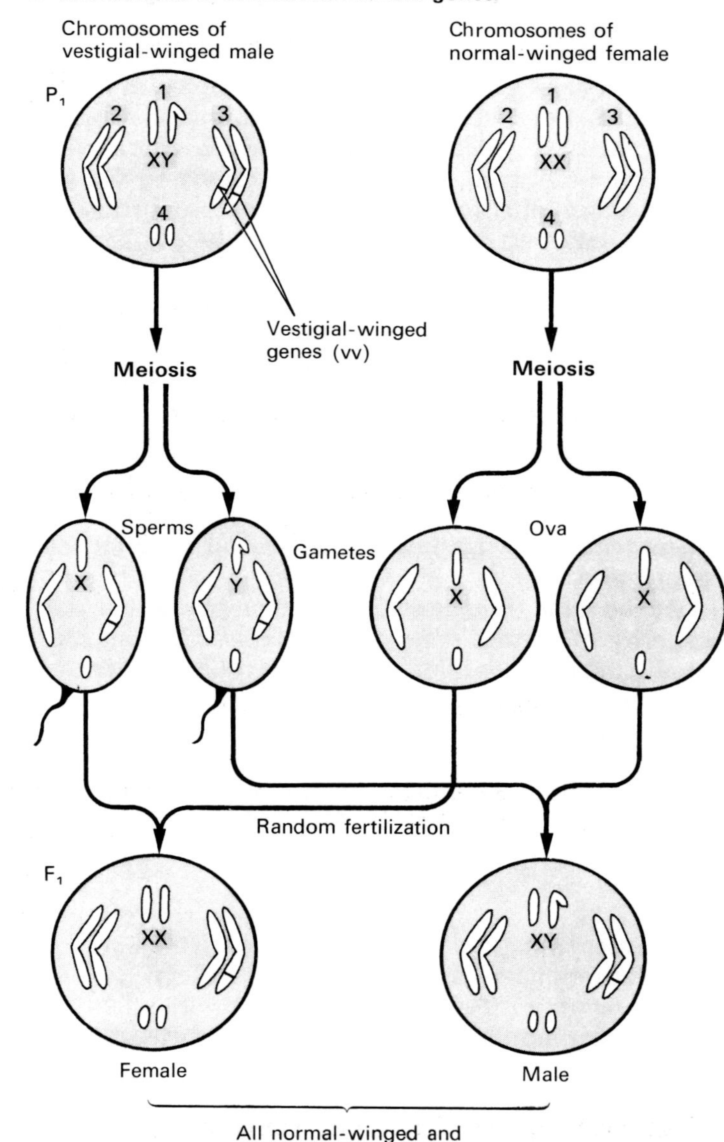

Fig. 19.8 Cross between normal-winged and vestigial-winged *Drosophila*.

A illustrates the phenotypes up to F_1 flies, together with segregation of genes in the F_2 flies. **B** illustrates the segregation of chromosomes and vestigial genes between parents and F_1 flies.

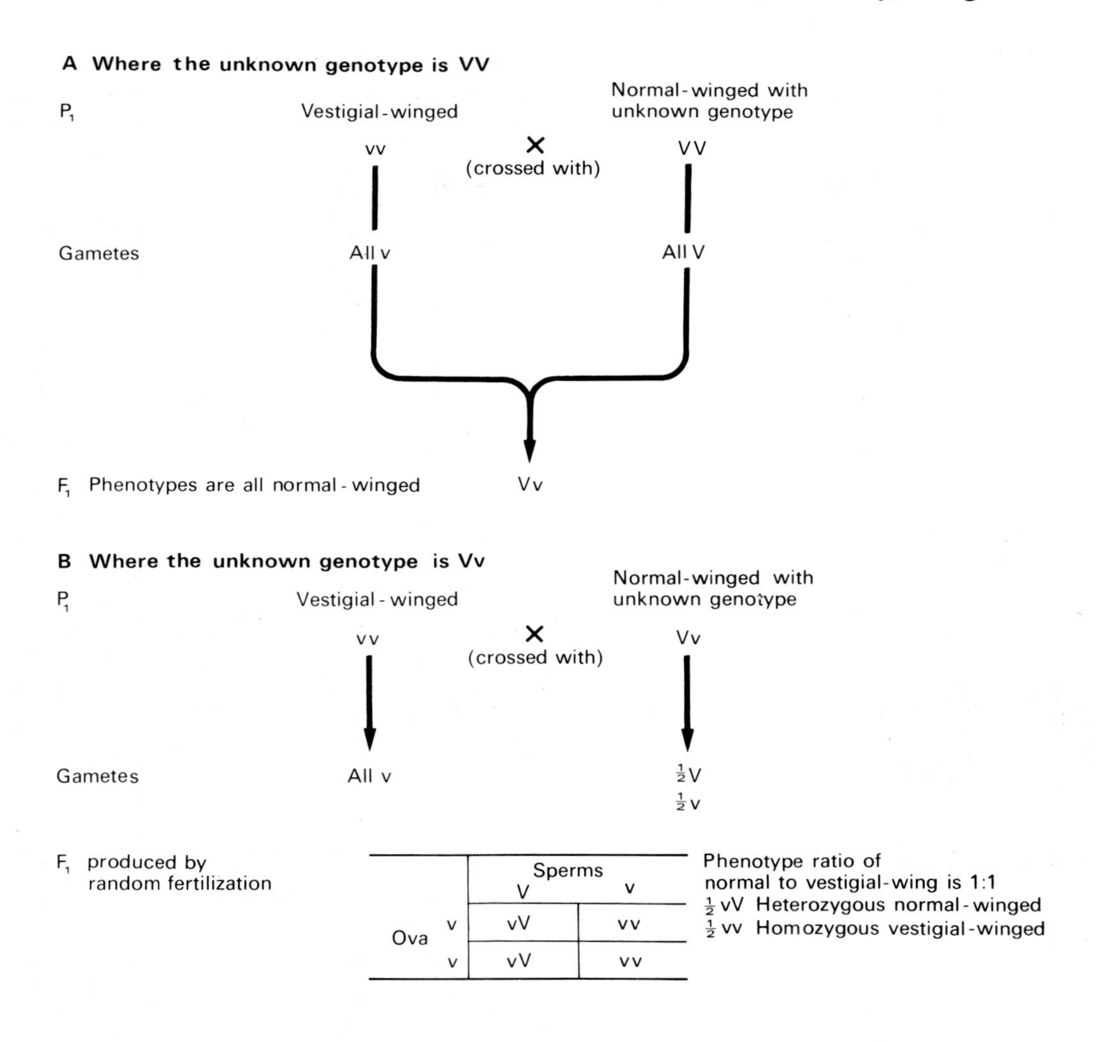

Fig. 19.9 Use of a test cross to distinguish homozygous dominant VV flies from those with a Vv genotype

wings (i.e. wings which are abnormally small).

Figure 19.8 uses modern terms to describe a cross between a mutant *Drosophila* with vestigial wings, and a normal-winged fly. As in Mendel's monohybrid crosses, the F_1 shows only the dominant character, in this case normal wings, and the phenotypes of the F_2 generation show the dominant and recessive characters in a ratio of 3:1. But there remains the problem of distinguishing between homozygous VV flies and heterozygous Vv flies, the phenotypes of which are both normal-winged. At this point Mendel self-pollinated his pea plants (Fig. 19.3) but this is impossible with *Drosophila* (and other animals). The problem is solved by a **test-cross**, or back-cross as it is sometimes called, between F_2 flies with an unknown genotype and homozygous vv flies whose genotype is clearly visible from their vestigial wing shape. Figure 19.9A shows that where the unknown genotype is VV, all the F_1 generation develop normal wings. Figure 19.9B shows that where the unknown genotype is Vv the phenotypes of the F_2 generation show the dominant and recessive characters in a ratio of 1:1.

Figure 19.8B illustrates the segregation of chromosomes and vestigial-wing genes from parent flies to the F_1 generation. In addition, this figure illustrates the presence of XX chromosomes in female flies, and XY chromosomes in males. These are sometimes called the **sex chromosomes** because they are concerned with determining the sex of the organism.

Sex determination

In *Drosophila*, and many other animals including humans, if sexual reproduction produces a zygote with an X and a Y chromosome that zygote will always develop into a male. But zygotes with an X and an X chromosome always develop into females. The two types of zygote are produced in the following way.

Sex chromosomes separate at meiosis (like all other chromosome pairs) and only one goes into each gamete. Therefore, all female gametes carry one X chromosome, while half the male gametes carry a Y chromosome and the other half an X chromosome. Since there is an equal chance of both X and Y male gametes fertilizing a female gamete there should be an equal number of male and female offspring. For some unknown reason, however, there are usually more males than females. In humans, for instance, the ratio of female to male babies born alive is about 100:106.

It is not clear why the XX combination produces females while the XY combination produces males, since the genes which control the development of sexual characteristics are not restricted to the sex chromosomes but are distributed throughout all homologous pairs. In fact, a zygote appears to have all the genes for producing *both* male and female characteristics. The final outcome as to which sex develops seems to depend on the Y chromosome: when the Y is present only the male characteristics develop, and when it is absent female characteristics develop.

Verification and inquiry exercises

A *To demonstrate that random fertilization between equal numbers of dominant and recessive genes produces a phenotype ratio of 3:1*

1. If animals with the genotype Aa are crossed, the F_1 offspring will be produced from random fertilization between equal numbers of A and a sperms and A and a ova. Thus:

		Sperms A	*Sperms* a
Ova	A	AA	Aa
Ova	a	aA	aa

$\frac{1}{4}$AA, $\frac{1}{4}$Aa, $\frac{1}{4}$aA } $\frac{3}{4}$

$\frac{1}{4}$aa } $\frac{1}{4}$

The phenotypes occur in the ratio of 3:1

It is possible to show that the 3:1 ratio results from random fertilization in the following way.

2. Take two coins of the same denomination (2p or 10p coins are suitable). Use white sticky paper to label one side of one coin Sperm A, and the opposite side Sperm a; then label one side of the other coin Ovum A, and the opposite side Ovum a. Copy this chart.

Sperm	*Ovum*	*Tally*	
A	A		Total = (A A, a A, A a)
a	A		
A	a		
a	a		Total =

3. Working in pairs spin an 'ovum' coin and a 'sperm' coin simultaneously noting how they fall, and enter the result in the appropriate part of the tally column. Repeat this at least 50 times.

4. Does the ratio of one total to the other amount to roughly 3:1?

5. Explain why coin tossing can be used in place of fertilization in this experiment.

6. How does this experiment show that, to be successful, all genetics experiments must involve large numbers of offspring?

B *Variation*

1. Every member of the class should record the following information:

a) The length, to the nearest millimetre, of one index finger.

b) The number of times the heart beats in one minute (after sitting at rest for at least five minutes).

Use this information to prepare histograms or line graphs showing variation in finger length and heart-beat rate in the whole class.

2. Collect at least 50 leaves from one tree.

a) Measure either the length of each leaf (including the petiole) or the width.

b) Use these measurements to construct a histogram or line graph showing variation in leaf size in the plant.

C *A monohybrid cross with maize plants*

1. Maize is useful in genetical experiments for two main reasons.

a) Pure line varieties are available with clearly visible differences such as dark and light coloured grains, and smooth and wrinkled grains.

b) One maize cob contains several hundred grains, each of which results from a single fertilization. Therefore, after a cross has been made, it is only necessary to count the types of grains on the cobs to obtain F_1 and F_2 ratios.

2. Sow seeds of a pure line dark grained variety close to seeds of a light grained variety.

a) Cover the male and female flower heads with plastic bags to prevent self-pollination and to prevent pollination by plants not included in the experiment.

b) When the male flowers are ripe, cross-pollinate the plants by transferring the plastic bags and pollen which has been shed into them from the flowers of the dark variety to the flowers of the light variety, and from the light flowers to the dark flowers.

c) What colour are the grains of the F_1 plants?

d) Which colour is dominant and which is recessive?

e) Sow F_1 grains, but this time ensure that only self-pollination occurs. What is the ratio of dark to light coloured grains in the F_2 plants?

Comprehension test

1. Mendel crossed pure line pea plants with green pods, and pure line plants with yellow pods. All the F_1 plants had green pods. Out of 580 F_2 plants, 428 had green pods and 152 had yellow pods. Which characteristic is dominant and which recessive? In the F_2 plants how many are: homozygous recessive, homozygous dominant, and heterozygous? Using G to represent the dominant gene, and g to represent the recessive gene, write out a plan showing the segregation of genes from the parents to the F_2 plants.

2. If two *Drosophila* flies, heterozygous for genes of one allelomorphic pair, were bred together and had 200 offspring, about how many would have the dominant phenotype? Of these offspring some will be homozygous dominant and some heterozygous. How is it possible to establish which is which?

3. Mendel believed that hereditary factors (genes) were always either dominant or recessive. How might he have altered this view had he performed the following crosses? When pure line sweet peas with red flowers are crossed with pure lines having white flowers, all the F_1 plants have pink flowers. When pure line shorthorn cattle with red coats are crossed with pure lines having white coats their offspring have coats with a mixture of both red and white hairs (this is called the **roan** condition).

4. Live bacteria which were all red in colour were placed under an ultra-violet lamp. After several days groups of white bacteria began to appear among the red. What conclusions, if any, can be made at this stage? What further experiments should be performed?

Summary and factual recall test

Mendel was looking for the (1) which govern the way in which (2) characteristics pass from parents to (3). To do this he chose characteristics with (4) variation, which means (5). His basic method involved hybridization which means (6). Study of *one* characteristic by this method is called a (7) cross.

Mendel crossed tall plants with dwarf plants and produced F_1 plants which were (8). From this he concluded that tallness must be the (9) characteristic because (10). Mendel now (11) the F_1 plants and produced F_2 plants in which three-quarters were (12) and one quarter (13), giving a ratio of (14). From this he concluded that dwarfness must be a (15) character because (16).

Mendel suggested that hereditary (17) are controlled by factors that operate together in (18), the members of which (19) during gamete formation so that each gamete receives (20). But at fertilization (21–describe what happens to the factors).

By comparing how factors behave during reproduction with the behaviour of (22) during (23) or reduction division it has become clear that (24). The behaviour of the two are similar in three ways which are: (25).

Factors are now called genes. Each gene has a partner which occupies the same (26) on the corresponding (27) chromosome. Genes of such a pair are said to be (28) of each other. When the members of these pairs are identical they are said to be (29), and when opposite they are said to be (30). The (31) of an organism is the nature and arrangement of its genes, and is contrasted with its phenotype which is (32).

A mutation is (33). Mutation rate has been artificially increased by (34–describe three ways). Only mutations carried in the (35) are inherited, and most of these do not appear in the phenotype because (36), but if they do, (37) selection usually operates against them because most of them are (38). Because of this (39) mutations disappear from a species while (40) ones survive and are multiplied. In time this process can lead to (41) change.

Genes are made of DNA, which is short for (42). DNA can be compared with an architect's plan because both (43). A DNA molecule consists of strands twisted into a double (44) joined by cross-bridges made of substances called (45), of which there are four types: (46). It is the (47) of these substances along the DNA molecule which make up the genetic (48).

Genes control the types of (49) made in a cell, and thereby control the structure and functions of the whole organism because (50).

20

Parasites and the body's defences against them

A healthy organism is one in which all the physical and chemical processes of life are working in harmony. Apart from accidental damage and starvation, one of the chief causes of disharmony, i.e. dis-ease, in an organism is the presence on or inside its body of another organism which lives as a **parasite**.

Parasites are organisms which obtain their food from the living body of another organism called the **host**. In doing so they often damage and poison the host, causing the symptoms of a disease. Parasites which harm their host are said to be **pathogenic**, and **pathology** is the study of the diseases which they cause. Not all parasites, however, harm their host. For example the bacterium called *Escherischia coli* lives harmlessly as a non-pathogenic parasite in the intestines of most humans and many other animals.

This chapter is concerned mainly with pathogenic parasites. It begins with a description of some bacteria and viruses which cause disease; goes on to describe tapeworms, which are much larger parasites; and concludes with an account of the human body's defences against parasites and the ill-effects which they cause.

20.1 Bacteria and viruses

Bacteria are among the smallest and simplest living things. Some biologists also consider viruses to be living organisms, but they do not exhibit all the normal characteristics of living things. Bacteria and viruses which cause disease are commonly called **germs**.

General features of bacteria

The commonest bacteria vary in size from 0.0005 mm to 0.005 mm, which means that a row of a hundred quite large bacteria would just reach across the full stop at the end of this sentence.

The variety of shape and form among bacteria is illustrated in Figure 20.1. Some types are unicellular, e.g. micrococci, while other types such as the staphylococci, which form clusters, and the streptococci, which form chains, are considered by some to be multicellular because they consist of thousands of cells connected to one another by fine strands of protoplasm.

Bacteria are responsible for some of the deadliest diseases of man such as tuberculosis, leprosy, cholera, diphtheria, pneumonia, and typhoid fever. In addition, bacteria cause many conditions which, though less deadly, are very debilitating such as boils, food poisoning, and bacillary dysentery. Bacteria are also responsible for several plant diseases, some of which affect crops. Pathogenic bacteria harm their host by destroying its tissues and by producing poisons, called **toxins**, which upset the host's metabolism.

It must be remembered, however, that only a small proportion of all bacteria are harmful, and many are extremely useful to man. For instance, bacteria can be used to ripen cheese, make sewage harmless, produce vinegar, alcohol, and acetone by fermentation, cure tobacco, produce silage (cattle food), and 'ret' flax (i.e. to soften the fibres so that they can be used in linen manufacture). In addition, the action of saprotrophic bacteria is essential for the continuance of life on earth (chapter 22).

Bacterial cells It is now generally accepted that bacteria do not have a nucleus like other cells. Bacterial cells do contain DNA molecules but there is no conclusive evidence that they have chromosomes or a nuclear membrane (Fig. 20.2). The protoplasm of each bacterial cell is enclosed within a cell wall of uncertain composition. It has been suggested that this wall consists of protein and fat molecules combined with a polysaccharide similar to cellulose.

At some time during their life cycle most if not all bacteria are enclosed within a layer of slime which their protoplasm secretes through the cell wall. This slime layer is certainly present around all parasitic bacteria while they are inside their host. It is believed that the slime provides some protection against harmful chemicals in the host's body fluids.

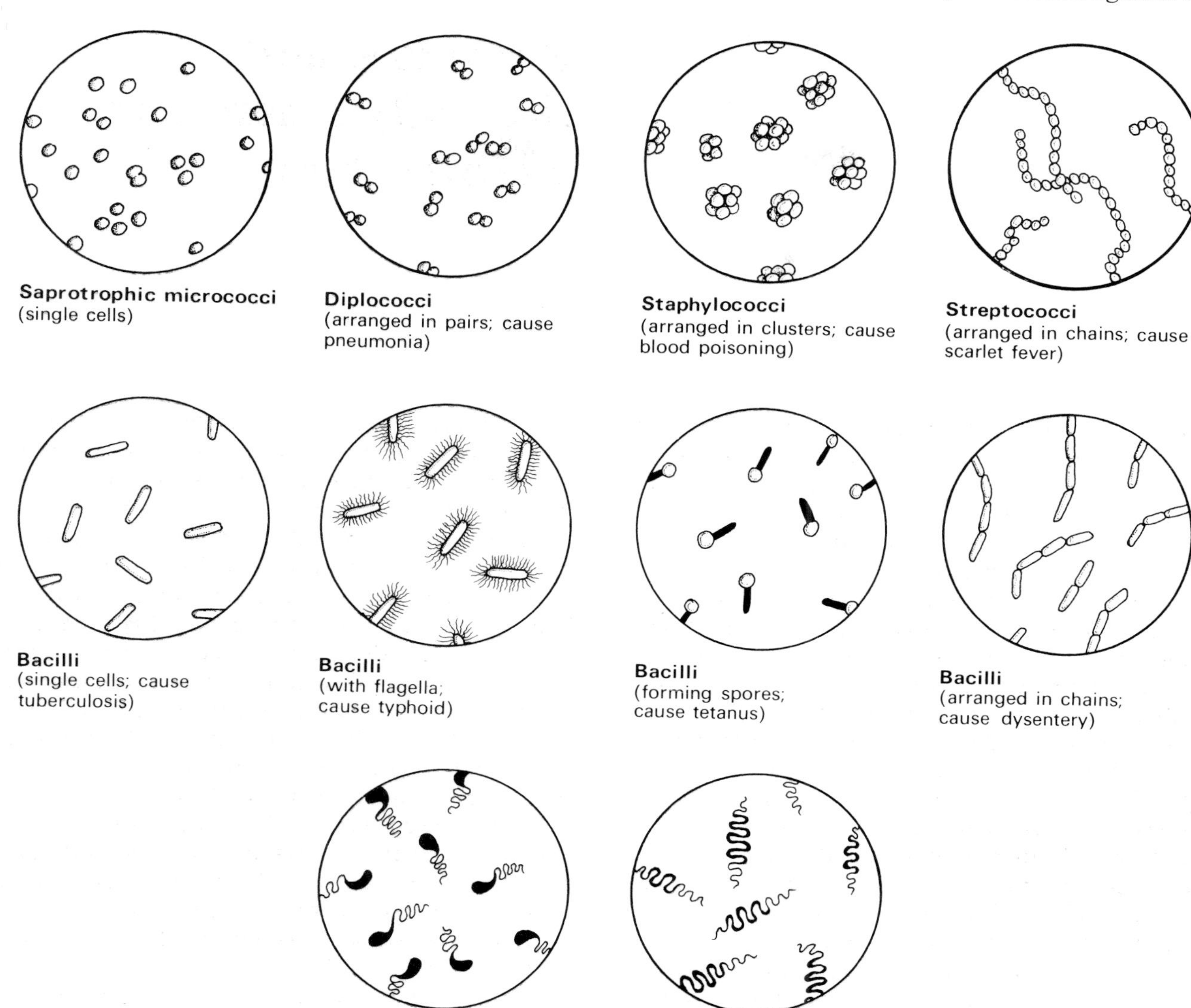

Fig. 20.1 Some types of bacteria

Movement Several types of bacteria can propel themselves through liquids by means of whip-like movements of one or more flagella. Estimates of their speed vary. One observer reported a type of bacillus moving for very short periods at a speed of 720 mm per hour, which is equivalent to a man travelling at more than 400 km per hour.

Respiration Some bacteria are aerobic (use oxygen in respiration); others are anaerobic (can respire without oxygen); while others can be either aerobic or anaerobic, depending on whether or not oxygen is available. Some anaerobic types are actually poisoned by oxygen.

Nutrition Apart from parasitic types, bacteria can be saprotrophic or autotrophic. Autotrophic bacteria obtain energy either from sunlight trapped by bacteriochlorophyll in their cells, or from energy released in chemical reactions which they control, e.g. nitrogen cycle bacteria described in chapter 22.

Reproduction and life cycle Under favourable conditions bacteria reproduce asexually by cell division every 20 or 30 minutes. Starting with a single cell this rate of division produces a slow initial build-up of numbers, i.e. 64 bacteria in $2\frac{1}{2}$ hours. But cell division can produce more than 16 million bacteria in 8 hours, and 4000 million million million in 24 hours. Clearly,

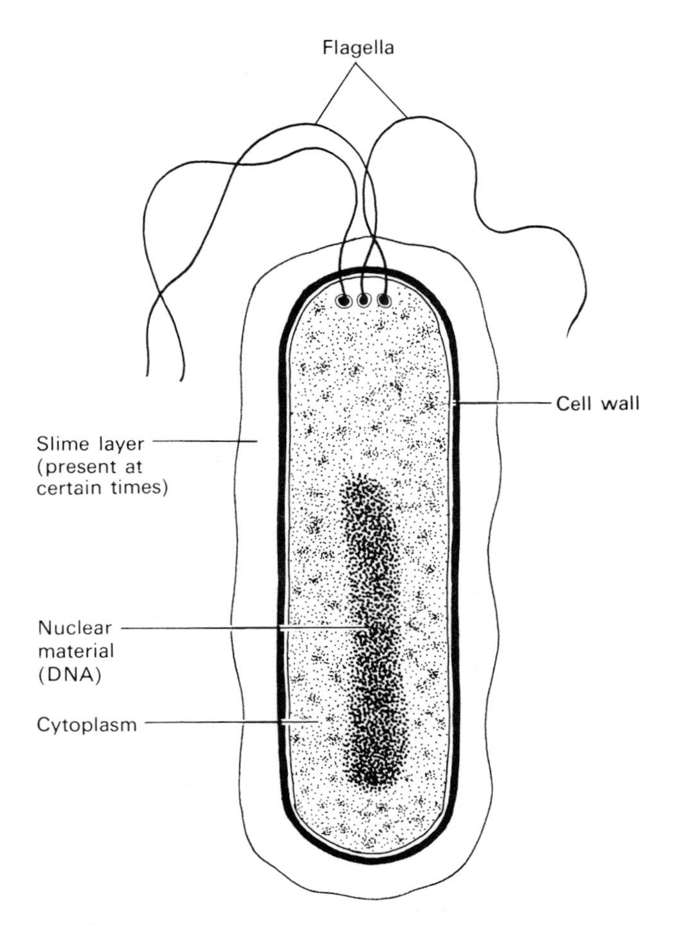

Fig. 20.2 A diagram of the structures found in bacterial cells

if the death rate did not eventually balance the rate of increase, bacteria would conquer the earth in a few days.

Sexual reproduction has not been clearly demonstrated in bacteria. Nevertheless, there is some indirect evidence that in certain species a type of cellular fusion takes place, roughly equivalent to fertilization, followed by segregation of genes. However, until more conclusive evidence is obtained it must be assumed that the majority of bacteria reproduce solely by asexual means.

When conditions are unfavourable for growth some bacteria form **spores**. This is a resting, or dormant stage in their life cycle (Fig. 20.3). In forming a spore, each cell loses its protoplasmic connections with surrounding cells, after which its contents round off and become enclosed in a thick protective wall. Unlike spore formation in fungi, bacterial spores are not a method of multiplication. Only one spore is formed in each bacterial cell, and only one bacterium emerges from a spore upon germination. Spores are extremely resistant to adverse conditions. Normal, actively growing bacteria are killed in minutes when temperatures rise above 60°C, but spores can survive boiling for up to 6 hours. Growing bacteria deprived of water die within a few hours, but spores can exist for years as dry dust blown about in the wind. Growing bacteria can be killed instantly by disinfectants and antiseptics, and yet their spores remain unscathed after soaking for several hours in such chemicals.

Classification of bacteria There is considerable disagreement about how to classify bacteria in the scheme of living things. At one time they were thought to be animals, later they were classified as plants, and then they were placed with fungi and algae in the now obsolete division Thallophyta. The modern view is that bacteria are neither plants nor animals but a distinct group with an independent origin. For this reason they are sometimes placed, along with other simple organisms of doubtful origin, into the kingdom Protista (chapter 1). More recently, however, biologists have decided to establish a group called **Procaryotic organisms**. This group includes bacteria, viruses, and the blue-green algae, all of which differ from other organisms in having no definite nucleus.

General features of viruses

In 1890 Professor Beijerinck of Delft University extracted juice from diseased tobacco plants and filtered it through porous (unglazed) pottery. He then discovered that the clear filtrate, which contained no visible particles under the highest magnification then available, caused the same disease when smeared on

Fig. 20.3 Life cycle of a spore-forming bacterium

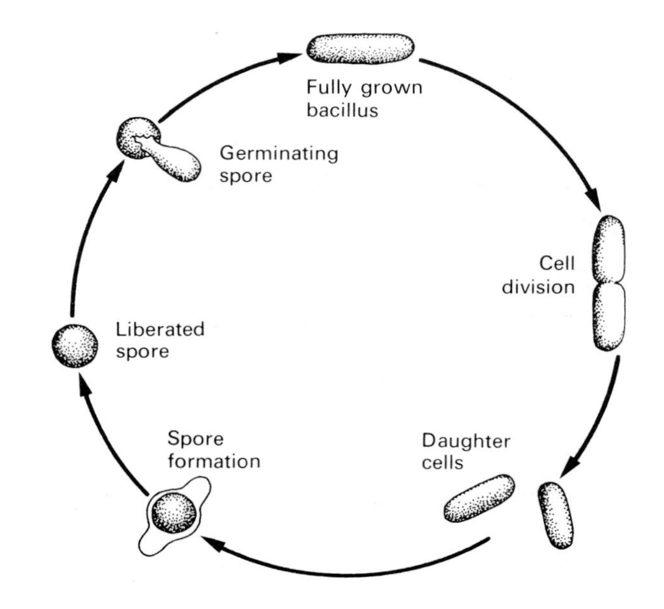

healthy plants. Moreover, filtered juices from these plants transmitted the disease to a third group, and so on indefinitely. Beijerinck was unable to accept that his filtrates contained disease organisms of any kind, and so he blamed the liquid itself for producing the disease. He called the liquid filtrate a **virus**, which means poison.

Later discoveries showed that many diseases are caused by filterable viruses. But in 1928 experiments were carried out using very fine filters which produced filtrate that no longer transmitted disease. This showed that sub-microscopic particles must be causing these diseases, and calculations made by comparing filters which allowed these particles to pass through with filters that stopped them revealed the incredible fact that virus particles measure from 10 to 3000 million millionths of a millimetre in diameter. (Figure 20.4 compares the relative size of viruses and bacteria.)

In 1935 the American biochemist W. Stanley produced a few grams of needle-like crystals by refining juice extracted from tens of thousands of diseased tobacco plants. Amazingly, this apparently dead, dry powder could still produce disease when dissolved in water and smeared on healthy tobacco plants, even after it had been stored for months or years in a bottle. This powder was the first pure sample of crystallized virus, and is now called tobacco mosaic virus because of the mottled pattern it produces on the leaves of tobacco plants. The preparation of this powder started a controversy among scientists which continues to this day about whether viruses are alive or dead. But it is certain that viruses are not cells. Furthermore, they show signs of life only when inside the living cells of another organism. Viruses are purely parasitic, and exist in a dormant state when outside their host.

Fig. 20.4 Comparative sizes of bacteria and viruses

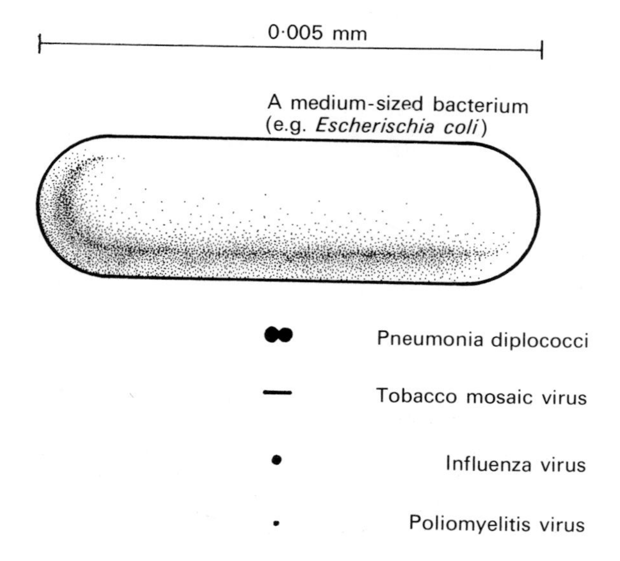

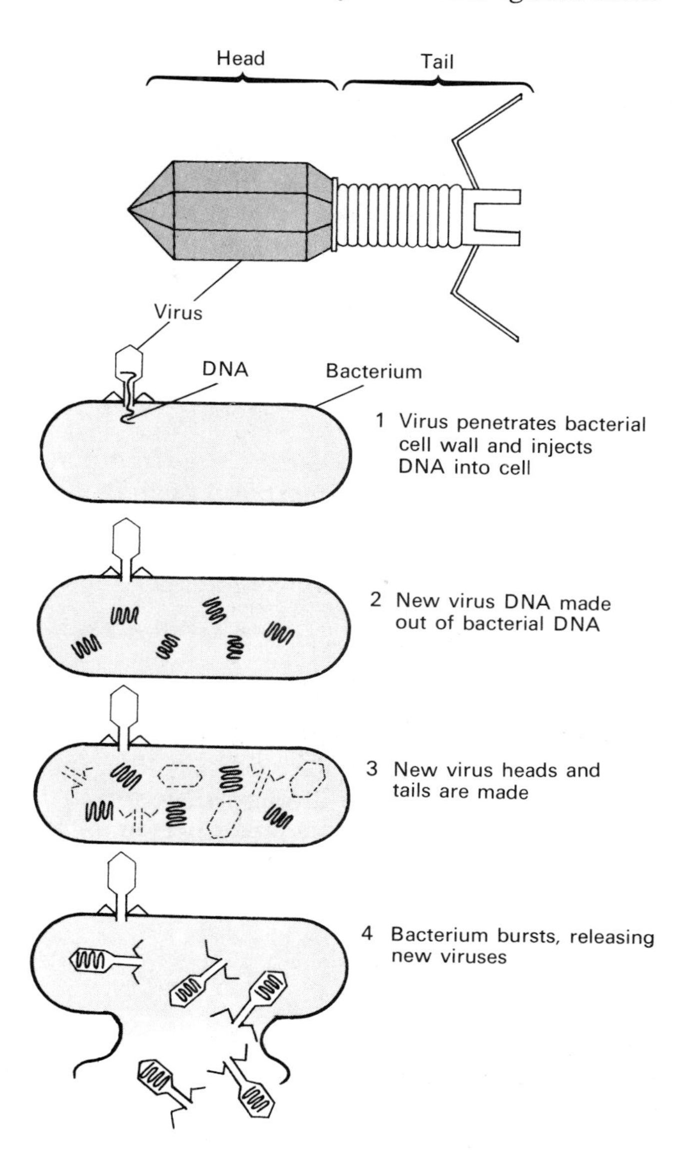

Fig. 20.5 Life cycle of a bacteriophage virus

Part of the mystery about what virus particles are, and how they multiply inside host cells has been solved by studying viruses which attack bacteria. Viruses of this type are called **bacteriophages**, which means bacteria-eaters. Figure 20.5 illustrates the structure and life history of a bacteriophage. The virus consists of a 'head' shaped like a many-sided crystal, with a rod-like projection, the 'tail', at one side. The head of such a virus contains a strand of DNA which is injected into a bacterial cell. Inside the bacterium the virus DNA makes use of bacterial DNA as a raw material to make replicas of itself, and then uses other cell materials to make new virus head and tail units. After about 30 minutes, the bacterium bursts open releasing about 200 perfect replicas of the virus, which repeat the process on other bacteria.

It is almost certain that the same process takes place inside cells of the human body when they are attacked by viruses such as those which cause chicken-pox, measles, poliomyelitis, the common cold, and influenza.

20.2 **The spread and prevention of infection**

Pathogenic bacteria and viruses can spread from person to person in many ways, the most important of which are: contact with infected droplets from someone with a disease; consumption of contaminated water and food; contact with infected people and contaminated objects; and contact with animals.

Droplet infection

Disease organisms present in the mouth, nose, and lungs can be carried out of the body in droplets of moisture whenever a person breathes out, talks, coughs, or sneezes. These droplets may spread the disease organisms directly to another person, or they may infect water and food which could later by consumed by others. Viral infections such as the common cold, influenza, and pneumonia are spread by droplets, as are bacilli which cause diseases such as whooping cough and diphtheria.

Contaminated water and food

Food and water can be contaminated with disease organisms by sewage, by contact with people suffering from disease, and by insects, birds, and many other animals. Typhoid fever, cholera, and food poisoning are spread in this way.

Direct contact

Diseases spread by contact with infected people and objects are said to be **contagious**. Infections can be spread by direct bodily contact, or indirectly by touching objects such as handkerchiefs, books, or coins previously handled by infected people. Germs may then be transferred from the hands to the mouth as, for instance, when eating hand-held foods such as sandwiches. Smallpox, measles, and tuberculosis are spread in this way.

Animals

Animals which spread infection are called **vectors**. They do this by carrying the disease organisms either inside or outside their bodies. House-flies, for example, carry typhoid, cholera, and dysentery germs both on their bodies and in their faeces. The yellow

Fig. 20.6 Diagram summarizing the spread of infection

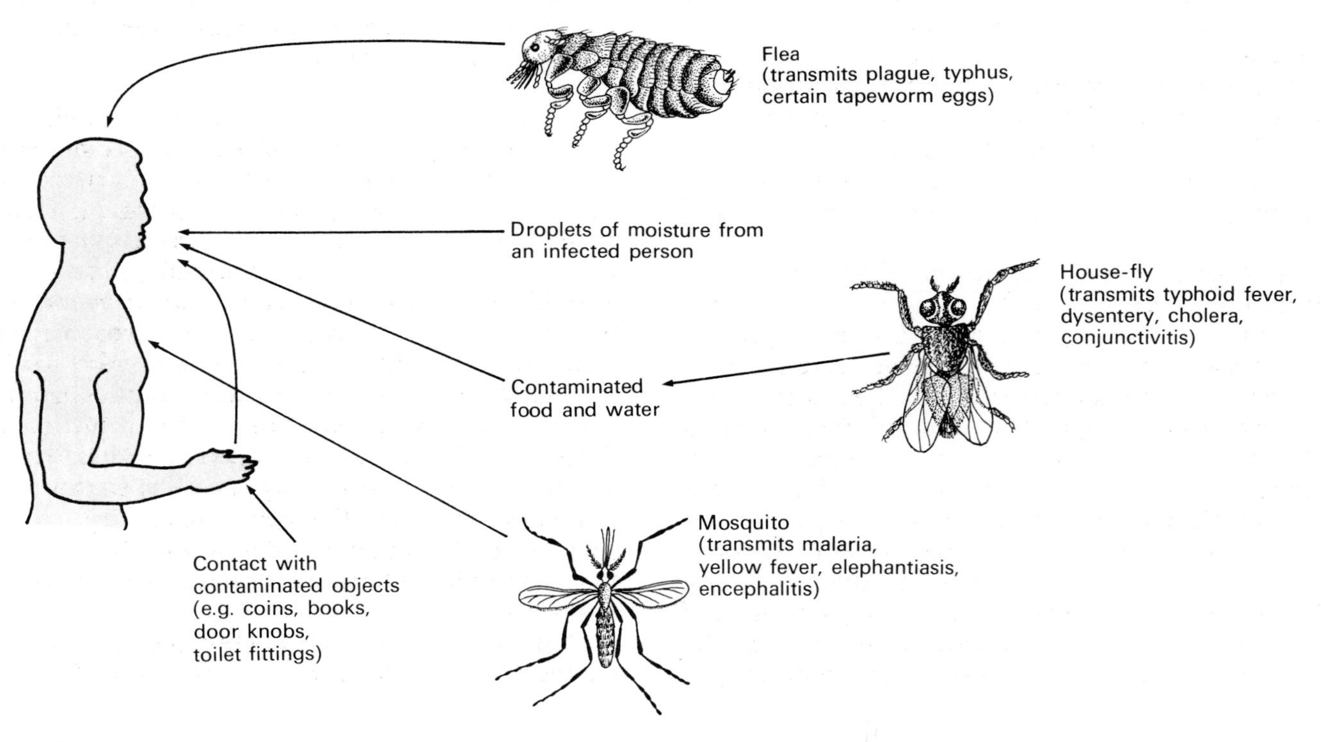

fever virus is carried from person to person, and from animals to people, inside the bodies of certain mosquitoes, which spread the virus when they suck blood.

Prevention of infection

The most effective ways of controlling the spread of disease have been found by studying the life cycles of pathogenic organisms and discovering how they enter the body. This research has led to the development of insecticides, more efficient sewage disposal methods, and methods for decontaminating water supplies and cleansing cities. Personal cleanliness is also necessary if the spread of infection is to be avoided. The skin and hair must be washed regularly to avoid the accumulation of bacteria, and the dirt in which they grow. It is especially important that the hands are kept clean to avoid contamination from, and contamination of, all the articles and objects which are handled daily by different people.

20.3 **The tapeworm**

Tapeworms are animals which live as parasites in the intestines of vertebrates. Figure 20.7 illustrates the structure and life cycle of a tapeworm called *Taenia solium*, commonly known as the pork tapeworm.

The adult tapeworm

The body of an adult tapeworm consists of a small round 'head' which is attached to a long flat tape-like body. The head bears hooks and suckers by which it attaches itself to the intestine wall of the host. The tape consists of a chain of flat segments. New segments are formed continuously by a growth region behind the head, and old segments drop off the hind end of the worm and are expelled from the host's body through the anus. In *Taenia*, the tape has been known to grow up to 8 m in length, when it consists of about 1000 segments.

Tapeworms have no mouth or digestive system, and do not produce digestive enzymes. They have a very simple nervous system and no sense organs or locomotory organs, although they have muscles which can make the tape undulate slightly. In fact the adult tapeworm is totally dependent on its host for protection and food, and so cannot survive in the outside world. In the host's intestine, a tapeworm is both protected by the host's body, and bathed in the host's digested food, which is absorbed over the tapeworm's entire body surface. The tapeworm is not digested along with its host's food because a thick enzyme-resistant cuticle covers its head and tape. It survives in the intestine without oxygen by respiring anaerobically.

Life cycle

The life cycle of the tapeworm involves two hosts: the first, or **primary host** carries the adult worm, and the **secondary host** carries a dormant stage of the life cycle called a **bladderworm**. The life cycle is completed when the secondary host, and its bladderworms, is eaten by the primary host. In *Taenia solium* the primary host is man, and the secondary host is a pig (hence the name pork tapeworm).

Egg production in the primary host A tapeworm egg has only the minutest chance of reaching the secondary host, but eggs are produced on an enormous scale. In fact, almost all of a tapeworm's body is concerned with egg production. A newly formed segment is hermaphrodite, i.e. it contains both male and female sex organs, but after fertilization the sex organs disappear leaving up to 50 000 developing eggs. The eggs are ripe by the time the segment drops off the end of the tape, and then they pass out of the host's body with its faeces.

The pork tapeworm produces about 100 million eggs a year and can continue to do so for about 25 years. But this is a slow rate of reproduction compared with star performers such as the beef tapeworm, *Taenia saginata*, which produces nearly 600 million eggs a year, and the fish tapeworm, *Diphyllobothrium latum*, which releases up to 36 million eggs a day!

Under unhygienic conditions, and where there are poor sewage disposal arrangements, tapeworm eggs may get into pig food and so be eaten by the secondary host. Unless this happens the eggs undergo no further development and eventually die.

Bladderworm development in the secondary host A tapeworm egg consists of a protective outer shell which contains an embryo 0.02 mm in diameter armed with six hooks. If a pig eats an egg the pig's gastric juice dissolves the shell and releases the embryo, which then uses its hooks to bore through the intestine wall. Here it enters a blood vessel and is carried around the pig's body in its blood stream. Embryos which come to rest inside muscles develop into bladderworms. These are fluid-filled bags containing a tapeworm head inside-out (Fig. 20.7). When fully developed a bladderworm is about 7 mm in diameter, and has no affect on the pig, even when it is in the muscles of its heart. Bladderworms are the dormant stage of the life cycle and develop no further unless the muscle which contains them is eaten by the primary host–man.

If pork is cooked thoroughly the heat kills any bladderworms which it may contain. But if man eats

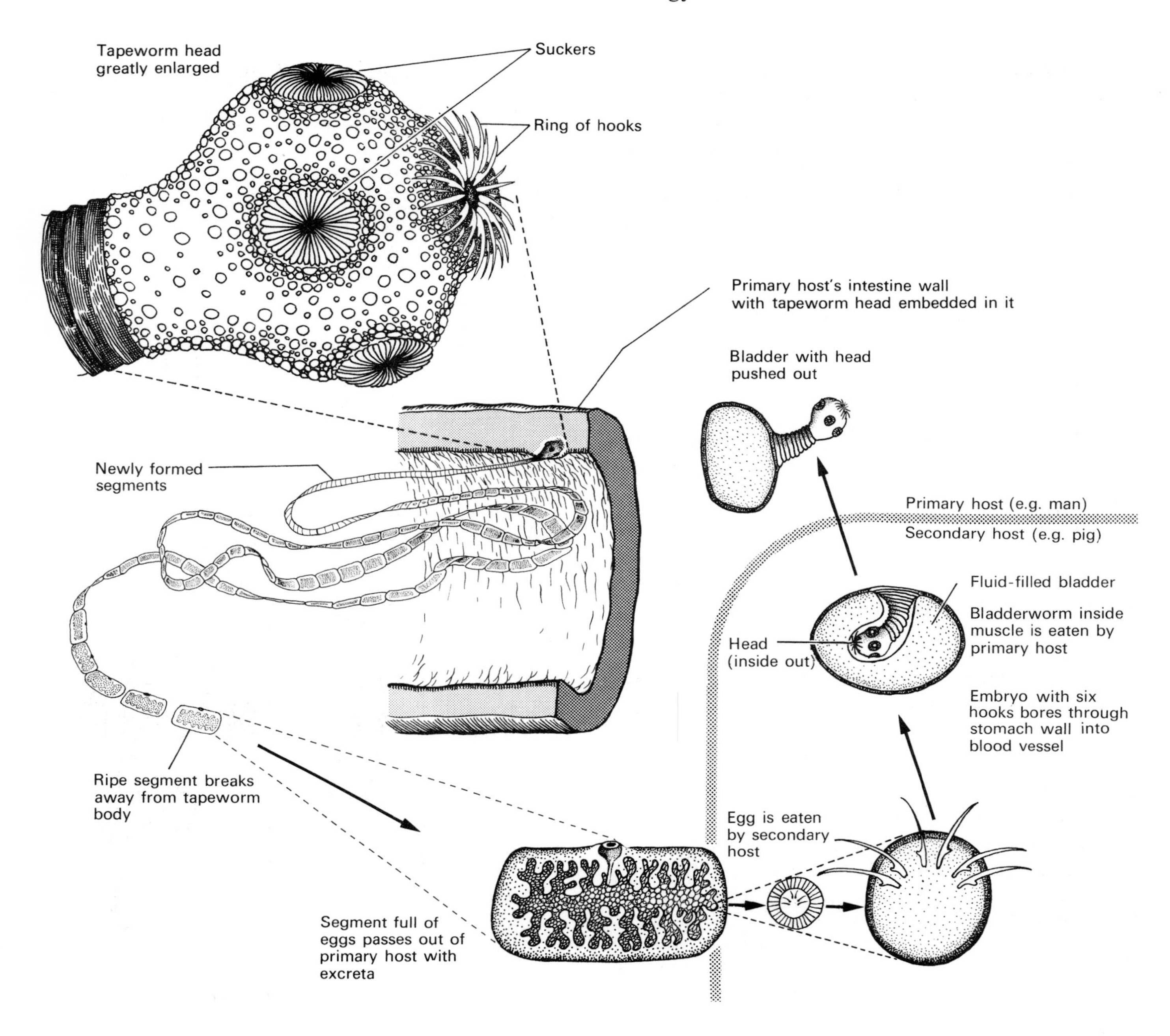

Fig. 20.7 Life cycle of the pork tapeworm

undercooked pork containing a live bladderworm his digestive juices release the tapeworm head, which unfolds so that its suckers and hooks are on the outside. The head now becomes attached to the intestine wall, and a tape develops behind it.

Pathological effects of tapeworms

On the whole, the effects of tapeworms on man are so slight that the victim may be unaware of its presence. Some people, however, suffer a loss of appetite, abdominal pains, loss of weight, nausea, and dizziness. Occasionally a tapeworm head may tear the wall of the intestine, causing wounds which can then be infected with bacteria, and become ulcerated.

Control of tapeworms

The most effective way of preventing infection by tapeworms is to construct proper sanitary arrangements which ensure that eggs do not contaminate pig food. This interrupts the life cycle, and the worms gradually die out. Tapeworms can also be controlled by careful inspection of imported meat for bladderworms, which are clearly visible to the unaided eye. Control measures such as these have almost completely removed the risk of tapeworm infection in Britain and most of Europe. All types of tapeworm infections in man can be treated medically with the drug quinacrine. This kills the worms, which then pass out of the body in the faeces.

20.4 The body's defences against parasites

The body possesses many defences against parasites, and together these give it what is called **natural immunity**. In humans, these natural defences include the skin, the ciliated membranes lining the respiratory system, stomach acid and enzymes of the digestive system, the type of white blood cells called phagocytes, and chemicals called **antibodies** in the blood and tissue fluid. In addition, medical science has developed ways of helping the body to develop additional defences against parasites. Together these defences are referred to as **artificial immunity**.

Natural immunity

The defences described below are summarized diagrammatically in Figure 20.8.

The skin The outer surface of the body is covered with a layer of dead cells known as the **cornified layer** of the skin (Fig. 10.3). As fast as they wear away or are damaged, these dead cells are replaced by a region of live growing cells beneath the surface, called the Malpighian layer. In addition, the dead cornified layer is kept supple, water-repellent, and mildly antiseptic by an oily substance called **sebum**, produced by sebaceous glands in the hair follicles. The human skin therefore forms a waterproof, germproof, self-repairing barrier preventing the entry of germs and dirt into the body.

The skin which covers the eyes, called the **conjunctiva** (Fig. 14.5), is extremely thin and delicate, but it is protected by a whole battery of natural defences, including antiseptic tear drops, and the blink reflex of the eyelids.

The respiratory system Dust and germs breathed in through the nose do not usually reach the lungs; they are trapped in sticky mucus which covers the membranes lining the nasal cavity and trachea. Trapped dirt and germs are then carried by cilia (Fig. 9.3) to the oesophagus, where they are swallowed and eventually passed out of the body in the faeces.

The digestive system Many germs are unavoidably consumed with food and drink. Fortunately they are usually harmless, but in any case the majority are killed by stomach acid and digested by enzymes.

The examples of natural immunity described so far may be thought of as the body's first line of defence.

Fig. 20.8 Diagram summarizing natural immunity

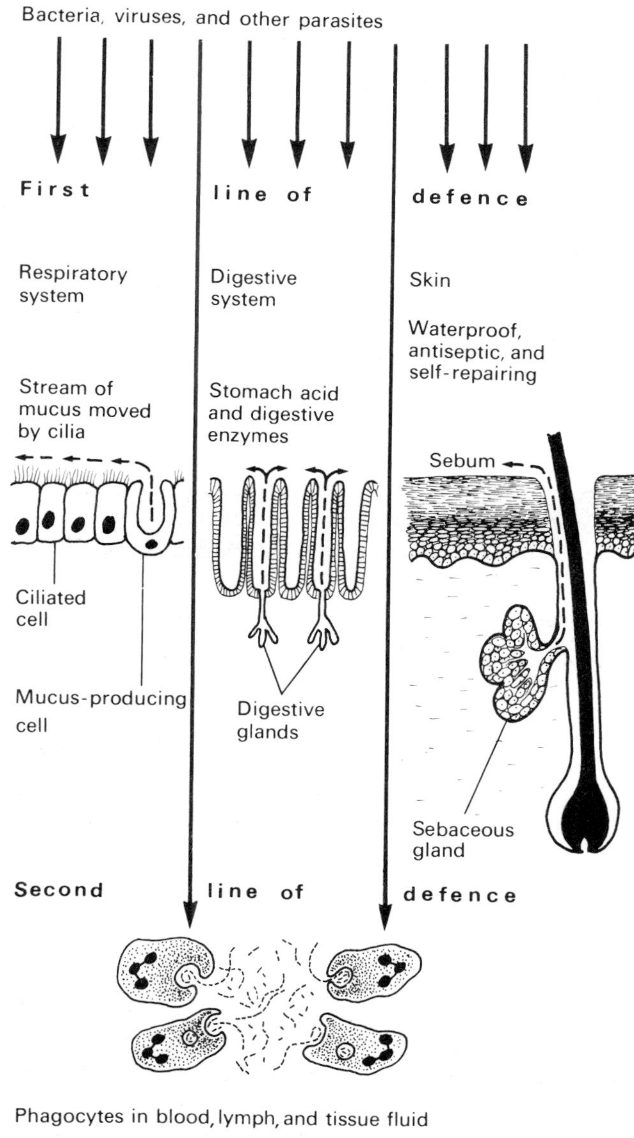

But if these defences are broken, as happens when the skin is cut, grazed, or burned, the body's second line of defence comes into operation, and this is controlled by the blood.

Bleeding and clot formation Whenever the skin is broken bleeding occurs for a time and washes the wound clean. Then a series of chemical reactions takes place causing blood to clot and form a solid plug in the wound. Blood clots when the platelets and surrounding damaged tissues release an enzyme which helps to convert a soluble blood-protein called fibrinogen into a network of insoluble threads made of fibrin. A network of fibrin threads forms across the wound and acts like a filter, trapping red cells which would otherwise have escaped from the damaged blood vessels. The fibrin and its entangled red cells eventually dry out forming a barrier, the clot, which prevents further loss of blood, and seals off the damaged area against dirt and germs until new tissue grows and heals the wound.

Sometimes bacteria enter a wound in such large numbers that the washing action of bleeding is insufficient to remove them all, and an infection develops. In this case, blood vessels around the wound dilate, permitting more blood to flow through the area. Millions of white blood cells of a type known as phagocytes flow out through the distended blood vessel walls, together with large quantities of tissue fluid. As a result, the wound swells and becomes hot and painful to the touch.

Action of phagocytes Phagocytes are white blood cells (Fig. 6.1) which destroy bacteria by engulfing and digesting them in the same way that an amoeba eats its prey (Figs. 4.2 and 4.4). At the same time, some phagocytes are killed by toxins produced by the invading bacteria, and the dead cells resulting from this 'battle' gather beneath the blood clot as a white fluid called pus. But unless the infection is of uncontrollable size, the bacteria are eventually destroyed and the pus absorbed, leaving the wound clean and free to heal.

Sometimes the body is invaded by large numbers of germs which overwhelm its outer defences. Infections of this size are dealt with in three ways. First, bacteria in the blood and tissue fluid are destroyed by free-swimming phagocytes of the type already described. Second, bacteria in the lymph vessels are destroyed by large phagocytes attached to the walls of the lymph nodes (Fig. 6.11). Third, bacteria and their toxins in any part of the body are attacked by chemicals known as antibodies, which are produced by the body during an infection.

Antibodies To understand what antibodies are and what they do, it is first necessary to understand that the substances which make up the bodies of parasites, and the toxins which they produce, are chemically different from the substances which make up the human body. The body is able to detect the presence of parasites and their toxins by 'recognizing' these chemical differences. Whenever this happens specialized cells in the lymph nodes, spleen, liver, and bone marrow react by producing antibodies. Antibodies are chemicals which combine with the 'foreign' chemicals in the bodies of parasites and in their toxins, either destroying these chemicals or rendering them harmless in various ways. Any substance, from whatever source, which stimulates the production of antibodies, is called an **antigen**. Parasites and most of their toxic products are antigens.

Antibodies are known to be specific, which means that each type of antibody can combine with one specific antigen and no other. The antibody against measles virus antigen will destroy this virus alone and have no effect on other disease organisms. The body must be able to produce different antibodies to combat each type of antigen, but it is not yet clear how this is done. Antibodies also vary in their effects on antigens. These effects are summarized below, and in Figure 20.9.

1. **Opsonins** are antibodies which combine with antigen material on the outer surface of bacteria. This seems to make the bacteria more likely to be destroyed by phagocytes. It is as if the antibody makes a bacterium more 'appetizing'.
2. **Lysins** are antibodies which kill bacteria and viruses by dissolving them.
3. **Agglutinins** are antibodies which cause bacteria and viruses to stick together in clumps. In this state the germs can neither penetrate host cells nor reproduce properly.
4. **Anti-toxins** are antibodies which combine with toxins produced by disease organisms, and render them harmless to the body.

Antibodies are produced slowly at first, but if the disease persists for more than three or four days they are produced in much larger quantities. Moreover, the antibodies against certain disease organisms remain in the blood for many years after the infection has disappeared, thereby giving the body a built-in resistance (immunity) to the disease in the event of re-infection. A child who has recovered from chicken-pox or measles for example, is unlikely to suffer from the disease again despite repeated exposure to infection, owing to the presence in his blood of specific antibodies against these germs.

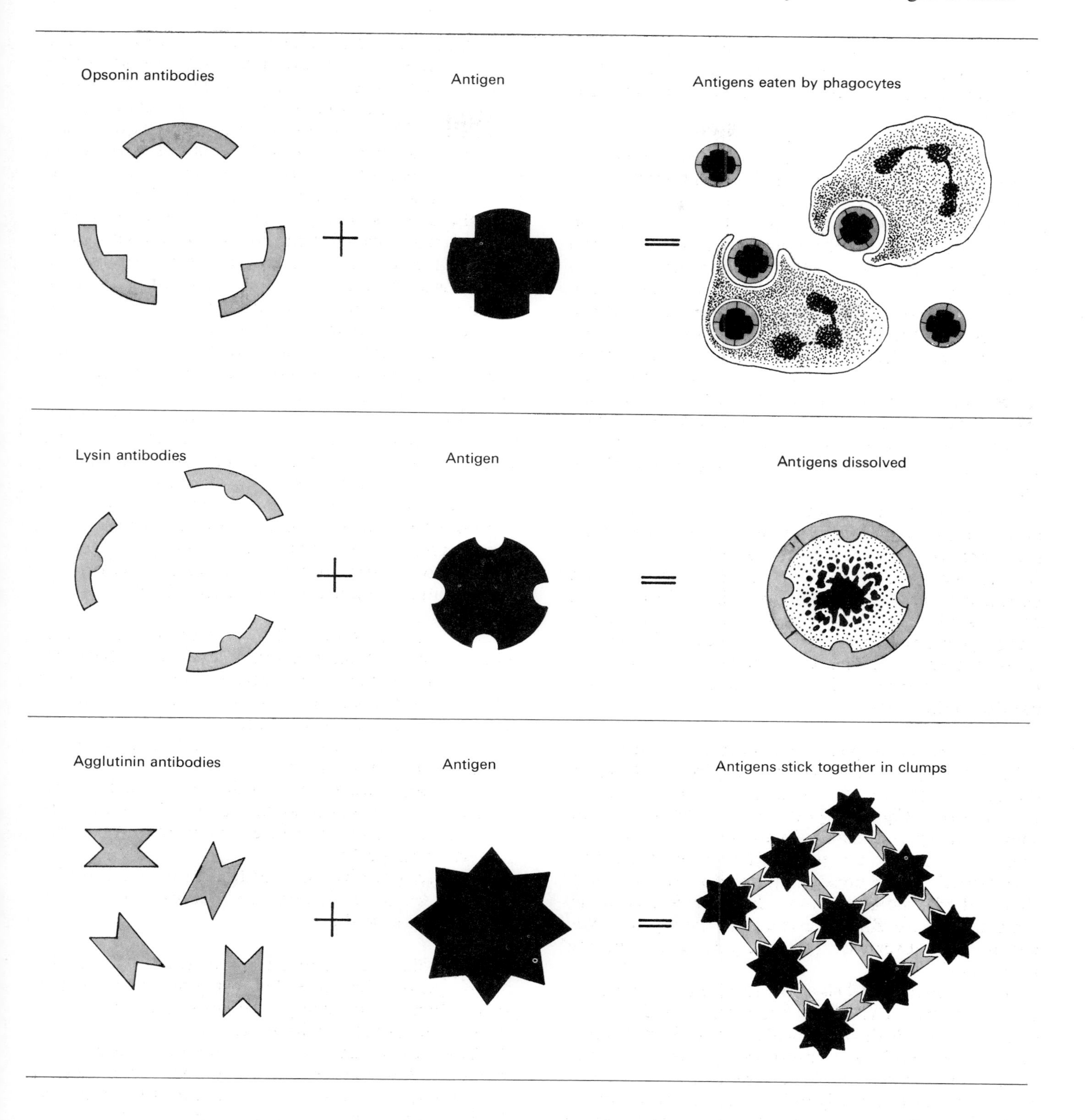

Fig. 20.9 Diagram showing how certain antibodies work.

Antibody molecules are not really shaped like those in the diagram. The shapes indicate that each antibody will react with only one specific antigen.

Artificial immunity

The production of antibodies in response to disease organisms is an example of **active immunity**, that is, 'active' in the sense that the antibodies are produced by the body in response to an antigen. It is now possible to make use of the body's capacity for active antibody production in order to produce immunity by artificial means so that the body can be prepared in advance to fight off infections. This is done by introducing **vaccines** to the body.

Vaccines A vaccine can be made either from a suspension of dead disease organisms – a suspension of germs which have been inactivated in various ways so that they no longer cause a disease – or from germs which are very similar to those which cause a serious disease but are actually harmless. When such vaccines are injected into the blood-stream the body responds by producing antibodies as if it were undergoing an attack from the actual disease organism. These antibodies remain in the blood and thereby make the body immune to the disease. The period of immunity so produced varies according to the disease from a few months to several years.

One of the first people to use a vaccine was Edward Jenner (1749–1823), a country doctor in a small village in Gloucestershire. Jenner became interested in a local legend that people who had recovered from a mild disease known as cowpox, because it is caught from cows, never thereafter suffered from the dreaded disease of smallpox. To test the truth of the legend Jenner performed an extremely dangerous experiment. He scratched the skin of a healthy eight-year-old boy and rubbed pus from the hand of a milk-maid suffering from cowpox into the wound. The boy caught cowpox and quickly recovered from it. Jenner then inserted pus from a patient suffering from smallpox into the boy in a similar manner. The boy did not catch smallpox on this occasion, nor on a subsequent occasion when Jenner again attempted to infect him.

Cowpox virus is called vaccinia after the Latin word for cow, and because of its association with Jenner's work it has given rise to the modern terms vaccine and vaccination. There are now vaccines available against many diseases, including typhoid fever, poliomyelitis, cholera, and bubonic plague.

Ready-made antibodies Medical science has devised a way of supplying ready-made antibodies, which are injected into the body to assist it in fighting germs before its own defence mechanisms have come into operation. This treatment is an example of **passive**

Gillray's cartoon illustrating the supposed effects of vaccination

immunity, which is given that name because the body acquires added immunity without doing any work.

Diphtheria antibodies, for example, are produced by injecting horses with toxins produced by diphtheria bacilli. The horses then produce antibodies of the anti-toxin variety to neutralize the diphtheria poison in their bodies. By taking blood from these horses and extracting serum from it, a supply of anti-toxin is obtained which is extremely effective in treating and helping to prevent the spread of diphtheria in humans. Horses are also used to produce anti-toxins against tetanus; and an attack of measles may be prevented, or its effects lessened, by injecting serum taken from humans who have recently recovered from this disease.

Verification and inquiry exercises

A *Making a culture of live bacteria*

1. Boil a few grams of finely chopped meat, carrots, and potatoes in water for 15 minutes, then filter off the solid matter to obtain a fairly clear broth.

2. Leave the broth in open test-tubes for a few hours. Plug the tubes with cotton wool and leave them in a warm place (at approximately 25 °C) until the broth has 'gone bad' owing to the growth of bacteria. This is an example of bacteria culture.

B *Stained preparations*

1. Smear some saliva on a slide and allow it to dry. Scrape a little food material from between the teeth, mix it with a drop of water on a slide, then smear it and allow it to dry.

2. The dried smears may be stained in either gentian-violet, crystal-violet, or methylene blue stain. Put a drop of stain on the smear and allow it to remain there for 2 to 3 minutes before washing it off under a slow-running tap. Dry the back of the slide and place a cover slip over the stained smear. Observe under low, then under high-power magnification. What shapes are the bacteria?

C *An investigation of the effect of heat, cold, and chemicals on bacteria*

1. *Heat*

a) Plug six test-tubes with non-absorbent cotton wool and sterilize them in an autoclave (pressure cooker) for 15 minutes at 103 kPa.

b) Prepare a quantity of broth as in A above, and pour about 2 cm into each of the sterile tubes, replugging them immediately afterwards. Place all the tubes upright in a beaker half-filled with water and boil them for another 15 minutes.

c) When the tubes are cool put a few drops of live bacteria culture in each from A2 above, replugging them immediately afterwards, and taking great care to wash the hands thoroughly afterwards, and sterilize the dropper used to transfer the bacteria. Leave the tubes in a warm place for a day.

d) Call the next day 'day one' and proceed as follows:

Day one–Boil all six tubes in a beaker of water for 5 minutes and allow to cool. Put them in a warm place.

Day two–Put two tubes on one side in a warm place and label them A and B. Boil the other four tubes for 5 minutes, allow them to cool and put them in a warm place.

Day three–Put two tubes on one side and label them C and D. Boil the other two for 5 minutes, and when they cool label them E and F. Put them with the others in a warm place. A and B are therefore heated once; D and C are heated twice; and E and F are heated three times. Leave the tubes in a warm place for two or three days and examine them, without unplugging, for traces of bacterial growth.

e) Heating to 100 °C kills growing and reproducing bacteria but not spores. How does this fact help explain the results of this experiment? Explain how this procedure could be used to sterilize food which is made inedible by pressure cooking (i.e. by temperatures in excess of 100 °C).

2. *Chemicals and cold*

a) Devise experiments making use of bacteria cultures prepared according to C1 above to verify: that strong salt and sugar solutions kill bacteria; and that refrigeration slows down the rate of bacterial growth.

b) Devise other experiments to test the effectiveness of various antiseptics and disinfectants in killing bacteria, when used at different strengths.

c) How do these results show how food may be preserved from decay owing to the action of bacteria, and diseases may be stopped from spreading?

d) *Never* touch live bacteria cultures with the hands. *Always* wash the hands after handling equipment containing live bacteria. *Always* dispose of bacteria cultures by boiling them and washing the equipment in strong disinfectant.

D *To demonstrate the presence of bacteria throughout the environment*

In the following experiment use either commercially prepared sterile nutrient agar and disposable petri dishes, or proceed as follows.

1. Prepare a quantity of broth as in A above, and while it is hot stir in 1.5 g of agar powder to every 100 cm^3 of liquid. Prepare about 12 sterile test-tubes as in C1(a) above, add 2 cm^3 of hot agar to each, then replug and boil them in a beaker of water for 15 minutes. Take the tubes from the water and prop them against something so that the agar solidifies at an angle–making what is known as an agar 'slope'. Finally place the tubes upright in a test-tube rack. (*Note:* if absolute sterility is required then the test-tubes must be autoclaved for 15 minutes after adding the agar.)

2. Leave one tube untouched as a control. Each of the remaining tubes should be contaminated in one of the

following ways, then replugged and labelled accordingly:

A with soil
B with dust from various parts of the room (using more than one tube if necessary)
C with aquarium or pond water
D with tap water
E leave open to the air for 1 hour indoors
F leave open out of doors
G, H, J, etc. Think of other ways to contaminate these tubes.

3. Watch for signs of bacterial growth.

Comprehension test

1. It was once thought that living things could form directly out of non-living matter: people believed, for example, that discarded food would eventually turn into bacteria and moulds. An experiment was prepared as illustrated in Figure 20.10. Tube A was left untouched. Tubes B, C, and D were prepared as illustrated, then boiled for 15 minutes in a beaker of water. A went bad in one day, B in two days, C in one week, and D remained fresh indefinitely, or until the bung and 'S' tube were removed. How do these results disprove the following statements:

a) Meat soup turns into bacteria.

b) The process of boiling meat soup alters it so that it cannot turn into bacteria.

c) Boiling the soup alters the air in the tube so that the soup cannot turn into bacteria.

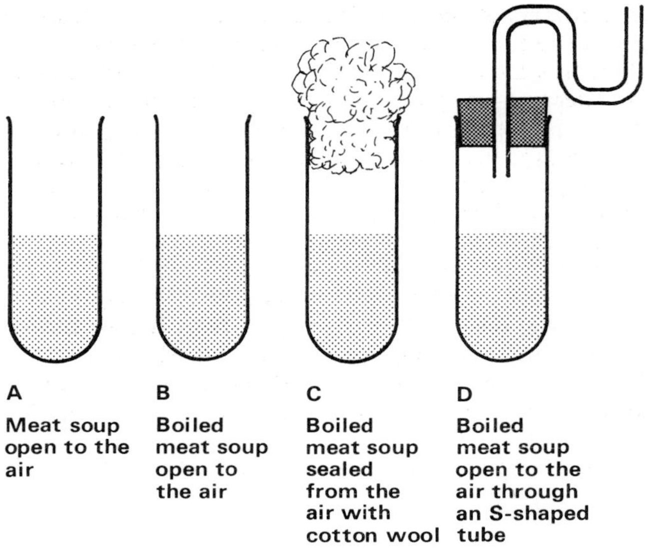

Fig. 20.10 Diagram for comprehension test 1

Why does the soup in C eventually go bad, whereas that in D remains fresh? How do these results help to show that bacteria are living creatures which can exist floating in the air around us?

2. Make a list of all the features of a tapeworm which enable it to live as it does. In what way do these features make tapeworms completely unfitted to an independent (non-parasitic) existence? What are the advantages and disadvantages of a parasitic way of life?

Summary and factual recall test

Parasites are organisms which obtain (1) from the (2) body of another organism called the (3). Parasites which are harmful are said to be (4).

Bacteria cause serious diseases such as (5–name five), and less serious conditions such as (6–name three). Other bacteria are useful to man in the following ways (7–name at least six). Under favourable conditions bacteria reproduce (8) about every (9) minutes. Under unfavourable conditions some bacteria form a resting stage called a (10), in the following way (11). These structures are extremely (12) to adverse conditions such as (13–describe three).

Viruses cause diseases such as (14–name five). Viruses show signs of life only when (15); at other times they exist in a (16) state. Viruses which attack bacteria are called (17). They inject (18) into a bacterial cell where it produces replica viruses in the following way (19).

Diseases spread by droplet infection are (20–name three). Droplet infection occurs in the following way (21). Food and (22) can be contaminated with germs by (23–name three ways). Diseases spread in this way are (24–name three). Contagious diseases are those spread by (25). Examples of how this happens are (26–describe two), and diseases spread in this way are (27–name three). Animals which spread infection are called (28). Examples are (29–name two vectors and the diseases they carry).

Adult tapeworms live in the (30) host, while a dormant stage called the (31) lives in the (32) host. Adult worms are hermaphrodite, which means (33). They produce

(34) numbers of eggs. Under unhygienic conditions such as (35), the eggs of pork tapeworms may be eaten by (36). When this happens the (37) of the egg is dissolved by (38) releasing an (39) which uses its six (40) to burrow into the (41) wall where it enters a (42). If they come to rest in a (43) they develop into (44). If (45) eats under-cooked (46) containing (47), the tapeworm life cycle is completed in the following way (48). Tapeworm infections can be prevented by (49–describe two ways).

Sebum, produced by (50) glands in the hair (51), make the skin (52–describe three effects). The (53) layer of the skin replaces dead cells of the (54) layer as fast as they are (55) or (56). (57) and germs entering the respiratory system are trapped in (58), and moved by (59) to the (60) where they are (61). Most germs in food are killed by (62) and (63).

When the skin is cut (64) occurs for a time which (65) the wound. Eventually a clot forms in the following way (66). The clot prevents further loss of (67), and helps prevent (68) and (69) entering the wound.

Germs inside the body are destroyed by a type of (70) blood cell called (71). In addition, antibodies made by cells in the (72–name four places) destroy germs and toxins in four different ways: (73–name the antibodies and give their functions).

Vaccines can be made from (74–name three things). The body reacts to a vaccine by producing (75), which make it (76) to attack from certain germs. Vaccines are available against (77–name four diseases).

Antibody production in response to germs is an example of (78) immunity, which is given this name because (79). Horses can be used to make antibodies in the following way (80). Treatment with these ready-made antibodies is an example of (81) immunity, so called because (82).

21
Soil

Soil is not just a dead substance. It is an environment which provides both food and shelter for a wide variety of organisms such as bacteria, fungi, worms, and insects, which carry out their whole life cycle within it. Man and many other animals also depend on soil, because it provides support, water, and minerals for the plants which are their food.

This chapter describes the composition of soil, its origin, cultivation, and a number of methods for studying its physical and chemical properties.

21.1 Composition and origin of soil

Shake a few grams of garden soil in a test-tube of water for about thirty seconds, allow it to settle for a few minutes and compare the result with Figure 21.1. This is called a **sedimentation test**, and is a method of separating the major components of soil, which are

Fig. 21.1 Results of a soil sedimentation test

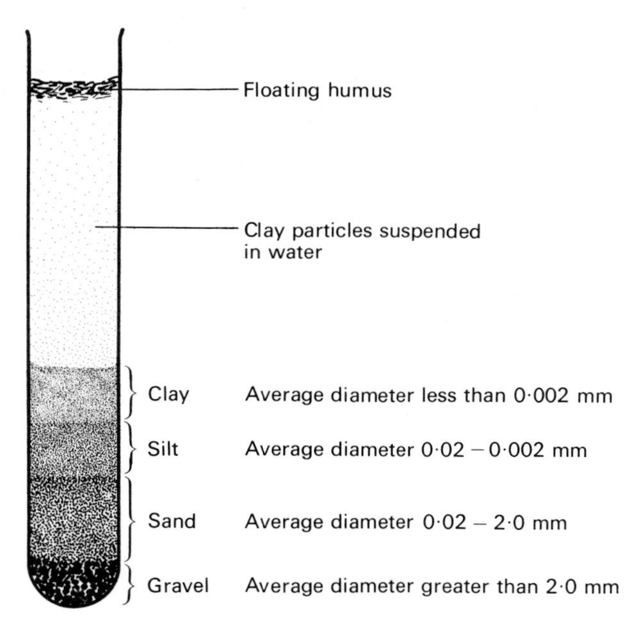

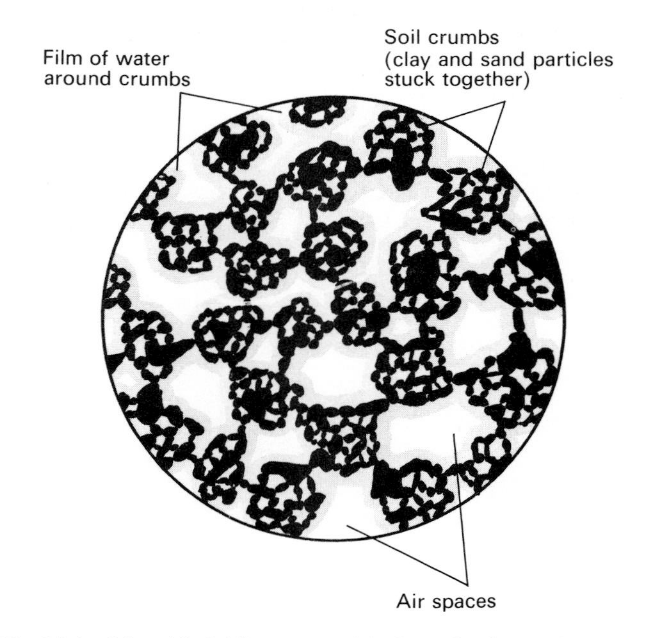

Fig. 21.2 Magnified (diagrammatic) view of soil crumbs

humus and **mineral particles**. Humus is the black material which floats to the top of the tube, and mineral particles are those which settle to the bottom. They settle in order of weight, the heaviest particles (which are usually the largest) settle first.

Humus

When animals and plants die their remains decompose owing to the saprotrophic activities of bacteria and fungi. One of the end-products of decay is the black fibrous humus material which floats to the surface in a sedimentation test. This fibrous humus comes mainly from cellulose and lignin fibres of plants and the hard parts of animals. The soft tissues of dead organisms decay into a liquid which passes into the soil forming a sticky coating around mineral particles. This glues the particles together into clumps called **soil crumbs** (Fig. 21.2). The soft decaying tissues also form nitrates, phosphates, and ammonium salts. These dissolve in soil water and are essential for the healthy growth of plants.

The presence of fibrous humus and well developed soil crumbs are characteristics of fertile soil. First, a soil with these features does not usually become waterlogged during rain because the air spaces between the crumbs provide adequate drainage. Second, these air spaces allow oxygen to reach plant roots, bacteria, fungi, and other soil organisms, permitting rapid respiration and growth. Third, humus retains moisture which would otherwise drain away after rainfall. Fourth, humus absorbs dissolved minerals from soil water thereby preventing them from being washed out of the soil. Fifth, humus binds the mineral particles together so that they are not easily blown away by high winds.

Mineral particles

Mineral particles originate from rocks which have been broken up by various physical and chemical processes known as **weathering**. These processes are summarized in Figure 21.3.

Physical weathering There are several ways in which physical weathering can take place. Water in cracks and crevices may shatter the rock as it freezes and expands. Cracks may be widened further by the pressure of plant roots which grow within them. Over many years these processes, together with the removal of soluble substances from rocks by rainwater, can reduce them to boulders and then smaller fragments. When this happens on a slope, loosened rocks roll or are washed downwards, undergoing further wear and tear as they knock and rub against each other. This kind of weathering is called mechanical breakdown. When rock fragments are washed into rivers they can be carried long distances, and during this time mechanical breakdown continues as the rocks are rolled over and pounded together. Eventually, this reduces them to microscopic particles of sand, silt, and clay.

Fig. 21.3 Diagram summarizing the ways in which soil is formed

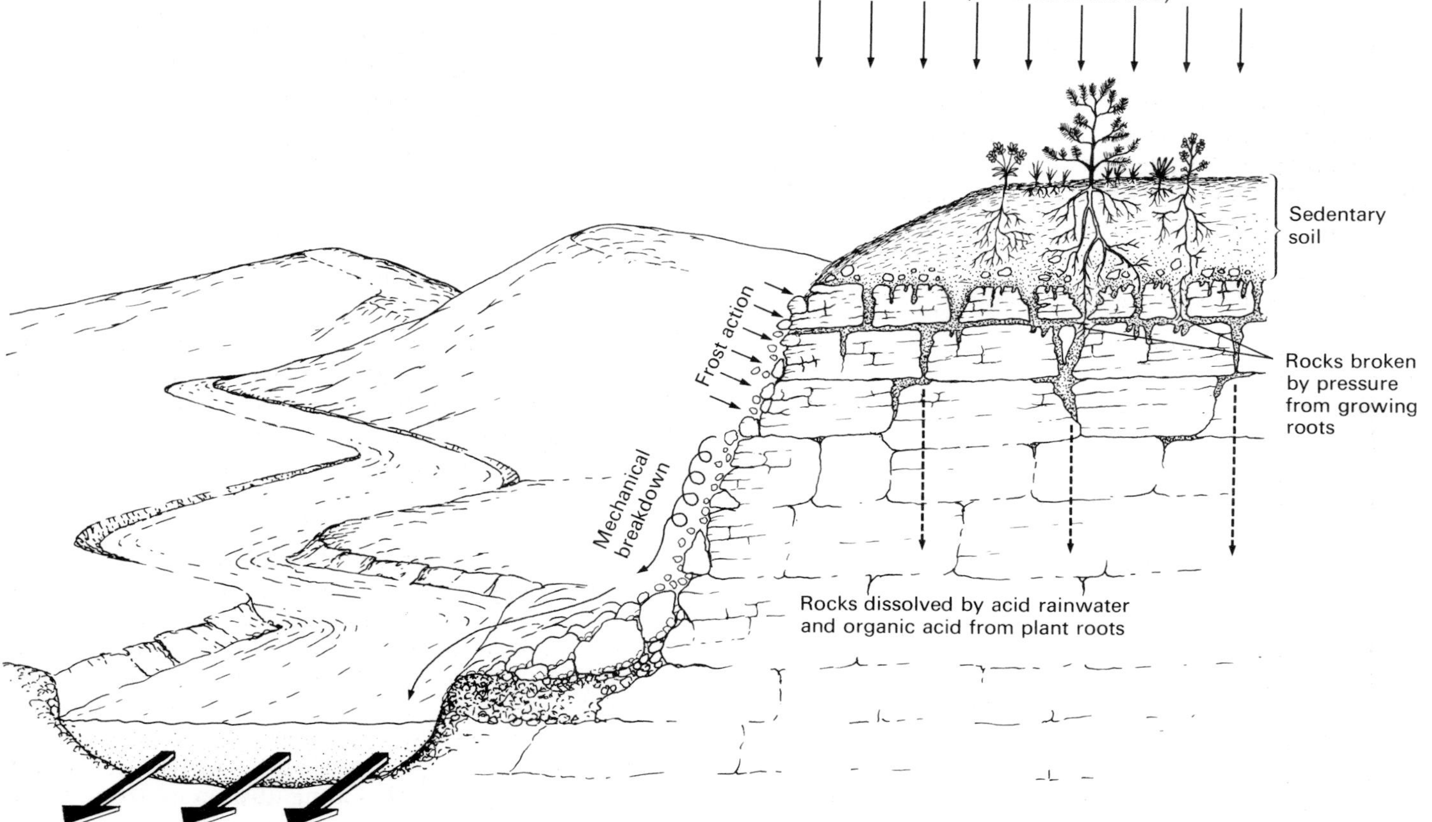

Chemical weathering Carbon dioxide in the air dissolves in falling rain to make a weak solution of carbonic acid. When this acid falls on rocks, especially limestone, it reacts chemically forming soluble substances which wash away, leaving insoluble fragments behind which develop into soil. As rainwater percolates through the soil it picks up more carbon dioxide from respiring animals and plant roots. The carbonic acid is therefore stronger by the time water reaches underlying rocks. In addition, plant roots produce weak organic acids which also contribute to the chemical decomposition of rocks below the surface, resulting in a gradual deepening of the soil layer.

Besides humus and minerals, air, water, and living organisms are essential components of fertile soils.

Air

Most soil organisms, including plant roots, are aerobic (i.e. use oxygen in respiration). Consequently, no soil is fertile unless oxygen can circulate through it. The amount of air, and therefore oxygen, in soil depends mainly on the size of its particles: the larger the particles the larger the air spaces between them.

Water

Soil water is essential for plant growth both on its own, and because it carries in solution the many different minerals required by plants.

Soil water comes mostly from rainfall. Some of this water evaporates back into the atmosphere, and some drains away. A large amount of water is held in the soil by forces called **adsorption** and **capillarity**. The force of adsorption causes water to gather in a thin film around the surface of the soil particles (Fig. 21.2). Consequently, the volume of water retained by a soil depends on the total surface area of the particles which it contains. The smaller the particles in a soil, the greater its total surface area (comprehension test 2). Therefore, in equal volumes of sand and clay, the clay will retain most water by adsorption. This adsorbed water attracts more water by a force called capillarity. The strength of capillarity depends on the distance apart of the soil particles. The closer they are together the stronger the force of capillarity (this is demonstrated in exercises B3 and B4). Owing to their small size, particles of clay are closer together than the larger particles of sand. Consequently, clay soils have greater capillary force than sandy soils. In fact, the capillary force of clay soils is great enough to overcome the force of gravity and so clay actually draws water upwards by capillarity from lower regions as fast as it evaporates from the surface. This is an important feature of clay soils; it means their surface layers are kept moist for a longer period than sandy soils during drought.

Soil which contains all the water it can possibly hold is said to be at **field capacity**. The water which it cannot retain drains away and may eventually form a completely waterlogged layer deeper down, the upper level of which is called the **water table**. Depending on the amount of rainfall and the rate at which water can seep away through underlying rocks, the depth at which the water table occurs may vary from a few centimetres to several hundred metres.

Soil life

Countless small and microscopic organisms inhabit soil. Among the most important as far as soil fertility is concerned are saprotrophic bacteria and fungi which decompose dead organisms, releasing nitrates and other minerals necessary for plant growth (nitrogen cycle, chapter 22).

Earthworms also increase soil fertility. They aerate the soil by burrowing through it, and some worms turn the soil over by taking it into their mouths, passing it through their digestive systems, and finally passing it to the surface as a worm cast. Most worms add to soil humus by dragging leaves down into their burrows as food, and leaving them there partly eaten.

Other burrowing animals which turn the soil over include rabbits, moles, badgers, certain millipedes, and insects. All animals increase soil fertility with their droppings, and ultimately with their dead bodies. Soil also contains many pests which destroy plant life, such as slugs, wire-worms, and parasitic fungi.

21.2 Types of soil

Soil can be classified in many different ways. For example, it can be classified as **alluvial** or **sedentary**, depending on how it was formed; as **topsoil** or **subsoil**, depending on the depth at which it is situated; and as **loam**, **sand**, or **clay**, depending on its composition. These are different ways of classifying what may be the *same* soil; it is quite possible, therefore, for a soil to be described as alluvial, and a topsoil, and at the same time a loam.

Sedentary and alluvial soils

A sedentary soil is one which is situated on top of the rock from which it developed (by weathering), like that on top of the rock formation in Figure 21.3. An alluvial soil is one composed of mineral particles which have been carried long distances from their place of origin by rivers or glaciers, and then deposited on flat areas such as valley bottoms and coastal plains.

Topsoil and subsoil

Both sedentary and alluvial particles slowly develop into fertile soil as they are colonized by plants and animals, and as they gather humus from the activities of bacteria and fungi. But these processes take place only in the upper regions of the soil, where they lead to the formation of a layer called topsoil. Good topsoil is dark in colour owing to the presence of humus, and is filled with organisms. Below the topsoil is subsoil, which is lighter in colour owing to the absence of humus, and contains very little life except the deeper roots of plants. Because of its humus content and the presence of organisms, topsoil is more fertile than subsoil. The best topsoil of all is called loam.

Sedentary and alluvial soils can be classified into different types according to the substances of which they are composed, e.g. loam, sand, and clay, and the proportions in which these substances exist. These substances in different proportions give a soil its particular physical and chemical properties.

Loam soils

Loams are the most fertile of all soils, because they possess all the characteristics which promote vigorous plant growth. The best loams consist of about 50 per cent sand, 30 per cent clay, and 20 per cent humus, with a little lime, all of which are mixed together and formed into well-developed soil crumbs. Soils of this type are well-aerated, drain freely, and yet retain plenty of moisture and dissolved minerals. Loams warm up quickly in the spring, and are easily cultivated by digging or ploughing.

Sandy soils

Sandy soils contain a high proportion of large mineral particles combined with little humus and hardly any clay or silt. Soils of this type have several advantages. There are large spaces between sand particles which give the soil good drainage and aeration, and this enables plant roots to penetrate quickly during germination. Sandy soils are easily cultivated all the year round because their particles do not stick together in large clumps, which is why they are generally called 'light' soils. But sandy soil does not retain water for long after rainfall, and this causes plants to wilt and die very quickly during drought. The rapid movement of water through sandy soil washes away its dissolved minerals which quickly reduces its fertility.

Clay soils

Clay soils contain a high proportion of minute mineral particles. Unlike sand, clay retains moisture during drought owing to its capillarity, and it loses minerals at a slow rate owing to its poor drainage. The disadvantages of clay are related to the smallness of clay particles. These are crowded together in compact masses which are sticky and heavy when wet, making the soil extremely difficult to cultivate. Wet clay is almost completely devoid of air, and in prolonged drought it dries and splits into rock-hard lumps.

21.3 Cultivation of soil

The soils of wild uncultivated areas are more or less balanced and unchanging. As fast as minerals are extracted from them by plants they are replaced by other minerals from animal droppings and the decaying remains of dead organisms. But this is not the case in garden and agricultural soils. Here, the natural sequence of life, death, and decay is replaced by intensive cultivation in which plants are removed from the soil for human consumption. Under these conditions, unless the minerals extracted by these plants are replaced artificially, the soil rapidly deteriorates and soon becomes incapable of supporting plant life.

However, it is possible to go further than simply maintaining soil fertility by replacing lost minerals; soils can be improved artificially so that they are capable of supporting a far larger crop of plants than in their natural state. In general, all types of soil improvement involve attempting, as far as possible, to turn poor soils into best quality loam.

Improving clay soil

Heavy clay soils can be made lighter by deep ploughing or digging in the autumn. This produces large clods which are broken down by winter frosts. Drainage of clay can be improved by adding the substances which it lacks in order to turn it into loam, such as sand, and humus in the form of peat or well rotted manure. An especially effective method of improving clay is to add lime. This reacts chemically with clay particles making them join together in clumps like soil crumbs. This is called **flocculation**, and is investigated in exercise C2. Clay also tends to be rather acid, a condition which reduces the growth of certain crop plants. The acid is neutralized by the addition of lime.

Improving sandy soil

Sandy soil can be improved by adding humus in the form of peat and manure. When this is ploughed into the soil it improves water-holding capacity and mineral content. In dry weather, evaporation from the surface of sandy soil can be reduced by laying wet rotting manure or peat on the soil surface. This process is called **mulching**.

Fertilizers

Of the many mineral elements present in soils, nitrogen, phosphorus, potassium, magnesium, calcium, and sulphur are required by plants in the greatest quantities. Consequently, these elements must be added to garden and agricultural soils at regular intervals if fertility is to be maintained. A **fertilizer** is a substance which supplies one or more of these essential elements.

An **organic fertilizer** is one which is derived from animals and plants, whereas an **inorganic fertilizer** is a factory-produced chemical. The main difference between the two is that organic fertilizers act more slowly because they must first decompose in the soil before they can release their mineral elements. Inorganic fertilizers, on the other hand, dissolve almost immediately in soil water and their minerals are readily available to plants.

Nitrate and phosphate, for instance, can be added to soil in slow-acting organic fertilizers such as dried blood, bone-meal, or farmyard manure, or in quick-acting inorganic chemicals such as sulphate of ammonia and superphosphate of lime. But for quick results it is more convenient to use compound inorganic fertilizers which contain balanced proportions of nitrate, phosphate, and potassium. However, these may in the long run be less beneficial or even harmful, because if used excessively they can lead to the breakdown of soil crumbs and the formation of inorganic acids in the soil.

Verification and inquiry exercises

A *An investigation into the composition of various soils*

Carry out the following tests on soils collected from different environments, e.g. meadows, marshes, woods, moorland. If this is not possible at least ensure that samples of clay, sand, and a good quality of loam soil are available. Relate the results of the following tests to the type of vegetation found on each type of soil. Finally, summarize the results in a chart so that the various soils can be compared at a glance.

1. *Sedimentation test*

Repeat the sedimentation test described on page 268 on equal volumes of a variety of soils, but use a measuring cylinder instead of a test-tube. After the sediments have settled roughly estimate the proportions of humus and each type of mineral particle.

2. *An estimation of water content*

a) Weigh a soil sample in an evaporating basin, heat it for 30 minutes at approximately 100 °C and weigh it when cool. (The soil may be heated in an oven, or over a beaker of boiling water, but it must not be heated to a temperature at which its humus content begins to burn.) Repeat until no further weight loss is observed. The difference between the original weight and the weight after heating represents the amount of water that the soil contained. Calculate the result as a percentage of the original weight.

b) Devise a way of discovering the field capacities of various soils. Try to account for any differences discovered.

3. *An estimation of humus content*

a) Weigh a quantity of dry soil from exercise A2 above in a crucible, and heat it strongly for about 20 minutes to burn away all its humus content.

b) Weigh again when cool and calculate the humus content as a percentage.

c) Find the difference between the humus content of topsoil and subsoil.

4. *An estimation of air content*

a) Find the volume of an empty tin (which has had one end removed cleanly with a can opener) by filling it with water and emptying it into a large measuring cylinder.

b) Punch one or two holes in the bottom of the tin and push it, open end first, into the soil until its bottom is level with the surface. Dig the tin out of the soil without disturbing its contents, turn it the right way up and smooth its surface level.

c) Empty *all* the soil from the tin into a large measuring cylinder containing a known volume of water. After bubbles have stopped rising from the soil note the amount by which the water level has risen up the measuring cylinder. This rise represents the volume of solid matter in the soil.

d) Calculate the volume of air in the soil sample. Calculate the result as a percentage.

B *An investigation into the physical properties of various soils*

Carry out the following tests on the same variety of soils and add the results to the chart.

1. *To compare the rate at which air percolates through various soils*

It is extremely difficult to measure air percolation through undisturbed soils. Nevertheless, useful comparisons can be made using dried powdered samples of soil.

a) Place a known volume of soil in the funnel of the apparatus illustrated in Figure 21.4. Tap the funnel to make the soil settle.

b) Start a stop-watch and simultaneously open the clip. Water will drain from the tube, pulling air through the soil sample. Stop the watch when the tube is empty.

c) Repeat, using an equal volume of a different soil.

d) What factors affect the rate at which air percolates through soil?

e) In which type of soil will roots receive the most air?

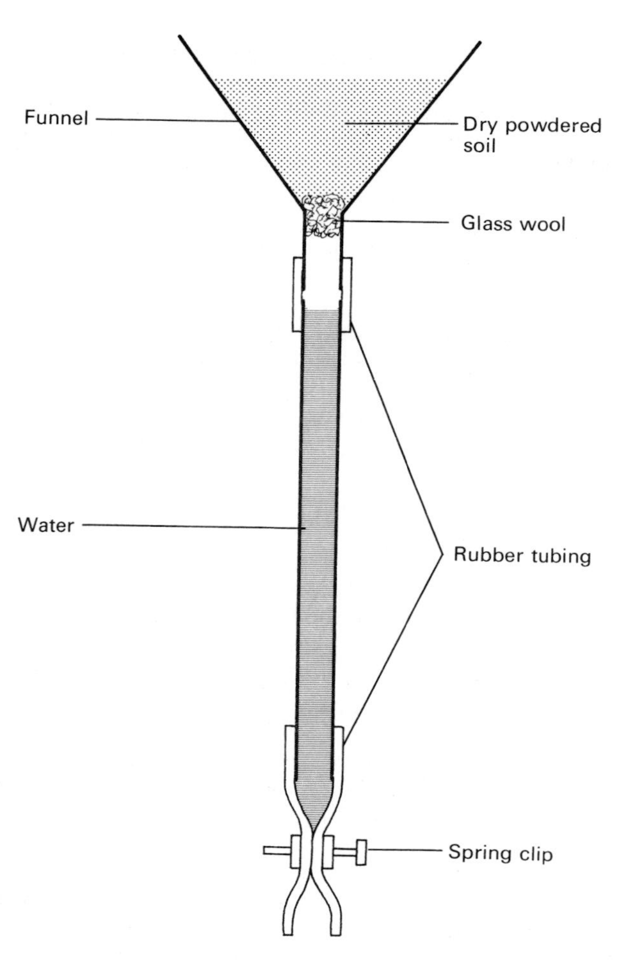

Fig. 21.4 Materials and apparatus for exercise B1

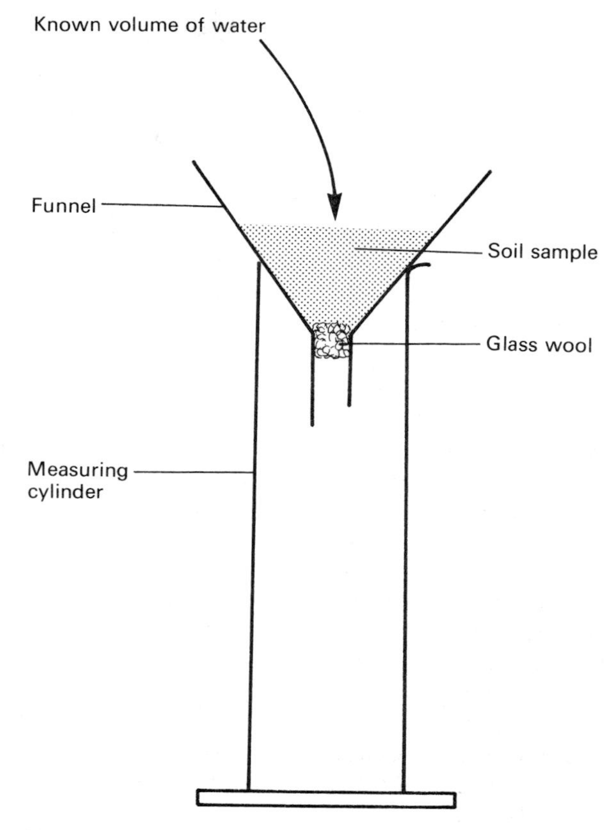

Fig. 21.5 Materials and apparatus for exercise B2

2. *To compare the water-retaining capacities of soils, and the rates at which water percolates through them*

a) Place a known volume of dry powdered soil in the funnel of the apparatus illustrated in Figure 21.5. Pour a known volume of water gently on to the soil, and note the volume of water which eventually drains into the measuring cylinder. The difference between the two volumes represents the amount of water retained by the soil.

b) Start a stop-watch, and simultaneously pour another known volume of water on to the wet soil. Note the length of time taken for this water to drain through the soil.

c) Repeat these two tests on different types of soil.

d) What factors affect the water-retaining and drainage characteristics of soil?

e) What type(s) of soil: retain so much water and drain so slowly that they are likely to flood during heavy rainfall; and drain so readily that they are likely to have the soluble minerals washed out of them?

3. *To compare the ability of various soils to take up water by capillarity*

Capillarity results in water moving upwards against the pull of gravity from the water table. The apparatus illustrated in Figure 21.6 can be used to measure this.

a) Obtain a number of wide glass tubes about 20 cm long and prepare them as illustrated using a different type of dry powdered soil in each.

b) Note the date on which the bases are immersed in water, and the dates on which the seeds germinate.

c) Account for any differences observed.

4. *To demonstrate the relationship between the height to which water rises by capillarity and the width of the channel through which it passes*

a) Strap two pieces of glass together with elastic bands, then slip a length of matchstick between them to hold two edges apart (Fig. 21.7).

b) Lower this apparatus into a shallow dish of coloured water. Tap the glass a few times, and observe the way in which water rises between the two sheets.

c) What is the relationship between channel width and the height of capillary rise?

d) How does the result of this test help to explain why water rises to a greater height by capillarity in clay soil than in sandy soil?

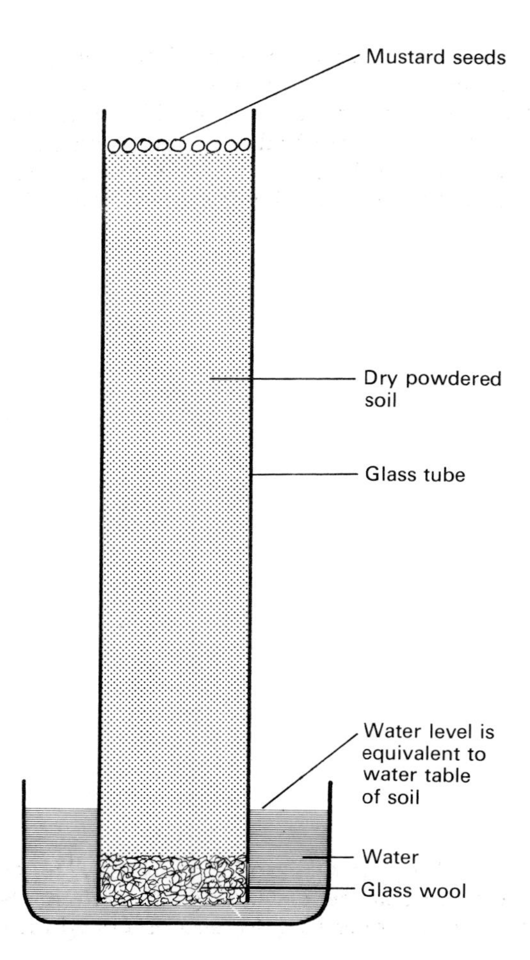

Fig. 21.6 Materials and apparatus for exercise B3

C *An investigation into the chemical properties of soil*

1. *To discover the degree of acidity or alkalinity of a soil sample (i.e. its pH value)*

Obtain a quantity of universal indicator, such as BDH or as prepared by a biological supplier. (An indicator is a substance which changes colour according to the pH of chemicals which are mixed with it.)

a) Place a few drops of indicator on a white tile, mix a little soil with it, then tilt the tile to make indicator run out of the soil.

b) Observe the indicator's colour and, from the colour chart provided by the supplier, read off the soil's pH value. To be of any use this test should be related to the type of vegetation found growing on a soil.

2. *To investigate the effects of lime on clay*

a) Two-thirds fill two large test-tubes with distilled water and add an equal volume of dry powdered clay to each. Shake both tubes thoroughly, add lime water to one tube and an equal volume of distilled water to the other. Shake the tubes again and put them in a test-tube rack to settle. What are the effects of flocculation?

b) Fill two funnels with wet clay soil and place each on a measuring cylinder (Fig. 21.5). Pour a known volume of lime water into one funnel and an equal volume of distilled water into the other. Compare the rates at which they drain through clay, and explain any difference observed.

D *An investigation of the living things found in soil*

1. *To verify that soil contains living organisms*

a) Hang a cloth bag containing a sample of moist garden soil inside a flask of lime water (as illustrated in Figure 8.4).

b) As a control, prepare another flask containing a bag of baked soil. Note any changes in the lime water, and explain.

2. *To verify the presence of bacteria and fungi in soil*

Prepare tubes of agar as described in exercise D of chapter 20, and sprinkle a little soil from different localities into each. Observe the growth of bacteria and fungi.

Fig. 21.7 Materials and apparatus for exercise B4

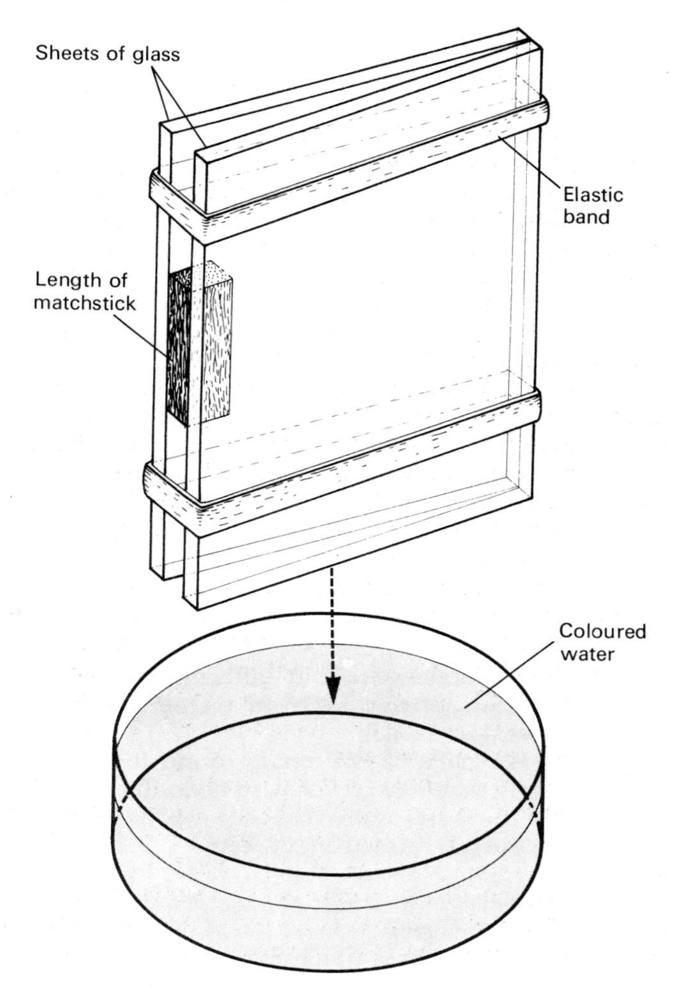

Comprehension test

1. A bottle was filled with water, sealed with a screw cap and placed in the freezing compartment of a refrigerator. A few hours later the bottle shattered. Explain how this result illustrates one of the factors responsible for soil formation.

2. A box 10 cm square will hold 1000 glass beads 1 cm in diameter, or 125 glass beads 2 cm in diameter. The surface area of a sphere = $4\pi r^2$. If $\pi = 3$, what is the total surface area (a) of all the 1 cm beads, and (b) all the 2 cm beads? How does the difference help to explain the different water-holding capacities of sand and clay?

Summary and factual recall test

When organisms die their remains (1) owing to the (2) action of bacteria and (3) in the soil. One end-product of this process is a black fibrous material called (4) which comes mainly from (5–name three sources). The soft parts of decaying organisms form a (6) which sticks soil particles together into clumps called soil (7). Soft decaying tissues also form chemicals such as (8–name three) which are essential for (9) growth. Humus and crumbs are an important part of soil because (10).

In order of increasing size the types of mineral particle in soil are: (11). These particles originate from rock by a number of (12) and (13) processes called weathering. Cracks in rocks are widened by (14) which freezes and (15); and by pressure from the growth of plant (16). Rocks are also broken down by mechanical weathering such as: (17–give two examples).

(18) dissolves in falling rain to make weak (19) acid. This reacts with rocks, especially (20), forming (21) substances which (22) away leaving small fragments that develop into (23). More of this acid is formed from (24), together with (25) acid from plant roots. These deepen the soil by (26).

Air is an essential part of soil because (27). The amount of air in soil depends on (28).

Water is an essential part of soil because it carries (29) in solution. A force called (30) retains water around soil particles. This water attracts more water by a force called (31), which gets (32–stronger/weaker) the smaller the distance between soil particles. This explains why (33) soils can draw water upwards from the (34) table against the force of gravity. The field capacity of a soil is (35).

Bacteria and fungi increase soil fertility mainly by (36). Worms (37) the soil by burrowing through it. Some worms turn the soil over by (38), and others add humus to the soil by (39).

A sedentary soil is one which is situated (40). An (41) soil is formed from particles carried long distances by (42) and (43), and then deposited in places such as (44–name two). Topsoil differs from subsoil in at least three ways: (45).

Loam soils consist of about 50 per cent (46), 30 per cent (47), and 20 per cent (48), with a little (49). All these are formed into well-developed (50). Loam is the most fertile of all soils for at least four reasons: (51).

Sandy soils consist mostly of (52–large/small) particles with little or no (53–name two other components). The agricultural advantages of sandy soil are (54–describe four). Its disadvantages are (55–describe two).

Clay soils consist mostly of (56–large/small) particles. The advantages of clay are (57–describe two). Its disadvantages are (58–describe three).

Uncultivated soils can be described as 'balanced' because (59). Garden and agricultural soils must have (60) added to them regularly because (61). Heavy clay soils can be improved by (62–describe four ways). Sandy soils can be improved by (63).

Three examples of organic fertilizer are (64). These release minerals to the soil (65–slowly/quickly) because (66). Inorganic fertilizers release their minerals (67–slowly/quickly). Two examples of inorganic fertilizers are (68).

22
The balance of nature

The planet Earth is a ball of rock hurtling through space. The rock has depressions in it filled with water and flat areas covered with soil, and it is entirely contained in an envelope of air. But this is only the physical world. The Earth is inhabited by countless millions of living organisms which make up the biological world.

This chapter is concerned with the continuous exchanges of material and energy which take place between and within these two worlds. These exchanges occur in cycles, which involve a continuous circulation of substances between organisms and their physical environment. Substances such as oxygen, carbon dioxide, water, and minerals are constantly absorbed by organisms, but as fast as these substances are lost from the physical world they are replaced from the natural processes of photosynthesis, respiration, excretion, and decay in the biological world. In other words, there is a **balance of nature** in which losses equal replacements, and in which materials are used and re-used over and over again.

The **nitrogen cycle** is an excellent example of how one element circulates around and through the physical and biological worlds (Fig. 22.1).

Fig. 22.1 The nitrogen cycle

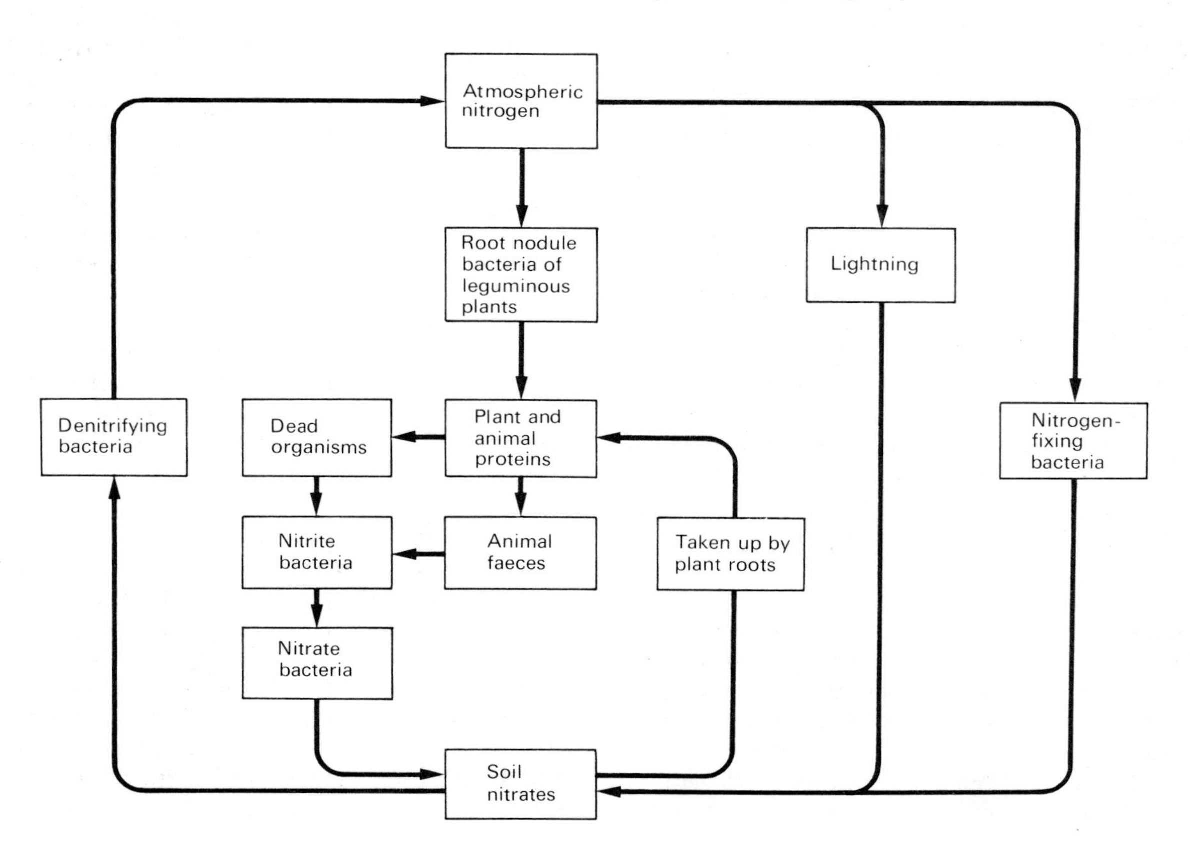

22.1 The nitrogen cycle

Life cannot exist without nitrogen. It is an essential component of all proteins. Nitrogen gas forms approximately four-fifths of the atmosphere but neither plants nor animals can make use of it in this form. Plants can take in nitrogen only in the form of nitrates which they absorb from the soil, and the only way animals can obtain nitrogen is by eating plants or animals which eat plants. Consequently, both plants and animals depend, directly or indirectly, on mechanisms which replenish the soil with nitrates as fast as they are removed from the soil by plant roots. These mechanisms which replace nitrates in the soil may be divided into two types: those which transform atmospheric nitrogen into soil nitrates, and those which transform plant and animal protein into nitrates.

Transformation of atmospheric nitrogen

Nitrogen is transformed into nitrates by lightning and by certain soil bacteria. (Nitrates are compounds of nitrogen and oxygen.)

Lightning During lightning flashes an extremely high temperature is generated. This results in nitrogen combining with oxygen to form nitrous and nitric oxide gases. These gases dissolve in rainwater and form nitrous and nitric acid. These acids soak into the soil where they react with other chemicals to form nitrates.

Nitrogen-fixing bacteria Bacteria of this type use carbohydrate and atmospheric nitrogen to make compounds which are eventually released into the soil as nitrates. This process is called **nitrogen-fixation.** Some of these bacteria obtain carbohydrate from soil humus. Others live inside the root cells of leguminous plants (e.g. peas, beans, clover, and vetches) where they cause tiny swellings called **root nodules** (Fig. 22.2). Here the bacteria obtain both protection and carbohydrate from the plant cells, and in return they release nitrates into the plant tissues and the soil. (An association of this kind in which two different organisms benefit from living together is called **symbiosis.**) Leguminous plants are often cultivated in order to increase the nitrate content of agricultural soils.

Transformation of plant and animal proteins

Saprotrophic, or putrefying bacteria and fungi decompose proteins in dead animals and plants. In this process ammonia is released and immediately dissolves in soil water. Here, the ammonia combines with soil chemicals to form various ammonium compounds which are converted into nitrates by nitrifying bacteria.

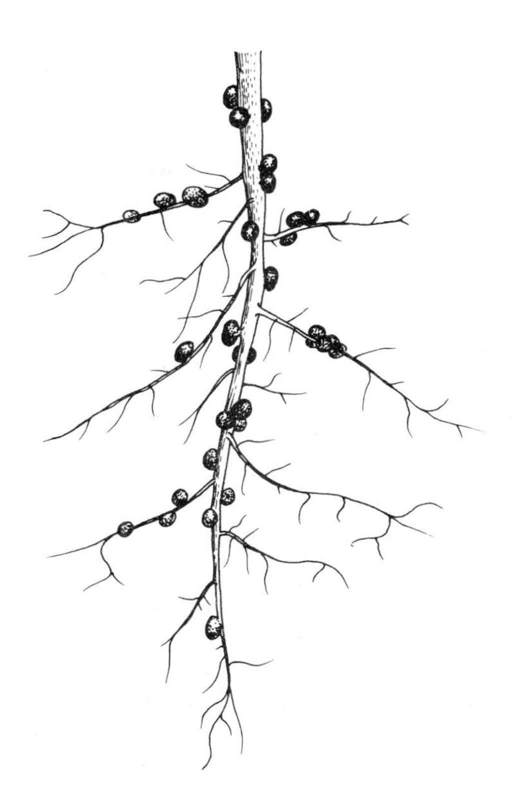

Fig. 22.2 Root nodules on a bean plant

Nitrifying bacteria There are at least two types of nitrifying bacteria in the soil. First, **nitrite bacteria** turn ammonium compounds into nitrites by combining them with oxygen. (Nitrites are chemicals with less oxygen in their molecules than nitrates.) Second, **nitrate bacteria** combine nitrites with more oxygen to form nitrates. By this sequence nitrogen in proteins is changed into a form which can be absorbed by plant roots. The same series of bacteria also form nitrates out of the nitrogen-containing compounds in animal droppings and urine.

Lightning and the nitrifying bacteria replace nitrates removed from soils by plant roots. But there are some bacteria which remove nitrogen from soil nitrates and return it to the atmosphere as gas. These are called **denitrifying bacteria.**

Denitrifying bacteria

Waterlogged and heavy clay soils contain very little oxygen and therefore from a biological point of view they are anaerobic. Denitrifying bacteria are anaerobic, whereas nitrifying bacteria are aerobic. Consequently, in waterlogged conditions and in heavy clay soils denitrifying bacteria are the more active of the two. Denitrifying bacteria obtain their energy by breaking down nitrites and nitrates in the soil into nitrogen and oxygen gases. These gases pass into the

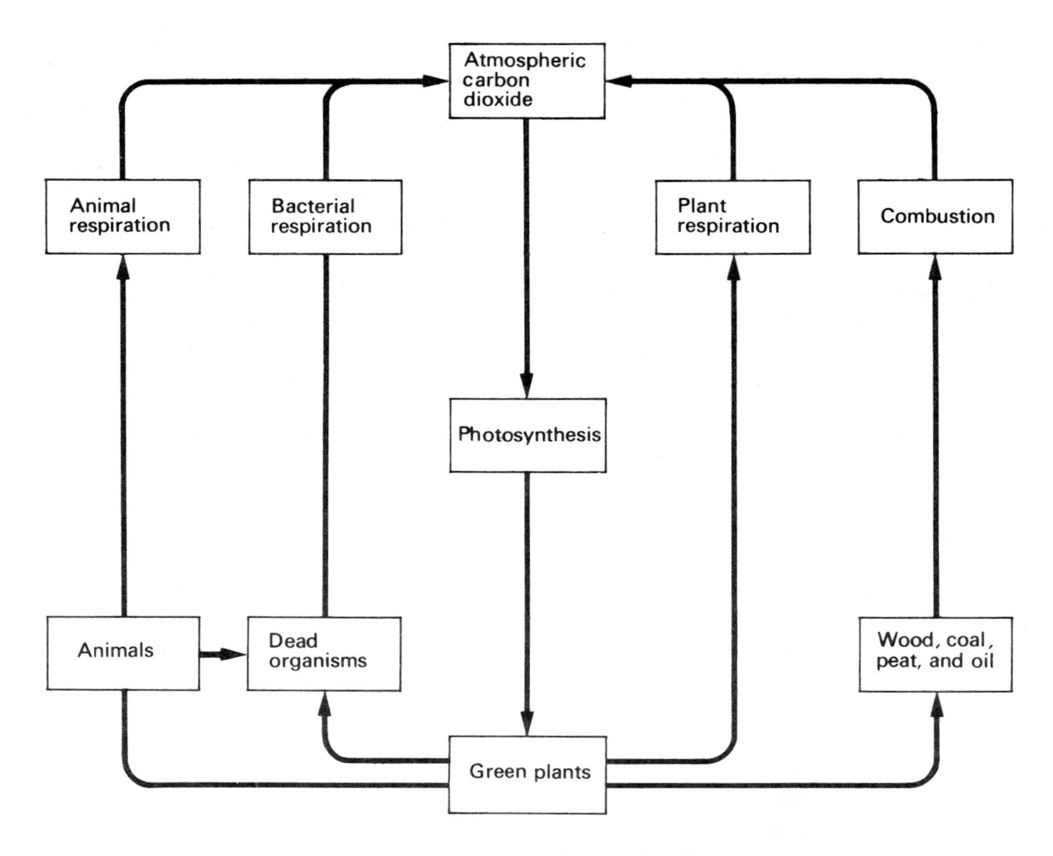

Fig. 22.3 The carbon cycle

atmosphere where the nitrogen is no longer available to plants. For this reason it is good agricultural practice to improve drainage of soils where flooding is likely, and to break up clay soils by the methods described in section 21.3.

22.2 The carbon cycle

The carbon cycle is another example of how an element circulates continuously within the physical and biological worlds (Fig. 22.3). Carbon dioxide is absorbed by green plants during daylight hours as a raw material of photosynthesis. Despite this process the amount of carbon dioxide in the atmosphere remains constant at about 0.033 per cent. This is because it is replaced in a number of ways as fast as it is absorbed by plants.

Respiration

Some of the carbohydrate produced by photosynthesis is used to build up the plant body, but the remainder is eventually respired for energy. This releases carbon dioxide back into the atmosphere.

Carbon atoms in the carbohydrates, fats, and proteins of plants are transferred to the bodies of herbivores when these animals feed on plant tissues. Later these carbon atoms may be transferred again if the herbivores are eaten by carnivores or omnivores. As these animals respire some of the carbon atoms are released to the atmosphere as carbon dioxide.

Decay

After death, the bodies of organisms are decomposed and absorbed as food by saprotrophic bacteria and fungi. Carbon atoms in this absorbed material are released to the atmosphere as the saprotrophs respire.

Combustion

Combustion, or burning, of inflammable materials results in the release of carbon atoms as carbon dioxide. Combustion can form part of the carbon cycle in the following ways.

Carbon absorbed by a tree during photosynthesis and used to build woody tissue in its trunk will be returned to the atmosphere if the tree is chopped down and burned as fuel. Over millions of years, the bodies of dead organisms have produced fossil fuels such as coal, oil, and natural gas. Some of these organisms were plants which took carbon from the air during photosynthesis, and some were animals which fed on the plants, or on animals. Therefore, when these fossil fuels are burned today, they release carbon atoms which were trapped by photosynthesis in plants that lived millions of years ago.

The circulation of carbon in each of these ways can give individual carbon atoms a long and varied history. For instance, carbon atoms in the breath of a gardener may be absorbed by photosynthesis into the leaves of his rose bushes where they are incorporated into sugar. Later, the same carbon may enter the bodies of greenfly as they suck sugar from rose phloem. It could then be transferred to ladybirds which eat greenfly; to insect-eating birds which feed on ladybirds; to hawks which feed on insect-eating birds; and then returned to the atmosphere by hawk respiration.

This sequence of events illustrates that the raw materials of life are never lost or created, but re-used endlessly as they circulate between organisms and their environment. But circulation is not restricted to carbon and nitrogen; there is a continuous circulation of all the chemicals which make up living things.

22.3 Food chains and food webs

The sequence of events starting with photosynthesis in rose bushes and ending in hawks illustrates another important biological principle: that the materials and energy needed for life pass from one organism to another in what are called **food chains**. Like every other food chain, this example begins with **producer organisms** (the rose bushes), which are always green plants producing food substances by photosynthesis. The 'links' in a food chain are made up of consumer organisms which occur in a certain order. The **first-order consumers** are the animals which feed on plants (e.g. greenfly); the **second-order consumers** (e.g. ladybirds) are those which eat the first-order consumers, and so on to the end of the chain.

The pyramid of numbers

Figure 22.4 illustrates an example of a pyramid of numbers, which is a characteristic of certain food chains. The base of this particular pyramid consists of

Fig. 22.4 A pyramid of numbers

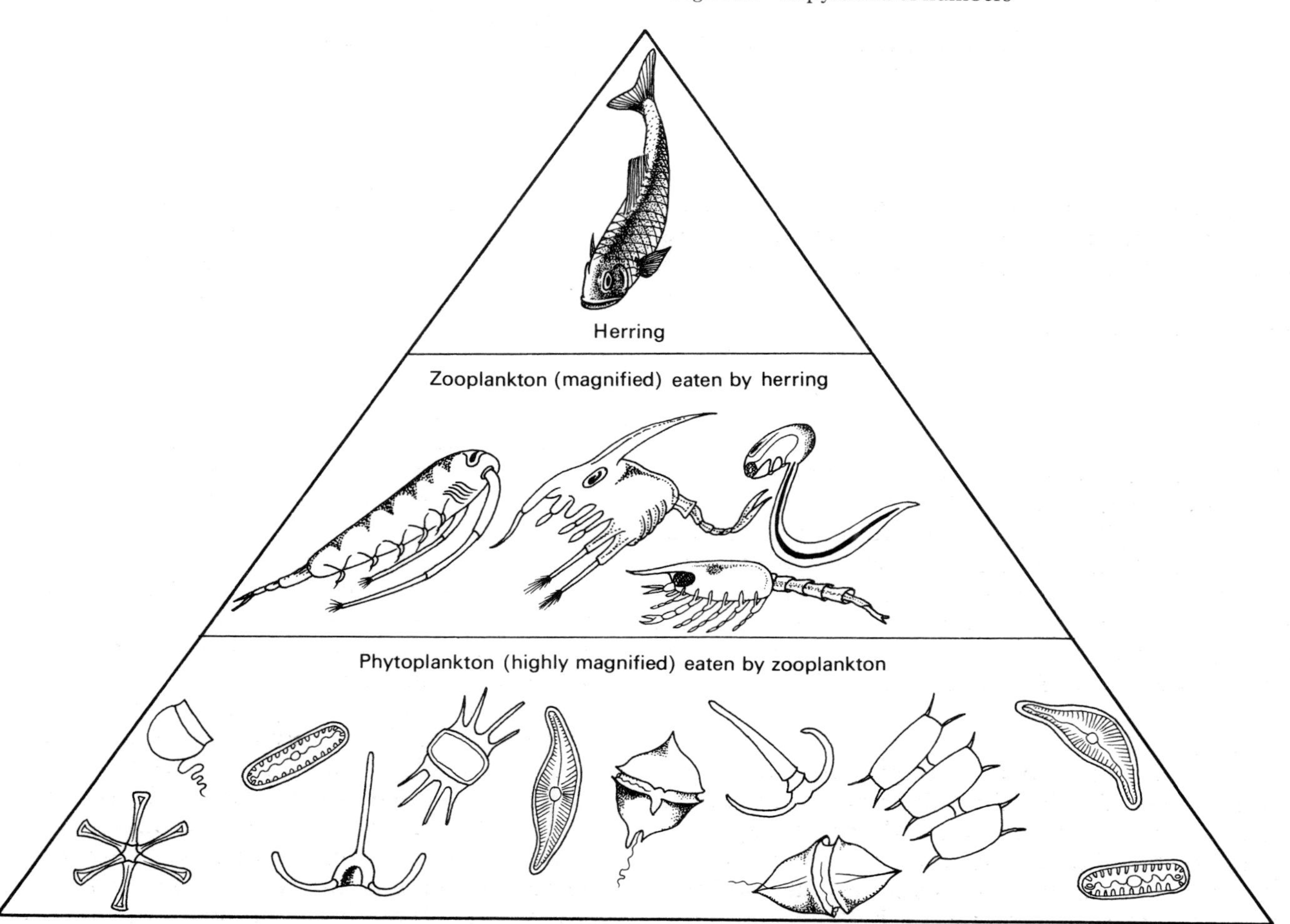

microscopic plants called phytoplankton. These plants are the producers in the food chain and are eaten by first-order consumers which are tiny animals known collectively as zooplankton. The zooplankton are eaten by second-order consumers which are fish such as herring.

This type of food chain can exist only if there are more producers than first-order consumers, and more first-order consumers than second-order consumers, and so on up the chain, with a reduction of population at each link. Only under these circumstances are there sufficient numbers in each link of the chain to reproduce at a rate which replaces losses to the organisms in the link above. For instance, if the zooplankton were able to increase to the point at which they devoured phytoplankton faster than these plants could reproduce, there would be a rapid decline in phytoplankton numbers and then the whole chain would collapse.

Food chains are another example of the balance of nature, since they depend upon a delicate balance between losses and gains. If anything happens to disturb this balance, such as disease or pollution which destroys a link in the chain, all the other organisms in the chain will be affected. Here is an imaginary example of such disruption.

Suppose that hawks were somehow removed from the top of the food chain described above. Without hawks to limit their numbers insect-eating birds would increase and could completely overwhelm the ladybird population. Similarly, without ladybirds to prey on them greenfly would increase in numbers and might damage the rose bushes so badly that both they and the greenfly would perish along with the other members of the food chain.

Food webs

In nature, food chains are not quite so simple as those described so far. Consumer organisms rarely depend on only one type of food, and often a particular food is eaten by more than one consumer. Zooplankton, for instance, are eaten by herring and by many other predators including prawns, shellfish, marine worms, and whalebone whales. In consequence, food chains are interconnected at many points forming what is best described as a **food web**.

Communities and ecosystems

Figure 22.5 illustrates part of a food web. It shows that many different organisms interact with one another in such a complicated way that each type comes to depend on many other organisms for its

Fig. 22.5 Part of a food web in an area of woodland surrounding a small pond

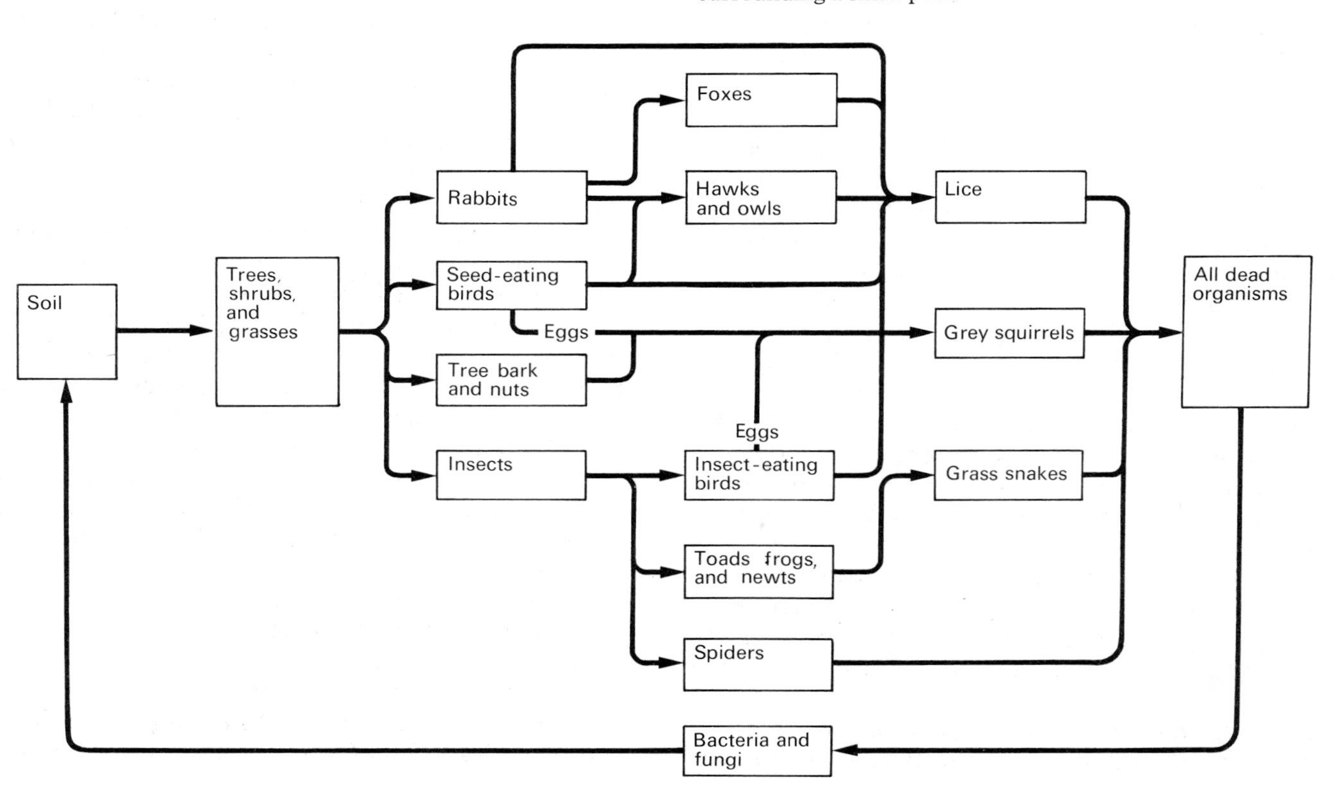

existence. This, and other groups of interdependent organisms which inhabit a common environment are known as a **community**.

The complex relationships between the members of a community, and between the community as a whole and its physical environment, are described as an **ecosystem**. An ecosystem is made up of all the producers and consumers in a community, including the parasites which live off these organisms; the rocks, soil, water, and air of their physical environment; and the circulation between this environment and the community of materials such as nitrogen, carbon, water, and oxygen.

It is very important to remember that ultimately the materials and energy which are distributed throughout a community through its food chains owe their origin to the ability of plants to trap the energy of sunlight and utilize it in photosynthesis.

Ecology

Ecology is the scientific study of ecosystems, and it involves tracing the complicated interrelationships which exist between organisms and their environment. This work has shown that ecosystems are rarely stable and unchanging; different organisms are always entering and leaving communities or dying out altogether, and this inevitably alters the balance of the community. When stability is eventually achieved a community is described as having reached a **climax**. When this happens the species of the community remain unchanged almost indefinitely. An example of a climax community is the oak forest found in parts of lowland Britain.

Ecologists often make a detailed study of a particular **habitat**, which is the name for part of an ecosystem that is occupied by its own community of organisms. A sea-shore rockpool is an example of a habitat, with a community usually made up of crabs, sea-anemones, marine worms, sponges, and many types of seaweed.

By studying individual species in a habitat it is possible to discover their **niche**: that is, the part they play in the over-all pattern of the community. In other words, niches are not the animals themselves, but include features such as their feeding habits, which give them their own special position in a food web. For example, flesh-eating, insect-eating, blood-sucking, grass-eating, and seed-eating are all different types of niche. The occupants of a particular niche vary from country to country. Foxes occupy a flesh-eating niche in Britain, whereas in Africa lions occupy a flesh-eating niche, and in North America pumas occupy a similar niche.

One of the most important aspects of ecology is the study of man's influence on the balance of nature. The earliest members of the human race were hunters and food gatherers and as such fitted into a niche which was in balance with their ecosystem. Unfortunately, this is no longer true. During the rapid technological advance which has taken place over the last two hundred years, man has developed the power to change his environment: by damming rivers; stripping away vegetation and soil to build roads and cities; removing natural communities and replacing them with agricultural crops and farm animals; and, perhaps most damaging of all, by producing poisons and other harmful substances known generally as **pollutants**.

22.4 Pollution

A substance is called a pollutant when its presence causes harm to living things. This means that a substance may be a pollutant or not according to circumstances. Carbon dioxide, for example, is harmless to animals and useful to plants when present in the air at its normal level of 0.033 per cent. But at much higher concentrations carbon dioxide is classed as a pollutant owing to its harmful effects on animals. Similarly, nitrates are vital to the healthy growth of plants, but they become pollutants when they accumulate in lakes and cause such a rapid growth of green algae that the water looks like pea soup and other organisms become endangered. Some substances are both useful and harmful at the same time. Examples are the chemicals applied to seeds such as wheat to prevent fungal parasites attacking the seedlings. These chemicals are useful insofar as they increase crop yields, but are harmful (pollutants) when they enter food chains starting with seed-eating birds and ending with hawks and owls.

In view of these facts a pollutant can also be defined as a substance present in the wrong amount, at the wrong place, at the wrong time.

The majority of pollutants are harmful for a while but soon become harmless after they react chemically with other substances in the environment. Examples are sewage discharged into the open sea, where it is quickly rendered harmless by bacteria and other organisms, and poisonous gases such as sulphur dioxide which are almost immediately diluted to harmless levels in the atmosphere and then change to a less poisonous form.

On the other hand, a number of pollutants retain their harmful characteristics for very long periods. These often cause special problems because there is a danger of them accumulating either in the physical environment or in the bodies of organisms in high concentrations. Lead and mercury compounds are examples of long-lasting pollutants. Lead compounds, for instance, may be inhaled from certain fumes or absorbed from contaminated water in minute quantities over a number of years with no ill-effects. But once inside the body lead cannot be excreted and may slowly accumulate to harmful levels.

Air pollution

Life on earth is supported by a layer of air called the **troposphere** which extends to a height of approximately 10 km. It is usual to describe pollution of the troposphere by stating the concentration of pollutants in parts per million (ppm), which means the number of cubic centimetres of pollutant in a cubic metre of air. The main cause of air pollution is the burning of fossil fuels such as coal and oil.

When coal is burned it produces smoke containing particles of various sizes and fumes made up of different gases–mainly carbon dioxide and sulphur dioxide. Smoke particles settle out of the atmosphere as dust, some of which covers the leaves of plants and harms or kills them by cutting off their light and so limiting photosynthesis. These particles also block their stomata. The dust is inhaled by people and, if present in high concentrations, may cause serious respiratory ailments.

Air pollution from a coal by-products plant

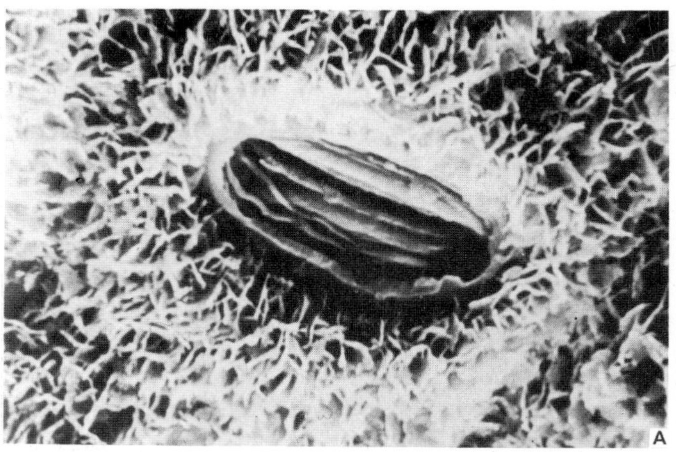

Leaf surfaces seen with the scanning electron microscope. **A** An oak leaf with stoma from an unpolluted area. **B** An oak leaf from a heavily polluted area, showing particles of dirt which clearly interfere with the opening and closing of the stoma

Sulphur dioxide is harmful both as a gas and when it has reacted with water vapour in the atmosphere, producing sulphuric acid. The gas and the acid vapour may be inhaled into the lungs. They cause no detectable harm in man at concentrations below 1 ppm except that they may aggravate ailments like bronchitis. But in plants concentrations of sulphur dioxide as low as 0.1 ppm have been shown to produce leaf damage and reduce yields of crops, and recent evidence suggests that crop yields may be reduced even when sulphur dioxide levels are too low to produce visible effects.

Plant-like organisms called **lichens** are useful indicators of sulphur dioxide pollution because certain varieties are killed by concentrations of only 0.05 ppm. If these sensitive lichens disappear, or are absent from an industrial area but still grow in surrounding rural districts, it indicates that sulphur dioxide pollution is having some biological effect.

Another cause of air pollution is the exhaust fumes from petrol and diesel engines. Such engines, especially when badly tuned, emit smoke and fumes which often gather in busy city streets forming a visible fog. Depending on the fuel used, exhaust gases consist of oxides of nitrogen and certain lead compounds, both of which are poisonous. However, even in a city street exhaust fumes appear to have little or no harmful effect on the human body unless a person is suffering from anaemia or lung disease. Even so, these gases are unpleasant to inhale and may have as yet unknown long-term effects.

In Britain atmospheric pollution seldom reaches a level which is harmful to man because, under normal weather conditions, smoke and fumes rise high into the atmosphere where they are dispersed by winds and transformed into relatively harmless substances. But pollutants are not dispersed in still air and may be trapped at a low level in the atmosphere when there is a climatic condition known as a **temperature inversion** (Fig. 22.6).

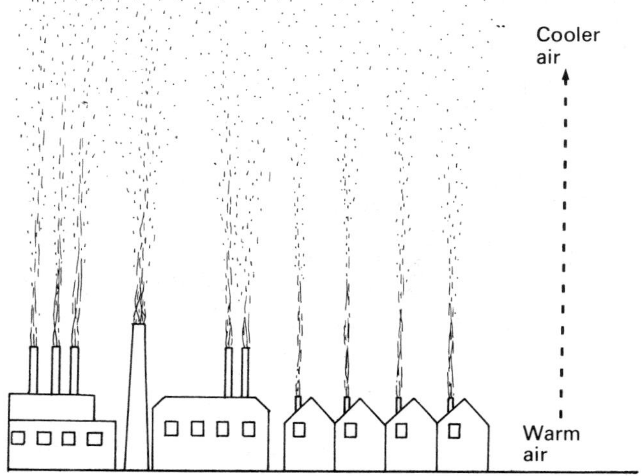

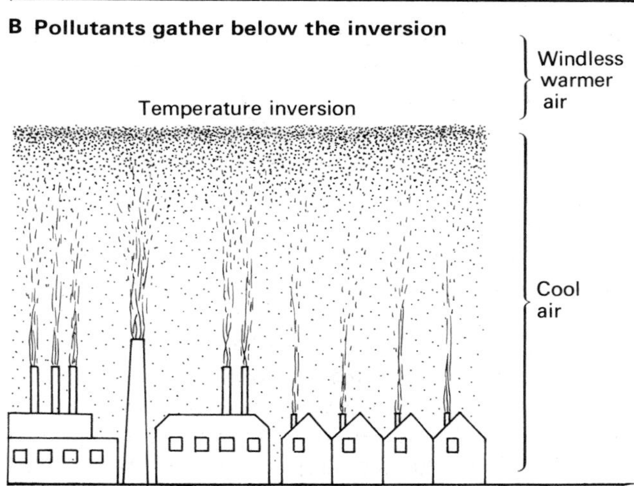

Fig. 22.6 Comparison between normal conditions and those which lead to smog formation

A temperature inversion and still air occurred simultaneously in London from 5–9 December 1952. During this time sulphur dioxide levels increased to 0.75 ppm, and smoke gathered in the atmosphere along with lead compounds, sulphuric acid, and oxides of nitrogen. All of these substances gave rise to a thick yellow fog, or 'smog'. Over this period the death rate in London increased rapidly; in one area there were about 4000 more deaths than would normally be expected at that time of year. It is not certain that pollution was the main cause but, in conjunction with other factors such as cold and damp, it almost certainly contributed to the increased death rate.

Under certain conditions oxides of nitrogen from cars and industrial areas are transformed by the energy of sunlight into peroxacyl nitrates, which gather in the atmosphere as visible clouds called **photochemical fog**. This substance can damage plants by attacking the spongy mesophyll layer of their leaves. At high concentrations it causes eye irritations in man. Photochemical fog is particularly dense over the Los Angeles area in the United States where it has reduced the yield of citrus fruits and damaged many other plants. Fortunately, this type of fog is rare in Britain, largely because there is much less sunshine.

Since 1952, legislation in Britain such as the Clean Air Act, combined with the increased use of oil instead of coal as a source of power in industry and the railways, has done a great deal to reduce air pollution.

Water pollution

At present British rivers are being polluted by domestic, industrial, and agricultural waste to such an extent that in certain regions purification processes cannot make the water fit for human consumption or, in severe cases, even for industrial use. This pollution has caused great damage to the animal life of inland waters. In England and Wales, for example, about 5 per cent of rivers are totally without fish.

Domestic waste Domestic waste can be made relatively harmless at sewage treatment plants, but in many areas population growth has temporarily overloaded these plants so that much untreated sewage is discharged into rivers and the sea. When solid organic matter in sewage enters a river it is decomposed by bacteria which multiply within it and begin to use oxygen at a rate that quickly deoxygenates the water, making it uninhabitable for most types of fish, insects, crustaceans, and aquatic worms.

The treatment of sewage removes the bulk of its organic matter. But this process creates a problem by releasing chemicals such as phosphate from the sewage, which is added to an even greater amount of phosphate released from the use of synthetic detergents. When sewage rich in phosphate is discharged into rivers it may be joined by nitrate which drains out of agricultural land that has been treated with quick-acting inorganic fertilizers. As all this phosphate and nitrate gathers in water it gives rise to a condition known as **eutrophication**, which is the accumulation of minerals that promote plant growth. When eutrophication occurs in rivers, lakes, and reservoirs the algae that live in the water multiply at a much faster rate than normal. In most cases the algae gather at the surface where they form a thick green mat that prevents light from reaching plants living lower down. Consequently, these submerged plants die and the water, deprived of their photosynthetic oxygen, quickly becomes uninhabitable by animal life.

Industrial waste One of the main problems with industrial waste is that it often contains highly poisonous and long-lasting pollutants such as compounds of cyanide, lead, mercury, copper, and zinc. These chemicals are dangerous even in low concentrations because they can accumulate in fish and other aquatic creatures and then spread along food chains to other animals such as water birds, otters, and even man.

Pollution of the river Thames

Agricultural waste The traditional agricultural methods in which animal manure is spread on land used to grow crops cause little or no pollution. However, many modern farmers keep poultry, cattle, and pigs in buildings and have no other land on which to use their manure. All too often, owing to labour shortages and the high cost of transport, this manure is disposed of by methods which pollute local streams and rivers when it could have been utilized as a valuable fertilizer. Once in the water untreated animal manure decomposes causing reduced oxygen levels as described above.

Other pollutants resulting from modern agricultural practice include chemicals in crop sprays used to kill insects and fungal pests. If these chemicals are washed into local rivers and ponds they disrupt food chains in the same way as industrial waste.

Pollution of the sea The oceans of the world contain nearly 200 million million million cubic metres of water, but this does not mean they are large enough to absorb all the waste that man is pouring into them without becoming polluted.

Until oil supplies run out, or until science discovers an alternative and cleaner source of energy, there will be a continual risk of oil spillage into the sea from tankers and offshore oil rigs. When this happens the oil kills sea birds either by poisoning them or by sticking their feathers together so that they cannot fly in search of food, and when washed ashore oil kills all forms of life on rocks and in sand.

Nevertheless, living things have a remarkable capacity for recovery. Provided these disasters do not occur too often sea bird colonies can replace their lost members, and shore life has been known to return to normal within only a year of heavy oil pollution. Unfortunately, in using detergents to clear oil pollution man has often caused more damage than if he had left the oil to be decomposed naturally by bacteria. It has been shown that these detergents, in concentrations of only 1 ppm, kill most sea-shore life, and that where they are used life takes at least two years to restore itself to normal. A less harmful alternative is to cover floating oil with powdered chalk. The oil then sinks to the bottom of the sea and is decomposed by bacteria.

A large quantity of untreated sewage is discharged into the sea, and this can cause pollution, especially where it gathers in restricted areas such as bays and inlets. Otherwise it is carried by tides and coastal currents into deeper water where it quickly decomposes and is then relatively harmless. A more serious problem arises from the discharge into the sea of poisonous industrial wastes, like those described above. When these gather in shallow-water fisheries they may cause a danger to human health by accumulating in fish and other sea creatures which are consumed as food.

Pollution by oil, sewage, and industrial waste is particularly advanced in the Mediterranean, because this is an inland sea whose only connection with the oceans is the Straits of Gibraltar. Movement of water through this gap is so slow that it takes at least a hundred years for all the water of the Mediterranean to be renewed. Hence, pollutants accumulate, particularly around large industrial areas and population centres such as Marseilles, Venice, and Haifa. Pollution of the Mediterranean is rapidly destroying aquatic life. Over the past twenty years marine life in Spain's territorial waters has been reduced by 40 per cent, and in the waters off Venice all but the ten most hardy species of fish have been destroyed. Pollution is also increasing the risk of certain diseases in the area. In 1973 there was a cholera outbreak in Naples which was probably caused by pollution of local mussel beds. There is also a risk of typhoid and hepatitis if water is accidentally swallowed while swimming in certain areas.

These are only a few of the results of pollution in the Mediterranean. They are examples of what might happen elsewhere if the oceans continue to receive waste materials at the present rate.

Pollution is a problem because man is clever but not always wise. He has been clever in learning how to exploit the world's resources to his own advantage, but not wise enough to foresee the environmental damage which results from this exploitation. Before it is too late man must learn how to reduce his interference with the balance of nature, otherwise he may endanger all forms of life on this planet, including his own.

Comprehension test

1. The number of organisms in each link of most food chains is limited by the numbers in each link below them. Explain the meaning, and the significance of this statement.

2. Study Figure 22.5 and answer the following questions.

a) Name the producer organisms.

b) How many complete food chains are to be found in the diagram?

c) Describe what might happen to the other organisms if a disease killed all the rabbits.

d) Why might the removal of all the grey squirrels improve the living conditions for the other animals?

e) If, after removing the grey squirrels, all hawks and owls were to be killed, why might this harm the living conditions of other animals?

f) Name those animals whose food supply would be reduced (both directly and indirectly) if the area were to be sprayed with insecticide.

g) What part is played by bacteria and fungi in maintaining this food web?

Summary and factual recall test

Nitrogen is an essential component of (1) and so is vital to all life. Plants cannot absorb nitrogen from the (2) but only in the form of (3) which they absorb from the (4). Some of the nitrogen absorbed by plant roots is replaced by nitrogen-(5) bacteria which use (6) nitrogen and (7) to make compounds which eventually release (8) into the soil.

Dead organisms are decomposed by (9). The process of decay releases (10) which is first converted into (11) by one set of bacteria and then into (12) by another set. Together these organisms are called (13) bacteria. The events described above are called the nitrogen cycle because (14).

(15) removed from the air by photosynthesis is replaced in three ways as follows (16). This process is called the (17) cycle because (18).

All food chains begin with producer organisms, which are (19). Next in line are the (20) consumers which are (21) that eat (22), and next are the (23) consumers which eat (24). Food chains are inter-connected at many points, forming a food (25).

A community is made up of inter-(26) organisms which inhabit a common (27). An ecosystem is made up of (28—summarize all its parts). The scientific study of ecosystems is called (29). This often involves the study of a (30), such as a rock pool. Flesh-eating and blood-sucking are examples of different types of (31), a study of which shows the position of each animal in a (32) web.

A substance is called a pollutant when it (33). The main sources of air pollution are (34—name two). These sources produce pollutants such as (35—name three).

Untreated sewage in a river is (36) by bacteria, which remove (37) from the water making it uninhabitable for animals such as (38—name four). Industrial waste often contains long-lasting pollutants such as (39—name five). These, and poisons used in agricultural (40) sprays can disrupt food chains in the following way (41).

Glossary

This glossary gives brief definitions of some of the most important biological terms used in the text. A word in *italics* within a definition can be found elsewhere in the glossary.

Absorption The movement of digested (soluble) food through the walls of the *alimentary canal* into the blood-stream.

Accommodation Alterations to the focus of the eye. In man this is done by changing the shape of the lens.

Adrenalin A *hormone* secreted by the adrenal glands. It prepares the body for instant action by increasing the rate of heart-beat, blood pressure, and blood sugar level.

Adsorption A force which holds water in a thin film around the surface of soil particles.

Aerobic respiration A type of *respiration* in which oxygen is consumed.

Aerofoil The shape of a bird's wing as seen in cross-section. It generates lift as the bird moves through the air.

Agglutinin A type of *antibody* which sticks bacteria together in clumps, thereby hindering their ability to reproduce.

Alimentary canal The intestine. A tube running from mouth to anus inside which *digestion* and *absorption* take place.

Alleles Two *genes* are called alleles of each other when they occupy the same relative position on *homologous chromosomes*, and control development of the same hereditary characteristic.

Alluvial soil Soil made up of particles which have been carried by rivers or glaciers to their present position.

Alternation of generations A type of life cycle in which a generation of organisms capable only of *sexual reproduction* alternates with a generation of organisms capable only of *asexual reproduction*. Typical of mosses and ferns.

Alveoli Bubble-like air pockets at the ends of the air passages in lungs. They are surrounded by blood vessels and are concerned with *gaseous exchange*.

Amino acids Organic compounds consisting of chemical units linked together in long chains to make protein molecules. These separate from one another when *protein* is digested.

Amnion The fluid-filled sac that surrounds and protects the *embryos* of reptiles, birds, and mammals.

Ampulla The swollen portion of the *semi-circular canal*. It contains sensory hairs and a *cupula*, and detects changes in the direction of movement.

Amylase A type of *enzyme* which digests *carbohydrates*.

Anabolism Chemical reactions of *metabolism* that build up complex substances from simple raw materials. This requires energy, which comes from *catabolism*.

Anaerobic respiration A type of *respiration* in which oxygen is not consumed.

Anaphase The stage of cell division in which either *chromatids* (*mitosis*) or *chromosomes* (*meiosis*) separate.

Androecium The collective name for the male sex organs (*stamens*) of a flower.

Antagonistic muscle system Two sets of muscles which oppose each other at each side of a joint. One set bends the joint (flexor muscles) and the other straightens it (extensor muscles).

Antennae Long narrow sense organs on the heads of insects, containing *receptors* of touch, taste, smell, humidity, and temperature.

Anther Terminal part of a *stamen*, which produces, and later releases, *pollen*.

Antibodies Chemicals made by the body in response to *parasites* and foreign substances called *antigens*. They destroy or neutralize *antigens*.

Antigens Bacteria, viruses, or foreign substances in the body which stimulate the production of *antibodies*.

Antitoxin A type of *antibody* which neutralizes poisonous substances, particularly those produced by *parasites*.

Aorta The main *artery* of the body.

Aqueous humour The liquid that fills the front chamber of the eye between the lens and the *cornea*.

Artery Blood vessel carrying blood away from the heart.

Asexual reproduction Reproduction involving one parent, without the fusion of *gametes*.

ATP Adenosine triphosphate A chemical that transfers energy released by *respiration* to other reactions in the body which absorb energy.

Atrium One of the thin-walled upper chambers of the heart. It receives blood from the *veins*.

Autotrophic (organisms) Those which manufacture all their own food requirements from simple substances, e.g. plants living by *photosynthesis*.

Auxin A *hormone* produced by plants which controls the rate of cell growth in roots and shoots.

Axon The nerve fibre of a *neurone* (nerve cell) which conducts nerve impulses away from the cell body.

Bile A greenish-yellow liquid made in the liver and passed into the *duodenum*, where its main function is to aid in the *digestion* of fats.

Binary fission A type of *asexual reproduction* in which the organism divides into two parts. Typical of protozoa, e.g. *Amoeba*.

Bladderworm A stage in the life cycle of a tapeworm during which it lives in the muscles of the secondary host. It consists of a fluid-filled bladder containing a tapeworm head.

Blind spot Point at which the optic nerve leaves the *retina* of the eye. It is insensitive to light.

Bowman's capsule A cup-shaped structure in a kidney 0.1 mm in diameter in man. It contains a *glomerulus* and leads to a *kidney tubule*.

Bronchial tree The branching air passages of the lungs, starting with the single *trachea* and ending in millions of fine bronchioles.

Budding A type of *asexual reproduction* in which offspring develop as outgrowths from the parent's body.

Bulb A organ of *vegetative reproduction* consisting of a very short stem surrounded by thick leaves swollen with water and stored food.

Caecum A blind-ended branch of the intestine at the junction between the *ileum* and *colon*. It has no function in man, but is concerned with *cellulose digestion* in herbivores.

Calyx Collective name for the *sepals* of a flower.

Cambium A region of unspecialized cells between the *xylem* and *phloem* of many *vascular bundles*. Cell division in the cambium produces new vascular tissue and increases the diameter of the stem.

Canine teeth Conical, dagger-like teeth at each side of of the jaws of some mammals. These teeth are well developed in carnivores, which use them to kill their prey and tear meat from their bones.

Capillarity The force that draws water upwards through narrow spaces, such as those between particles of soil.

Capillary Narrow, thin-walled blood vessel conveying blood from *arteries* to *veins*. The main exchanges of gases and dissolved substances between the blood and body cells take place through capillary walls.

Carbohydrates Compounds containing carbon, hydrogen, and oxygen, e.g. sugar, starch. These are the main source of energy for *metabolism*.

Carbon cycle The continuous circulation of carbon atoms between atmospheric carbon dioxide and the bodies of living organisms.

Carnivore Flesh-eating animal, e.g. lion, tiger.

Carpel Female sex organ of a flower.

Catabolism Chemical reactions of *metabolism*, which break down complex substances into simpler ones and release energy.

Catalysts Substances that increase the speed of a chemical reaction without themselves being used up in the reaction, e.g. *enzymes*.

Cell A unit of living matter, consisting of a *nucleus*, *cytoplasm*, and a *cell membrane*.

Cell membrane The *semi-permeable* membrane which forms the outer surface of all cells.

Cellulose A *carbohydrate* made up of long fibres. It forms the rigid cell wall which surrounds all plant cells.

Central nervous system Brain and spinal cord.

Centromere The part of a *chromosome* where its two *chromatids* are joined together, which becomes attached to the *spindle apparatus* during cell division.

Cerebellum The part of the brain which controls balance and muscular co-ordination.

Cerebral cortex *Grey matter* which forms the outer layer of the *cerebral hemispheres*. It controls voluntary movements, and is concerned with memory, thinking, and learning.

Cerebral hemispheres Two swellings at the front of the brain. These form the largest region of the human brain and are concerned with conscious sensation, learning, and memory.

Chlorophyll Green substance in plants. It absorbs light energy for use in *photosynthesis*.

Chloroplasts *Organelles* in plant cells containing *chlorophyll*.

Chromatid One of the two identical strands of a *chromosome*. These separate during cell division and move to opposite ends of the cell.

Chromosomes Rod-like structures visible in the *nucleus* of a cell during cell division, consisting of *genes*, and containing the *hereditary information* of the cell.

Chyme Semi-liquid, partly digested food which results from the action of the *stomach* and its *gastric juice*.

Cilia Minute hair-like projections from the surface of certain cells, e.g. *Paramecium*. Cilia flick back and forth, causing movement of surrounding fluids.

Ciliary muscles Muscles in the eye which change the shape of the lens during *accommodation*.

Cochlea A spiral tube within the inner ear containing sensory nerve endings which respond to sound vibrations.

Coleoptile Sheath-like protective covering over the first-formed leaves of grasses and other cereals.

Collecting tubule A tube within a kidney which collects *urine* from the *kidney tubules* and passes it into the *ureter*.

Colon Part of the large intestine. Its function is to absorb water and mineral salts from *faeces*.

Community A group of interdependent organisms that share a particular environment.

Companion cells Elongated cells situated alongside a *sieve tube*.

Compensation point The point at which *photosynthesis* and *respiration* in a plant are exactly balanced, and one process uses up the products of the other.

Cones Cone-shaped, light-sensitive cells in the *retina* of the eye. They work only in bright light and are sensitive to colour.

Conjugation A type of sexual reproduction in which cells join in pairs and exchange *hereditary information*.

Conjunctiva Transparent skin that covers and protects the front of the eye.

Consumers Organisms in a *food-chain* which live by consuming (eating) other organisms.

Co-ordination Control of the body so that its tissues and organs work together at the correct speed and in the correct sequence, thereby serving the body as a whole.

Corm An organ of *vegetative reproduction* consisting of a short stem swollen with stored food.

Cornea Transparent, circular window at the front of the eye.

Corolla The collective name for the petals of a flower.

Cupula A cone of jelly attached to sensory hair cells in the *ampulla* of a *semi-circular canal*. The cupula moves and stimulates the hair cells during changes in the direction of movement.

Cytoplasm All the contents of a cell except its *nucleus*.

Deamination Breakdown of unwanted *amino acids* in the liver by removal of the nitrogen-containing part of the molecules.

Denitrifying bacteria *Anaerobic* bacteria in soil which break down nitrates into nitrogen and oxygen.

Dentine A substance similar to bone which forms the inner part of a tooth, beneath the *enamel*.

Dentition The characteristic arrangement and types of teeth in a mammal.

Dermis The layer of skin beneath the *epidermis*, consisting of connective tissue, blood vessels, nerves, hair roots, and cells filled with fat.

Diaphragm Dome-shaped sheet of muscle at the base of the *thorax*. It is part of the mechanism that ventilates the lungs.

Differentiation The process by which cells become specialized in performing a particular function.

Diffusion Movement of molecules of liquids and gases from regions where they are highly concentrated to regions where they are less concentrated until they are equally distributed.

Digestion The process by which food is made soluble by the action of digestive juices (*enzymes*).

DNA Deoxyribonucleic acid A chemical within *chromosomes* which contains the *hereditary information* of the cell.

Dominant characteristic One which appears in the *phenotype* when crossed with a contrasting *recessive characteristic*.

Dormancy A resting, inactive condition, in which *metabolism* almost stops.

Duodenum The part of the *alimentary canal* between the *stomach* and the *ileum*.

Ear-drum A membrane of skin and muscle situated at the bottom of the ear canal of the outer ear. It is vibrated by sound waves in the air and transmits these vibrations to the ear *ossicles*.

Ecdysis The periodic shedding (moulting) of an insect's cuticle as it grows.

Ecosystem All the interactions between the members of a *community* and its environment.

Effectors Tissues or organs which respond to *stimuli*, e.g. muscles and glands.

Egestion Removal from the body of indigestible materials (*faecal matter*) which have passed through the intestine.

Egg nucleus The female *gamete* of flowering plants. One of the nuclei within an *ovule*.

Embryo The stage of development between the fertilized egg (*zygote*) and the newly formed organism.

Embryo sac Part of the *ovule* of a plant which contains the female *gamete* (*egg nucleus*).

Enamel The extremely hard, white substance which forms the outer surface of a tooth.

Encystment Formation of the dormant stage in an organism's life cycle. The organism becomes enclosed in a hard outer covering. In this state it can survive conditions which would kill it in the free-living stage.

Endocrine system A system of organs which produce *hormones*.

Endoplasmic reticulum A system of fluid-filled channels within the *cytoplasm* of most cells.

Endoskeleton A skeleton which forms inside the body of an organism.

Enzymes Protein substances which act as *catalysts* and control the rate of chemical reactions in cells, generally speeding them up.

Epidermis The outer layer of cells in an animal or plant.

Eustachian tube An air-filled tube running from the back of the mouth to the middle ear. It permits air pressure to be equalized on either side of the *ear-drum*.

Eutrophication Excessive growth of algae in river and lakes owing to accumulation in the water of minerals such as nitrates, which promote plant growth.

Evolution The sequence of gradual changes over millions of years in which new *species* may be produced.

Excretion Removal from the body of waste substances produced by *metabolism*, and substances which are in excess of the body's requirements.

Exoskeleton A skeleton which forms on the outside of an organism.

F_1 generation The first filial generation. Organisms produced by crossing animals or plants which form the starting point of a genetical experiment.

F_2 generation Organisms produced by crossing or selfing members of an F_1 generation.

Faeces The indigestible material which remains in the *colon* after *digestion* has taken place.

Fatty acids Chemical components released when fats are broken down during *digestion*.

Fermentation The breakdown of sugar by organisms such as yeast and bacteria which takes place under *anaerobic* conditions.

Fertilization The fusion of male and female *gametes* during sexual reproduction, resulting in the formation of a *zygote*.

Fibrinogen A *protein* in blood which is transformed into fibres of fibrin which block damaged blood vessels and form a blood clot.

Foetus The *embryo* of a mammal at the stage of development in which the main features are visible.

Food chain A sequence of organisms through which energy is transferred. The first organisms are *producers* (usually green plants), and succeeding 'links' are *consumers*.

Food web A number of interconnected *food chains*.

Fovea Region of the *retina* immediately opposite the lens. It consists of densely packed *cones*, and provides the clearest vision.

Gall bladder A small bladder inside the liver in which bile is stored.

Gametes Cells involved in *sexual reproduction*, e.g. *pollen grains*, *ova*. Gametes fuse together at *fertilization*, forming a *zygote* which develops into an *embryo*.

Gaseous exchange The process by which an organism absorbs oxygen from the air in exchange for carbon dioxide, which is released into the air. This takes place in respiratory organs such as lungs.

Gene Part of a *chromosome* which controls the appearance of a set of hereditary characteristics.

Genetics The scientific study of *genes*.

Genotype The genetic make-up of an organism, or the set of *genes* which it possesses.

Geotropism Growth movement of a plant in response to gravity.

Gland A collection of cells which manufactures and releases into the body useful substances such as *enzymes* or *hormones*.

Glomerular filtrate Fluid which results from the filtration of blood in *Bowman's capsules*. It consists of *urine* and many useful substances such as glucose.

Glomerulus A group of capillaries inside a *Bowman's capsule* in a kidney. Blood is filtered as it passes through the glomerulus and Bowman's capsule walls into the *kidney tubule*.

Glycogen A carbohydrate similar to starch. It is stored in the liver and muscles of mammals, and converted into glucose as the body requires energy for *metabolism*.

Goblet cells Cells which produce *mucus*. They are situated in the intestine walls and in the air passages of the lungs.

Graafian follicle A fluid-filled space in a mammalian *ovary* containing a cell which develops into a female *gamete* (ovum).

Grey matter Nervous tissue in the brain and spinal cord consisting mainly of *neurone* cell bodies.

Guard cells Crescent-shaped cells in the *epidermis* of plants which surround and control the diameter of stomatal pores (*stomata*).

Gynoecium The collective name for the female sex organs (*carpels*) of a flower.

Habitat A region of an environment containing its own particular *community* of organisms, e.g. marsh, sand dune.

Haemoglobin Red substance in *red blood cells*. It combines with oxygen and transports it to the tissues.

Hepatic portal vein Vessel in which blood containing absorbed food is carried from the intestine to the liver.

Herbivore An animal which eats only plants, e.g. horse, sheep.

Hermaphrodite An organism which possesses both male and female reproductive organs.

Heterotrophic (organisms) Those which feed on other organisms or their products.

Heterozygous An organism is said to be heterozygous for a characteristic when the *alleles* (two *genes*) which control it are opposite in nature.

Homeostasis Maintenance of a constant *internal environment*.

Homoiothermic (organisms) Those which are capable of maintaining a constant body temperature.

Homologous chromosomes *Chromosomes* form pairs in the early stages of *meiosis*. The two chromosomes of each pair are identical in shape and size and are described as homologous.

Homozygous An organism is said to be homozygous for a characteristic when the *alleles* (two *genes*) which control it are identical in nature.

Hormone A chemical produced in small amounts which helps to co-ordinate processes such as *metabolism*, growth, and reproduction.

Host An organism in or on which a *parasite* lives.

Humus Part of the soil which consists of the decayed remains of plants and animals.

Hydrostatic skeleton Support which comes from water in the cells of an organism.

Hydrotropism A growth movement in plants in response to water.

Hyphae Fine hollow threads which make up the body of many fungi.

Ileum The region of the *alimentary canal* between the *duodenum* and *colon*, where *digestion* is completed and *absorption* takes place.

Imago Fully developed adult insect.

Immunity The ability of an organism to resist infection from *parasites*.
Incisors Chisel-shaped teeth at the front of the jaws.
Inclusion A non-living substance, such as stored food, within the *cytoplasm* of a cell.
Insertion (of a muscle) The end of a muscle attached to the bone which moves when the muscle contracts.
Insulin A *hormone* produced by the *pancreas*. It helps to control the amount of sugar in the blood.
Intercostal muscles Muscles between the ribs that raise the rib cage during inspiration (breathing in).
Internal environment The *tissue fluid* that bathes every cell of the body, and supplies all their food and oxygen requirements.
Iris The coloured part of the eye, consisting of radial and circular muscles which alter the size of the *pupil* and control the amount of light entering the eye.

Key A sequence of statements about the characteristics of a group of organisms which can be used to identify an organism belonging to that group.
Kidney tubule A narrow tube leading from a *Bowman's capsule* in the kidney. It reabsorbs water, glucose, and other useful substances from *glomerular filtrate*.

Lacteal Part of the *lymphatic system* which extends through the centre of a *villus*. It absorbs digested fat from the *ileum*.
Lamina The flat part of a leaf on either side of the *mid-rib*.
Larva An early stage in the life cycle of certain organisms which bears little or no resemblance to the adult.
Larynx Voice box.
Lateral line A fluid-filled tube along each side of a fish containing sensory hairs which respond to vibrations in the water.
Ligament Band of fibres around a joint in the skeleton which holds the bones in place and helps prevent dislocation.
Lignin A hard rigid substance which forms in the walls of cells which make up *xylem tissue*.
Lipase An *enzyme* which digests fats and oils.
Loam A soil containing sand, clay, and *humus* in ideal proportions for healthy plant growth.
Locomotory organs Organs that enable animals to move, e.g. fins.
Lymph A liquid derived from *tissue fluid* after it has passed between the cells of the body and drained into the *lymphatic* system.
Lymphatic system A system of vessels that transport *lymph* from the tissues to the circulatory system.
Lymph node Part of the *lymphatic system* containing *phagocytes*, which remove germs and dead cells from the *lymph*.
Lymphocytes White cells of the blood produced in the *lymphatic* system. They produce *antibodies*, which destroy *antigens*.
Lysins *Antibodies* that destroy germs by dissolving them.

Malnutrition The effects on the body of an unbalanced diet.
Malpighian layer A region of rapidly dividing cells beneath the *epidermis* of the skin. It replaces cells worn away from the skin surface.
Meiosis Cell division that produces *gametes*, and results in cells with half the number of chromosomes found in the parent cell.
Menstruation Breakdown and removal from the body of the lining of the *uterus*. This occurs if an *ovum* has not been fertilized.
Metabolism All the chemical and physical processes necessary for life.
Metamorphosis The sequence of changes by which a *larva* becomes an adult organism.
Metaphase The stage of cell division in which *chromosomes* gather at the equator of the cell and become attached to the *spindle apparatus*.
Mid-rib The rigid rib in the centre of a leaf, containing *xylem* and *phloem*.
Mitochondria (singular mitochondrion) Rod-shaped *organelles* in the *cytoplasm* of cells, concerned with *respiration*.
Mitosis Type of cell division resulting in cells with the same number of *chromosomes* as the parent cell.
Molars Large teeth with four cusps, situated at the back of the jaw. They are used to crush and grind food into small pieces.
Monohybrid cross Cross between organisms which show contrasting variations of only one characteristic.
Motor end-plate Part of a *motor neurone* which is embedded in a muscle.
Motor neurone A *neurone* which conducts impulses from the *central nervous system* to a muscle or gland.
Mucus A sticky fluid produced by *goblet cells*.
Mutation A sudden unpredictable change in a *gene* or *chromosome*, which alters the characteristics that it controls.
Myopia Short-sightedness. Usually results from an abnormally elongated eyeball.

Natural selection A theory proposed by Charles Darwin, suggesting how evolutionary change could have occurred.
Nectary An organ in a flower that produces the sugary fluid nectar. It aids *pollination* by attracting insects.
Neurone A nerve cell consisting of a cell body and nerve fibres. It conducts nerve impulses.
Niche The position of an organism in a *food web*. This is determined mainly by its feeding habits.
Nitrifying bacteria Bacteria in soil which convert the decaying remains of organisms into soil nitrates.
Nitrogen-fixing bacteria Bacteria in soil and *root*

nodules which convert nitrogen in the air into soil nitrates.

Nucleus Part of a cell which contains *chromosomes*. It controls cell *metabolism* and division.

Nymph An early stage in the life cycle of certain insects that resembles the adult except that it is smaller, usually wingless, and sexually immature.

Oesophagus Tube through which food passes from the mouth to the *stomach*.

Oestrogen Female sex *hormone*, controlling conditions in the *uterus* before and during *pregnancy*.

Ommatidia Light-sensitive units that make up the compound eyes of insects.

Omnivore An animal that eats both animals and plants, e.g. man.

Opsonins *Antibodies* which combine with chemicals on the surface of bacteria, making them more likely to be attacked by *phagocytes*.

Organ A structure consisting of several *tissues* which work together in performing a particular function, e.g. the heart.

Organelles Structures in the *cytoplasm* of a cell which take an active part in the life and function of the cell, e.g. *chloroplasts*.

Origin (of a muscle) The anchorage point of a muscle, i.e. the end which does not move during contraction.

Osmoregulation Control of the movement of water in and out of cells by *osmosis*.

Osmosis Diffusion of water molecules through a *semi-permeable membrane* from a weak to a strong solution.

Osmotic potential Solutions are said to have an osmotic potential when they are prevented from taking in water by *osmosis*.

Osmotic pressure The pressure required to prevent water entering a solution by *osmosis*.

Ossicles (of the ear) Three bones in the middle ear which transmit vibrations from the *ear-drum* to the inner ear.

Otolith Minute grains of chalk embedded in blobs of jelly attached to sensory hairs in the *utricles* and saccules of the inner ear. They are displaced by body movements, thereby causing the hair cells to send impulses to the brain.

Ova (singular ovum) Female gametes of animals.

Ovary An organ which produces females gametes (*ova* or *ovules*).

Oviduct A tube leading from a funnel-shaped opening near an *ovary* to the outside of the body.

Ovulation The release of an *ovum* from a ripe *Graafian follicle*.

Ovule The part of a *carpel* containing the female gamete, or *egg nucleus*. Ovules develop into seeds after *fertilization*.

Oxygen debt This occurs in muscle tissue during strenuous exercise, when oxygen is consumed faster than it can be supplied by the blood.

Oxyhaemoglobin *Haemoglobin* which has combined with oxygen in the *red blood cells*.

Palisade mesophyll A layer of cylindrical cells at right angles to the upper *epidermis* of leaves. They contain more chlorophyll than other plant cells, and are the main cells concerned with *photosynthesis* in plants.

Pancreas An organ situated between the *stomach* and the *duodenum*, producing *trypsin*, *amylase*, and *lipase*.

Parasite An organism that obtains food from the living body of another organism called the *host*.

Pathology The scientific study of the effects on the body of disease organisms.

Pedicel Flower stalk.

Pepsin An *enzyme* produced by the *stomach* which begins the digestion of *proteins*.

Peristalsis Wave-like contractions of tubular organs such as the *alimentary canal* that propel the contents of the tube in one direction.

Petiole Leaf stalk.

Phagocytes *White blood cells* that engulf and digest germs.

Pharynx Area at the back of the mouth immediately above the *trachea* and *oesophagus*.

Phenotype The visible hereditary characteristics of an organism, as opposed to its *genotype* or genetical characteristics.

Phloem A plant tissue that transports the products of *photosynthesis* from the leaves to the growing points and food storage organs. It consists mainly of *sieve tubes* and *companion cells*.

Photosynthesis The process by which plants use light energy trapped by *chlorophyll* to form sugar out of carbon dioxide and water.

Phototropism The growth movement of a plant in response to the direction of light.

Placenta The organ through which the *foetus* of a mammal obtains food and oxygen from its mother's blood, and passes waste substance into the mother's blood.

Plasma The liquid part of blood.

Plasmolysis The shrinkage of cell *cytoplasm* owing to loss of water by *osmosis*.

Platelets Particles in the blood which are concerned with clot formation in wounds.

Pleural cavity The fluid-filled space between the outer surface of the lungs and the inner surface of the rib cage.

Poikilothermic (organisms) Those which cannot maintain a constant body temperature, but vary according to the temperature of their surroundings.

Pollen Male *gametes* of flowering plants.

Pollination Transfer of *pollen* grains from *stamens* to *stigmas*.

Pollutant A substance present in large enough quantities in the environment to be harmful to living things.

Polysaccharide A *carbohydrate* whose molecules are made up of long chains of simple sugar molecules, e.g. *cellulose*.

Pregnancy The period during which a female mammal carries a developing *embryo* in her *uterus*.

Presbyopia Old sight. A condition resulting from old

age in which the lens loses its ability to change shape during *accommodation*.

Producers Organisms such as green plants that produce food. Producers form the starting points of *food chains*.

Prophase The first stage of cell division, when *chromosomes* become visible in the *nucleus*.

Proprioceptor A sensory nerve ending which picks up *stimuli* originating inside the body, e.g. a stretch receptor in a muscle.

Protandrous The condition in a flower when the *stamens* ripen before the *carpels*. This prevents self-fertilization.

Protease An *enzyme* which *digests* protein.

Proteins The main body-building foods such as meat, eggs, and fish.

Protogynous The condition in a flower which occurs when the *carpels* ripen before the *stamens*. This prevents self-fertilization.

Pseudopodia Projections from the cytoplasm of certain cells, e.g. *Amoeba*. They are used in locomotion and feeding.

Puberty The stage of development at which men and women become sexually mature, i.e. able to reproduce.

Pulmonary circulation The system of vessels that conveys blood from the right *ventricle* to the lungs and back to the left *atrium* of the heart.

Pupa A stage in the life cycle of an insect during which *metamorphosis* takes place.

Pupil The hole in the *iris* of the eye through which light enters.

Receptacle The surface at the top of a flower stalk to which all the parts of a flower are attached.

Receptors The regions of sensory nerve fibres where *stimuli* are received and converted into nerve impulses, e.g. the *rods* and *cones* of the eye.

Recessive characteristic One that does not appear in the *phenotype* when crossed with a *dominant characteristic*.

Rectum The last part of the *alimentary canal*.

Red blood cells Disc-shaped cells containing *haemoglobin*, which transport oxygen from the lungs to the body tissues.

Reflex A response that does not have to be learned, and occurs very quickly with no conscious thought, e.g. withdrawal from a painful *stimulus*.

Rennin A substance produced in the stomach of a young mammal. It is concerned with the *digestion* of milk.

Respiration A sequence of chemical reactions which release energy from food.

Response An activity in the body that results from a *stimulus*.

Retina A layer of light-sensitive cells at the back of the eye on which images are projected.

Rhizome A horizontal underground stem filled with stored food. It enables a plant to survive the winter and make new growth in the spring, and produces new plants from branches which grow from underground buds.

Rods Rod-shaped light-sensitive cells found in the *retina*. They work in dim light but do not respond to differences in colour.

Root hairs Hair-like outgrowths from single cells in the *epidermis* of a root in a zone near the root apex.

Root nodules Swellings on the roots of certain leguminous plants (such as peas and clover), containing *nitrogen-fixing bacteria*.

Root pressure Pressure causing water to pass up the *xylem* from the living cells of the root.

Saliva Fluid produced and released into the mouth by three pairs of salivary glands in response to food. It contains the *enzyme* salivary amylase, mucin, and various minerals.

Saprotrophs (saprophytes) Organisms which feed on organic matter, such as the dead remains of animals and plants, by releasing *enzymes* that *digest* the food externally, reducing it to a liquid which is absorbed into the saprotroph body, e.g. certain bacteria and fungi.

Sebaceous gland A *gland* in the hair follicles of the skin. It *secretes* an oily substance called sebum, which makes skin supple, waterproof, and mildly antiseptic.

Secretion The production by *glands* of substances such as *enzymes* which are useful to the body.

Sedentary soil Soil which overlies the rocks from which it was formed.

Segregation The separation of *genes* which are *alleles* of each other at *meiosis* and their movement into separate *gametes*.

Semen Fluid produced by the *testes* of mammals. It consists of *sperms* and chemicals which nourish them and stimulate their swimming movements.

Semi-circular canals Three semi-circular tubes in the inner ear. They contain fluid and sensory hair cells attached to the *cupula*. They respond to changes in the direction of movement.

Semi-lunar valves Pocket-like valves in the main *arteries* at the point where they leave the heart. They stop blood flowing back into the *ventricles*.

Semi-permeable membrane A membrane which allows certain substances to pass through but prevents the passage of others, e.g. the *cell membrane*.

Sensory neurone A neurone which conducts impulses from a *receptor* (e.g. sense organ) to the *central nervous system*.

Sepals Leaf-like structures at the outer region of a flower. They cover and protect the flower in bud.

Sexual reproduction Reproduction usually involving two parents, which produce *gametes*. These fuse together making a *zygote*, which develops through an *embryo* stage into a new organism.

Sieve tube A tube which forms part of *phloem* tissue. It transports food from regions of *photosynthesis* to growth and food storage regions.

Species A group of organisms which can mate together and produce fertile offspring.

Sperms The male *gametes* of animals.

Sphincter A ring of muscle found in the walls of tubular organs such as the *alimentary canal*. Its contraction slows or stops movement of substances through the tube.

Spindle apparatus The arrangement of fine fibres which are thought to draw the *chromosomes* apart during cell division.

Spiracle An opening on an insect's body through which air moves in and out of its *tracheal system*.

Spleen An organ immediately below the *stomach* which produces *white blood cells*, and destroys old, worn-out *red blood cells*.

Spongy mesophyll The layer of cells in a leaf immediately below the *palisade*. It contains large intercellular air spaces.

Spontaneous generation The now discredited theory that organisms can arise spontaneously from non-living matter.

Spore A microscopic reproductive cell released from an organism during *asexual reproduction*. Typical of fungi, mosses, and ferns. In bacteria, a spore is a resting or dormant stage of the life cycle, usually formed when conditions are unfavourable.

Stamens The male reproductive organs of a flower. The *anthers* of stamens produce *pollen* grains.

Stigma The part of a *carpel* to which pollen grains become attached during *pollination*.

Stimulus Anything which produces a *response* in an organism, e.g. a painful burn on the skin.

Stomach A bag-like organ at the end of the *oesophagus*.

Stomata Pores in the *epidermis* of plants through which air enters and leaves, and water evaporates during *transpiration*.

Suspensory ligaments Fibres which hold the lens in position within the eye.

Sweat gland A *gland* in the skin. It produces water which evaporates into the air and cools the body.

Swim bladder An air-filled bladder in the bodies of certain fish. It is used to adjust the density of the body until it equals that of water, making the fish weightless.

Symbiosis A close association between two different organisms in which both benefit.

Synapse A microscopic gap over which nerve impulses pass when moving from one nerve cell to the next.

Synovial joint Any freely moveable joint in the skeleton, e.g. the elbow.

Systemic circulation The series of vessels which carries blood from the left *ventricle* around the body and back to the heart at the right *atrium*.

Taste bud A collection of sensory nerve endings in the tongue which respond to certain chemicals in food, producing the sensation of taste.

Telophase The final stage of cell division during which the parent cell separates into two daughter cells.

Tendon A strong band of fibres which attaches muscles to bones.

Testa The protective outer covering of a seed.

Testis Male reproductive organ of animals producing *gametes* called *sperms*.

Thorax (of insects) The middle segment of the body.

Thorax (of mammals) The cavity within the chest that contains the lungs, heart, and main blood vessels.

Threshold The level of stimulation at which nerve impulses begin to pass from a sense organ, or to cross over a *synapse*.

Thyroid An *endocrine gland* in the neck. It produces a *hormone* called thyroxin which has a major influence on physical and mental development.

Tissue A collection of similar cells which work together to perform a particular function.

Tissue fluid Fluid which is forced through capillary walls and moves between all cells of the body, providing them with food and oxygen and removing their waste products.

Toxin A poisonous substance.

Trace elements Minerals which are essential for the healthy growth of plants but which are required only in minute quantities, e.g. boron.

Trachea The windpipe.

Tracheal system A system of tubes through which air passes in and out of an insect's body.

Translocation In general, the movement of substances within a plant. More often used to describe the movement of sugar through *phloem*.

Transpiration Evaporation of water from plant cells and out of their *stomata*.

Tropism A movement in a plant in which the direction of root and shoot growth alters according to the direction of a *stimulus*.

Trypsin An *enzyme* produced by the *pancreas*. It digests *proteins*, converting them into *amino acids*.

Turgor pressure Pressure within plant cells which results from the absorption of water by *osmosis*.

Umbilical cord A tube containing blood vessels connecting a developing *embryo* with its *placenta*.

Ureter A tube which carries *urine* from a kidney to the bladder.

Urethra A tube which carries *urine* out of the body from the bladder.

Urine Liquid containing waste materials removed from the blood as it is filtered by the kidneys. It consists of water, urea, and various mineral salts.

Uterus A bag-like organ of the female reproductive system. It contains, protects, and nourishes the developing *embryo*.

Utricles Fluid-filled spaces of the inner ear containing sensory hair cells and *otoliths*. They detect acceleration and deceleration of the body and changes in its position relative to the pull of gravity.

Vaccine A suspension of dead, inactivated, or relatively harmless germs which, when introduced into the blood-stream, stimulate the production of *antibodies* and make the body *immune* to attack from harmful disease organisms.

Vacuole Fluid-filled space in the *cytoplasm* of a cell.

Vascular bundle Strand of *xylem* and *phloem* tissues running from the roots into the leaves. It transports food and water throughout the plant and supports the softer tissues.

Vascular system (of animals) The heart and blood vessels.

Vascular system (of plants) *Xylem* and *phloem* tissues.

Vector An animal which carries disease organisms.

Vegetative reproduction A form of *asexual reproduction* in which outgrowths from a plant eventually separate and continue an independent existence.

Vein (of animals) A vessel which carries blood towards the heart.

Vein (of plants) A strand of *xylem* and *phloem* tissue in a leaf.

Vena cava The main *vein* of the body.

Ventilation The movement of air or water across a respiratory surface such as a lung or gill, which enables *gaseous exchange* to take place.

Ventricle One of the large, thick-walled lower chambers of the heart that pump blood into *arteries*.

Vertebral column The backbone or spine. A chain of small bones called vertebrae that support the body, protect the spinal cord, and permit bending movements.

Vestigial structure The remains of an organ that is thought to have lost its function in the course of evolution, e.g. the appendix in man.

Villi Minute finger-like structures on the inner surface of the *duodenum* and *ileum*. These occur in millions, greatly increasing the surface area available for *absorption*.

Vitamins Chemicals required in small amounts to maintain health.

Vitreous humour The jelly-like substance that fills and supports the chamber of the eye behind the lens.

Wall pressure A cell taking in water by *osmosis* inflates until its cellulose wall can stretch no further. The restraining force of the cell wall is called wall pressure.

Water potential The ability of a cell to take in water by *osmosis*. It is calculated by finding the difference between *wall pressure* and *osmotic pressure*.

Weathering The process by which rocks are broken down into small fragments, e.g. by wind, rain, and frost.

White blood cells (leucocytes) The general name for a number of different colourless cells in the blood, e.g. *phagocytes*, *lymphocytes*.

White matter Nervous tissue in the brain and spinal cord which consists of nerve fibres.

X chromosome A *chromosome* which, when present in a *zygote* either alone or with another *X chromosome*, causes that zygote to develop into a female organism.

Xylem A plant tissue which transports water and dissolved minerals from the soil to the leaves, and also supports the softer plant tissues. It consists of xylem vessels and fibres.

Y chromosome A *chromosome* which, when present in a *zygote*, results in that zygote developing into a male organism.

Zygote The cell which results from the fusion of a male and a female *gamete* (a fertilized egg).

Index

Page numbers in bold type indicate major references in the text

Abscission 200
Absorption:
  in animals 35, 36, 37, **50**
  in plants 83, **86**
Accommodation **171–2**, 180
Adenine 246
Adipose tissue 27, **118**
Adrenal gland 162–3
Adrenalin 162–3
Adsorption 270
Adventitious roots 201, 203
Aerobic respiration **94**, 96–7
Aerofoil **140**, 150
After-birth 225
Agar 128-9, 252, 265
Agglutinin 262–3
Agranulocytes 67–8
Air:
  pollution of 282–3
  in soil 270, 272
Albinism 248
Albumen 215–17
Alcohol 95, 254
Algae 5, 6, 186, 284
Alimentary canal 36, **47–50**
Allele 247
Alluvial soil 270
Alternation of generations 189–90
Alveoli **102–4**
Amino acids **28**, 33, 35, 48, 49, 50, 114
Ammonia 114
Amnion 224
Amoeba 5, 6, 13, 36–7, 66, 67, 68
  feeding and digestion 37
  life cycle 184–5
Amphibia 10 (*see also* frog)
Ampulla 178
Amylase:
  pancreatic 49
  salivary 45
Anabolism 2
Anaemia (pernicious) 31
Anaerobic respiration **94–6**, 99–100
Anaphase:
  meiosis 244–5
  mitosis 20–1
Androecium 191
Angiospermae (*see* flowering plants)
Annelida (true worms) 5
Annual 199
Annual ring 80
Antagonistic muscles **138**, 145, 148
Antennae 168–9
Anther 191, 193, 194
Antheridia 189
Antibody **262**, 264–5
Antigen 262
Antiseptic 256
Anti-toxin 262, 265
Anus 50
Aorta 70
Appendix 50
Aqueous humour **171**, 172
Archegonia 189
Aristotle 52–3
Arterioles 73
Artery 69, 70, **73**
Arthropoda **5**
Artificial propagation 202
Artificial selection 235
Ascorbic acid (vitamin C) 31
  test for 33
Asexual reproduction 182–3
Assimilation 35
Association area 160
Atlas vertebra 137, 138
ATP (adenosine triphosphate) 96–7
Atrium 70
Auxin **128**, 130, 131
Axillary bud 201, 202, 203
Axis vertebra 137
Axon 153–5

Bacilliarophyta (diatoms) 5, 6
Bacillus 254–6
Back-cross 249
Bacteria 32, 45, 48, 50, 67, 95, **254–6**, 261, 262, 281, 284
  culture of 265–6
  in nitrogen cycle 277
  in soil 268, 274
Bacteriophage 257–8
Balance of nature 276–85
Balance, sense of 178–9
Balanced diet 24, 32–3, 34
Ball-and-socket joint 137
Barb and barbule (*see* feather)
Bee 174–5
  life cycle 209–13
Beri-beri 30, 31
Biceps muscle 138, 139
Bicuspid valve 70
Biennial 199
Bile 48–9, 115
Bile duct 48
Binary fission 184
Bird 9, 10
  flight 139–42
  infection by 258
  reproduction 216–17
Birth 224
Birth control 225–6
Bladder 119
Bladderworm 259–60
Bleeding 262
Blind spot 173, 180, 181
Blood 50
  and immunity 262–5
  circulation 69–74
  composition 66–9
  functions 75–7
  pressure 73–4
  purification 114
Blood clot 68, 78, 114, 225, **262**
Bolus 46
Bone 133–5, 136, 137
Bone marrow 66, 67, 68, 262
Bowman's capsule 121
Brain 153, 155, 156, 157, 158, **159–61**, 165, 166, 167, 169, 173
Breathing 94, 101, **105–7**
Bronchial tubes 102
Bronchiole 102
Bronchitis 282
Bronchus 102
Bryophyta 5, 7, 189–90
Bubonic plague 264
Budding:
  Hydra 207
  yeast 185
Buds (of plants) 199, 200, 202, 203, 204–5
Bulb: tulip 202
Buttercup:
  flower structure 191–2
  pollination 192
  vegetative reproduction 201

Butterfly:
feeding 39, **42**
life cycle 208, **210**

Caecum 50
Calciferol (vitamin D) 31
Calcium:
in animal nutrition 29
in plant nutrition 61, 272
Calorimeter 25
Calyx 191
Cambium 82, 200
Canine teeth 43
Capillarity 270, 271, 273
Capillary (blood vessel) 50, 70, **74**, 104
Capillary bed 74, 75
Carbohydrate **25**, 32, 45, 52, 277
test for 33
Carbon cycle 278–9
Carbon dioxide **77**, 270, 278, 281
in photosynthesis 52, 53, 54, 55, 58, 61–3, 64
in respiration 2, 94, 95, 101, 104, 105, 107, 110
transport in blood 75–7
Carbonic acid 77, 269
Cardiac muscle 70
Carnassial teeth 43
Carnivore 1, 43, 59, 278
Carotene 30
Carpel 191
Cartilage 102, 136
Castle's intrinsic factor 31, 48
Catabolism 2
Catalyst 35
Caterpillar 133, 208, 210
Cell division 16–18, 20–1, 244–5 (*see also* mitosis and meiosis)
Cell membrane **13**, 16, 85
Cell wall 13, 85
Cells 11–23
comparison between plant and animal 13
growth 18–19
osmosis 85–6
size 16
structure 12–13
Cellulose 13, 21, 25, 44, 48, 50, **55**, 85, 254
test for 33
Central nervous system 153, 156–61
Centromere 21
Cerebellum 160
Cerebral cortex 160–1
Cerebrum 160–1
Cervix 219, 224, 225
Chewing 44–5
carnivore 43
herbivore 44
man 44
Chicken pox 258, 262
Chitin 133, 145
Chlorophyll **14**, 15, 52, 53–6, 58, 63, 126
Chlorophyta (green algae) 5, 6
Chloroplasts 14
Cholera 258, 264
Cholesterol 48
Chondricthyes 8, 10
Chordata 5, 8–10
Choroid 171
Chromatids 21
Chromatin 13, 20
Chromosomes 13, 20–1
and heredity 18–20, 182, 244–6
structure of 245–6
Chrysalis 208, 210
Chyle 50
Chyme 47
Cilia **14**, 37, 101, 261
Ciliary muscles 171–2
Ciliophora (ciliates) 5
Circulatory system 70–4
Classification 3–10
Clay soil 269, 270, **271**, 274, 277–8
Climax community 281
Clinostat 129–30
Cloaca:
bird 217
frog 215
Clone 182–3, 207
Cobalamine (vitamin B_{12}) 31, 48
Cochlea 177
Cockroach 208–9
Coelenterata 5, 8, 15, 207
Cold blooded 115
Coleoptile 127, 128, 129
Collecting duct 121
Colon 50
Colour vision 173
Common cold 258
Community 280–1
Companion cells 80
Compensation point 58
Complete metamorphosis 208
Composite flowers 191
Compound eyes 173–5
Conditioned reflex 158–9
Condom 226
Cones (of the eye) 173
Conjugation:
Mucor 186
Spirogyra 188
Conjunctiva 169, 261
Constipation 32
Consumer 279–80
Contagious disease 258
Continuous variation 240
Contraceptive diaphragm 225
Contraceptive pill 225
Contractile root 203
Controlled experiments 64
Converging lens 172
Co-ordination **153–63**, 166
Copper 284
Copulation 184
birds 217
mammals 222
Corm 202
Cornea 169
Cornified layer 261
Corolla 191
Coronary blood vessels 72
Corpus luteum 220
Cortex:
brain 160–1
kidney 120–1
root 81, 86
stem 82
Cotyledon 5, 196
Courtship:
birds 216–17
stickleback 214
Cowpox 264
Cretinism 162
Cross pollination 191–3
Crumbs (soil) 268–9, 271
Crustacean 5, 133
Cultivation of soil 271–2
Culture:
bacteria 265–6
Drosophila 252
Cupula 178
Cuticle:
insects 145, 208
plants 57
tapeworm 259
Cuvier, George 228
Cyanide 284
Cyst 185
Cytoplasm 13
Cytosine 246

Dandelion:
flower 191, 195
fruit 198
Dark reaction 54–5
Darwin, Charles 126–7, 233–5
Deamination 114
Defecation 118
Deficiency diseases:
animals 30
plants 60, 61
Dendrites 155
Dendron 153–4
Denitrifying bacteria 277–8
Dentine 42
Dentition:
carnivores 43
herbivores 44
omnivores 44
Deoxygenated blood 70, 104
Depressor muscle 141, 146–7
Dermis 116
Detergent 283–4
Development:
bird 217
frog 215–16
insect 208–13
man 224
Diabetes 163
Dialysis tubing 84, 90
Diaphragm (of mammal) 101, 106
Diatoms 5, 6
Dicotyledon 5, 56

Index

Diet 24–34
  balanced 24, 32–3, 34
  slimming 34
Differentiation 18
Diffusion 66, 83–4, 105
Digestion 35
  Amoeba 37
  extracellular 36–7
  Hydra 37–8
  intracellular 36–7
  mammals 44–9
Diphtheria 254, 258, 265
Diplococcus 255
Disaccharide 25
Discontinuous variation 239–40
Disease 254–66
  carriers of 258
  infection 258
  prevention of 259
  (*see also* immunity)
Disinfectant 256
Dispersal of fruits and seeds 196–9
Distance judgement 173
Diverging lenses 172
Division of labour 14–15
DNA (deoxyribonucleic acid) 18, 20, 21, 245–6, 254, 257–8
Dominant characteristic 241
Drone bee 209, 212, 213
Droplet infection 258
Drosophila 248–9
  culture of 252
Duodenum 48–9
Dwarfism 161
Dysentery 258

Ear 175–7
Ear drum 175
Earthworm:
  movement 148–9
  in soil 270
  support 148
Ecdysis 145, 208
Echinodermata **10**
Ecology 281
Ecosystem 281
Egg:
  bird 216, 217
  frog 215
  tapeworm 259
Egg nucleus 191, 194–5
Ejaculation 217, 222
Elbow joint 137, 138, 139
Elements:
  major 61
  trace 61
Elevator muscle 141, 146
Embryo:
  bird 217
  frog 215
  mammal 223–4
Embryo sac 191, 194–5
Emulsification of fats 48–9
Enamel (teeth) 42
Endocrine system 153, **161–3**
Endolymph 177
Endoplasmic reticulum 13
Endoskeleton 133, 149–50
Endosperm 196
Energy:
  release from food 94
  sunlight 52, 54
Environment:
  and heredity 247–8
  internal 112
Enzymes, digestive 28, 30, **35–6**, 51, 259, 261–2
Ephemerals 199
Epidermis:
  mammals 116, 261
  plants 15, 57–8
Epididymis 222
Epiglottis 46
Erepsin 49
Erythrocyte (*see* red blood cells)
Essential amino acids 28
Euglena 5, 6
Eustachian tube 176, 181
Eutrophication 283–4
Evolution 50, **228–37**
Excretion 2, 32, 77, **118–23**
Exhaust fumes 282–3
Exoskeleton 133, 145, 149
Expiration 106–7
Extensor muscle 138, 145
Extrinsic muscles 169
Eye 169–73
  accommodation 171–2
  compound (of insects) 173–5
  correction of defects 172
Eyelid 169

F_1 generation 241
F_2 generation 241
Faeces 50, 118, 259
False fruit 196
Fat **25–7**, 33
  digestion of 36, 49, 50
  storage of 27, 118
  tests for 33
Fatty acids 27, 36, 50, 68
Feather 140–2, **151**
Feed-back 112–13, 162
Feeding 1
  Amoeba 37
  Hydra 37–8
  insects 39–42
  mammals 42–4
  Paramecium 37
  tapeworm 259
Fermentation 95–6
Ferns 5, 7, 190
Fertilization:
  birds 216, 217
  external 184
  flowering plants 194–5
  frog 215
  Hydra 207
  internal 184, 216, 217
  mammals 222
Fertilizer 197, 272, 284
Fibrin 262
Fibrinogen 68, 114, 262
Field capacity (soil) 270
Fins 144–5
Fish 8, 10
  life cycle 213–14
  sound detection 178
  swimming 142–5
Flaccid cells 85
Flexor muscle 138, 145
Flight:
  birds 139–42
  insects 147–8
Flocculation 271, 274
Floret 191
Flower:
  fertilization in 194–6
  pollination of 191–4
  shape and symmetry 191
  structure 191
Flowering plants:
  classification of 5
  life cycle of 199–200
  vegetative reproduction of 200–5
Fog 282–3
Follicle (Graafian) 219
Food 203
  energy value of 25
  produced by photosynthesis 59
  release of energy from 94
  tests for 33
  types of 25–32
Food chain 279–80, 281, 284
Food web 280, 281
Fossil fuel 59–61, 278, 282
Fossils 229–30
Fovea 173, 180, 181
Frog 215–17
Fructose 25
Fruit 191, 196
  dispersal of 196–9
Fungi 5, 6, 185–6, 270 (*see also Mucor* and yeast)

Galapagos Islands 236–7
Gall bladder 48
Gametes 183–4, 191, 194–5, 218, 244–5, 247, 248, 249
Gametophyte 189–90
Gaseous exchange 101–9
Gastric glands 47
Gastric juice 45, 47
Gene 245–7
Genetic code 18, 245–6
Genetics 18, 239, **245–6**
Genotype 247
Geotropism 125, 129–30
Germination 95, 130, 196, 205
Gills:
  fish 101, 108
  tadpoles 216
Glands:
  adrenal 162–3
  digestive 47–8

Glands (*continued*):
endocrine 161–3
pituitary 161–2
thyroid 162
Glomerular filtrate 121
Glomerulus 121
Glucose 25, 33, 36, 53, 54–6, 94–6, 113, 163
Glycogen 25, 113, 163
Gowland Hopkins, Frederick 30, 34
Graafian follicle 219
Grafting (plants) 203–4
Granulocytes 67
Grass flower 193, 196
Grey matter 155, 160
Gristle (cartilage) 136
Growth 2
cells 16–18
hormones 161
plants 22, 23
Guanine 246
Guard cells 57, 88
Gymnospermae 5
Gynoecium 191

Habitat 281
Haemoglobin 66–7, 75, 104–5
Hair 116–17, 118, 167
Haptotropism 131
Harvey, William 69–70
Hearing 175–7
Heart 66, 69, 70–2
Heat stroke 116
Heavy soil 271
Hepatic portal vein 50, 77, 113
Hepaticae (liverworts) 5, 189
Herbaceous perennial 200
Herbivores (*see also* ruminants) 1, 44, 48, 50, 59, 278
Hereditary characteristics 18, 239–53
Hereditary information 18, 182, 189, 193
Heredity 239–53
Hermaphrodite 184
Heterozygous 247
Hinge joint 43–4, 137
Hip joint 137
Homeostasis 112–24
Homoiothermic 115
Homologous chromosomes 244–5
Homozygous 247
Honey 209–12
Hooke, Robert 11–12
Hormones 31
animals 161–3
plants 128–30, 131
Horse 265
evolution of 230, 235
Horse chestnut twig 200
Host 254, 259
Housefly:
feeding 40
spread of infection by 258
Humerus bone 136
Humidity and transpiration 87
Humus 268–9, 270, 272, 277

Hybridization 240
Hydra:
feeding and digestion 37–8, 50
reproduction 207
structure 15
Hydrochloric acid 47
Hydrostatic skeleton 133, 148
Hydrotropism 125, 131
Hypermetropia (long sight) 172
Hypha 185–6
Hypopharynx 40

Ileum 49–50
Image formation 169–73
Imago 208
Immunity:
active 264
artificial 264–5
natural 261–2
passive 264–5
Impulse (nervous) 155–6
Incisors 43
carnivore 43
herbivore 44
Inclusion 13
Incomplete metamorphosis 208
Incubation 217
Infection 258
by animals 258–9
by contact 258
by contaminated food 258
by droplets 258
prevention of 259
Influenza 258
Ingenhousz, Jan 53
Inheritance (*see* heredity)
Inner ear 175–7
Insects:
antennae 168–9
compound and simple eyes 173–5
feeding 39–42
flight 147–8
life cycle 208–13
pollination by 194
respiratory organs 108–9
sound detection by 177–8
vectors 258–9
walking 145–6
Inspiration (breathing) 105–6
Insulin 113–14, 163
Integuments 196
Intercostal muscles 105, 106, 107
Internal environment 112
Interphase 20
Intervertebral disc 136
Intestine 36, **47–50**
Iodine:
in diet 29
in starch tests 61
Iris (eye) 169, 171
Iris (plant) 202–3
Iron 29
Isotopes 54, 90

Jaw 43–4
Jenner, Edward 264
Joints 135–7

Keys 4
Kidney 77, 114, **118–22**, 161, 163
Kidney tubules 121
Knee-jerk reflex 158
Knee joint 137
Kwashiorkor 29

Labium:
locust 39
mosquito 40
Labrum 39
Lactase 49
Lacteal 50
Lactic acid 94, 96
Lamarck, Jean Baptiste 233
Lamina (leaf) 57
Large intestine 50
Larva 208, 213
Larynx (voice box) 101
Lateral line 178
Lead 282, 284
Leaf 52
photosynthesis 58–9
structure 55–8, 63
Leaf-fall 200
Leguminous plants 277
Lens 169, 171–2, 173
Leprosy 254
Leucocytes (*see* white blood cells)
Levers in skeletons 138–9
Lichens 282
Life, characteristics of 1–2
Life histories (cycles):
bacteria and viruses 255–8
bird 216–17
butterfly 208, 210
fish 213–14
flowering plants 199–200
frog 215–16
insects 208–13
tapeworm 259–60
Ligaments 133, 165
Light, in photosynthesis 54–5, 58, 61, 64
Light reaction 54–5
Light soil 271
Lightning 277
Lignin 80
Lime 271, 274
Lipase 49
Liver 66, 77, **113–15**, 262
Loam soil 271
Locomotion (*see* movement)
Locus of gene 247
Locust (feeding) 39, 50–1
Long sight 172–3
Lungs 70, **75–7**, 282
structure 101–4
ventilation 105–7
Lymph node 75
Lymphatic system 50, **75**, 262
Lymphocytes 67–8
Lysin 262

Magnesium 61, 272
Maize fruit 197
Malnutrition 24
Malpighi, Marcello 70, 121
Maltase 49
Maltose 45
Mammals 10
  reproduction 217–24
  sound detection 175
  support and movement 133–9
Mammary glands 161, 218, 224
Mandibles (insect):
  locust 39
  mosquito 40
Manure 271, 284
Marrow (bone) 66, 67–8, 135
Marsupials 10
Mastigophora (flagellates) 5
Mating:
  birds 217
  fish (stickleback) 214
  frogs 215
  mammals 222
Maxilla 39
Measles 258
Mediterranean 284
Medulla (kidney) 121
Medulla oblongata 159–60
Meiosis 183, **244–5**
Membrane:
  cell 13
  semi-permeable 84, 85
Mendel, Gregor 239–43
Mendelian factors 242
Mendel's First Law 242
Menstrual cycle 219–20
Menstruation 219
Mercury 281–2, 284
Meristems 18
Mesophyll 58
Metabolism 2, 13, 97, 118, 162
Metamorphosis:
  complete 208
  frog 215–16
  incomplete 208
  insects 208
Metaphase:
  meiosis 244–5
  mitosis 20–1
Micrococci 254
Micro-organisms (*see* bacteria and viruses)
Micropyle 195
Micro-villi 50
Middle ear 175
Mid-rib 57, 58
Milk 48, 161, 218, 224
Milk teeth 43
Minerals:
  absorption by plants 86
  in animal nutrition 29
  in bones 133–4
  in plant nutrition 53, **61**
  in soil 269–70, 271–2
Mitochondria 13
Mitosis **19–21**, 182, 196
Molar teeth 43–4
Molluscs **10**
Monocotyledons 5
Monohybrid cross 240–3, 248
Monosaccharide 25
Monotremes 10
Mosquito:
  feeding 40
  spread of infection 259
Moss 5, 7, 189–90, 205
Motor area 160
Motor end-plate 155
Motor neurones 155
Mould (*see Mucor* and fungi)
Moulting (insects) 145, 208
Mouth:
  digestion in 44–5
  teeth 43–4
Mouth parts of insects 39–42
Movement:
  birds 139–42
  earthworm 148–9
  fish 142–5
  insects 145–8
  mammals 133–9
Mucin 45, 47
Mucor 185–6
Mucus 46, 50, 101–2, 261
Mulching 271
Multiple fission 185
Muscle 138–9, 143–4, 148–9
Muscle fibres 95–6
  cardiac 70
  in *Hydra* 15
Mutation 234, **247–8**
Mycelium 185–6
Mycophyta (fungi) 5, 6
Myopia (short sight) 172

Nasal passages 101, 261
Nastic movements 132
Natural selection **233–7**, 248
Nectar 191, 209, 212
Nectary 191
Nematocyst 38
Nerve cell 153–5
Nerve fibre 153–5
Nerve impulse 153, **155–6**, 165
Nervous system 153–61
Nest building:
  bird 216–17
  stickleback 214
Neurone (*see* nerve cell)
Niche 281
Nicotinic acid (niacin) 31
Nitrate 268, 272, 277, 283–4
Nitrifying bacteria 277
Nitrite 277
Nitrite bacteria 277
Nitrogen 61, 277 (*see also* nitrate)
Nitrogen cycle 277–8
Nitrogen-fixing bacteria 277
Nitrogenous excretion 119, **121–2**
Nucleus **13**
  meiosis of 244–5
  mitosis of 19–21
Nutrition:
  autotrophic 1, 52–61
  heterotrophic 1–2, 24–33
Nymph (insect) 208

Obesity 24, 34
Ocellus 175
Oesophagus 46
Oestrogen 163, 219
Offset 202
Oil **25–7**, 36, 49, 50
  test for 33
Oil pollution 284
Old sight 172
Olfactory organs 168
Ommatidium 173–5
Omnivore 1, 44, 59
Operculum 107, 216
Opsonin 262
Optic nerve 169, 173
Orbit 169
Organelles 13, 14
Orgasm 222
Osmoregulation 123–4
Osmosis 74–5, **83–6**, 123
Osmotic potential **84**
Ossicles (ear) 175–7
Osteichthyes 8, 10
Otolith 178–9
Outer ear 175
Oval window 175–7
Ovary:
  flowering plants **191**, 194–5, 196, 197, 199, 205–6
  Hydra 207
  mammals 163, **218**
Oviduct:
  birds 217
  mammals 219
Ovulation 219–20
Ovule 191, 195, 196
Ovum 183, 189, 207, 218–20, 222
Oxygen 269, 270, 277
  from photosynthesis 52, 53, 54, 58–9, 63
  in respiration 94, 101, 104–5, 107, 112
  transport by blood 66–7, 74, 75
Oxygen debt 96
Oxygenated blood 70, 104–5
Oxyhaemoglobin 66, 78

Pain 167
Palisade layer 58
Palp (locust) 39, 51
Pancreas **48–9**, 113–14, 163
Pancreatic juice 48–9
Paramecium 14, 66, 101
  feeding and digestion 37, 50

Parasites 1–2, 59, 261–5
  bacteria 254–6
  tapeworm 259–60
  viruses 256–8
Pavlov, Ivan 158
Peat 271
Pedicel 191
Pellagra 31
Penis 222
Peppered Moth **235–6**, 248
Pepsin **47**, 51
Peptide 48, 49
Perennating organs 200
Perennials 199–200
Perianth 191
Perilymph 177
Peristalsis 32, **46**
Permanent teeth 43
Perspiration (sweating) 29, 32, **115–16**
Petals **191**, 193–4
Petiole 55, 89, 91
Phaeophyta (brown algae) 5, 6
Phagocytes **67**, 75, **262**
Pharynx 101
Phenotype 247
Phloem 58, **80**, 88
Phosphate 133, 268, 272, 283–4
Phosphorus (*see also* phosphate) 61, 272
Photosynthesis 1, 2, 13, **52–61**, 88, 278–9, 282
Phototropism **125–9**, 130
Phylloquinone (vitamin K) 32
Phylum 3
Phytoplankton 280
Pinna 175
Pitch (of sound) 175
Pitching 144
Pituitary gland **161–2**, 219
Placenta 223–4
Plasma 66, **68–9**, 77
Plasmolysis 85
Platelets 66, **68**, 262
Pleural cavity 105–6
Plumule 196
Pneumonia 258
Poikilothermic 115
Poliomyelitis 258, 264
Pollen **191–4**, 211–12
Pollen basket 209
Pollen comb 209
Pollen tube 195
Pollination **191–4**, 240
  insect 194
  self 193
  wind 193–4
Pollution 280, **281–5**
  air 282–3
  rivers 283–4
  sea 284–5
Polysaccharide 25, 56, 254
Poppy (fruit) 198
Posture 138
Potassium **29**, 45, 61, 272
Potato tuber 201
Potometer 90–1
Pregnancy 223–4
Premolars 43–4
Presbyopia (old sight) 172
Pressure, sense of 166
Priestley, Joseph 53
Primary feathers 140
Primates 10
Proboscis (insect) **40**, **42**, 208
Producer 279–80
Proleg 210
Prophase:
  meiosis 245
  mitosis 19
Proprioceptor 165–6
Prostate gland 222
Protandry 193
Protein 114, 247, 277
  adequate 28
  digestion of 48, 49
  essential 28–9
  functions in diet 28, 33
  structure of 28
  tests for 33
Protista 5, 6
Protogyny 193
Protozoa 5, 6
Pseudopodia 37
Pteridophyta (ferns) 5, 6, 190
Ptyalin 45
Puberty 219
Pulmonary artery and vein 69, 70
Pulmonary circulation 69, 70
Pulse 73
Pupa 208, 213, 226
Pupil 171, 180
Pus 262
Pyramid of numbers 279–80

Radicle 196
Radius bone 134, 137
Reabsorption (in kidney) 121–2
Receptacle 191
Recessive characteristic 241
Rectum 50
Red blood cells:
  formation 66
  functions 75, 104–5, 262
Reduction division (*see* meiosis)
Reflex 101, **157–9**, 164, 222
Regulation:
  amino acids and protein 114
  blood sugar 113–14
  body temperature 115–18
Renal artery and vein 119, 120, 121
Rennin 48
Reproduction 2, **182–4**
  Amoeba 184–5
  asexual 182–3
  bacteria 255–6
  birds 216–17
  fish 213–14
  flowering plants 190–200
  frog 215–16
  Hydra 207
  insects 208–13
  mammals 217–24
  moss 189–90
  Mucor 185–6
  sexual 183
  Spirogyra 186–8
  vegetative 200–5
  virus 257–8
  yeast 185
Reptile 9, 10
Residual air 107
Respiration 2, 25, 66, 86, **94–7**, 162, 269, 270, 278
  aerobic 94, 101, 255
  anaerobic 94–6, 225, 259, 277
Respiratory organs:
  fish 107
  insects 108–10
  mammals 101–5, 261
Respiratory surface 101
Retina 169, **172–3**
Retinol (vitamin A) 30
Rhabdom 173
Rhizoid 189
Rhizome 201
Riboflavine (vitamin B_2) 31
Ribosomes 13
Ribs 106, 136
Rickets 31
Rods 173
Rolling 144
Root:
  adventitious 201, 203
  contractile 203
  functions of 83, 86
  response to gravity 125, **129–30**
  response to water 131
  uptake of minerals 86–7
  uptake of water 83, **86–7**
Root cap 81
Root hairs 86
Root nodules 277
Roughage 32
Royal jelly 211, 212
Rumen 44
Ruminants 44
Runner 201

Saccule 178–9
Sachs, Julius 53, 61
Saliva:
  housefly 40
  mammal 45, 51
  mosquito 40
Sandy soil 270, **271**
Saprotroph (saprophyte) 2, 59, 185, 225, 277
Sclerotic 170
Scrotum 220
Scurvy 30, **31**
Sebaceous gland 116, 261
Sebum 261
Secondary feathers 140
Secondary sexual characteristics 163

Secretion 118
Sedentary soil 270
Sedimentary rock 229
Seeds (*see also* fruit) 191, 199
  dispersal 196–9
  germination 205
  structure 196
Segregation:
  chromosomes and genes 245, 256
  Mendelian factors 242–3
Self-fertilization 184
Semen 222
Semi-circular canal 175, 178
Semi-lunar valves 70, 73
Seminal vesicle 222
Semi-permeable membrane **83**, 85
Sensation 165
Sense organs 156
  ear 175–7
  eye 169–73
  inside body 165
  in nose 168
  in the skin 166
  taste buds 167–8
Sensitivity of plants 1, **125–30**
Sensory adaptation 165
Sensory area (of brain) 160
Sensory neurones 153
Sepals 191
Seta 148
Sewage 258, 259, 281, 283, 284
Sex chromosomes 245
Sex determination 249
Sex hormones 163, 219, 221
Sexuality 184
Shell (of bird egg) 217
Shivering 117
Shoot 196
  response to gravity 129–30
  response to light 125–9
Short sight 172
Sieve tube 58, 80, 88
Silt 269
Skeleton:
  animals 133–7
  endo- 133
  exo- 133
  hydrostatic 133, 148
Skin 115–18
  and immunity 261
  receptors in 166–7
  structure 116
Slimming diet 34
Small intestine 48
Smallpox 258, 264
Smell 168
Smog 282, 283
Smoke 282
Sodium bicarbonate 49, 77
Sodium chloride 29, 115–16
Soil 86
  components of 268–70
  formation of 268–9
  types of 270–1
Special creation, theory of 228
Species:
  definition of 3–4
  origin of 233–5
Sperm duct 222
Spermatophyta (seed plants) 5
Sperms 183, 189, 207, 212, 214, 215, 217, 221, 222
Sphincter muscle 47, 119
Spikelet (of grass flower) 196
Spinal column (*see* vertebral column)
Spinal cord 153, 154, 155, 157, 158, 160
Spindle apparatus 21
Spiracle (of insects) 109
Spirillum 255
Spirogyra 186–8
Spleen 66, 262
Spontaneous generation 2, 266
Sporangium 185–6
Spores:
  bacteria 256
  mosses 189
  Mucor 185–6
Sporophyte 5, 189–90
Stamen 191, 193
Staphylococci 254, 255
Starch 25
  in leaves 55–6
  test for 33, 61
Stem 86, 125, 199–200
  structure of 80
  underground 202–3
Stereoscopic vision 173
Sternum 136, 151
Stickleback 213–14
Stigma 191, 193–4
Stimulus 165
Sting cell (nematocyst) 15, 38
Stomach:
  human 47, 261
  ruminant 44
Stomata (stoma) 53, 57–8, 86, **88**, 282
Storage (of food) 27, 55–6, 118, 196, 199–205
Strawberry 198
Streptococci 254–5
Stretch receptors 160, **165–6**
Style 191, 194–5
Subsoil 271
Sucker 202
Sucrase 49
Sucrose 25
Sugar (*see also* glucose) 25
Sulphur 61, 272
Sulphur dioxide 282
Sulphuric acid 282
Sunlight:
  and vitamin D 31
  in fog formation 283
  in photosynthesis 52, **54–5**
Surface area:
  cells 16
  intestine 50
  lung 102
  roots 83
Survival of the fittest 234–5
Suspensory ligaments 171–2
Swallowing 46, 101
Swarming 213
Sweat glands 115, 118
Sweating 115–16
Swim bladder (fish) 143
Symbiosis 277
Synapse 156, 159–60
Synovial fluid 136, 137
Synovial joints 136, 137
Systemic circulation 70

Tadpole 215–16
Tail:
  fish 143–4
  tadpole 216
Tapeworm 259–60
Taste, sense of 167–8
Tear glands 169
Teeth 42–4
Telophase:
  meiosis 245
  mitosis 21
Temperature control 77, 113, **115–18**
Temperature inversion 282
Temperature, sense of 166–7
Tendon 138, 165
Terminal bud 201
Territory (of stickleback) 213–14
Testa 196
Testis 183
  Hydra 207
  mammals 220–1
Testosterone 163, 221
Tetanus 265
Thallophyta 256
Thiamine (vitamin B_1) 30–1
Thigmotropism 131
Thorax:
  insects 145, 147
  mammals 101, 105
Threshold level 156
Thymine 246
Thyroid gland 162
Thyroxin 162, 164
Tidal air 107
Tissue 14, 18, 20
  animals 15
  plants 16
Tissue fluid 68, **74–5**, 112, 113, 122, 124, 262
Tongue 45, 167–8
Topsoil 271
Touch, sense of 166
Toxin 254, 262, 265
Trace elements 61
Trachea (wind-pipe) 46, **101–2**, 261
Tracheal system (insects) 109–10
Tracheole (insects) 109–10
Translocation 88–9
Transpiration 86–8
Tricuspid valve 70
Tropism 125
Troposphere 282

True leg 210
Trypsin 49
Tuber 201, 202
Tuberculosis 254, 258
Turgor pressure 85, 86
Typhoid fever 258

Ultra-violet light 174–5
Umbilical cord 223, 224
Urea 77, 114, 118, 121, 122
Ureter 119
Urethra 119, 222
Urinary system 119
Urine 32, 77, 121, 122, 123, 124, 161, 277
Uterus 161, 219, 222, 223, 224
Utricle 178–9

Vaccine 264
Vacuole **14**, 36–7, 38, 50
Vagina 219, 224
Valve:
- heart 70
- vein 69, 74

Van Helmont, Jean Baptiste 53
Variation 234, 235, 239
- continuous 240
- discontinuous 240

Vascular bundle 80
Vascular tissue:
- in animals 69–74
- in plants 80–2

Vasoconstriction 118, 161
Vasodilation 116
Vector 258
Vegetative reproduction (propagation) 200–5
Veins:
- circulatory system 69, **74**
- leaves 57, 58, 87

Vena cava 70
Ventilation 101
- in fish 107
- in insects 109
- in mammals 105–6

Ventricle 70, 73
Venules 73
Vertebra 136, 143
Vertebral column (backbone) 136
Vertebrata (vertebrates) 10
Vibrio 255
Villus 50, 77, 223
Virus 256–8
Vision 169–75
Vitamins **30–2**, 33, 48, 114
Vitreous humour 171, 172
Vocal cords 101
Voice 101
Vulva 219

Wall pressure 85
Wallace, Alfred Russel 233
Warm blooded 115
Water:
- in animal diet 32
- in photosynthesis 54
- pollution of 283–5
- in soil 270, 271, 272, 273
- support by 133

Water (*continued*):
- uptake by plants 86

Water culture 61
Water potential 86, 87
Water table 270
Weathering 269–70
White blood cells (leucocytes) 67–8, 262
White dead nettle 194
White matter 155
Whooping cough 258
Wilting 86
Wind dispersal 197
Wind-pipe (trachea) 47, 101–2
Wind pollination 193–4
Wings:
- birds 139–42
- insects 146–8

Winter twigs 200, 206
Worker bees 209–12

X-chromosomes 249
Xylem 80, 86, 88–9

Y-chromosomes 249
Yawing 144
Yeast 95
- life cycle 185

Yellow fever 258–9
Yolk 215, 217

Zinc 284
Zooplankton 280
Zygospore 186, 188
Zygote 18, 20, 183, 186, 207, 243

Answers to factual recall tests

The following are provided as one set of possible answers for the chapter summaries. It is not suggested that these are the only correct answers; in many cases there are other equally appropriate words, phrases, or sentences.

Chapter 1
(1) fins, wings, legs (2) plants do not move about from place to place (3) leaves turning towards the light, roots growing into the soil (4) light, sound, touch, smell, taste (5) co-ordinate (6) light, gravity, water (7) energy (8) growth (9) repair (10) autotrophic (11) carbon dioxide (12) water (13) sunlight (14) photosynthesis (15) herbivores, carnivores, omnivores, parasites, saprotrophs (16) oxygen (17) energy (18) carbon dioxide (19) gills, lungs (20) excretion (21) egestion (22) other living organisms (23) spontaneous (24) living matter can arise spontaneously out of non-living matter (25) animal growth is limited, whereas plants can grow throughout their whole life span (26) all the chemical and physical processes of life (27) respiration (28) photosynthesis (29) balanced (30) *Homo sapiens* (31) genus (32) species (33) they help to prevent confusion between different organisms (34) fertile

Chapter 2
(1) cell membrane (2) selectively (3) cytoplasm (4) starch, fat, excretory substances (5) inclusions (6) organelles (7) mitochondria, endoplasmic reticulum, ribosomes, chloroplasts (8) chromatin (9) chromosomes (10) deoxyribonucleic acid (11) hereditary (12) genetic (13) plant cells are enclosed in a cellulose wall; they contain vacuoles and chloroplasts (14) tissues (15) labour (16) muscles, nerves (17) organ (18) the heart (19) surface area (20) volume (21) all cells absorb food and oxygen, and remove waste products through their surface (22) number (23) size (24) replica (25) chromosomes (26) chromatids (27) centromere (28) anaphase (29) opposite (30) prophase (31) same (32) it ensures that information in the genetic code reaches every cell of an organism

Chapter 3
(1) sweet fruit, jam, treacle, bread, potato, rice (2) energy (3) seventeen (4) glycogen (5) muscles (6) liver (7) fat (8) butter, lard, suet, dripping, olive oil (9) thirty-eight (10) adipose (11) under the skin, around muscles, around the heart (12) skin (13) heat (14) it has double the energy yield, and is less bulky (15) meat, liver, kidney, eggs, fish (16) building (17) they supply the materials from which body tissues are made (18) enzymes (19) minute (20) speed up reactions which would otherwise proceed very slowly (21) carbon, hydrogen, oxygen, and nitrogen (22) amino (23) twenty (24) eggs, meat, fish (25) they contain the eight essential amino acids from which the human body makes all its proteins (26) children (27) pregnant women (28) people recovering from illness (29) they require body-building materials (30) fifteen (31) meat, eggs, milk, green vegetables (32) minute (33) scurvy, beri-beri, pellagra, rickets (34) water (35) metabolism (36) food (37) excretory waste (38) sufficient carbohydrate, fat, protein, vitamins, and minerals for the body's energy, growth, and repair requirements (39) body size, sex, occupation, health, age

Chapter 4
(1) simpler (2) soluble (3) insoluble (4) absorbed (5) enzymes (6) starchy (7) sugars (8) glucose (9) fats (10) oils (11) fatty acids (12) glycerol (13) protein (14) amino acids (15) inside (16) unicellular (17) *Amoeba* (18) outside (19) intestine (20) mouth (21) anus (22) It breaks down food into easily swallowed pieces; it increases the surface area of food and so speeds digestion; it mixes food with saliva (23) starch (24) maltose (25) wave (26) muscular (27) peristalsis (28) sphincters (29) gastric (30) hydrochloric (31) pepsin (32) peptides (33) bile (34) droplets (35) emulsion (36) trypsin (37) amylase (38) lipase (39) erepsin (40) peptides (41) maltase, sucrase, and lactase (42) lipase (43) fatty acids (44) glycerol (45) villi (46) increase (47) absorption (48) water (49) salt (50) faecal

Chapter 5
(1) carbon dioxide (2) stomata (3) lower (4) water (5) roots (6) chlorophyll (7) light (8) $6CO_2$ (9) $6H_2O$ (10) light energy (11) chlorophyll (12) $C_6H_{12}O_6$ (13) $6O_2$ (14) water (15) hydrogen (16) oxygen (17) 'light' (18) CO_2 (19) $2H_2O$ (20) $C(H_2O)$ (21) H_2O (22) O_2 (23) chloroplasts (24) palisade (25) mesophyll (26) xylem (27) water (28) sieve (29) sugar (30) growth (31) food (32) Photosynthesis produces oxygen which is necessary for respiration of all living things (33) photosynthesis converts light energy into chemical energy, thereby producing substances necessary for plant growth. These are used as food, directly or indirectly, by all heterotrophic organisms

Chapter 6
(1) plasma (2) water (3) glucose, fatty acids, glycerol, amino acids, vitamins, minerals, albumin, fibrinogen, antibodies, hormones, urea, and carbon dioxide (4) capillary (5) tissue fluid (6) food (7) oxygen (8) waste (9) bi-concave (10) bone marrow (11) nucleus (12) haemoglobin (13) iron (14) oxygen (15) oxyhaemoglobin (16) irregular (17) lobed

(18) granules (19) phagocytic (20) engulf disease-causing bacteria (21) large round (22) granules (23) antibodies (24) disease (25) clot (26) heart (27) arteries (28) thick (29) capillaries (30) permeable (31) veins (32) valves (33) heart (34) pulmonary circulation (35) systemic circulation (36) atrium (37) right (38) left (39) tricuspid (40) bicuspid (41) ventricles (42) right (43) pulmonary (44) left (45) aorta

Chapter 7
(1) vessels (2) water (3) minerals (4) leaves (5) sieve (6) sugar (7) leaves (8) storage areas (9) growing points (10 diffusion (11) water (12) semi-permeable (13) weak (14) strong (15) cellulose (16) turgor (17) wall (18) turgid (19) wall pressure prevents further entry of water molecules (20) seedlings, herbaceous plants, leaves, flowers (21) lose (22) plasmolysis (23) water (24) evaporation (25) spongy (26) stomata (27) under (28) guard (29) temperature, humidity, wind (30) brings water and minerals to leaves, essential for absorption of carbon dioxide, cools leaves (31) drought (32) xylem (33) transpiration

Chapter 8
(1) breakdown (2) energy (3) life (4) oxygen (5) carbon dioxide (6) water (7) energy (8) oxygen (9) energy (10) lactic acid (11) alcohol (12) yeast (13) bacteria (14) ethanol, citric acid, oxalic acid (15) the fermentation process of yeast produces carbon dioxide which forms bubbles as it escapes from the dough (16) alcohol (17) poisonous (18) lactic acid (19) contraction (20) debt (21) it transfers energy from the chemical reactions of respiration to the other reactions of metabolism which use energy

Chapter 9
(1) diaphragm (2) flatter (3) intercostal muscles (4) backbone (5) sternum (6) cage (7) increase (8) thoracic (9) air pressure in the pleural cavity is always lower than in the lungs, and this pressure difference stretches the elastic lung tissue until it almost fills the thorax, and as the volume of the thorax increases so does the volume of the lungs (10) pressure (11) germs (12) dust (13) mucus (14) goblet (15) cilia (16) mouth (17) swallowed (18) trachea (19) bronchi (20) bronchioles (21) alveoli (22) capillaries (23) gaseous (24) larger (25) As red cells sequeeze through the capillaries they expose a large surface area to the capillary walls for oxygen absorption (26) Friction slows down blood flow thereby increasing the time available for oxygen absorption (27) opercula (28) gill bar (29) lamellae (30) capillaries (31) valves (32) mouth (33) oxygen (34) carbon dioxide (35) tracheal (36) blood (37) spiracles (38) tracheoles (39) exchange

Chapter 10
(1) maintenance (2) internal environment (3) feed (4) man (5) mammals (6) birds (7) blood (8) tissue (9) carbon dioxide, oxygen, temperature, food, urea, poisons, osmotic pressure (10) sugar (11) amino acids (12) protein (13) A, D, B_{12} (14) iron, copper, potassium (15) removing harmful substances produced by disease organisms, and those taken into the body such as certain drugs and alcohol (16) clotting (17) fibrinogen (18) heat (19) warms (20) bile (21) can maintain a constant body temperature (22) man, dog (23) poikilothermic (24) they have no temperature control mechanism (25) warm-blooded animals can be cooler than their surroundings, and cold-blooded animals can be warmer than their surroundings (26) sweat (27) evaporates (28) heat (29) superficial (30) vasodilation (31) over-heated (32) close (33) radiation (34) shivering (35) heat (36) erector (37) by stopping cold air from reaching the skin (38) by producing a layer of still air around the body which insulates it (39) waste (40) metabolism (41) excess (42) urea (43) urine (44) urea contains nitrogen derived from the breakdown of excess amino acids in the body (45) glomeruli (46) Bowman's capsules (47) tubules (48) glucose, amino acids, vitamins, minerals (49) reabsorption

Chapter 11
(1) growth movements (2) the movement occurs only at a plant's growing points (3) phototropism (4) A (5) A (6) B (7) C (8) green (9) the features of sets A and B were due to lack of light, and growth in light from only one direction respectively (10) positively (11) chlorophyll (12) geotropism (13) upwards (14) negative (15) downwards (16) positive (17) auxin (18) direction (19) tip (20) increases (21) elongation (22) towards

Chapter 12
(1) caterpillars, earthworms (2) water (3) hydrostatic (4) exo- (5) outside (6) chitin (7) wax (8) cuticle (9) the periodic shedding of the cuticle and its replacement by a new, larger one (10) synovial (11) elbow, knee, finger knuckles (12) they move in one plane only (13) shoulder, hip (14) they consist of a ball-shaped end of a bone which fits into a hollow socket in another bone (15) ligaments (16) tendons (17) does not move when the muscle contracts (18) moves when the muscle contracts (19) antagonistic (20) flexors (21) extensors (22) fulcrum, effort, and load (23) the atlas which is the fulcrum, the head which is the load, and the neck muscles which are the effort (24) the elbow joint which is the fulcrum, the hand which is the load, and the biceps which is the effort (25) the ball of the foot which is the fulcrum, the body which is the load, and the calf muscles which are the effort (26) aerofoil (27) air flowing across its upper surface is at a lower pressure than air flowing across its lower surface (28) leading (29) trailing (30) backwards (31) forwards (32) 'wrist' (33) there is a sudden build-up of air pressure beneath it (34) swim (35) density (36) equal (37) weightless (38) muscle (39) backbone (40) waves (41) backwards (42) press (43) tail fin (44) yawing, rolling, pitching (45) elevator (46) downwards (47) depressor (48) lengthwise (49) pressure (50) upwards (51) figure-of-eight (52) horizontal (53) vertical (54) changing (55) longer (56) thinner (57) anchored (58) setae (59) shorter (60) fatter (61) front

Chapter 13
(1) the brain and spinal cord (2) fibres (3) insulated (4) eyes, ears, nose, taste buds, touch receptors (5) dendron (6) body (7) axon (8) central nervous system (9) cell (10) central nervous system (11) dendrites (12) axon (13) a muscle or gland (14) threshold (15) all-or-none principle (16) the frequency at which they pass along a nerve (17) synapse (18) behaviour in which a stimulus results in a response which is unlearned, and occurs quickly without conscious thought (19) the knee-jerk reflex (20) base (21) spinal cord (22) regulation of temperature, blood pressure, rate of heartbeat and breathing (23) organs of balance, stretch receptors in muscles (24) balance (25) muscular (26) walking, cycling (27) grey (28) cerebral cortex (29) neurone cell bodies (30) white (31) nerve fibres (32) sensory (33) glands (34) motor (35) it is here that associations take place between information from the senses and memory, to produce conscious awareness

and thought (36) ducts (37) hormones (38) blood (39) brain (40) it controls other endocrine glands (41) to influence growth, cause vasoconstriction, and contract the uterus during childbirth (42) neck (43) sugar (44) respiration (45) physical (46) mental (47) sudden, violent effort (48) escape from predators (49) stress, fear, anger (50) glucose (51) increasing the rate of conversion of glucose into glycogen in the liver (52) oestrogens (53) to control development of secondary sexual characteristics; prepare the uterus for pregnancy; and maintain the uterus during pregnancy (54) testes (55) testosterone (56) secondary sexual characteristics in males

Chapter 14

(1) muscles, tendons, ligaments (2) tension (3) muscular (4) angle (5) joints (6) the direction and speed of his movements, and the position of his limbs relative to the body. (7) hard (8) soft (9) rough (10) smooth (11) temperature (12) compare temperatures (13) they give warning of damage to the body (14) taste buds (15) sweet, sour, salt, bitter (16) olfactory organs (17) they give warning of poisons or other harmful substances (18) touch, sound, smell, taste, temperature, humidity (19) feet (20) mouth parts (21) by the skull, the conjunctiva, tears, and the blink reflex (22) extrinsic (23) movements (24) hole (25) pupil (26) light (27) depth (28) radial muscles contract making the pupil larger, and circular muscles make it smaller (29) ciliary (30) contract (31) releases (32) suspensory (33) more convex (34) increases (35) bend (refract) (36) increases (37) aqueous (38) vitreous (39) stretch (40) ciliary (41) less convex (42) presbyopia (43) elasticity (44) near (45) hypermetropia (46) the distance between the lens and the retina is less than normal (47) converging (48) myopia (49) the eyeball is abnormally elongated (50) diverging (51) retina (52) rods and cones (53) Cones are sensitive to colour and work in bright light. Rods are insensitive to colour and work in dim light (54) 20 (55) 20 000 (56) ear drum (57) ossicles (58) oval (59) twenty (60) perilymph (61) cochlea (62) endolymph (63) hair cells (64) vibrate up and down (65) auditory (66) semi-circular (67) ampullae (68) cupula (69) sensory hair cells (70) movements of the head (71) head movements disturb fluid in the semi-circular canal in the same plane as the movement, which displaces the cupula and stimulates the sensory hairs (72) chalk (73) otoliths (74) acceleration, deceleration, and position of the body relative to gravity (75) movements or changes of position displace the jelly and its otoliths, which stimulates the sensory hair cells

Chapter 15

(1) asexual (2) mitosis (3) hereditary (4) copy (5) change (6) generation (7) sexual (8) gametes (9) meiosis (10) fertilization (11) zygote (12) it has inherited two different sets of hereditary information: one set from each parent (13) that there are separate male and female sexes (14) that one organism contains both male and female sex organs (15) external (16) internal (17) reproductive system (18) copulation (19) that fertilization is more certain because the eggs and sperms are in a confined space (20) binary fission (21) the nucleus divides into two by mitosis, followed by the cytoplasm, thus forming two daughter cells (22) water and food (23) spherical (24) hard (25) cyst (26) heat and cold (27) multiple fission (28) the nucleus divides many times and each receives an equal share of the cytoplasm, thus forming many small amoebae (29) fungus (30) budding (31) the cytoplasm forms a bud, which receives a nucleus from the parent cell where mitosis has taken place (32) hyphae (33) mycelium (34) upwards (35) cytoplasm (36) nuclei (37) swell (38) sporangium (39) multiple fission (40) elliptically (41) spores (42) nuclei (43) by wind and insects (44) conjugation (45) plus (46) minus (47) zygospore (48) nuclei (49) spores (50) sporangium (51) fragmentation into smaller filaments (52) conjugation (53) tubular (54) adjacent (55) cytoplasm (56) nucleus (57) plus (58) minus (59) fertilization (60) zygospore (61) the spore breaks open and a new *Spirogyra* filament grows out of it.

Chapter 16

(1) antheridia (2) sperms (3) female (4) archegonia (5) fertilize (6) spore case (7) spores (8) wind (9) warm, damp (10) spores produce a branching green thread which develops buds that become new moss plants (11) it entails the alternation of a gamete-producing gametophyte generation, and a spore-producing sporophyte generation (12) pedicel (13) receptacle (14) whorls (15) female (16) gynoecium (17) ovary (18) ovules (19) embryo sac (20) egg (21) gamete (22) ovule (23) ovary wall (24) male (25) androecium (26) filament (27) anther (28) pollen sacs (29) pollen (30) gametes (31) corolla (32) by their colour, scent, and the nectar produced by nectaries at their bases (33) pollinate (34) calyx (35) to protect the other floral parts during the bud stage of development (36) its petals are the same size and shape (37) radially (38) it can be cut in two along many vertical planes to produce identical halves (39) its petals are different shapes and sizes (40) bilaterially (41) it can be cut into two identical halves along only one vertical plane (42) transfer of pollen from anthers to stigmas in the same flower or between flowers on the same plant (43) transfer of pollen from one plant to another of the same species (44) cross (45) hereditary information of two different plants is intermixed (46) evolutionary (47) unisexual flowers, protandry, protogyny (48) petals (49) nectaries (50) abundant pollen; small light pollen grains; large anthers hanging outside flowers; 'feathery' stigmas; flowers above the leaves or opening before them (51) increase (52) pollen (53) stigmas (54) stinging nettle, hazel, willow (55) large, coloured and scented petals, often with honey guides; sticky or spiky pollen grains; anthers and stigmas situated where insects will brush against them (56) attract (57) they pollinate the flowers (58) white dead nettle, sweet pea, buttercup (59) tubular (60) pollen tube (61) ovule (62) micropyle (63) embryo (64) bursts open (65) male (66) female (67) radicle (68) plumule (69) seed (70) cotyledons (71) cotyledons (72) endosperm (73) testa (74) it avoids over crowding, and increases the chances of a species spreading into new territory (75) by wind, animals, and self-dispersal (76) the plant increases in size and weight and stores food (77) reproductive (78) flowers, fruit, and seeds are produced (79) grow and reproduce several times in one season (80) chickweed, groundsel (81) grow and reproduce (82) shepherd's purse, marigold (83) two (84) carrot, radish (85) they produce enough food in their vegetative phase to reproduce and still have some to spare (86) asexual (87) stolons, runners, suckers, rhizomes, bulbs, corms, tubers (88) they are food storage organs which allow a plant to survive from one growing season to the next, i.e. to perennate (89) rhizomes, bulbs, corms, tubers

Chapter 17

(1) budding (2) body wall (3) that testes and ovaries occur on the same animal (4) the testes release sperms before the ovaries on the same animal are ripe (5) ball of cells (6) hard outer wall (7) warm (8) dragonflies, cockroaches, locusts (9) that they are smaller, wingless, and sexually immature (10) incomplete (11) there are relatively few changes between young and adult stages (12) butterflies, bees, flies (13) complete

(14) the young change completely in form (15) eating (16) ecdysis (17) pupae (18) larval cells disintegrate, giving up their stored food to unspecialized cells which construct the adult organs inside the pupa (19) blood (20) crumpled and folded (21) strips of vegetation (22) glue produced by his kidneys (23) swollen with eggs (24) lays eggs (25) fertilizes them (26) male (27) he guards them from predators, and fans the nest with his tail to aerate it (28) hibernating (29) mud and dense vegetation near water (30) mating (31) pads on his thumbs (32) sperms (33) eggs (34) cloaca (35) ten (36) their mouths are closed and they still have some yolk left in their intestinal cells (37) two or three (38) they scrape algae from plants with their horny lips (39) internal (40) operculum (41) spiracle (42) two (43) lungs, and an intestine suited to a carnivorous diet (44) metamorphosis (45) three (46) the tail is absorbed; forelimbs appear; limbs are used for locomotion; and the tongue is used to catch food (47) they have internal fertilization (48) it enables birds to recognize the opposite sex of the same species; enables males to drive away competitors; initiates nest building; and makes copulation possible (49) ejaculated (50) oviduct (51) albumen (52) a shell (53) the embryo is cold-blooded (54) female (55) brood (56) they cannot fly or feed themselves at first (57) that the embryo develops inside the female's reproductive system (58) the developing embryos are kept warm, fed, protected, and the female is free to lead a normal life (59) milk (60) mammary (61) puberty (62) ovulation (63) month (64) Graafian (65) corpus luteum (66) stimulate development of the uterus lining in preparation for a fertilized ovum (67) oviduct (68) uterus (69) thirty-six (70) dies (71) menstruation (72) fourteen (73) testes (74) epididymis (75) seminal vesicle (76) prostate (77) sperm ducts (78) penis (79) semen (80) stimulate swimming movements of sperm tails, and nourish the sperms (81) pregnancy (82) nine (83) embryo (84) uterus (85) it produces enzymes which digest the cells in the uterus wall (86) placenta (87) food (88) oxygen (89) blood (90) carbon dioxide (91) nitrogenous waste (92) it is suddenly cut off from its supply of food and oxygen (93) uterus (94) abdominal (95) head (96) umbilical (97) bleeding

Chapter 18

(1) change (2) evolution of many different organisms from one type of ancestor (3) limb (4) pattern (5) they evolved from a common ancestor with less specialized limbs (6) acquired (7) young inherit characteristics acquired by parents in the course of their lives (8) natural selection (9) double (10) generation (11) constant (12) struggle (13) reproductive (14) survival (15) they help an organism to survive in the struggle for existence (16) inheritance (17) variation (18) natural selection (19) inheritance (20) features with survival value (21) species (22) artificial (23) breeds of horse, sheep, many varieties of plants (24) they cause corresponding changes in the process of natural selection, and so different features of organisms may suddenly gain survival value (25) ice ages, mountain building, spread of deserts, continental drift

Chapter 19

(1) laws (2) hereditary (3) offspring (4) discontinuous (5) they were distinctly different and produced no intermediate forms when crossed (6) crossing organisms which differ in one or more characteristics (7) monohybrid (8) all tall (9) dominant (10) it had 'dominated' the dwarf characteristic (11) self-pollinated (12) tall (13) dwarf (14) 3:1 (15) recessive (16) it had 'receded' during the F_1 generation (17) characteristics (18) pairs (19) separate (20) one member of each pair (21) the factors come together again and the pairs are restored (22) chromosomes (23) meiosis (24) the factors are carried by the chromosomes (25) Both factors and chromosomes occur in pairs. Factors and chromosomes separate at meiosis and one from each pair goes into each gamete. Factor and chromosome pairs are restored at fertilization (26) locus (27) homologous (28) alleles (29) homozygous (30) heterozygous (31) genotype (32) its visible appearance (33) a change in a gene, or a chromosome, usually involving an alteration in a DNA molecule (34) heat, chemicals such as mustard gas, radiation (35) gametes (36) mutations are usually recessive (37) natural (38) harmful to the organism (39) unfavourable (40) favourable (41) evolutionary (42) deoxyribonucleic acid (43) contain the instructions for building something (44) helix (45) bases (46) adenine, thymine, cytosine, guanine (47) sequence (48) code (49) protein (50) proteins are the building materials of life, and form enzymes which control the chemistry of life

Chapter 20

(1) food (2) living (3) host (4) pathogenic (5) tuberculosis, cholera, diphtheria, pneumonia, typhoid fever (6) boils, food poisoning, dysentery (7) ripening cheese; processing sewage; producing vinegar, alcohol, acetone and silage; curing tobacco (8) asexually (9) twenty to thirty (10) spore (11) the cell rounds off and forms a protective wall (12) resistant (13) drying up, high temperatures, poisonous chemicals (14) chicken pox, measles, poliomyelitis, common cold, influenza (15) inside living organisms (16) dormant (17) bacteriophages (18) DNA (19) the virus DNA uses bacterial DNA to make replicas of itself, while other cell materials are used to make head and tail units (20) colds, influenza, pneumonia (21) droplets of moisture containing germs are carried out of the body during breathing, coughing, sneezing, and speaking (22) water (23) contact with infected people, insects, birds (24) typhoid fever, cholera, food poisoning (25) contact with infected people and objects (26) by contact with the body, or handling infected objects such as books, coins, and handkerchiefs (27) smallpox, measles, tuberculosis (28) vectors (29) house-flies which carry typhoid fever, cholera, and dysentery; mosquitoes which carry yellow fever (30) primary (31) bladderworm (32) secondary (33) their bodies contain both male and female sex organs (34) large (35) poor sewage disposal methods (36) pigs (37) shell (38) stomach acid and enzymes (39) embryo (40) hooks (41) intestine (42) blood vessel (43) muscle (44) bladderworms (45) man (46) pork (47) live bladderworms (48) a tapeworm head emerges from the bladderworm, becomes attached to the intestine wall, and grows a tape (49) by efficient sanitary arrangements, and inspection of meat for bladderworms (50) sebaceous (51) follicles (52) supple, water-proof, and partly germ-proof (53) Malpighian (54) cornified (55) worn away (56) damaged (57) dust (58) mucus (59) cilia (60) oesophagus (61) swallowed and passed out of the body in the faeces (62) stomach acid (63) digestive enzymes (64) bleeding (65) cleans (66) fibrinogen is converted into a network of fibrin fibres, which trap red blood cells and dry out forming a solid clot (67) blood (68) dirt (69) germs (70) white (71) phagocytes (72) lymph nodes, liver, spleen, and bone marrow (73) opsonins make germs more likely to be eaten by phagocytes; lysins dissolve germs, agglutinins stick germs together, and antitoxins neutralize poisons formed by germs (74) a suspension of dead germs, inactivated germs, or harmless germs similar to pathogenic types (75) antibodies (76) immune (77) typhoid fever, poliomyelitis, cholera, bubonic plague (78) active (79) the body actively produces the antibodies (80) toxins injected into the horse cause it to make

antibodies, which are later extracted in the form of blood serum (81) passive (82) the body gains immunity without doing any work

Chapter 21

(1) decay (2) saprotrophic (3) fungi (4) humus (5) the lignin and cellulose of plants, and the hard parts of animals (6) liquid (7) crumbs (8) nitrates, phosphates, ammonium salts (9) plant (10) they provide air spaces which increase drainage and oxygen levels in the soil; they help retain water and soil minerals; and help prevent wind erosion (11) clay, silt, sand, gravel (12) physical (13) chemical (14) water (15) expands (16) roots (17) being pounded against each other while rolling down a hill; being washed along a river bottom (18) carbon dioxide (19) carbonic (20) limestone (21) soluble (22) wash, or drain (23) soil (24) carbon dioxide produced by respiring soil organisms (25 organic (26) decomposing (dissolving) underlying rocks (27) most soil organisms and plant roots need oxygen for respiration (28) the size of its particles; the larger the particles the greater the amount of air (29) minerals (30) adsorption (31) capillarity (32) stronger (33) clay (34) water (35) the maximum amount of water which it can retain (36) causing organisms to decay and producing humus and minerals (37) aerate (38) passing it through their bodies and ejecting it at the soil surface as a worm cast (39) dragging leaves into their burrows where they are left to decay (40) on top of the rocks from which it was formed (41) alluvial (42) rivers (43) glaciers (44) valley bottoms and plains (45) topsoil is situated above subsoil, it is black, and contains more humus and organisms (46) sand (47) clay (48) humus (49) lime (50) crumbs (51) it is well aerated; drains freely yet retains plenty of moisture and dissolved minerals; it warms quickly in spring; it is easily cultivated (52) large (53) humus, clay (54) good drainage and aeration; germinating roots penetrate it easily; and it is easily cultivated (55) it loses water and minerals quickly (56) small (57) it is able to draw water from below by capillarity, and it retains its minerals (58) its heavy consistency; lack of air; tendency to dry out into hard lumps (59) their mineral supplies are continually replaced by animal droppings and decaying organisms (60) minerals (61) they receive no droppings and contain few decaying organisms (62) ploughing to expose lumps to frost; adding sand, humus, and lime (63) adding humus (peat and manure) (64) manure, dried blood, bone-meal (65) slowly (66) they must first decay (67) quickly (68) sulphate of ammonia, superphosphate of lime

Chapter 22

(1) protein (2) atmosphere (3) nitrates (4) soil (5) fixing (6) atmospheric (7) carbohydrates (8) nitrates (9) bacteria and fungi (10) ammonia (11) nitrites (12) nitrates (13) nitrifying (14) they are involved in the continuous circulation of nitrogen between and within the physical and biological worlds (15) carbon dioxide (16) by respiration of animals and plants, and by decay bacteria, and by combustion, all of which release carbon dioxide gas (17) carbon (18) it involves the circulation of carbon between the atmosphere and living organisms (19) green plants, i.e. producing food by photosynthesis (20) first-order (21) animals (22) plants (producers) (23) second-order (24) first-order consumers (25) web (26) dependent (27) environment (28) producers, consumers, and their parasites; their physical environment; and the circulation between the two of carbon, nitrogen, oxygen, water, etc. (29) ecology (30) habitat (31) niche (32) food (33) harms living things (34) smoke (e.g. from burning coal) and car exhaust fumes (35) soot particles, sulphur dioxide, oxides of nitrogen, carbon monoxide (36) decomposed (37) oxygen (38) fish, crustaceans, insects, aquatic worms (39) compounds of cyanide, lead mercury, copper, zinc (40) crop (41) the chemicals enter organisms such as fish, which are at the bottom end of many food chains, and from here the chemicals spread up through many other animals often reaching humans.